Lecture Notes in Computer Science 773

Edited by G. Goos and J. Hartmanis

Douglas R. Stinson (Ed.)

Advances in Cryptology – CRYPTO '93

13th Annual International Cryptology Conference
Santa Barbara, California, USA
August 22-26, 1993
Proceedings

Springer-Verlag
Berlin Heidelberg New York
London Paris Tokyo
Hong Kong Barcelona
Budapest

Series Editors

Gerhard Goos
Universität Karlsruhe
Postfach 69 80
Vincenz-Priessnitz-Straße 1
D-76131 Karlsruhe, Germany

Juris Hartmanis
Cornell University
Department of Computer Science
4130 Upson Hall
Ithaca, NY 14853, USA

Volume Editor

Douglas R. Stinson
Computer Science and Engineering Department
and Center for Communication and Information Science
University of Nebraska, Lincoln, NE 68588-01115, USA

CR Subject Classification (1991): E.3-4, D.4.6, G.2.1

ISBN 3-540-57766-1 Springer-Verlag Berlin Heidelberg New York
ISBN 0-387-57766-1 Springer-Verlag New York Berlin Heidelberg

Printed in Germany

Typesetting: Camera-ready by author
SPIN: 10131900 45/3140-543210 - Printed on acid-free paper

PREFACE

The CRYPTO '93 conference was sponsored by the International Association for Cryptologic Research (IACR) and Bell-Northern Research (a subsidiary of Northern Telecom), in co-operation with the IEEE Computer Society Technical Committee. It took place at the University of California, Santa Barbara, from August 22–26, 1993. This was the thirteenth annual CRYPTO conference, all of which have been held at UCSB. The conference was very enjoyable and ran very smoothly, largely due to the efforts of the General Chair, Paul Van Oorschot. It was a pleasure working with Paul throughout the months leading up to the conference.

There were 136 submitted papers which were considered by the Program Committee. Of these, 38 were selected for presentation at the conference. There was also one invited talk at the conference, presented by Miles Smid, the title of which was "A Status Report On the Federal Government Key Escrow System."

The conference also included the customary Rump Session, which was presided over by Whit Diffie in his usual inimitable fashion. Thanks again to Whit for organizing and running the Rump session. This year, the Rump Session included an interesting and lively panel discussion on issues pertaining to key escrowing. Those taking part were W. Diffie, J. Gilmore, S. Goldwasser, M. Hellman, A. Herzberg, S. Micali, R. Rueppel, G. Simmons and D. Weitzner.

These proceedings contain revised versions of the 38 contributed talks, as well as two talks from the Rump Session. Please remember that these papers are unrefereed, and many of them represent work in progress. Some authors will write final versions of their papers for publication in refereed journals at a later time. Of course, the authors bear full responsibility for the contents of their papers.

I am very grateful to the members of the Program Committee for their hard work and dedication in the difficult task of selecting less than 30% of the submitted papers for presentation at the conference. The members of the program committee were as follows:

Mihir Bellare (IBM T. J. Watson)
Eli Biham (Technion, Israel)
Ernie Brickell (Sandia Laboratories)
Joan Feigenbaum (AT&T Bell Laboratories)
Russell Impagliazzo (UCSD)
Andrew Odlyzko (AT&T Bell Laboratories)
Tatsuaki Okamoto (NTT, Japan)
Birgit Pfitzmann (Hildesheim, Germany)
Rainer Rueppel (R^3, Switzerland)
Scott Vanstone (Waterloo, Canada)

As has been done since 1989, submissions to CRYPTO '93 were required to be anonymous. As well, we followed recent tradition which dictates that Program

Committee members could be an author or co-author of at most one accepted paper. Papers submitted by members of the Program Committee underwent the normal reviewing process (and, of course, no Program Committee member reviewed his or her own paper).

Thanks to Jimmy Upton for help with the pre-proceedings that were distributed at the conference (incidentally, this is the last year that CRYPTO will have both pre-proceedings and proceedings — starting in 1994, the proceedings will be available at the conference). Thanks also to Gus Simmons and Carol Patterson, who helped out with registration at the conference. And I would also like to convey my gratitude to Deb Heckens and my student, K. Gopalakrishnan, for their assistance.

Finally, I would like to thank everyone who submitted talks for CRYPTO '93. It goes without saying that the success of the conference depends ultimately on the quality of the submissions — CRYPTO has been and remains a leading conference in the discipline due the the high quality of the papers. I am also grateful to the authors for sending me final versions of their papers for publication in these proceedings in a timely fashion.

Douglas Stinson
Program Chair, CRYPTO '93
University of Nebraska
November, 1993

CONTENTS

Cryptosystems
Chair: Scott Vanstone

Stream Ciphers and Cryptographic Functions
Chair: Rainer Rueppel

Proof Systems and Zero-knowledge
Chair: Joan Feigenbaum

Secret Sharing

Chair: Mihir Bellare

Number Theory and Algorithms

Chair: Andrew Odlyzko

Differential Cryptanalysis

Chair: Spyros Magliveras

Complexity Theory

Chair: Joe Kilian

Applications

Chair: Birgit Pfitzmann

Authentication Codes

Chair: Doug Stinson

Hash Functions

Chair: Ivan Damgård

Cryptanalysis

Chair: Eli Biham

Key Distribution

Chair: Tatsuaki Okamoto

Efficient Signature Schemes Based on Birational Permutations

Adi Shamir

Dept. Computer Science, The Weizmann Institute of Science, Rehovot 76100, Israel

Abstract: Many public key cryptographic schemes (such as cubic RSA) are based on low degree polynomials whose inverses are high degree polynomials. These functions are very easy to compute but time consuming to invert even by their legitimate users. To overcome this problem, it is natural to consider the class of birational permutations f over k-tuples of numbers, in which both f and f^{-1} are low degree rational functions. In this paper we develop two new families of birational permutations, and discuss their cryptographic applications.

Remark: At the rump session of CRYPTO 93, Coppersmith Stern and Vaudenay presented two elegant and powerful attacks on the two signature schemes suggested in this paper. The attacks are quite specific, and thus it is conceivable that other signature schemes based on birational permutations will not be affected. The reader is thus encouraged to study the underlying mathematical structure of the schemes and the attacks, but to excercise great caution in implementing the ideas in practice.

1 Introduction

The original proposal for public key cryptography (Diffie and Hellman [1976] was based on the notion of trapdoor permutations, i.e., invertible functions which are easy to compute but apparently difficult to invert, unless some trapdoor information (which makes the inversion easy) is known. The best known implementation of this idea is the RSA scheme (Rivest, Shamir and Adleman [1978]), which can solve in a unified way the problems of key management, secure transmission, user identification, message authentication, and digital signatures. In one of the variants of this scheme, the encryption function is the low degree polynomial $f(x) = x^3 \pmod{n}$, which can be efficiently computed with two modular multiplications. Unfortunately, its inverse $f^{-1}(v) = v^d \pmod{n}$ is a very high degree polynomial, and thus its evaluation is quite slow (especially in software implementations).

In spite of extensive research in the last 16 years, there had been no fundamentally new constructions of trapdoor permutations which are faster than the RSA scheme. To overcome this difficulty, researchers have developed specialized solutions to various cryptographic needs which are not based on this unifying notion. For example, Diffie and Hellman [1976] proposed a key management scheme which is based on the one-way permutation of exponentiation modulo a prime. Since this function cannot be efficiently inverted, it is neither an encryption nor a signature scheme. The cryptosystem of Merkle and Hellman [1978] is invertible, but its mapping is not onto and thus it can not generate digital signatures. The Fiat-Shamir [1986] and DSS [1991] signature schemes are not one-to-one mappings, and thus they can not be used as cryptosystems.

A natural approach to this problem is to search for low degree algebraic mappings (polynomials or rational functions) whose inverses are also low degree algebraic mappings. Such mappings are called birational functions. We are particularly interested in multivariate mappings $f(x_1, \ldots, x_k) = (v_1, \ldots, v_k)$ in which the x_i and the v_i are numbers modulo a large $n = pq$, since the solution of general algebraic equations of this type is at least as hard as the factorization of the modulus. In this context, we say that a polynomial modulo n is low degree if its degree is a constant which does not grow with n, and a rational function is low degree if it is the ratio of two low degree polynomials. For example, in the case of cubic RSA, the function is considered low degree, but its inverse is not. Nonlinear algebraic mappings do not usually have unique inverses, when they do have inverses they usually cannot be written in closed form, and when the closed forms exist they are usually based on root extractions (radicals) or exponentiations whose computation modulo a large n is very slow. The construction of nonlinear birational mappings is thus a non-trivial task.

One attempt to construct birational permutations, due to Fell and Diffie [85], used the following DES-like idea. Let $(x_1, x_2, \ldots, x_k)$ be an initial k-vector of variables, and let g be a secret nonlinear multivariate polynomial. Alternately replace the current k-vector of multivariate polynomials $(p_1, p_2, \ldots, p_k)$ by $(p_1 + g(p_2, \ldots, p_k), p_2, \ldots, p_k)$, and rotate the k-vector to the right. After sufficiently many iterations, expand and publish the resultant k-vector of multivariate polynomials as your public key. When the trapdoor information g is known, the inverse of f can be computed by undoing the transformations (i.e., by alternately subtracting p_1 and rotating the k-vector to the left). Unfortunately, even when g is a quadratic function, the degree (and thus the size of the public key) grows exponentially with the number of iterations, which cannot be too small for security reasons. As the authors themselves conclude, "there seems to be no way to build such a system that is both secure and has a public key of practical size".

The problem is not accidental, due to the following generic attack: If f is known, the cryptanalyst can prepare a large number of input-output pairs for this function. Since f is invertible, these pairs (in reverse order) can be used to interpolate the unknown low-degree function f^{-1} by solving a small number of linear equations relating its coefficients. We do not know how to solve this problem in the context of public key cryptosystems. However, in the context of public key signature schemes there is a simple way to avoid this specific attack with the following modification:

Key Generation: Each user in the system chooses a particular birational permutation $f(x_1,\ldots,x_k)=(v_1,\ldots,v_k)$ consisting of k rational functions $f_i(x_1,\ldots,x_k) = v_i$, discards the first $s > 0$ of these f_i functions, describes the other $k-s$ f_i functions in his public key, and keeps the inverse of f as his private key.

Signature Generation: Given a digital message m, the signer chooses $v_i = r_i$ for $i = 1,\ldots,s$, and computes $v_i = h(m,i)$ for $i = s+1,\ldots,k$, where r_i are newly chosen secret random values, and h is a publicly known cryptographic hash function. He then uses his knowledge of the secret f^{-1} to compute a signature $(x_1,\ldots,x_k)$ satisfying $f(x_1,\ldots,x_k) = (v_1,\ldots,v_k)$.

Signature Verification: The verifier checks that $f_i(x_1,\ldots,x_k) = h(m,i)$ for $i = s+1,\ldots,k$, where the f_i's are taken from the signer's public key.

This modified scheme can no longer be used as a cryptosystem, since the cleartext $(x_1,\ldots,x_k)$ cannot be uniquely recovered from the shorter ciphertext $(v_{s+1},\ldots,v_k)$. It can be used as a signature scheme, since messages can have multiple signatures. The cryptanalyst cannot interpolate f^{-1} since it is not uniquely defined by the public key: He cannot generate complete input-output pairs for f^{-1} by himself, and cannot use input-output pairs generated by the legitimate signer since each one of them is based on new unknown values r_i. The recommended choice of s is 1, which makes the verification condition hardest to satisfy.

The security of this scheme depends on the choice of birational permutations. In Section 2 we introduce a simple family of birational permutations based on sequentially linearized equations, and in Section 3 we introduce a more sophisticated family of birational permutations based on algebraic bases in polynomial rings. Both families yield signature schemes with very low computational complexity. Unfortunately, both families also turn out to be breakable by the new attacks of Coppersmith Stern and Vaudenay[1993]. The reader is encouraged to look for other families of birational permutations and to study their efficiency and security.

2 A Birational Permutation Based on Sequentially Linearized Equations

Let n be the public product of two large secret primes p and q. verify Consider the triangular birational permutation $g(y_1, \ldots, y_k) = (w_1, \ldots, w_k) \pmod n$ in which the i-th output depends only on the first i inputs via the mapping $g_i(y_1, ..., y_i) = w_i \pmod n$, where g_i is a low degree polynomial which is linear in its last input y_i (the other inputs can and should occur non-linearly). Given the values of inputs $y_1, \ldots, y_k$, we can easily compute the values of the outputs $w_1, \ldots, w_k$ by evaluating k low degree polynomials. Given the values of the outputs $w_1, \ldots, w_k$, we can easily recover the values of the inputs $y_1, \ldots, y_k$ by solving a series of linear equations: First we solve for y_1 in $g_1(y_1) = w_1 \pmod n$. Then we substitute the computed value of y_1 into its (non-linear) occurrences in $g_2(y_1, y_2) = w_2 \pmod n$, and solve the remaining linear equation in y_2. We proceed in this order until we compute the last y_k. Each y_i is thus a low degree rational function of the w_j's, which is easy to compute with a small number of arithmetic operations modulo n.

To hide the easy solvability of the g_i's, the user has to transform them into more random looking polynomials before publishing them as his public key. We recommend the following two transformations:

1. Let A be a randomly chosen invertible $k \times k$ matrix, and consider the variable transformation $Y = AX$, where Y is the column vector of original variables $(y_1, \ldots, y_k)^t$ and X is a column vector of new variables $(x_1, \ldots, x_k)^t$. When the resultant polynomials are expanded, they contain all the variables in a non-linear way.
2. Let B be a randomly chosen invertible $k \times k$ matrix, and consider the mixing transformation $F = BG$, where G is the column vector of polynomials $(g_1, \ldots, g_k)^t$ and F is a column vector of new polynomials $(f_1, \ldots, f_k)^t$. Each f_i is thus a polynomial whose coefficients are random linear combinations of the corresponding coefficients of the given $g_1, \ldots, g_k$.

When A and B are known, it is easy to solve the resultant system of equations $f_i(x_1, ..., x_k) = v_i \pmod n$ for $i = 1, ..., k$ by inverting these transformations. of them change be To minimize the size of the public key, we recommend using g_i's which are homogeneous quadratic expressions of the form:

$$g_i(y_1, \ldots, y_i) = l_i(y_1, \ldots, y_{i-1}) \cdot y_i + q_i(y_1, \ldots, y_{i-1}) \pmod n$$

where l_i is a randomly chosen linear function of its inputs and q_i is a randomly chosen homogeneous quadratic function of its inputs. The only exception is g_1 in which l_1 and q_1 have no inputs. Since the coefficients of the linear g_1 cannot be mixed with the coefficients of the quadratic $g_2, \ldots, g_k$, and since we have to eliminate at least one of the polynomials in order to overcome the interpolation attack, we recommend the elimination of g_1 from the user's public key.

Without loss of generality, we can assume that $g_1(y_1) = y_1$ and $g_2(y_1, y_2) = y_1 y_2$ (since they can always be brought to this form by linear transformations). The case $k = 2$ is thus equivalent to the OSS scheme (Ong, Schnorr, Shamir [STOC83]), where the variable transformation A is $y_1 = x_1 + ax_2$, $y_2 = x_1 - ax_2$ and the mixing transformation B is the identity (all the arithmetic operations are carried out modulo a composite n). solves the signature The OSS scheme was successfully attacked by Pollard [1984], who showed that one quadratic equation in two variables can be solved even when the factorization of the modulus is unknown. A typical example of the extended signature scheme for $k = 3$ with the toy modulus $n = 101$ is:

Example: Consider the following sequentially linearized system of equations:

$$y_1 = w_1 \pmod{101}$$
$$y_1 y_2 = w_2 \pmod{101}$$
$$(29y_1 + 43y_2)y_3 + (71y_1^2 + 53y_2^2 + 89y_1y_2) = w_3 \pmod{101}.$$

Apply the linear change of variables:

$$y_1 = x_1 + 25x_2 + 73x_3 \pmod{101}$$
$$y_2 = x_1 + 47x_2 + 11x_3 \pmod{101}$$
$$y_3 = x_1 + 83x_2 + 17x_3 \pmod{101}$$

to obtain the new expressions:

$$x_1 + 25x_2 + 73x_3 = w_1 \pmod{101}$$
$$x_1^2 + 64x_2^2 + 96x_3^2 + 72x_1x_2 + 84x_1x_3 + 70x_2x_3 = w_2 \pmod{101}$$
$$83x_1^2 + 55x_2^2 + 16x_3^2 + 28x_1x_2 + 97x_1x_3 + 74x_2x_3 = w_3 \pmod{101}.$$

Mix these three expressions g_1, g_2, and g_3 by computing: $f_1 = g_1 \pmod{101}$, $f_2 = (39g_2 + 82g3) \pmod{101}$, $f_3 = (93g_2 + 51g3) \pmod{101}$. The resultant expressions are:

$$x_1 + 25x_2 + 73x_3 = v_1 \pmod{101}$$
$$78x_1^2 + 37x_2^2 + 6x_3^2 + 54x_1x_2 + 19x_1x_3 + 11x_2x_3 = v_2 \pmod{101}$$
$$84x_1^2 + 71x_2^2 + 48x_3^2 + 44x_1x_2 + 33x_1x_3 + 83x_2x_3 = v_3 \pmod{101}.$$

Discard the first equation, publish the last two equations as your public key, and retain the variable transformation A and the mixing transformation B as your secret key.

To sign a message m, hash it into two numbers v_2, v_3 in $[0, 101)$. Assume that $v_2 = h(m, 2) = 12$ and $v_3 = h(m, 3) = 34$. Its signature is any triplet of numbers x_1, x_2, x_3 which satisfies the two published equations for these values of v_2, v_3. To find such a solution, choose a random value for v_1 (each choice will yield a different signature). For example, let $v_1 = 99$. Apply the inverse mixing transformation B^{-1} by computing $w_1 = v_1 = 99 \pmod{101}, w_2 = 8v_2 + v_3 = 29 \pmod{101}, w_3 = 27v_2 + 18v_3 = 27 \pmod{101}$. The equations in terms of the y_i's are thus:

$$y_1 = 99 \pmod{101}$$
$$y_1 y_2 = 29 \pmod{101}$$
$$(29y_1 + 43y_2)y_3 + (71y_1^2 + 53y_2^2 + 89y_1 y_2) = 27 \pmod{101}.$$

To solve it, substitute $y_1 = 99$ into the second equation and solve the linear equation $99y_2 = 29 \pmod{101}$ to obtain $y_2 = 36$. Substitute both values into the third equation, simplify it into $76y_3 + 45 = 27 \pmod{101}$, and solve it to obtain $y_3 = 29$. Finally, use the inverse of the variable transformation A to change this Y solution into the X solution $x_1 = 40, x_2 = 27, x_3 = 22$ which is a valid signature of m since it satisfies the two published equations. ∎

Two general homogeneous quadratic equations in three unknowns have 12 coefficients, but they can be changed without loss of generality into a normal form which reduces the size of the public key to five numbers. The size of each signature is 196 bytes (when n is a 512-bit modulus). The generation and verification of each signature requires about 20 modular multiplications, compared to about 760 modular multiplications in the RSA scheme and about 50 modular multiplications in the Fiat-Shamir Scheme.

3 A Birational Permutation Based On Algebraic Bases

In this section we introduce a different family of birational permutations, which is inherently non-linear in all its components. The construction is quite unexpected, and seems to have many possible extensions and modifications.

Let $F_d[y_1, y_2, \ldots, y_k]$ denote the set of all the homogeneous polynomials of degree d in the k variables $y_1, y_2, \ldots, y_k$. We consider in particular the case of quadratic polynomials ($d = 2$), which are linear combinations of the $k(k+1)/2$ elementary quadratics y_iy_j for $i \leq j$ with coefficients a_{ij}. Such a set of quadratics is called a linear basis of $F_2[y_1, \ldots, y_k]$.

If we allow the additional operations of multiplication and (remainder-free) division of polynomials, we can express some of the elementary quadratics by other elementary quadratics. For example, the standard linear basis of $F_2[y_1, y_2, y_3]$ is $\{y_1^2, y_2^2, y_3^2, y_1y_2, y_1y_3, y_2y_3\}$. However, the first three elementary quadratics can be expressed by the last three in the following way:

$$y_1^2 = (y_1y_2)(y_1y_3)/(y_2y_3), \; y_2^2 = (y_1y_2)(y_2y_3)/(y_1y_3), \; y_3^2 = (y_1y_3)(y_2y_3)/(y_1y_2).$$

We can thus reduce the six linear generators $\{y_1^2, y_2^2, y_3^2, y_1y_2, y_1y_3, y_2y_3\}$ into the three algebraic generators $\{y_1y_2, y_1y_3, y_2y_3\}$. Another triplet of algebraic generators is $\{y_1^2, y_1y_2, y_2y_3\}$. However, $\{y_1^2, y_2^2, y_2y_3\}$ does not algebraically generate $F_2[y_1, y_2, y_3]$ since it cannot express y_1y_2.

To formalize this notion, consider an arbitrary set G of polynomials over a ring R. Its algebraic closure $[G]$ of G is defined as the smallest set of polynomials which contains G and R, and is closed under addition, subtraction, multiplication, and remainder-free division. Note that $[G]$ is not necessarily an ideal in the ring of polynomials over R, since we do not allow arbitrary polynomials as coefficients of the generators. For example, when $G = \{y^2\}$, $[G]$ is the set of all the polynomials in y whose monomials have even degrees. Note further that root extractions (radicals) are not allowed as basic operations, since they cannot be carried out efficiently in some rings R. G is called an algebraic basis of F if the polynomials in G are algebraically independent and F is contained in $[G]$.

Theorem: *1. The set of polynomials $\{y_1^2, y_1y_2, y_2y_3, \ldots, y_{k-1}y_k\}$ is an algebraic basis of $F_2[y_1, y_2, \ldots, y_k]$ for any k.*

2. The set of polynomials $\{y_1y_2, y_2y_3, y_3y_4, \ldots, y_ky_1\}$ is an algebraic basis of $F_2[y_1, y_2, \ldots y_k]$ for any odd $k > 1$.

Proof (sketch): We show that the specified sets can generate all the other elementary quadratics y_iy_j for $i \leq j$. Assume first that $j - i$ is an odd number. Then we can use the telescoping formula:

$$y_iy_j = (y_iy_{i+1})(y_{i+2}y_{i+3}) \cdots (y_{j-1}y_j)/(y_{i+1}y_{i+2}) \cdots (y_{j-2}y_{j-1}).$$

If $j-i$ is an even number, this approach will yield y_i/y_j instead of y_iy_j. To turn the former into the later, we have to multiply it by y_j^2. In case 1, we are given y_1^2 as a generator, and we can turn each y_t^2 into y_{t+1}^2 by using the formula:

$$y_{t+1}^2 = (y_ty_{t+1})^2/y_t^2.$$

In case 2, we use the fact that k is odd, and therefore the distance from j to j via the cyclic order $y_j, y_{j+1}, \ldots, y_k, y_1, \ldots, y_j$ is odd. The telescoping formula will thus yield the desired y_j^2 if we cycle through all the variables y_t. ■

The Theorem specifies two types of algebraic bases of size k for $F_2[y_1, \ldots, y_k]$. To get other bases, we can use the variable transformation A and the mixing transformation B described in Section 2. Due to the invertible and algebraic nature of these transformations, it is easy to translate the representation of each f in the original basis into a representation of f in the new bases.

Example: Consider the algebraic basis $G = \{y_1y_2, y_2y_3, y_3y_1\}$ of $F_2[y_1, y_2, y_3]$, and the randomly chosen linear transformations:

$$A = \begin{pmatrix} 37 & 62 & 71 \\ 89 & 45 & 68 \\ 50 & 17 & 93 \end{pmatrix} \qquad B = \begin{pmatrix} 41 & 73 & 51 \\ 89 & 12 & 60 \\ 37 & 94 & 19 \end{pmatrix}$$

Then the linear change of variables $Y \leftarrow AY \pmod{101}$ changes the generators g_i in G into:

$$\begin{aligned} g_1' &= (37y_1 + 62y_2 + 71y_3)(89y_1 + 45y_2 + 68y_3) \\ &= 12y_1y_2 + 38y_2y_3 + 48y_3y_1 + 61y_1^2 + 63y_2^2 + 81y_3^2 \pmod{101} \\ g_2' &= (89y_1 + 45y_2 + 68y_3)(50y_1 + 17y_2 + 93y_3) \\ &= 26y_1y2 + 89y_2y_3 + 62y_3y_1 + 6y_1^2 + 58y_2^2 + 62y_3^2 \pmod{101} \\ g_3' &= (50y_1 + 17y_2 + 93y_3)(37y_1 + 62y_2 + 71y_3) \\ &= 93y_1y_2 + 4y_2y_3 + 22y_3y_1 + 32y_1^2 + 44y_2^2 + 38y_3^2 \pmod{101} \end{aligned}$$

and the linear transformation $G'' \leftarrow BG' \pmod{101}$ changes the generators g_i' in G' into the new generators:

$$\begin{aligned} g_1'' &= 41g_1' + 73g_2' + 51g_3' \\ &= 63y_1y_2 + 78y_2y_3 + 41y_3y_1 + 26y_1^2 + 72y_2^2 + 89y_3^2 \pmod{101} \\ g_2'' &= 89g_1' + 12g_2' + 60g_3' \\ &= 92y_1y_2 + 44y_2y_3 + 74y_3y_1 + 48y_1^2 + 55y_2^2 + 32y_3^2 \pmod{101} \\ g_3'' &= 37g_1' + 94g_2' + 19g_3' \\ &= 9y_1y_2 + 51y_2y_3 + 43y_3y_1 + 96y_1^2 + 34y_2^2 + 53y_3^2 \pmod{101} \end{aligned}$$

Since G'' is an algebraic basis, it can be used to represent any given quadratic polynomial such as $f = y_1^2$, but the representation is not obvious unless the invertible transformations A and B are known. ■

To simplify our notation, we assume without loss of generality that $k > 1$ is odd and G is always the symmetric basis $\{y_i y_{i+1}\}$ (where $i+1$ is computed mod k). Given the invertible linear transformation $Y \leftarrow AY$, we define $A\&A$ as the $k \times k(k+1)/2$ matrix whose i-th row represents the quadratic polynomial $y_i y_{i+1}$ after the change of variables. The coefficients of the final basis G'' can thus be compactly represented by the $k \times k(k+1)/2$ matrix $B(A\&A)$.

Assume now that we are given an arbitrary assignment of values $x_1, \ldots, x_k$ to the elementary quadratics $y_1y_2, y_2y_3, \ldots, y_ky_1$ in the standard basis of $F_2[y_1, \ldots, y_k]$. We can use the telescoping formulas to compute the values of all the $k(k+1)/2$ elementary quadratics y_iy_j for $i \leq j$. We denote this extended version of X by $E(X)$, and note that it increases the height of the column vector from k to $k(k+1)/2$. The values of the k generators in G'' for this assignment X can be computed by the matrix-vector product $V = [B(A\&A)][E(X)]$. Note again the non-linearity of this transformation, which is due to the quadratic computation of the coefficients of A&A from the coefficients of A, and the multiplications and divisions required to extend X into $E(X)$.

This function maps the k-vector X into the k-vector V. Our goal now is to invert this function and recover the original values in X when A and B are known. First, we can undo the effect of B by computing $W = B^{-1}V$, and obtain the relationship $W = [A\&A][E(X)]$. By definition, the values w_i in W are the values of the generators g'_i in the intermediate G'. Since G' is an algebraic basis, it can represent any quadratic polynomial in $F_2[y_1, \ldots, y_k]$ as an algebraic expression in the generators g'_i. In particular it can represent the k elementary quadratics y_iy_{i+1} in such a way. However, each x_i was defined as the value of y_iy_{i+1}, and thus it can be recovered by evaluating an appropriate algebraic expression in the values $w_1, w_2, \ldots, w_k$. It is easy to check that these algebraic expressions can be compactly represented as $X = [A^{-1}\&A^{-1}][E(B^{-1}V)]$.

Example (continued): Consider the algebraic bases G, G' and G'' of the previous example. Let x_i denote the value of y_iy_{i+1}, i.e.: $y_1y_2 = x_1$, $y_2y_3 = x_2$, $y_3y_1 = x_3$ (mod 101). The values of the other three elementary quadratics can be expressed by as $y_1^2 = x_3x_1/x_2$, $y_2^2 = x_1x_2/x_3$, $y_3^2 = x_2x_3/x_1$ (mod 101). The values v_i of the g''_i are computed via $[B(A\&A)][E(X)]$, i.e.:

$$v_1 = 63x_1 + 78x_2 + 41x_3 + 26x_3x_1/x_2 + 72x_1x_2/x_3 + 89x_2x_3/x_1 \pmod{101}$$
$$v_2 = 92x_1 + 44x_2 + 74x_3 + 48x_3x_1/x_2 + 55x_1x_2/x_3 + 32x_2x_3/x_1 \pmod{101}$$
$$v_3 = 9x_1 + 51x_2 + 43x_3 + 96x_3x_1/x_2 + 34x_1x_2/x_3 + 53x_2x_3/x_1 \pmod{101}$$

In particular, when the input is $x_1 = 1$, $x_2 = 2$, $x_3 = 3$, the output is $v_1 = 54$, $v_2 = 63$, $v_3 = 85$. To invert this mapping, we reverse the transformation $G'' \leftarrow BG'$ (mod 101) by computing $W = B^{-1}V$ (mod 101). When $v_1 = 54$,

$v_2 = 63$, $v_3 = 85$, this matrix-vector product yields $w_1 = 94$, $w_2 = 69$, $w_3 = 1$, which are the values of the intermediate generators g_1', g_2', g_3'. We then extend this W into $E(W) = (w_1, w_2, w_3, w_3 w_1/w_2, w_1 w_2/w_3, w_2 w_3/w_1)$, and compute $X = [A^{-1} \& A^{-1}][E(W)] \pmod{n}$, i.e.:

$$x_1 = 29w_1 + 75w_2 + 45w_3 + 74w_3w_1/w_2 + 42w_1w_2/w_3 + 45w_2w_3/w_1 \pmod{101}$$

$$x_2 = 46w_1 + 72w_2 + 14w_3 + 99w_3w_1/w_2 + 61w_1w_2/w_3 + 58w_2w_3/w_1 \pmod{101}$$

$$x_3 = 1w_1 + 58w_2 + 46w_3 + 87w_3w_1/w_2 + 31w_1w_2/w_3 + 77w_2w_3/w_1 \pmod{101}.$$

For $w_1 = 94$, $w_2 = 69$, $w_3 = 1$, the extension of W to $E(W)$ yields the column vector $(94, 69, 1, 16, 22, 19)^t$. When these values are substituted into the expressions above, we get the original inputs $x_1 = 1$, $x_2 = 2$, $x_3 = 3$. ∎

Note that the application and inversion of the function have essentially the same form. They use a small number of modular additions multiplications and divisions but no root extractions. ∎

The new signature scheme can now be formalized in the following way:

Key Generation: Pick a set F of rational functions in k variables, and a standard algebraic basis G of F with the property that the representation of any f in F as an algebraic expression in terms of the generators g_i in G can be easily computed. Transform the easy basis G into an apparently hard basis G'' by using randomly chosen invertible algebraic transformations. Eliminate $s > 0$ generators from G'', publish the other $k - s$ generators in G'' as the public key, and keep the algebraic transformations as the private key.

Signature Generation: To sign a given message m, assign to each published g_i'' the hashed value $v_i = h(m, i)$ of m, and to each eliminated g_i'' a newly chosen random number r_i. Use the secret algebraic transformations to express each g_i in the easy basis G in terms of the generators g_j'' in the hard basis G''. The values x_i of the easy generators g_i for this assignment of values to the hard generators g_j'' form the signature X of m.

Signature Verification: Assign the values x_i from the signature X to the easy generators g_i, and compute the values v_i of the $k - s$ hard generators $g''i$ which appear in the signer's public key. Evaluate the $k - s$ hashed forms $h(m, i)$ of m, and accept the validity of the signature if $v_i = h(m, i)$ for all the $k - s$ values of i.

The recommended choice for F is the set $F_d[y_1, \ldots, y_k]$ of all the homogeneous polynomials of degree d over the ring Z_n. The factorization of $n = pq$ can be destroyed as soon as n is chosen and published. The recommended choice for G is some set of monomials $y_1^{e_1} y_2^{e_2} \ldots y_k^{e_k}$ with $e_1 + e_2 + \ldots + e_k = d$ such that any other monomial in $F_d[y_1, \ldots, y_k]$ can be generated by a sequence of multiplications and divisions. It is not difficult to show that for any d and k, $F_d[y_1, \ldots, y_k]$ has an algebraic basis of this type which consists of exactly k monomials of degree d (for example, $G = \{y_1^3, y_2^2 y_3, y_1 y_2 y_3\}$ is an algebraic basis for $F_3[y_1, y_2, y_3]$). The problem with large choices of k and d is that the number of coefficients in the published generators G'' grows as $O(k^{d+1})$. However, for fixed d this key size grows only polynomially in k, and the recommended choice of d is 2. The recommended bases for $d = 2$ are the standard bases $G = \{y_1 y_2, \ldots, y_{k-1} y_k, y_k y_1\}$ for odd k and $G = \{y_1^2, y_1 y_2, \ldots, y_{k-1} y_k\}$ for arbitrary k. The recommended choice of invertible algebraic transformations is a pair of randomly chosen $k \times k$ matrices A (the variable transformation) and B (the mixing transformation). Note that bases G with $\mathrm{GCD}(g_1, g_2, \cdots, g_k) \neq 1$ should be avoided — such a GCD remains nontrivial under the transformations $Y \leftarrow AY$ and $G'' \leftarrow BG'$, and its computation can reveal some of the rows of A. This condition is automatically satisfied in our standard bases G for $k \geq 3$. For example, when $k = 3$, two random linear combinations of $y_1 y_2$, $y_2 y_3$ and $y_3 y_1$ will almost surely be relatively prime. This signature scheme is comparable in its key size, signature size, and computational complexity to the signature scheme described in Section 2. By using the techniques of Shamir[1993], we can show that recovering the secret matrices A and B from the public key $B(A\&A)$ is at least as hard as factoring the modulus n by using the symmetry of the problem when the rows of A and the coloumns of B are cyclically rotated. However, the attack of Coppersmith Stern and Vaudenay overcomes this obstacle by computing signatures directly without extracting the original secret keys.

Bibliography

1. D. Coppersmith J. Stern and S. Vaudenay [1993]: "Attacks on the Birational Permutation Signature Schemes", Proceedings of CRYPTO 93 (this volume).

2. W. Diffie and M. Hellman [1976]: "New Directions in Cryptography", IEEE Trans. Information Theory, Vol IT-22, No 6, pp 644-654.

3. DSS [1991]: "Specifications for a Digital Signature Standard", US Federal Register Vol 56 No 169, August 30 1991.

4. H. Fell and W. Diffie [1985]: "Analysis of a Public Key Approach Based on Polynomial Substitution", Proceedings of CRYPTO 85, Springer-Verlag Vol 218, pp 340-349.

5. A. Fiat and A. Shamir [1986]: "How to Prove Yourself: Practical Solutions to Identification and Signature Problems", Proceedings of CRYPTO 86, Springer-Verlag Vol 263, pp 186-194.

6. R. Merkle and M. Hellman [1978]: "Hiding Information and Signatures in Trapdoor Knapsacks", IEEE Trans. Information Theory, Vol IT-24, No 5, pp 525-530.

7. R. Rivest, A. Shamir and L. Adleman [1978]: "A Method for Obtaining Digital Signatures and Public Key Cryptosystems", Comm. ACM, Vol 21. No 2, pp 120-126.

8. A. Shamir [1993]: "On the Generation of Multivariate Polynomials Which Are Hard To Factor", Proceedings of STOC 1993.

A new identification scheme based on syndrome decoding *.

Jacques Stern

Laboratoire d'informatique, École Normale Supérieure

Abstract. Zero-knowledge proofs were introduced in 1985, in a paper by Goldwasser, Micali and Rackoff ([6]). Their practical significance was soon demonstrated in the work of Fiat and Shamir ([4]), who turned zero-knowledge proofs of quadratic residuosity into efficient means of establishing user identities. Still, as is almost always the case in public-key cryptography, the Fiat-Shamir scheme relied on arithmetic operations on large numbers. In 1989, there were two attempts to build identification protocols that only use simple operations (see [11, 10]). One appeared in the EUROCRYPT proceedings and relies on the intractability of some coding problems, the other was presented at the CRYPTO rump session and depends on the so-called Permuted Kernel problem (PKP). Unfortunately, the first of the schemes was not really practical. In the present paper, we propose a new identification scheme, based on error-correcting codes, which is zero-knowledge and is of practical value. Furthermore, we describe several variants, including one which has an *identity based* character. The security of our scheme depends on the hardness of decoding a word of given syndrome w.r.t. some binary linear error-correcting code.

1 The signature scheme

Since the appearance of public-key cryptography, basically all practical schemes have been based on hard problems from number theory. This has remained true with zero-knowledge proofs, introduced in 1985, in a paper by Goldwasser, Micali and Rackoff ([6]) and whose practical significance was soon demonstrated in the work of Fiat and Shamir ([4]). In 1989, there were two attempts to build identification protocols that only use simple operations (see [11, 10]). One relied on the intractability of some coding problems, the other on the so-called Permuted Kernel problem (PKP). Unfortunately, the first of the schemes was not really practical.

In the present paper, we propose a new identification scheme, based on the syndrome decoding problem for error-correcting codes (SD), which is zero-knowledge and seems of truly practical value. The proposed scheme uses a fixed $(n\text{-}k)$-matrix H over the two-element field. This matrix is common to all users and is originally built randomly. Thus, considered as a parity-check matrix, it should provide a linear binary code with a good correcting power.

---

* PATENT CAUTION: This document may reveal patentable subject matter

Any user receives a secret key s which is an n-bit word with a prescribed number p of 1's. This prescribed number p is also part of the system. The public identification is computed as

$$i = H(s)$$

The identification scheme relies heavily on the technical notion of a *commitment.* If u is an sequence of bits, a *commitment* for u is the image of u via some cryptographic hash function. A commitment will be used as a one-way function: in order to disclose it, one announces the original sequence from which it was built. Once this is done, anyone can check the correctness of the commitment.

We now describe the basic interactive zero-knowledge protocol that enables any user (which we will call the prover as usual) to identify himself to another one (which we call the verifier). The protocol includes r rounds, each of these being performed as follows:

1. The prover picks a random n-bit word y together with a random permutation σ of the integers $\{1 \cdots n\}$ and sends commitments c_1, c_2, c_3 respectively for $\langle \sigma, H(y) \rangle$, $\langle y.\sigma \rangle$ and $\langle (y \oplus s).\sigma \rangle$ to the verifier. Note that $\langle , \rangle$ denotes the action of the hash function on the concatenation of all the bits of information of its arguments. A permutation σ is being considered in this setting as a vector of bits which encodes it; also note that $y.\sigma$ refers to the image of y under permutation σ.
2. The verifier sends a random element b of $\{0, 1, 2\}$.
3. If b is 0, the prover reveals y and σ. If b is 1, the prover reveals $y \oplus s$ and σ. Finally, if b equals 2, the prover discloses $y.\sigma$ and $s.\sigma$.
4. If b equals 0, the verifier checks that commitments c_1 and c_2, which were disclosed in step 2, have been computed honestly.
   If b equals 1, the verifier checks that commitments c_1 and c_3, were correct: note that σ is known from step 3 and that $H(y)$ can be recovered from $H(y \oplus s)$ by the equation

   $$H(y) = H(y \oplus s) \oplus i$$

   wher i is the user's public key.
   Now, if b is 2, the verifier checks commitments c_2 and c_3 and the property that the weight of $s.\sigma$ has the prescribed value p.

The number r of consecutive rounds depends on the required level of security and will be discussed further on as well as the values of the parameters n, k, p.

2 Security of the scheme

Of course, the security of the scheme relies on the difficulty of inverting the function

$$s \longrightarrow H(s)$$

when its arguments are restricted to valid secret keys. In order to give evidence of this difficulty, let us recall from [1] that it is NP-complete to determine whether a code has a word s of weight p whose image is a given k-bit word i. Let us also observe that, if p is small enough, finding s is exactly equivalent to finding the codeword w minimizing the weight of $t \oplus w$, when an element t of $H^{-1}(i)$ is chosen. But this is the problem of decoding unstructured codes which is currently believed to be unsolvable. Algorithms known for this problem (such as those described in [7, 12]) have a computing time that grows exponentially. A correct asymptotic evaluation of those algorithms has been recently given in [2] and confirmed by experiments in moderate sizes. Thus, as is the case for factoring, it is possible to state an *intractability assumption for the Syndrome Decoding Problem*, upon which the security of our scheme rests. Several versions of this assumption can be given, depending on the underlying model of security: for example, one may claim that no polynomial time probabilistic algorithm that takes as an input the parity check matrix H of an $(n, n-k)$ code, together with a binary k-bit vector i, can output the minimum-weight solution of the equation $H(s) = i$, with a non-negligible probability of success and a for a significant part of the inputs that correspond to a ratio k/n lying in a fixed interval. As usual, the precise formulation of these intractability assumptions is rather cumbersome: we keep it for the final version of our paper.

In order to counterfeit a given signature without knowing the secret key, various strategies can be used.

- Having only y and σ ready for the verifier's query and replacing the unknown s by some arbitrary vector t of weight p. In this case, the false prover hopes that b is 0 or 2 and the probability of success is $(2/3)^r$, where r is the number of rounds. A similar strategy can be defined with $y \oplus s$ in place of y.
- Having both y and $y \oplus t$ ready where t is some element such that $H(t) = i$, distinct from s and whose weight is not p. This yields the same probability of success.

It is fairly clear that shifting beetween one strategy to another has also the same probability of success. Furthermore, it can be proven formally that any probabilistic interactive algorithm that is accepted by the verifier with a probability of success that is significantly above $(2/3)^r$ can be turned into another that discloses the secret key or else outputs collisions for the cryptographic hash functions used for commitments. This is along the lines of the proof of theorem 3 of [5] and will appear in the full paper. Thus, if secure cryptographic hash functions do exist and if the intractability assumption for the Syndrome Decoding Problem holds, our system actually provides a proper proof of knowledge.

It can also be proved formally that the scheme is zero-knowledge. We will only give a brief hint. As we observed above, anyone can be ready to answer two queries among the three possible ones at each round. Hence, by using the standard idea of resettable simulation (see [6]), one can devise a polynomial-time simulation algorithm that mimics the fair communication between the prover and the verifier in expected time $O(2/3.r)$. Speaking more informally, the only information that comes out of a round is either a random word, y or $y \oplus s$

together with a random permutation, or else one random word y together with another random word of fixed weight p. Therefore, it is virtually impossible to undertake any statistical analysis that might reveal s.

We now discuss the size of the various parameters. Although a large variety of choices are opened, we recommend that n equals $2k$. Thus possible sizes are:

- $n = 512, k = 256$
- $n = 1024, k = 512$

As for the value of p, we observe that, although it is tempting to lower it, this would be rather dangerous: using the arguments of [12], one can see that secret keys of small weight (e.g. $p = 20$) will presumably be found. We recommend that p is chosen slightly below the value given by the so-called Warshamov Gilbert bound ([9]), which provides a theoretical limit value for the minimal weight d of a (n, k) random code, namely:

$$\frac{d}{n} = H_2(\frac{k}{n})$$

where H_2 is the entropy function defined by:

$$H_2 = -x \log_2(x) - (1 - x) \log_2(1 - x)$$

When n is $2k$, d is approximately $0.11n$ and values of p corresponding to the three possible data mentionned above are about 56 in the first case and 110 in the second. From the estimations of [2], the workfactor for the known algorithms that might reveal the secret is above $O(\Omega^k)$ where Ω is about 1.18 for the chosen values. This yields a value of 2^{61} for $k = 256$, which can be considered as unfeasible, especially as this is an estimate of a work factor, which means that the constant under the big-O notation is fairly large.

3 Performances of the scheme.

We will restrict ourselves to various remarks.

1. It might be thought that the proposed scheme requires a large amount of memory. THis is not accurate: on one hand, because the operations to perform are very simple, they can be implemented in hardware in a quite efficient way; on the other hand, if the scheme is implemented, partially or totally, in software, it is not neccessary to store all of H. One can only store words corresponding to some chosen locations and extend these by a fixed software random number generator.
2. The communication complexity of the protocol is comparable to what it is in the Fiat-Shamir scheme. If we assume that permutations come from a a seed of of 120 bits via a pseudo-random generator and that hash values are 64 bits long, which is perfectly acceptable from a security point of view, we obtain an average number of bits per round which is close to 950 bits when $n = 512$. Using a trick that saves space for sending $s.\sigma$, this can be lowered to approximately 860 bits.

3. The security of the scheme can be increased by taking n, k and p larger. The figures $n = 512$, $k = 256$ are to be considered as a minimum.
4. The heaviest part of the computing load of the prover (which is usually a portable device with a limited computing power) is the computation of $H(y)$, which is done in step 1. This load can be drastically reduced by extending the protocol to a 5-pass version which will be discussed in the next section
5. Considering that the probability of any cheating strategy is bounded above by $(2/3)^r$, where r is the number of rounds, we see that the basic protocol has to be repeated 35 times in order to achieve a level of security of 10^{-6}. A key difference between the proposed schemes and previous proposals is the fact that a single round offers only security 1/3 instead of 1/2. There is a variant of our protocol that achieves security 1/2. It will be discussed further on in this paper
6. As is the case for Shamir's PKP, our scheme is not identity based. This means that public keys have to be certified by the issuing authority. We will consider below a variant of this scheme with an identity-based character.

4 A variant that minimizes the computing load.

In order to minimize the computing load, we introduce a 5-pass variant. this variant depends on a new parameter q.

Step 1 is the same except that commitment c_1 is replaced by $\langle\sigma\rangle$. Thus $H(y)$ is not computed at this stage.

After step 1, the verifier sends back a choice of q indices from $\{1 \cdots k\}$ (these refer to a choice of q rows of the matrix H).

The prover answers by sending the list of bits $b_1, \cdots, b_q$ corresponding to the selected indices of the vector $H(y)$.

The rest of the protocol is similar (with obvious changes for the checking step 6).

Of course this opens up new strategies for cheating: basically, one will try to have both y and $y \oplus t$ ready where t is some element of weight p such that $H(t)$ differs from i on a small number of bits, say h. This will increase the probability of success by an amount which is close to $\frac{1}{3}(1-\frac{h}{k})^q$. In the case $n = 512$, $k = 256$, $p = 56$, $q = 64$, $h = 15$, this extra amount is roughly 0.007 and the loss can be compensated by adding only one extra round of the protocol.

Of course, the new strategy becomes more and more successful as h decreases; for example, making $h = 4$ and keeping all other figures unchanged increases the probability of cheating successfully to 0.78. But it can be shown that finding a t as above is equivalent to finding a word s' of weight at most $p + h$ with a given syndrome $H(s') = i$ and it is believed that, when h is very small, this remains unfeasable. Of course, many other trade-offs between n, k, p, h, q are possible.

5 A variant that minimizes the number of rounds.

In this variant, the secret key s is replaced by a simplex code generated by $s_1, \cdots, s_m$. Recall that a simplex code of dimension m has all its non zero codewords of weight 2^{m-1} (see [9]). It is easy to construct such a code with length $2^m - 1$ and to extend the length to any larger value n. The corresponding public key is the sequence $H(s_1), \cdots, H(s_m)$.

It is unknown whether or not it is easier to recover the family of secret vectors than to recover a single one. As a set of minimal values, we recommend $m = 7$ together with $n = 576$ and $k = 288$. This ensures consistency with our previous estimates.

We now describe one 5-pass round of a protocol that achieves identification.

1. The prover picks a random n-bit word y together with a random permutation σ of the integers $\{1 \cdots n\}$ and sends commitments c_1, c_2 respectively for $\langle \sigma, H(y) \rangle$, $\langle y.\sigma, s_1.\sigma, \cdots, s_m.\sigma \rangle$ to the verifier.
2. The verifier sends a random binary vector $b_1, \cdots, b_m$.
3. The prover computes

$$z = (y \oplus \bigoplus_{j=1}^{m} b_j s_j).\sigma$$

   and sends z to the verifier
4. The verifier responds with a one bit challenge b.
5. If b is 0, the prover reveals σ. If b is 1, the prover discloses $y.\sigma$ as well as the full sequence $s_1.\sigma, \cdots, s_m.\sigma$.
6. If b equals 0, the verifier checks that commitment c_1 has been computed honestly.Note that $H(y)$ can be recovered from $H(z.\sigma^{-1})$, the sequence of public keys and the binary vector issued at step 2.
   If b equals 1, the verifier checks that commitment c_2, was correct, that the computation of z is consistent and that $s_1, \cdots, s_m$ actually form a simplex code of the required weight.

This basic round can be repeated and it can be shown that the probability of success of a single round, when no information about the secret keys is known, is at most $\frac{1+2^{m-1}}{2^m}$, which is essentially 1/2. On the other hand, it is clear that the communication complexity is worse than in the single-key case.

6 An identity based version.

One attractive feature of the Fiat-Shamir scheme is that the public key can be derived from the user's identity, thus avoiding the need to link both by some signature from the issuing authority. Neither Shamir's PKP scheme nor our basic scheme have this feature. We now investigate various modifications that can turn our scheme into an identity-based scheme.

A first possibility is to use a set of t simplex codes of dimension m. If $s_1, \cdots, s_m$ is the first of these codes, then m bits can define a specific key

$$\bigoplus_{j=1}^{m} b_j s_j$$

and therefore, assuming that the identity of a user is given by tm bits, one can define t secret keys for each user. Now, these secret keys can be used randomly to perform identification. The verifier has to store tm vectors of k bits (the images of the basis vectors of the codes), which is much less than a full directory of users. We suggest $m = 7$ $t = 6$ as a reasonable implementation.

The other possibility we describe is a bit more intricate. It uses a "master code" consisting of t vectors $s_1, \cdots, s_t$ whose one-bits only cover a subset T of the possible n locations. Given the identity of a user as a sequence of t bits $e_1, \cdots, e_t$ it is easy, by Gaussian elimination, to find a linear combination s of the s_j's such that

$$H(s) = \sum_{j=1}^{t} e_j H(s_j)$$

and whose weight p is approximately $|T| - t/2$. This will be the secret key of the user, computed by the issuing authority. The security of this variant is more difficult to analyze. Typically, the existence of the master code implies that some code that can de defined from the public data has a vector with a small number of ones located within the (unknown) set T. The dimensions should be designed in order that the weight of this vector is large enough. We suggest, as an example, $t = 56$, $p = 95$, $n = 864$, $k = 432$.

7 An analogous scheme based on modular knapsacks.

In this section, we briefly mention an analogous scheme that can be devised by replacing the $\{0, 1\}$-matrix H by a matrix over a finite field with an extremely small number of elements (typically 3, 5 or 7). In this situation, the weight constraint is replaced by the constraint that the secret solution s to the equation $H(s) = i$ consists entirely of zeros and ones. Thus the underlying difficult problem is a modular knapsack. Although it is known that knapsacks can be attacked by methods based on lattice reduction (see [8, 3]), it is clear also that these methods do not apply to the modular case, at least when the modulus m is very small. Possible values for the scheme are (with the same notations as above)

- $n = 196$, $k = 128$, $m = 3$
- $n = 384$, $k = 256$, $m = 3$
- $n = 128$, $k = 64$, $m = 5$
- $n = 192$, $k = 96$, $m = 5$

One round of the protocol is performed as follows:

1. The prover picks a random vector y with coefficients from the m-element field, together with a random permutation σ of the integers $\{1 \cdots n\}$ and sends commitments c_1, c_2, c_3 respectively for $\langle \sigma, H(y) \rangle$, $\langle y.\sigma \rangle$ and $\langle (y \oplus s).\sigma \rangle$.
2. The verifier sends a random element b of $\{0, 1, 2\}$.
3. If b is 0, the prover reveals y and σ. If b is 1, the prover reveals $y + s \bmod m$ and σ. Finally, if b equals 2, the prover discloses $y.\sigma$ and $s.\sigma$.
4. The verifier makes the obvious checks.

8 Conclusion

We have defined a new practical identification scheme based on the syndrome decoding problem (SD). We have also described several variants of this scheme. The scheme only uses very simple operations and thus widens the range of techniques that can be applied in crytography. We welcome attaccks from readers and, as is customary when introducing a new cryptographic tool, we suggest that the scheme should not be adopted prematurely for actual use.

References

1. E. R. Berlekamp, R. J. Mc Eliece and H. C. A. Van Tilborg. On the inherent intractability of certain coding problems, *IEEE Trans. Inform. Theory,* (1978) 384-386.
2. F. Chabaud, Asymptotic analysis of probabilistic algorithms for finding short codewords, *Proceedings of EUROCODE 92*, Lecture Notes in Computer Science, to appear.
3. M. J. Coster, A. Joux, B. A. LaMacchia, A. M. Odlyzko, C. P. Schnorr and J. Stern, Improved low-density subset sum algorithms, *Computational Complexity*, to appear.
4. A. Fiat and A. Shamir, How to prove yourself: Practical solutions to identification and signature problems, *Proceedings of Crypto 86*, Lecture Notes in Computer Science 263, 181-187.
5. U. Feige, A. Fiat and A. Shamir, Zero-knowledge proofs of identity, *Proc. 19th ACM Symp. Theory of Computing*, 210-217, (1987) and *J. Cryptology*, **1** (1988), 77-95.
6. S. Goldwasser, S. Micali and C. Rackoff, The knowledge complexity of interactive proof systems, *Proc. 17th ACM Symp. Theory of Computing*, 291-304, (1985).
7. J . S. Leon, A probabilistic algorithm for computing minimum weights of large error-correcting codes. *IEEE Trans. Inform. Theory*, IT-34(5): 1354-1359.
8. J. C. Lagarias and A. M. Odlyzko, Solving low-density subset sum problems, *J. Assoc. Comp. Mach.* **32** (1985), 229-246.
9. F. J. MacWilliams and N. J. A. Sloane, The theory of error-correcting codes, North-Holland, Amsterdam-New-York-Oxford (1977).
10. A. Shamir, An efficient identification scheme based on permuted kernels, *Proceedings of Crypto 89*, Lecture Notes in Computer Science 435, 606-609.
11. J. Stern, An alternative to the Fiat-Shamir protocol, *Proceedings of Eurocrypt 89*, Lecture Notes in Computer Science 434, 173-180.
12. J. Stern, A method for finding codewords of small weight, *Coding Theory and Applications*, Lecture Notes in Computer Science 388 (1989), 106-113.

The Shrinking Generator

Don Coppersmith   Hugo Krawczyk   Yishay Mansour

IBM T.J. Watson Research Center
Yorktown Heights, NY 10598

Abstract. We present a new construction of a pseudorandom generator based on a simple combination of two LFSRs. The construction has attractive properties as simplicity (conceptual and implementation-wise), scalability (hardware and security), proven minimal security conditions (exponential period, exponential linear complexity, good statistical properties), and resistance to known attacks. The construction is suitable for practical implementation of efficient stream cipher cryptosystems.

1 Introduction

We present a new construction of a pseudorandom generator that uses as basic modules a pair of LFSRs. The inherent simplicity of LFSRs, the ease and efficiency of implementation, some good statistical properties of the LFSR sequences, and the algebraic theory underlying these devices turn them into natural candidates for use in the construction of pseudorandom generators, especially, targeted to the implementation of efficient stream cipher cryptosystems. Indeed, many such constructions were proposed in the literature. (See Rueppel's comprehensive survey on LFSR-based constructions of pseudorandom generators and related analysis tools [18]). On the other hand, some of the attractive properties listed above are also the reason for the failure of many of these constructions to meet a good cryptographic strength. In particular, the inherent linearity of LFSRs and the algebraic structure are many times the basis for breaking these systems.

Nevertheless, due to their technological advantages for simple hardware implementation of fast cryptosystems, LFSRs are still studied (and used!) as basic modules for these systems. In particular, the increasing speeds of transmitted information and the simple methods for LFSR parallelization and pipelining indicate that this interest is plausible to persist in the (visible) future. In addition, there is no reason to believe that good *and* simple constructions are impossible.

This paper presents a construction which is attractive in the sense that it is very simple (conceptually and implementation-wise) and passes the *minimal* tests that such constructions require to be worth being considered. We can prove that both the period and linear complexity of the resultant sequences is exponential in the LFSR's length, and that these sequences have some nice distributional statistics (measured in a rigorous way). The construction appears to be free of traditional weaknesses and has stood (up to now) several potential attacks. As said, these are just minimal conditions for the construction to deserve the attention of the cryptographic community, not a "proof" of their ultimate strength.

The practical strength of such a construction can be determined only after public scrutiny. (A desirable goal is to have a real proof of the strength of such a system - i.e., a proof of the unpredictability by efficient means of the generated sequences. Unfortunately, such a proof is not known for any efficient pseudorandom generator and, moreover, such a proof will require a breakthrough in complexity theory. On the other hand, the theoretically well-founded approach of relating the strength of a pseudorandom generator to the hardness of generic or specific problems [4, 20] has led to a beautiful theory and constructions but these are still too impractical for many real-world applications).

1.1 The Construction

Our construction uses two sources of pseudorandom bits to create a third source of pseudorandom bits of (potentially) better quality than the original sources. Here quality stands for the difficulty of predicting the pseudorandom sequence. (In general, through this paper, we use the notion of pseudorandomness and predictability in a rather informal way, although we rigorously analyze and prove some of the random-like properties of the resultant sequences). The sequence we build is a subsequence from the first source where the subsequence elements are chosen according to the positions of '1' bits in the second source. In other words, let $a_0, a_1, \ldots$ denote the first sequence and $s_0, s_1, \ldots$ the second one. We construct a third sequence $z_0, z_1, \ldots$ which includes those bits a_i for which the corresponding s_i is '1'. Other bits from the first sequence are discarded. (Therefore, the resultant sequence is a "shrunken" version of the first one). Formally, for all $k = 0, 1, \ldots$, $z_k = a_{i_k}$, where i_k is the position of the k-th '1' in the sequence $s_0, s_1, \ldots$. We call the resultant pseudorandom generator, *the shrinking generator (SG)*.

This generic idea can be applied to any pair of pseudorandom sources. Here we analyze the construction where the two sources are generated using Linear Feedback Shift Registers (LFSR). LFSRs are very well known structures consisting of a shift register controlled by a clock, which at each clock pulse outputs its most significant bit, shifts its contents in the most significant direction and inputs a bit to its less significant position. This *feedback bit* is computed as a linear combination (over GF(2)) of the bits in the shift register. This linear combination can be fixed (e.g. wired in a hardware implementation) or variable. In the latter case, the linear combination (or *connections*) is defined by a binary vector of the length of the LFSR. (In a hardware implementation this is achieved using, in addition to the shift register, a programmable control register which determines the shift register cells that are connected to the XOR circuit).

We denote by A the first LFSR in our construction, and by S (for *S*elector) the second one. $|A|$ and $|S|$ denote their lengths and the sequences they produce (after fixing the connections and initial contents of the registers) are denoted $a_0, a_1, \ldots$ and $s_0, s_1, \ldots$, respectively. We also refer to these sequences as *A-sequence* and *S-sequence.* Finally the resultant shrunken sequence is denoted by $z_0, z_1, \ldots$, and called the *Z-sequence.*

This construction is well defined for both fixed and variable connection LFSRs. In general, we recommend the use of variable connections both for security and flexibility. This issue is discussed throughout this paper in the appropriate places. Let us mention that in the case of a fixed connection implementation only the seeds (i.e., the initial contents of the shift registers) for the LFSRs A and S constitute the *secret key* for the pseudorandom generator (or the encryption/decryption key, when used as a stream cipher). If variable (programmable) connections are used then the value of these connections is also part of the key.

1.2 Properties

We analyze some of the properties of the resultant LFSR-based shrinking generator. We show that the period of the Z-sequences is exponential in both $|A|$ and $|S|$, and that its linear complexity is exponential in $|S|$. The linear complexity of a sequence is the length of the shortest LFSR that generates that sequence (or equivalently, the shortest recursive linear dependence over GF(2) satisfied by the sequence bits). The importance of this property is that sequences with low linear complexity are easily predictable (see section 2) and constructions based on LFSRs tend to preserve much of the linearity inherent to LFSRs. The above properties equally hold for fixed or variable connections. On the other hand, our statistical analysis of these sequences takes into account variable connections (chosen with uniform probability over the set of primitive connections[1]). We show that the space of resultant sequences has some of the necessary statistical properties for a pseudorandom generator: low correlation between the sequence bits, normalized appearance of 0's and 1's, and balanced distribution of subpatterns. Our statistical analysis uses Fourier analysis and ε-biased distributions as the main tools. The period and linear complexity bounds are proven mainly through algebraic techniques.

We stress, again, that all these properties are only *necessary* (but far from sufficient) conditions on the cryptographic strength of the pseudorandom generator. They just show that the elemental goals for an LFSR-based construction are achieved, namely, the destruction of the linearity while preserving the good statistical properties.

In section 4 we present some attacks and analyze their effect on our construction. These attacks work in time exponential in the length of register S, and indicate an effective key length bound of about half of the total key length.

Practical considerations regarding the implementation and practical use of our generator are discussed in section 5. In particular, we show how the problem

[1] Connection vectors for LFSRs are closely related to polynomials over GF(2). Best connections for LFSRs are those which correspond to primitive polynomials of the same degree as the LFSR's length. In that case, the sequence generated by the LFSR has *maximal length*, namely, a period of $2^n - 1$, where n is the length of the register [9]. Throughout this paper we implicitly assume a construction of the SG using primitive connections. Such connections are easy to find using probabilistic methods, e.g. see [17].

of irregular rate of the output bits present in our basic scheme can be solved at a moderate cost in hardware implementation.

Finally, in section 6 we discuss some existing alternative constructions and their relation to the shrinking generator.

2 Period and Linear Complexity

In this section we prove exponential bounds on the period and linear complexity of sequences produced by the shrinking generator. In the case of the period this bound is tight; for the linear complexity there is a gap by a factor of 2 between the lower and upper bound.

The importance of a long period is to avoid the repetition of the sequence after short period of times. An exponentially large linear complexity avoids one of the more generic attacks on pseudorandom sequences and/or stream ciphers. There is no need to even know the way a sequence is generated in order to break it through its linear complexity. Any sequence of linear complexity ℓ can be entirely reconstructed out of 2ℓ known bits by using the Berlekamp-Massey algorithm, which in time $O(\ell^2)$ finds the shortest linear dependency satisfied by the sequence (a-priori knowledge of the value of ℓ is not necessary). See, e.g. [3]. (On the other hand, high linear complexity by itself is far from being an indication of the sequence unpredictability. It suffices to mention that the sequence 00...001 has linear complexity as the length of the sequence).

Our results on the period and linear complexity of sequences generated with the shrinking generator are stated in the next theorems.

Theorem 1. *Let A and S form a shrinking generator Z. Denote by T_A , T_S, the periods of the A- and S- sequences respectively. If*

- *A and S are maximal length (i.e. have primitive connections)*
- $(T_A, T_S) = 1$

then the shrunken sequence Z has period $T_A \cdot 2^{|S|-1} = (2^{|A|} - 1) \cdot 2^{|S|-1}$.

Note: S must not be of maximal length. In the general case the period of the Z-sequence is $T_A \cdot W_S$, where W_S is the number of 1's in a full period of S. If both A and S are of maximal length then the condition $(T_A, T_S) = 1$ is equivalent to $(|A|, |S|) = 1$. For the next theorem S may also not be a maximal length sequence but we do need that W_S be a power of 2.

Theorem 2. *Under the conditions of Theorem 1, the shrunken sequence Z has linear complexity LC, where $|A| \cdot 2^{|S|-2} < LC \leq |A| \cdot 2^{|S|-1}$*

In the following proofs of theorems 1 and 2 we use some well-nown algebraic facts about sequences produced by LFSRs. These properties can be found in many textbooks (e.g. [9, 14]).

Notation: For the sake of readiness we use the following notation through these proofs: $a(i)$ denotes the A-sequence, $s(i)$ the S-sequence, and $z(i)$ the shrunken sequence Z. By k_i we denote the position of the i-th '1' in the S sequence. In other words, $\forall i, z(i) = a(k_i)$. We denote by W_S the number of 1's in a full period of S. For a maximal length sequence S this number is $2^{|S|-1}$.

Proof of Theorem 1:

Assumption: For simplicity of the proof we assume

$$|S| \leq T_A \ \ (i.e.\ |A| > log|S|). \tag{1}$$

The following fact is immediate from the definition of the shrunken sequence Z. **Fact 1:** Advancing W_S elements in the sequence z results in advancing T_S elements in the sequence a. Formally, $z(i + W_S) = a(k_i + T_S)$.
In general, for all $j = 0, 1, \ldots,$

$$z(i + jW_S) = a(k_i + jT_S). \tag{2}$$

Fact 2: Let k and k' be any pair of indices. If for all j: $a(k+jT_S) = a(k'+jT_S)$, then T_A divides $k - k'$.

Proof: Because of the A-sequence being of maximal length and $(T_A, T_S) = 1$ then the sequence $a(k + jT_S)$, $j = 0, 1, \ldots,$ is also maximal length and thus its period is T_A. ◇

Denote by T the (minimal) period of the sequence z. Clearly, the sequence z becomes periodic after $T_A \cdot W_S$ elements (since then both sequences a and s simultaneously complete a period). Therefore, T divides $T_A \cdot W_S$. We now proceed to show that $T_A \cdot W_S$ divides T.
By definition of T, for all i , $z(i) = z(i + T)$. In particular, for all i and j, $z(i + jW_S) = z(i + T + jW_S)$. Using (2) we get, for all i and j : $a(k_i + jT_S) = a(k_{i+T} + jT_S)$. Using Fact 2, we have

$$\forall i,\ T_A \ divides\ k_{i+T} - k_i. \tag{3}$$

Next step is to show, that (3) is possible only if W_S divides T. We reformulate (3) as:

$$\forall i, \exists j_i : k_{i+T} = k_i + j_i T_A \tag{4}$$

Putting $i + 1$ instead of i in (4) we get

$$k_{i+1+T} = k_{i+1} + j_{i+1} T_A \tag{5}$$

Subtracting (4) from (5) we get:

$$\forall i,\ k_{i+T+1} - k_{i+T} = k_{i+1} - k_i + (j_{i+1} - j_i) T_A. \tag{6}$$

Notice that k_{i+T} and k_{i+T+1} , as well as k_i and k_{i+1} , are the positions of consecutive 1's in the S-sequence. If $j_{i+1} - j_i$ would be different than zero, it would imply the existence of at least T_A consecutive zeros in the S-sequence, which is impossible by assumption (1). Therefore we get $j_{i+1} - j_i = 0$, and then for all i, $k_{i+T+1} - k_{i+T} = k_{i+1} - k_i$.
The later implies that the subsequence of s starting at $s(k_i)$ is identical to the subsequence starting at $s(k_{i+T})$. This means that T_S divides $k_{i+T} - k_i$, or

equivalently, that the number of elements in the S-sequence between $s(k_i)$ and $s(k_{i+T})$ is a multiple of its period . But then the number of 1's in this segment is a multiple of W_S. On the other hand, the number of 1's is exactly T, thus proving that W_S divides T.

Let t be such that

$$T = tW_S. \tag{7}$$

We have, for all j:

$$a(k_0) = z(0) = z(jT) = z(jtW_S) = a(k_0 + jtT_S). \tag{8}$$

Last equality follows from (2). We got that for all $j : a(k_0) = a(k_0 + jtT_S)$. This implies that T_A divides tT_S , and since $(T_A, T_S) = 1$, then T_A divides t. From (7) we get $T_A \cdot W_S$ divides T. □

The lower bound in the following proof of Theorem 2 is derived using the proven exponential period through an elegant technique from Gunther [10].

Proof of Theorem 2:

Upper bound on the linear complexity: Let z denote the variable corresponding to the sequence Z. To show an upper bound on the linear complexity of the sequence Z it suffices to present a polynomial $P(\cdot)$ for which $P(z) = 0$ (i.e. the coefficients of P represent a linear relation satisfied by the elements of Z). The variable z^{W_S} denotes the sequence Z decimated by W_S, i.e. the sequence $z(jW_S), j = 0, 1, \ldots$ Fact 1 in the proof of Theorem 1 states that this decimation, written in terms of the A-sequence, results in a sequence of the form $a(i + jT_S)$. Since we assume $(T_S, T_A) = 1$, the latter is a maximal length sequence with same linear complexity as the original A-sequence, and then it satisfies a polynomial $Q(\cdot)$ of degree $|A|$. But then also the decimated sequence z^{W_S} satisfies the polynomial, i.e. $Q(z^{W_S}) = 0$. Therefore, we have found a polynomial $P(z) = Q(z^{W_S})$ of degree $|A| \cdot W_S$, such that $P(z) = 0$, and then the linear complexity of the Z-sequence is at most $|A| \cdot |W_S| = |A| \cdot 2^{|S|-1}$.

Lower bound on the linear complexity: Let $M(z)$ denote the minimal polynomial of z. Since the sequence Z satisfies $Q(z^{W_S}) = 0$, we have that $M(z)$ must divide $Q(z^{W_S})$. Since $W_S = 2^{|S|-1}$, we have $Q(z^{W_S}) = Q(z^{2^{|S|-1}}) = (Q(z))^{2^{|S|-1}}$, and then $M(z)$ must be of the form $(Q(z))^t$ for $t \leq 2^{|S|-1}$. Assume $t \leq 2^{|S|-2}$. Then, $M(z)$ divides $(Q(z))^{2^{|S|-2}}$. Since $Q(z)$ is an irreducible polynomial of degree $|A|$ it divides the polynomial $1 + x^{T_A}$. Therefore, $M(z)$ divides $(1 + x^{T_A})^{2^{|S|-2}} = 1 + x^{T_A \cdot 2^{|S|-2}}$, but then the period of Z is at most $T_A \cdot 2^{|S|-2}$ contradicting Theorem 1. Therefore, $t > 2^{|S|-2}$ and the lower bound follows. □

3 Statistical Properties

3.1 Background

In this subsection we bring the required background on the techniques used in our analysis of the statistical properties of the shrinking generator; specifically, the notions of Fourier Transform (for Boolean domains) and ε-bias distributions.

Fourier Transform Boolean functions on n variables are considered as real valued functions $f: \{0,1\}^n \rightarrow \{-1,1\}$. The set of all real functions on the cube is a 2^n-dimensional real vector space with an inner product defined by:

$$< g, f >= 2^{-n} \sum_{x \in \{0,1\}^n} f(x)g(x) = E(gf)$$

(where E is expectation) and as usual the *norm* of a function is defined: $\|f\| = \sqrt{< f, f >}$, which is the Euclidean norm.

The basis of the cube $\mathbf{Z_2^n}$ is defined as follows: For each subset S of $\{1, \cdots, n\}$, define the function χ_S:

$$\chi_S(x_1, \cdots, x_n) = \begin{cases} +1 & \text{if } \sum_{i \in S} x_i \text{ is even} \\ -1 & \text{if } \sum_{i \in S} x_i \text{ is odd} \end{cases}$$

The following properties of this basis functions can be easily verified:

- For every A, B: $\chi_A \chi_B = \chi_{A \Delta B}$, where $A \Delta B$ is the symmetric difference of A and B.
- The family $\{\chi_S\}$ for all $S \subset \{1 \cdots n\}$ forms an orthonormal basis, i.e., if $A \neq B$, then $< \chi_A, \chi_B >= 0$, and for every A, $< \chi_A, \chi_A >= 1$.

Any real valued function on the cube can be uniquely expressed as a linear combination of the basis functions, i.e. $\sum_S c_S \chi_S$, where c_S are real constants. The *Fourier transform* of a function f is the expression of f as a linear combination of the χ_S's. For a function f and $S \subset \{1, \cdots, n\}$, the S'th Fourier coefficient of S denoted by $\hat{f}(S)$ is what was previously called c_S, i.e., $f = \sum_S \hat{f}(S) \chi_S$.

Since the χ_S's are an orthonormal basis, Fourier coefficients are found via:

$$\hat{f}(S) =< f, \chi_S >$$

For boolean f this specializes to:

$$\hat{f}(S) = Pr[f(x) = \oplus_{i \in S} x_i] - Pr[f(x) \neq \oplus_{i \in S} x_i]$$

where $x = (x_1, x_2, \ldots, x_n)$ is chosen uniformly at random.

ε-biased Distributions We consider a distribution function as a function from $\{0,1\}^n$ to the interval $[0,1]$. Given a probability distribution μ, then $\sum_x \mu(x) = 1$ and $\mu(x) \geq 0$. We can treat μ as any other function, and consider its Fourier coefficients. For example the uniform distribution is $U(x) = \frac{1}{2^n}$, which implies that $\hat{U}(S) = 0$, for $S \neq \emptyset$, and $\hat{U}(\emptyset) = \frac{1}{2^n}$.

A distribution is ε-bias if it is "close" to the uniform distribution in the following sense.

Definition 3. A distribution μ over $\{0,1\}^n$ is called an ε-bias distribution if for every subset $S \subset \{1 \ldots n\}$, $|\hat{\mu}(S)| \leq \varepsilon 2^{-n}$.

The notion of ε-bias distribution was introduced in [16], the main motivation being the derandomization of randomized algorithms, and the construction of small sample spaces that approximate the uniform distribution.

The following theorem from [1] connects LFSRs and ε-bias distributions.

Theorem 4. *([1]) Consider the distribution $\mathcal{D}(m,n)$ of strings of length n output by a LFSR A of length m, where the connections for A are chosen with uniform probability among all primitive polynomials over GF(2) of degree m, and the seed for A is chosen uniformly over all non-zero binary strings of length m. Then, $\mathcal{D}(m,n)$ is an $\frac{n-1}{2^m}$-bias distribution.*

Definition 5. Let f be a function from $\{0,1\}^n$ to the real numbers. Define $L_1(f) = \sum_S |\hat{f}(S)|$.

The following lemma relates ε-bias distributions and the norm $L_1(f)$. (See [13].) The function f can be seen as a test for distinguishing the distribution μ from the uniform distribution. The lemma states an upper bound on the quality of distinction, and therefore it is useful for tests of pseudorandomness.

Lemma 6. *([13])*

$$|E_U[f] - E_\mu[f]| \leq \varepsilon L_1(f)$$

where U is the uniform distribution and μ is an ε-bias distribution.

Proof. By simple arithmetic

$$E_\mu[f] = \sum_S \hat{f}(s)\hat{\mu}(S) = \hat{f}(\emptyset)\hat{\mu}(\emptyset) + \sum_{S\neq\emptyset} \hat{f}(S)\hat{\mu}(S).$$

Note that by definition $\hat{f}(\emptyset) = E_U[f]$. Since $\hat{\mu}(\emptyset) = 1/2^n$,

$$|E_U[f] - E_\mu[f]| = \sum_{S\neq\emptyset} \hat{f}(S)\hat{\mu}(S) \leq \varepsilon L_1(f).$$

Here we used the fact that each $\hat{\mu}(S)$ is bounded by ε. □

L_1 norm Lemma 6 is useful if we can upper bound the value $L_1(f)$. In this section we present some methods for bounding the L_1 norm of a function. The following technical Lemma gives a tool for doing that.

Lemma 7. *Let f and g be functions from $\{0,1\}^n$ to the real numbers. Then, $L_1(fg) \leq L_1(f)L_1(g)$ and $L_1(f+g) \leq L_1(f) + L_1(g)$.*

For many simple functions we can show that the L_1 is small. Here are a few examples.

Lemma 8.

- *Let* $sum(x) = \sum_{i=1}^{n} x_i$, *then* $L_1(sum) = n$.
- *Let* $AND(x) = \prod_i x_i$, *then* $L_1(AND) = 1$.
- *For* $B \in \{0,1,*\}^n$ *we define a template* $\texttt{template}_B(x) = 1$ *iff* x *and* B *agree on each* 0 *or* 1 *in* B, *i.e. for each* $b_i \neq *$ *then* $b_i = x_i$. *(For example* $\texttt{template}_{10*1*}(10110) = 1$ *while* $\texttt{template}_{10*1*}(00110) = 0$.*) For any* $B \in \{0,1,*\}^n$ *then* $L_1(\texttt{template}_B) = 1$.

Proof. For the *sum*, we can rewrite it as $n/2 + \sum_i \chi_{\{i\}}(x)/2$. Using the additivity of the L_1 the claim follows.

Note that the *AND* function is either 0 or 1 (and not ± 1). We rewrite the *AND* to be

$$\prod_{i=1}^{n} \frac{1 - \chi_{\{i\}}(x)}{2}$$

Note that $L_1(\frac{1-\chi_{\{i\}}(x)}{2}) = 1$, and the claim follows from the multiplicative properties of L_1 (see Lemma 7).

For $\texttt{template}_B(x)$ the proof is the same as for the AND function. We rewrite the function as,

$$\texttt{template}_B(x) = \left(\prod_{i:b_i=1} \frac{1 - \chi_{\{i\}}(x)}{2}\right)\left(\prod_{j:b_j=0} \frac{1 + \chi_{\{i\}}(x)}{2}\right)$$

and again we use the multiplicative property of the L_1. □

3.2 Applications to LFSR with variable connections

In this subsection we show that LFSR with variable connections have many properties that resemble random strings. In fact the only property that we use is that LFSR where the connections are chosen at random generates an ε-bias distribution, with exponentially small ε (see Theorem 4).

Theorem 9. *Let* A *be an LFSR where the connections for* A *are chosen with uniform probability among all primitive polynomials of degree* m *over* $GF(2)$. *Let* X *be the sum of* n *different bits* $i_1, \ldots, i_n$ *in the A-sequence (we assume that* $i_j \leq 2^m - 1$*). Let* $Y = \sum_{i=1}^{n} y_i$ *where* y_i *are i.i.d.* $\{0,1\}$*-random variables and* $Prob[y_i = 1] = 1/2$. *Then, the expected value of* X *is at most* $\frac{n^2}{2^m}$. *The difference between the variance of* X *and* Y *is bounded by* $\frac{n^3+n^2}{2^m}$. *Furthermore,* $|E[X^k] - E[Y^k]| \leq \frac{n^k}{2^m}$.

Proof. By definition $X = sum(a_{i_1}, \ldots, a_{i_n})$. By Lemma 8, $L_1(X) = n$. Also, $L_1(X^2) \leq n^2$ and $L_1(X^k) \leq n^k$. The theorem follows from Theorem 4 and Lemma 6. □

The following theorem applies the ideas of a general template to an LFSR sequence and shows that the probability that the template appears is close to the probability it appears in a random string.

Theorem 10. *Let A be an LFSR where the connections for A are chosen with uniform probability among all primitive polynomials of degree m over GF(2). Let X be the first n output bits of A and Y a random string of n bits. Let $B \in \{0,1,*\}^n$ be a template. Then*

$$|E[\texttt{template}_B(X)] - E[\texttt{template}_B(Y)]| \leq \frac{n}{2^m}.$$

Proof. By Lemma 8, we have that $L_1(template_B) = 1$. By Theorem 4 the string X is an ε-bias distribution, with $\varepsilon \leq \frac{n}{2^m}$. The theorem follows from Lemma 6. □

When we will consider the selector register S of the shrinking generator it would be important to argue how many bits we should consider in order to generate k output bits. The following theorem shows that the expected number is $O(k)$.

Theorem 11. *Let S be an LFSR where the connections for S are chosen with uniform probability among all primitive polynomials of degree m over GF(2). Let $i_k(S)$ be the location of the kth 1 bit in S, then the expectation*

$$E_S[i_k(S)] = O(k).$$

The proof of the above theorem will be given in the final version.
Remark: Note that all the proofs in this subsection were based only on the ε-bias properties, and therefore would hold for any ε-bias distribution.

3.3 Applications to the Shrinking Generator

In this subsection we apply the results in the previous subsection to the shrinking generator. Basically we show that the good random-like properties that existed in LFSR with variable connection remain in the shrinking generator. (Clearly, the shrinking generator has other essential properties not present in LFSR sequences, e.g. the exponential linear complexity.)

The following is a simple corollary of theorem 9, which states that the moments of the output of the shrinking generator are very close to the moments of a random string.

Corollary 12. *Let Z be a sequence generated by a shrinking generator with registers A and S. Let X be the sum of consecutive n bits in the Z-sequence (we assume that $n|S| \leq 2^{|A|}$). Let $Y = \sum_{i=1}^{n} y_i$ where y_i are i.i.d. $\{0,1\}$-random variables and $Prob[y_i = 1] = 1/2$. Then, the expected value of $|X|$ is at most $\frac{n^2}{2^{|A|}}$. The difference between the variance of X and Y is bounded by $\frac{n^3+n^2}{2^{|A|}}$. Furthermore, $|E[X^k] - E[Y^k]| \leq \frac{n^k}{2^{|A|}}$.*

Proof. Fix a specific S-sequence. The consecutive n bits in Z were generated by some n non consecutive different bits in the A-sequence, denote their indeces in this sequence by $i_1, \ldots, i_n$. Since $n|S| \leq 2^{|A|}$, we are in the same period of A, i.e. $i_j \leq 2^{|A|} - 1$. Since X is the sum of those bits, the corollary follows from Theorem 9. □

The following theorem shows that each template is distributed similarly in the output of the shrinking generator and a random string.

Theorem 13. *Let Z be a sequence generated by a shrinking generator with registers A and S. Let X be the first n bits in Z and Y be a random string of n bits. Let $B \in \{0,1,*\}^n$ be a template. Then*

$$|E[\mathtt{template}_B(Z)] - E[\mathtt{template}_B(Y)]| = O(\frac{n}{2^{|A|}}).$$

Proof. The bits of X come from the first $i_n(S)$ bits of A, where $i_n(S)$ is the index of the nth '1' bit in the S-sequence. Given S and B we can create a template B_S of size $i_n(S)$ for A (we simply put $*$ in any location that S is 0, and copy B in the locations where S is 1).

Note that $\mathtt{template}_B(X) = \mathtt{template}_{B_S}(A)$, once we fix S. Therefore it is sufficient to bound

$$\sum_S Prob[S] \left| E_A[\mathtt{template}_{B_S}(A)] - E_Y[\mathtt{template}_B(Y)] \right|.$$

By Theorem 10 the difference between the expectation is bounded by $i_n(S)/2^{|A|}$. Therefore,

$$\sum_S Prob[S] |E[\mathtt{template}_{B_S}(A)] - E[\mathtt{template}_B(Y)]| \leq$$

$$\leq \sum_S Prob[S] \frac{i_n(S)}{2^{|A|}} = \frac{E_S[i_n(S)]}{2^{|A|}} = O(\frac{n}{2^{|A|}}).$$

The last identity follows from Theorem 11. □

We now show some interesting applications of the above theorem. First we consider correlation between pairs of output bits. The correlation between two bit positions is the difference (in absolute value) between the probability that the two bits are equal and the probability that they differ.

Corollary 14. *Let Z be a sequence generated by a shrinking generator with registers A and S. Let X_1, X_2 be two bits in the Z-sequence that are at distance ℓ. The correlation between X_1 and X_2 is bounded by $O(\frac{\ell}{2^{|A|}})$.*

Proof. Simply use the four templates $\sigma_1 \overbrace{* \cdots *}^{\ell} \sigma_2$, where $\sigma_1, \sigma_2 \in \{0,1\}$, and apply Theorem 13. □

The next corollary shows that the distribution of patterns is almost uniform.

Corollary 15. *Let P be any binary string (pattern) of k bits and let X_k be the k consecutive bits in the Z-sequence. The probability that $X_k = P$ is in the range $2^{-k} \pm O(\frac{k}{2^{|A|}})$.*

Note that this corollary is a special case of Theorem 13.

4 Attacks

In this section we present some attacks on the shrinking generator. These attacks indicate an effective key length of the length of register S, or about twice this length if the connections for the registers are part of the key (i.e. the connections are variable and secret). More details on these and other attacks will be presented in the final version of this paper.

4.1 Attacking through S

If the connections for both S and A are known then one can exhaustively search for S's seed; each such seed can be expanded to a prefix of the S-sequence using the connection of S. Let $n = |A|$ and suppose we expand the S-sequence until its n-th '1' is produced. From this prefix, and from knowledge of a corresponding n-long prefix of the Z-sequence, one derives the value of n (non-consecutive) bits in the A-sequence. Since A's connections are known then A's seed can be recovered given these n bits by solving a system of linear equations (in general, the dimension of this system is about $n/2$ since about half of the seed bits – corresponding to 1's in S – are known). Therefore the attack's complexity is exponential in $|S|$ and polynomial in $|A|$, or more precisely, $O(2^{|S|} \cdot |A|^3)$.

If the connections of A are secret as we recommend, then the above procedure does not work since in order to write the system of equations one needs to know these connections. In this case the following attack avoids doing an exhaustive search on A's connections. This attack tries all possible seeds and connections for S (assuming S's connections are secret). Each pair of seed and connections for S is used to expand the seed into a t-long prefix of the S-sequence, for some integer t. With this prefix and sufficiently many bits (about $t/2$ bits) from the Z-sequence (known plaintext) it is possible to generate the first t bits of the product sequence $p_i = a_i \cdot s_i$. (Notice that bits from the A-sequence corresponding to positions of 1's in the S-sequence are known using the known part of the Z-sequence, and positions in which the $s_i = 0$ are also 0's in the product sequence). The interesting property of this product sequence is that its linear complexity is at most $|A| \cdot |S|$ (see [18]) and therefore having $t = 2 \cdot |A| \cdot |S|$ in the above attack suffices to find the whole product sequence p_i. The cost is quadratic in $|A| \cdot |S|$. This information together with the S-sequence, which is known, permits deriving the full sequence Z_i. Therefore the cost of the attack is the number of seeds and connections to be tried for S (about $2^{2|S|}/|S|$) times the complexity of recovering p_i through its linear complexity (i.e. $O((|A| \cdot |S|)^2)$). The necessary amount of plaintext (i.e. bits from Z) is $|A| \cdot |S|$. As before this attack indicates an effective key length of at most twice the length of S, or about half of the total key length.

4.2 Linear Complexity

Attacking the SG through its linear complexity requires the knowledge of an exponential in $|S|$ number of bits from the sequence, more precisely, $2^{|S|-2} \cdot |A|$

bits at least (see Theorem 2). On the other hand, the typically quadratic work that takes to derive the sequence from a prefix of that length is not necessary here. Having $2^{|S|} \cdot |A|$ consecutive bits from the sequence one can derive the whole sequence. The proof of Theorem 1 indicates that a decimation of the Z-sequence by factors of $W_S = 2^{|S|-1}$ implies the decimation of the A-sequence by a factor of $T_S = 2^{|S|} - 1$. Therefore, from $z(i + jW_S)$, $j = 0, 1, \ldots, 2 \cdot |A| - 1$ one derives $z(i + jW_S)$, for all j.

The complexity to break the whole sequence in this way is $O(2^{|S|} \cdot |A|^2)$ (even if the connections are secret). In addition to this computational complexity this attack requires $2^{|S|} \cdot |A|$ consecutive bits from the sequence. In any case, the parameters for the SG should be chosen such that collecting this many number of sequence bits be infeasible.

4.3 Other Attacks

The more traditional attacks on LFSR-based construction seem not to apply to our construction due to its different nature. These attacks include the analysis of boolean functions used for the combination of LFSR outputs, the correlation of generated bits relative to subcomponents in the system, and others (See [18] for more details on these attacks and their applications).

It is worth mentioning that a typical weakness of LFSR-based systems is encountered in implementations where the connection polynomials are chosen to be very sparse (i.e. only a few coefficients chosen to be non-zero). In this case, special attacks can be mounted taking advantage of this fact. We recommend not to implement any of these systems in such a way, including ours. (In a hardware implementation having sparse connections may be advantageous only if the connections are fixed). On the other hand, most of these attacks will work not only if the connection polynomial itself is sparse, but also if this polynomial has a multiple of moderately large degree which is sparse. We can mount special attacks on our system against such sparse multiples, although they are all exponential in $|S|$. Again these attacks are more relevant to fixed connection implementations, where heavy preprocessing can be done against the particular connections, than in the case of variable connections.

5 Practical Considerations

5.1 Overcoming Irregular Output Rate

The way the SG is defined, bits are output at a rate that depends on the appearance of 1's in S output. Therefore, this rate is on *average* 1 bit for each 2 pulses of the clock governing the LFSRs. This problem has two aspects. One is the reduced throughput relative to the LFSRs speed, the other the irregularity of the output. We show here that this apparently practical weaknesses can be overcome at a moderate price in hardware implementation (on the other hand, these "weaknesses" give most of the cryptographic strength to this construction).

We stress that this hardware cost is usually less than the required for adding more LFSRs (even one) to the construction (as many constructions do).

In order to achieve an average of 1 bit per clock pulse, the LFSRs can be easily speeded up with a very moderate cost in hardware: only the XOR tree is to be replicated (this is true also if the connections are variable!). Notice that whether this speed-up is necessary depends on the relation between the LFSR clock speed and the required throughput from the SG (e.g., when used in a stream cipher system this throughput depends on the data speed). If the clock is fast enough this speed-up may be not necessary at all. On the other hand, for fast data encryption a speedup mechanism may be necessary regardless of the reduced throughput of our construction.

The problem of irregular output rate can be serious in real-time applications where repeated delays are not acceptable. Fortunately, this problem can be also solved at a moderate cost. The solution is to use a short buffer for the SG output intended to gather bits from the SG output when they abound in order to compensate for sections of the sequence where the rate output is reduced. In [11] Markov analysis is applied to analyze the influence of such a buffer for the output rate of the SG. It is shown that even with short buffers (e.g., 16 or 24 bits) and with a speed of the LFSRs of above twice the necessary throughput from the SG the probability to have a byte of pseudorandom bits not ready in time is very small. (Examples are a probability of $5 \cdot 10^{-3}$ for buffer of size 16 and speedup factor of 9/4, or a probability of $3 \cdot 10^{-7}$ for a buffer of size 24 and speedup factor of 10/4. These probabilities decrease exponentially with increasing buffer sizes and speedup factors). We note that in most implementations of stream ciphers, some buffering naturally exist because of data coming in blocks of a given size (e.g depending on the bus width). Therefore the above technique may add none or very little bits to the buffer size. In many cases the above small probabilities of delayed pseudorandom bits is affordable. In cases it is not, we propose filling the missing bits with arbitrary values (e.g. alternate 0's and 1's) which can hardly hurt with a miss probability of $3 \cdot 10^{-7}$ or so. An alternative (but somewhat less simple) heuristic solution is to periodically buffer some bits of the A-sequence corresponding to 0's in S in order to use them for filling the missing bits in case of need.

5.2 Fixed vs. Variable Connections

Throughout the paper we have recommended several times the use of variable connections for the LFSRs A and S. Although variable connection do not influence the period and linear complexity of the resultant sequences, their advantage is apparent from the attacks discussed in section 4 (e.g., to avoid attacks using heavy precomputation for analyzing the particular connections, or the preparation of big preprocessing tables), and from the statistical analysis of section 3. They may be also beneficial in standing future attacks to the system.

In addition to these security advantages, using variable connections provides a large degree of flexibility to the construction (this is true for other LFSR-based constructions as well). Through the programming of these connections the security of the sytem can be tuned down or up with no change in the hardware. This is most important for systems where versions of different security levels use the same physical device (e.g. cryptographic systems sold in different countries with different levels of permitted security). Tuning down the security is done through a virtual shortening of the registers by loading zeros into the most significant locations of the connection registers.

We stress that while there is a cost in hardware associated with the connection registers, this cost is compensated with the possible choice of shorter registers when using variable connections, and by the above advantages. Moreover, having shorter registers implies having shorter seeds. The latter are the part of the key which keeps changing with bit generation while the connections are kept unchanged for long periods. Having shorter seeds help the key management and synchronization aspects (especially, when used in a stream cipher cryptosystem).

6 Discussion and Related Work

LFSR-based constructions are encountered today in many practical systems, especially for implementation of stream ciphers. Because of their conceptual and implementation simplicity they will keep being attractive; in particular, since they are simple to parallelize and pipeline they are natural candidates for high speed encryption, too. Moreover, LFSRs are widely used in non-cryptographic applications (coding, CRCs, whitening, etc), and then it's plausible to have new technologies supporting the construction of efficient LFSRs. In addition, LFSR-based constructions have the important practical property that the amount of required hardware can be traded-off against different levels of security; on the other hand, same hardware can handle different levels of security (see Section 5.2). From a theoretical point of view, it is puzzling whether such simple constructions may have a good cryptographic strength. For all these reasons it seems important to have some good construction(s) well evaluated by the cryptographic community. The one presented in this paper may be a good candidate for evaluation, as it compares to the best existing alternatives, and may have the potential to prove better.

Interesting examples of existing LFSR-based constructions for comparison with the shrinking generator are Gunther's *alternating step generator* [10], and some of the clock-controlled generators discussed in [8], in particular the *1-2 generator*. They have similar proven properties as ours, but both are developments of the weak "stop-and-go generator" [2]. This generator uses two LFSRs where the first one is used to control the clock of the second LFSR. Therefore, a '1' output by the first LFSR causes the second one to shift its state, while a '0' implies that the state keeps unchanged (but still a bit, same as the previous one, is output). The output of this second LFSR is then the output of the stop-and-go generator; and the weakness of the repeated bit is clear. The 1-2 generator solves this problem by shifting one bit of the second LFSR when the first LFSR outputs '0', and shifting two bits when the first LFSR outputs '1'.

Gunther's construction uses three registers and outputs the bitwise XOR of two stop-and-go sequences controlled by the same third LFSR. Actually, Gunther's generator is equivalent to a generator that merges two LFSR sequences S_0 and S_1 according to the '0's and '1's output by a third LFSR (a '0' implies taking next bit from S_0 a '1' implies taking next bit from S_1). This construction has the nice property that each bit in the output may (a-priori) correspond to any of the two sequences; on the other hand, it lacks the property of omitting bits from these sequences.

One advantage of Gunther's generator is that it guarantees one output bit per LFSR clock pulse, but it pays for it with a third LFSR. In our construction, the hardware prize we pay in order to regulate the output rate (see section 5.1) is usually lower than introducing a third LFSR (this is due to the fact that XOR gates usually cost significantly less than memory elements). Moreover, this third LFSR brings the effective key length of Gunther's scheme to one third of the total length (it can be broken through exhaustive search on only one of the three registers). The 1-2 generator has the effect of omitting bits through its irregular clocking, but this omission is by nature very local, e.g. one of any two consecutive bits originally output by one of the LFSRs appears in the generator's output sequence.

Locality appears in other versions of clock-controlled generators as well. In our construction the uncertainty about omission of bits is significantly superior (e.g., in clock-controlled constructions t bits from the control sequence determine the original locations in the other register of t output bits; in the shrinking generator, however, $2t$ bits in the selecting register S are necessary (on average) to determine the original locations of t bits in the Z-sequence). In particular, notice that the shrinking generator is *not* a special case of a clock-controlled generator (e.g., its output is not synchronized with the selecting register as it is the case in any clock-controlled scheme). Moreover, the general techniques on clock-controlled generators [8] do not directly apply to our construction.

Finally, the work by Golic and Zivkovic [7] shows that most *irregularly decimated* LFSR-sequences have high linear complexity; however, their result is non-constructive by nature and has no implication on our construction.

We stress that the omission of bits is important not only in LFSR-based constructions but also in other constructions as well. On the other hand, not every scheme for omission of bits is effective (e.g. a decimated LFSR sequence is as bad as the original sequence itself). For the linear congruential number generator outputting all of the bits of a generated number makes the task of breaking it a very easy one [5]. Even if some bits are omitted but a block of consecutive bits are output, efficient predicting methods are known [6, 19]. The extended family of congruential generators is efficiently predictable if sequence elements are output with no omission [12], but no efficient methods are reported for these sequences if part of the bits are omitted. It is an interesting open problem what can be proven for a shrinking generator based on congruential generators. Finally, let us mention that the idea of outputting individual bits of a sequence, is best captured by the notion of *hard bits* of a one-way function, a notion that plays a central role in the construction of complexity-theory based pseudorandom generators (see [4, 20] and subsequent works). It would be interesting to know whether the shrinking generator applied to two ε-predictable sequences guarantees, in general, a third sequence which is ε'-predictable for $\varepsilon' < \varepsilon < \frac{1}{2}$. (Roughly speaking, a sequence is ε-predictable if no polynomial-time algorithm can predict it with probability greater than $\frac{1}{2} + \varepsilon$).

Acknowledgement

We owe special thanks to Celso Brites, Amir Herzberg and Shay Kutten for their help and involvement during the development of the shrinking generator. Many people have contributed in different ways to this investigation; they include: Aaron Kershenbaum, Ilan Kessler, Ronny Roth, Kumar Sivarajan, and Moti Yung. To all of them many thanks.

References

1. Noga Alon, Oded Goldreich, Johan Hastad, and Rene Peralta. Simple constructions of almost k-wise independent random variables. In 31th *Annual Symposium on Foundations of Computer Science, St. Louis, Missouri*, pages 544–553, 1990.
2. Beth, T., and Piper, F., "The stop-and-go Generator", in *Lecture Notes in Computer Science 209; Advances in Cryptology: Proc. Eurocrypt '84*, Berlin: Springer-Verlag, 1985, pp. 88-92.
3. Blahut, R., *Theory and Practice of Error Control Codes*, Addison-Wesley, 1984.
4. Blum, M., and Micali, S., "How to Generate Cryptographically Strong Sequences of Pseudo-Random Bits", *SIAM Jour. on Computing*, Vol. 13, 1984, pp. 850-864.
5. Boyar, J. "Inferring Sequences Produced by Pseudo-Random Number Generators", *Jour. of ACM*, Vol. 36, No. 1, 1989, pp.129-141.
6. Frieze, A.M., Hastad, J., Kannan, R., Lagarias, J.C., and Shamir, A. "Reconstructing Truncated Integer Variables Satisfying Linear Congruences", *SIAM J. Comput.*, Vol. 17, 1988, pp. 262-280.
7. Golic, J.DJ., and Zivkovic, M.V., "On the Linear Complexity of Nonuniformly Decimated PN-sequences", *IEEE Trans. Inform. Theory*, Vol 34, Sept. 1988, pp. 1077-1079.
8. D. Gollmann and W.G. Chambers, "Clock-controlled shift registers: A review", *IEEE J. Selected Areas Commun.*, vol. 7, pp. 525-533, May 1989,
9. S.W. Golomb, *Shift Register Sequences*, Aegean Park Press, 1982.
10. Gunther, C.G., "Alternating Step Generators Controlled by de Bruijn Sequences", in *Lecture Notes in Computer Science 304; Advances in Cryptology: Proc. Eurocrypt '87*, Berlin: Springer-Verlag, 1988, pp. 88-92.
11. Kessler, I., and Krawczyk, H., "Buffer Length and Clock Rate for the Shrinking Generator", preprint.
12. Krawczyk, H., "How to Predict Congruential Generators", *Journal of Algorithms*, Vol. 13, 1992. pp. 527-545.
13. E. Kushilevitz and Y. Mansour. Learning decision trees using the fourier spectrum. In *Proceedings of the* 23rd *Annual ACM Symposium on Theory of Computing*, pages 455–464, May 1991.

14. Lidl, R., and Niederreiter, H., "Finite Fields", in *Encyclopedia of Mathematics and Its Applications*, Vol 20, Reading, MA: Addison-Wesley, 1983.
15. Yishay Mansour. An $o(n^{\log\log n})$ learning algorihm for DNF under the uniform distribution. In 5th *Annual Workshop on Computational Learning Theory*, pages 53–61, July 1992.
16. Joseph Naor and Moni Naor. Small bias probability spaces: efficient construction and applications. In *Proceedings of the* 22nd *Annual ACM Symposium on Theory of Computing, Baltimore, Maryland*, pages 213–223, May 1990.
17. Rabin, M.O., "Probabilistic Algorithms in Finite Fields", *SIAM J. on Computing*, Vol. 9, 1980, pp. 273-280.
18. Rueppel, R. A., "Stream Ciphers", in Gustavos J. Simmons, editor, *Contemporary Cryptology, The Science of Information*, IEEE Press, 1992, pp. 65-134.
19. Stern, J., "Secret Linear Congruential Generators Are Not Cryptographically Secure", *Proc. of the 28rd IEEE Symp. on Foundations of Computer Science*, 1987.
20. Yao, A.C., "Theory and Applications of Trapdoor Functions", *Proc. of the 23rd IEEE Symp. on Foundation of Computer Science*, 1982, pp. 80-91.

An Integrity Check Value Algorithm for Stream Ciphers

Richard Taylor

Telematic and System Security, Telecom Australia Research Laboratories,
P.O. Box 249, Clayton, Victoria 3168, AUSTRALIA, r.taylor@trl.oz.au

Abstract. A method of calculating an integrity check value (icv) with the use of a stream cipher is presented. The strength of the message integrity this provides is analysed and proven to be dependent on the unpredictability of the stream cipher used. A way of efficiently providing both integrity and encryption with the use of a single stream cipher is also explained. Note that the method of providing message integrity, used with or without encryption, is not subject to a number of attacks that succeed against many conventional integrity schemes. Specifically any legitimate message-icv pair that is copied or removed and subsequently replayed will have an appropriately small small chance of deceiving the receiver. Furthermore, any message-icv pair generated by an attacker and injected into the communication channel will have an appropriately small chance of escaping detection unless the attacker has actually broken the stream cipher. This is the case even if the attacker has any amount of chosen messages and corresponding icvs or performs any number of calculations.

1 Introduction

An integrity check value (icv) refers to a function of a secret key and variable length input messages. For a given key the function maps variable length input messages into a fixed length output. The secret key is shared between the message sender and receiver. The icv of a message is calculated by the sender and appended to the message before being sent. The receiver calculates the icv for the received message and compares the calculated icv with the received one. The message is deemed to be intact if the received and calculated icvs match. Assume that a hostile third party (or attacker) can tap into the communication channel and so can accumulate messages and their corresponding icvs. The integrity protection afforded against such an attacker is a measure of their difficulty in generating a new message and legitimate icv. Note that the term integrity check value as used here is also referred to in the literature as keyed hash function, cryptographic checksum or message authentication code.

The majority of integrity check value algorithms that have appeared in the literature are based on the use of block ciphers (see for example [10], [8] or the standards document [4]). Other work on the related notion of hash functions without secret keys has been motivated by the desirability of digital signatures for public-key cryptosystems (see [3], [12], [5]). Unfortunately there appears to be a lack of methods to efficiently extend the functionality of stream ciphers used for encryption so that both encryption and integrity are provided. Motivated by the need for such a method we propose an integrity check value algorithm based on the use of stream ciphers that is both efficient and secure. This algorithm is particularly suitable as part of an integrated method of providing both encryption and integrity with the use of a single stream cipher. Note that an alternative method of using a stream cipher for calculating an icv has recently been proposed (see [9]). This method can be successfully attacked with high probability however by altering a message and icv in transit (for example if the last bit of a message is altered then one of only two bits in the corresponding icv need be altered to create the matching icv).

Note that the integrity check value algorithm presented is related to certain unconditionally secure authentication schemes (see [1], [2] and [14]). However the use of the polynomial function (2) is new as far as the author is aware, as is the proposed use of the integrity method together with encryption from a single stream cipher.

2 Integrity

Consider any stream cipher with output stream $Z = (z_0, z_1, z_2, ...)$ where each output z_i consists of w bits. Thus each z_i may be considered as an integer from 0 to $2^w - 1$. We show how the output of the stream cipher may be used to provide message integrity by the calculation of an icv that is appended to a message and sent with it. In the following for positive integers t and u we shall write $t[u]$ to represent the unique integer satisfying

$$t[u] \equiv t(mod u) \text{ and } 0 \leq t[u] \leq u - 1.$$

To calculate the icv select a message block length b and prime number $p > 2^w$. It is suggested that p be chosen to be close to a power of 2 for efficient calculation of products modulo p (see [6] for example). Some examples are $p = 2^{31} - 1, 2^{61} - 1, 2^{89} - 1, 2^{107} - 1$ and $2^{127} - 1$. Let a message string $M = (m_0, m_1, ...)$ consisting of integers between 0 and $2^w - 1$ be partitioned into blocks $M_0, M_1, ..., M_s$ each containing at most b integers so that

$$M_j = m_{bj}, m_{bj+1}, ..., m_{b(j+1)-1}, j = 0, 1, ..., s - 1,$$
$$M_s = m_{bs}, m_{bs+1}, ..., m_{bs+t}, \text{ for some } t \leq b - 1.$$

Use the stream cipher to generate $s + 2$ outputs $z_i, z_{i+1}, z_{i+2}, ..., z_{i+s+1}$. The icv is calculated as

$$icv(M, b, p, z_i, z_{i+1}, ..., z_{i+s+1})$$
$$= (f(M_0, z_i) + f(M_1, z_{i+1}) + ... + f(M_s, z_{i+s}) + z_{i+s+1})[p] \quad (1)$$

where for any message string $N = (n_0, n_1, ..., n_r)$, and integer x

$$f(N, X) = (...((n_0 x + n_1)x + n_2)x + ... + n_r)x[p]. \quad (2)$$

Equivalently $f(N,x)$ may be expanded as

$$f(N,x) = (n_0 x^{r+1} + n_1 x^r + n_2 x^{r-1} + \dots + n_r x)[p].$$

Example 1. Let $w = 30, p = 2^{31} - 1, b = 20$. Let the cipher be in state 56 (the last output produced being z_{55}). Let $M = (m_0, m_1, \dots, m_{108})$ be a message string of length 109 that requires integrity. M is divided into blocks $M_0, M_1, \dots, M_4$ of length 20 where $M_i = (m_{20i}, m_{20i+1}, \dots m_{20i+19}), i = 0, 1, \dots, 4$, and one block of 9 integers $M_5 = (M_{100}, m_{101}, \dots, m_{108})$. The cipher is used to generate 7 outputs $z_{57}, z_{58}, \dots, z_{62}$, and the integrity check value

$$icv(M, 20, 2^{31} - 1, z_{56}, z_{57}, z_{58}, \dots, z_{62})$$

is calculated according to (1) and (2). The transmitted message is then

$$m_0, m_1, \dots, m_{108}, icv.$$

3 Analysis

The following Theorem and Corollaries establish a clear link between the strength of the integrity mechanism and the strength of the stream cipher from which it is constructed.

Theorem 1. *Let $p > 2^w$ be prime, and the function f be defined by (2). Let M and M' be any two unequal message strings of length b, and y any fixed integer. Then if x is a uniformly distributed random variable in the range 0 to $2^w - 1$,*

$$Probability(f(M,x) - f(M',x) \equiv y(mod p)) \leq \frac{b}{2^w}.$$

Proof. Let $M = (m_0, m_1, \dots, m_{b-1})$ and $M = (m'_0, m'_1, \dots, m'_{b-1})$. By expanding (2),

$$\begin{aligned} &f(M,x) - f(M',x) \\ &\equiv ((m_0 - m'_0)x^b + (m_1 - m'_1)x^{b-1} + \dots + (m_{b-1} - m'_{b-1}x)(mod p). \end{aligned}$$

Thus

$$f(M,x) - f(M',x) \equiv y(mod p)$$

if and only if

$$(m_0 - m'_0)x^b + (m_1 - m'_1)x^{b-1} + \dots + (m_{b-1} - m'_{b-1})x - y \equiv 0(mod p).$$

By a standard result of elementary number theory (see [11] p58) such an equivalence has at most b solutions in x, from which the result follows. □

Corollary 2 is a straightforward consequence of this theorem.

Corollary 2. *Let M and M′ be any two unequal message strings, and y any fixed integer. Let the function icv() be defined as in (1) and (2). Then if* $z_i, z_{i+1}, z_{i+2}, \ldots, z_{i+s}$ *are independent and uniformly distributed random variables in the range 0 to* $2^w - 1$,

$$Probability(icv(M, b, p, z_i, z_{i+1}, \ldots, z_{i+s+1}) \\ -icv(M', b, p, z_i, z_{i+1}, \ldots, z_{i+s+1}) \equiv y(mod p)) \leq \frac{b}{2^w}.$$

Corollary 3 indicates the strength of the integrity mechanism in terms of the likelihood of replacing, in transit, a message and the corresponding icv with a legitimate, but different, message-icv pair.

Corollary 3. *Let M and M′ be any two unequal message strings, and y, g any fixed integers. Let the function icv() be defined as in (1) and (2). Then if* $//z_i, z_{i+1}, z_{i+2}, \ldots, z_{i+s+1}$ *are independent and uniformly distributed random variables in the range 0 to* $2^w - 1$,

$$Probability(icv(M', b, p, z_i, z_{i+1}, \ldots, z_{i+s+1}) \equiv y(mod p) \\ \mid icv(M, b, p, z_i, z_{i+1}, \ldots, z_{i+s+1}) \equiv g(mod p)) \leq \frac{b}{2^w}. \quad (3)$$

Proof. Clearly the left hand side of the inequality in (3) is equal to

$$Prob(icv(M, b, p, z_i, z_{i+1}, \ldots, z_{i+s+1}) - icv(M', b, p, z_i, z_{i+1}, \ldots, z_{i+s+1}) \\ \equiv g - y(mod p) \mid icv(M, b, p, z_i, z_{i+1}, \ldots, z_{i+s+1}) \equiv g(mod p)).$$

However from (1)

$$icv(M, b, p, z_i, z_{i+1}, \ldots, z_{i+s+1}) - icv(M', b, p, z_i, z_{i+1}, \ldots, z_{i+s+1}) \\ \equiv (f(M_0, z_i) + f(M_1, z_{i+1}) + \ldots + f(M_s, z_{i+s}) \\ -f(M'_0, z_i) + f(M'_1, z_{i+1}) + \ldots + f(M'_s, z_{i+s}))(mod p)$$

is independent of z_{i+s+1} while

$$icv(M, b, p, z_i, z_{i+1}, \ldots, z_{i+s+1}) \equiv g(mod p))$$

if and only if

$$z_{i+s+1} \equiv (g - f(M_0, z_i) - f(M_1, z_{i+1}) - \ldots - f(M_s, z_{i+s}))(mod p).$$

Thus the events described in the conditional probability of (3) are independent and so the left hand side of the inequality (3) is equal to

$$Prob(icv(M, b, p, z_i, z_{i+1}, \ldots, z_{i+s+1}) \\ -icv(M', b, p, z_i, z_{i+1}, \ldots, z_{i+s+1}) \equiv (g - y)(mod p)).$$

The result now follows by Corollary 2. □

It follows from Corollary 3 that with an idealised stream cipher (with outputs that are uniformly distributed independent random variables) if any message and its integrity check value were to be altered in transit the new message and integrity check value would register as valid by the receiver with probability at most $b/2^w$. Moreover this is quite independent of whatever calculating ability or amount of chosen messages and corresponding icvs the party altering the message has. Thus we refer to $log_2(2^w/b)$ as the *effective icv length.* For the effective icv length to be a meaningful indicator of integrity strength with a practical deterministic cipher however, clearly the stream cipher key length must be at least as large as the effective icv length. On the other hand the contrapositive of Corollary 3 shows that if there is some way of altering or substituting message-icv pairs that goes undetected with a probability of more than $b/2^w$ then there must be some corresponding level of predictability in the stream cipher output. We illustrate this point with an example.

Example 2. Consider a stream cipher with a set V of initial vectors that determine the starting state of the cipher. For v in V let z_i^v denote the ith output of the stream cipher with initial vector v. Let the block length b equal 1. Assume that for some cipher position i there are a pair of unequal single integer messages m and m' and a function F that can calculate the icv of m' from that of m for all initial vectors v (so the integrity mechanism can be successfully attacked with probability 1). Furthermore assume that F can be evaluated in a reasonable amount of time (for example F should not embody a search through all initial vectors). Thus

$$icv(m', 1, p, z_i^v, z_{i+1}^v) = F(icv(m, 1, p, z_i^v, z_{i+1}^v)), \text{ for all } v \in V. \qquad (4)$$

Subtracting the icv of m from both sides of (4) and expanding the l. h. s. according to (1) and (2)

$$\begin{aligned}(m'z_i^v + z_{i+1}^v)[p] - (mz_i^v + z_{i+1}^v)[p] = \\ F(icv(m, 1, p, z_i^v, z_{i+1}^v)) - icv(m, 1, p, z_i^v, z_{i+1}^v)\end{aligned}$$

which implies that

$$(m' - m)z_i^v[p] = F'(icv(m, 1, p, z_i^v, z_{i+1}^v)) \qquad (5)$$

for a suitable function F'. Let S be the set of numbers s from 0 to $2^w - 1$ for which the equation

$$(m' - m)s[p] = F'(t)$$

has exactly one solution for t among $0, 1, ..., p - 1$. Then since $(m' - m)s[p]$ is a one to one function of s on $0, 1, ..., 2^w - 1$, it follows that S contains at least $2^w - k - 2$ integers. We may then define the inverse of F' on S, and by rearranging (5)

$$icv(m, 1, p, z_i^v, z_{i+1}^v) = F'^{-1}((m' - m)z_i^v[p]),$$

provided z_{i+1}^v is in S. By expanding the icv function from (1) and (2) and isolating z_{i+1}^v

$$z_{i+1}^v = (F'^{-1}((m' - m)z_i^v[p]) - mz_i^v)[p] \text{ for all } v \in V, z_i^v \in S.$$

Thus z_{i+1}^v can be obtained from z_i^v unless z_i^v happens to be one of at most $k+1$ values not in S. If the function F'^{-1} can be evaluated in a reasonable amount of time this amounts to an effective attack on the stream cipher.

4 ICV Length

In many conventional integrity mechanisms the icv is calculated as a function of a fixed key and the message. In such systems different messages that share the same icv (called collisions) may be found by an attacker who collects sufficiently many message-icv pairs. According to the so called Birthday Paradox (see [15]) the number of message-icv pairs required is approximately equal to the square root of the number of possible icvs. For this reason the length of the icv is usually chosen to be quite large (typically 128 bits) so that finding collisions is not feasible. In the method described in this article however collisions do not exist in the sense described above because the integrity function (1) and (2) uses fresh output from the stream cipher for each new message. Because of this it is suggested that the icv length may be much smaller than in conventional integrity systems.

Notwithstanding these remarks we describe how to modify the integrity method to increase the effective icv length by any required factor h. As in Section 2 let a message string $M = (m_0, m_1, ...)$ be divided into blocks $M_0, M_1, ..., M_s$ each containing at most b integers. Use the stream cipher to generate $h(s+2)$ outputs $z_i, z_{i+1}, z_{i+2}, ..., z_{i+h(s+2)-1}$. The integrity check value icv_h is defined as the concatenation of h integers between 0 and $p-1$ calculated according to

$$\begin{aligned}
&icv_h(M, b, p, z_i, z_{i+1}, ..., z_{i+h(s+2)-1}) = \\
&(f(M_0, z_i) + f(M_1, z_{i+1}) + ... + f(M_s, z_{i+s}) + z_{i+s+1})[p] \| \\
&(f(M_0, z_{i+s+2}) + f(M_1, z_{i+s+3}) + ... + f(M_s, z_{i+2s+2}) + z_{i+2s+3})[p] \| \\
&\cdot \\
&\cdot \\
&(f(M_0, z_{i+(h-1)(s+2)}) + ... + f(M_s, z_{i+h(s+2)-2}) + z_{i+h(s+2)})[p],
\end{aligned}$$

where the function f is given by (2). If $p < 2^{w+1}$, this provides an icv of length $h(w+1)$ bits with an effective icv length of

$$log_2\left(\left(\frac{2^w}{b}\right)^h\right) = hlog_2\left(\frac{2^w}{b}\right) = h(w - log_2 b). \tag{6}$$

Thus, for example, if $b = 20, w = 30, p = 2^{31} - 1, h = 4$ then the icv has length 124 bits while the effective icv length is

$$4(30 - log_2 20) \approx 102.7.$$

Note that this method of increasing the icv length by a factor of h requires the same increase factor in the amount of icv calculations required. As well the output from the stream cipher must be increased by a factor of h in order to complete the icv calculations. This however may be compensated for by increasing the block length b so the number of stream cipher outputs per message length remains approximately constant.

The other obvious way to increase the icv length by a factor of h is to increase the message and cipher integers to $w' = hw$ bits and take a prime $p' > 2^{w'}$. If p' can be chosen close to a power of 2 (say $p' = 2^v - 1$) then this method enjoys several advantages over the above method. Firstly the effective icv length becomes $hw - log_2 b$ which is larger than the corresponding $h(w - log_2 b)$ from (6). Also the output from the stream cipher required to process a given amount of message is not increased since a given message is divided into $1/h$ as many integers as before. Finally the multiplication of numbers modulo p' may take as much as h^2 times as much calculation as multiplication modulo p. However only $1/h$ as many multiplications are required to process a given amount of message. Thus the amount of calculations required to calculate the icv is increased by a factor of at most h (this figure may be further reduced for large h, see [7] pp 278-301 for a discussion of the complexity of multi-precision multiplication).

5 Integrity and Confidentiality

To provide both integrity and confidentiality the stream cipher can generate different outputs for both the integrity calculation and for message encryption. In order to prevent the message integrity from being undermined by a known plaintext attack it is important that cipher output that is used in icv calculations by the receiver could never be used for message encryption by the sender. To overcome this problem it is suggested that some integer $d < b$ is chosen and that only those z_i for which i is a multiple of d are used in the icv calculation, the remaining z_i being used for encryption.

Example 3. As in Example 1 let $w = 30, p = 2^{31} - 1, b = 20, M = (m_0, m_1, ..., m_{108})$ and the cipher be in state 56. Further let $d = 10$. To apply integrity and confidentiality the cipher is used to generate 121 outputs $z_{56}, z_{57}, z_{58}, ..., z_{176}$, and the icv is calculated as

$$icv(M, 20, 2^{31} - 1, z_{60}, z_{70}, ..., z_{120})$$

where the icv function is defined by (1) and (2). The transmitted message is

$$(m_0 + z_{56})[2^{30}], (m_1 + z_{57})[2^{30}], ..., (m_3 + z_{59})[2^{30}], (m_4 + z_{61})[2^{30}], \\ ..., (m_{108} + z_{176})[2^{30}], icv.$$

Finally we report on the performance of the combined confidentiality and integrity scheme using a stream cipher and implemented in computer software. The stream cipher used was constructed from three linear feedback shift registers and a combining function with memory (see [13] for designs of this type). In the icv calculation $w = 30, p = 2^{31} - 1$ and $b = 16$. This led to an effective icv length of 26 (thus the attackers ability to attack the integrity without attacking the stream cipher itself is one chance in 2^{26} or 68288512). A throughput of approximately 4 Mbits/second was attained for the entire system using a PC containing the Intel 486 processor running at 33 Mhz.

Acknowledgement

The permission of the Managing Director, Research and Information Technology, Telecom Australia to publish this paper is hereby acknowledged. Also the Author would like to thank the referees for several constructive comments on the draft version of this paper.

References

1. Brassard, G.: On Computationally Secure Authentication Tags. Advances in Cryptology-CRYPTO'82, proceedings, Springer-Verlag (1983) 79–86
2. Carter, J. L., Wegman, M. N.: Universal Classes of Hash Functions. Journal of Computer and Systems Sciences **18** (1979) 143–154
3. Damgaard, I. B.: A Design Principle for Hash Functions. Advances in Cryptology-CRYPTO'89, proceedings, Springer-Verlag (1990) 416–427
4. ISO/IEC 9797: Data Cryptographic Techniques-Data Integrity Mechanism using a Cryptographic Check Function employing a Block Cipher Algorithm. International Organisation for Standardisation (1989)
5. Jueneman, R. R.: A High-Speed Manipulation Detection Code. Advances in Cryptology-CRYPTO'86, proceedings, Springer-Verlag (1987) 327–346
6. Knobloch, H. J.: A Smart Card Implementation of the Fiat-Shamir Identification Scheme. Advances in Cryptology-EUROCRYPT'88, proceedings, Springer-Verlag (1989) 87–96
7. Knuth, D.: The Art of Computer Programming. Vol. 2, 2nd edition, Addison-Wesley, Reading, Mass. (1981)
8. Lai, X., Massey, J. L.: Hash Functions based on Block Ciphers. EUROCRYPT'92, extended abstracts (1992) 53–67
9. Lai, X., Rueppel, R. A., Woollven, J.: A Fast Cryptographic Checksum Algorithm based on Stream Ciphers. AUSCRYPT'92, abstracts (1992) 8-7–8-11
10. Merkle, R. C.: One Way Hash Functions and DES. Advances in Cryptology-CRYPTO'89, proceedings, Springer-Verlag (1990) 428–446
11. Niven, I., Zuckerman, H. S.: The Theory of Numbers (fourth edition). John Wiley and Sons, New York-Chichester-Brisbane-Toronto (1980)
12. Rivest, R. L.: The MD4 Message Digest Algorithm. Advances in Cryptology-CRYPTO'90, proceedings, Springer-Verlag (1991) 303–311
13. Rueppel, R. A.: Analysis and Design of Stream Ciphers. Springer-Verlag, Berlin (1986)

14. Wegman, M. N., Carter, J. L.: New Hash Functions and their use in Authentication and Set Equality. Journal of Computer and System Sciences **22** (1981) 265–279
15. Yuval, G.: How to Swindle Rabin. Cryptologia **3** (1979) 187–189

Nonlinearly Balanced Boolean Functions and Their Propagation Characteristics (Extended Abstract)

Jennifer Seberry * Xian-Mo Zhang ** and Yuliang Zheng ***,

Department of Computer Science
University of Wollongong, Wollongong, NSW 2522, Australia
{jennie, xianmo, yuliang}@cs.uow.edu.au

Abstract. Three of the most important criteria for cryptographically strong Boolean functions are the balancedness, the nonlinearity and the propagation criterion. This paper studies systematic methods for constructing Boolean functions satisfying some or all of the three criteria. We show that concatenating, splitting, modifying and multiplying sequences can yield balanced Boolean functions with a very high nonlinearity. In particular, we show that balanced Boolean functions obtained by modifying and multiplying sequences achieve a nonlinearity higher than that attainable by any previously known construction method. We also present methods for constructing highly nonlinear balanced Boolean functions satisfying the propagation criterion with respect to *all but one or three* vectors. A technique is developed to transform the vectors where the propagation criterion is not satisfied in such a way that the functions constructed satisfy the propagation criterion of high degree while preserving the balancedness and nonlinearity of the functions. The algebraic degrees of functions constructed are also discussed, together with examples illustrating the various constructions.

1 Preliminaries

Let f be a function on V_n. The $(1,-1)$-sequence defined by $((-1)^{f(\alpha_0)}, (-1)^{f(\alpha_1)}, \ldots, (-1)^{f(\alpha_{2^n-1})})$ is called the *sequence* of f, and the $(0,1)$-sequence defined by $(f(\alpha_0), f(\alpha_1), \ldots, f(\alpha_{2^n-1}))$ is called the *truth table* of f, where α_i, $0 \le i \le 2^n-1$, denotes the vector in V_n whose integer representation is i. A $(0,1)$-sequence ($(1,-1)$-sequence) is said *balanced* if it contains an equal number of zeros and ones (ones and minus ones). A function is balanced if its sequence is balanced.

* Supported in part by the Australian Research Council under the reference numbers A49130102, A9030136, A49131885 and A49232172.

** Supported in part by the Australian Research Council under the reference number A49130102.

*** Supported in part by the Australian Research Council under the reference number A49232172.

The *Hamming weight* of a $(0,1)$-sequence (or vector) α, denoted by $W(\alpha)$, is the number of ones in α. The *Hamming distance* between two sequences α and β of the same length, denoted by $d(\alpha,\beta)$, is the number of positions where the two sequences differ. Given two functions f and g on V_n, the Hamming distance between them is defined as $d(f,g) = d(\xi_f, \xi_g)$, where ξ_f and ξ_g are the truth tables of f and g respectively. The *nonlinearity* of f, denoted by N_f, is the minimal Hamming distance between f and all affine functions on V_n, i.e., $N_f = \min_{i=0,1,\ldots,2^{n+1}-1} d(f,\varphi_i)$ where $\varphi_0, \varphi_1, \ldots, \varphi_{2^{n+1}-1}$ denote the affine functions on V_n.

A $(1,-1)$-matrix H of order n is called a *Hadamard* matrix if $HH^t = nI_n$, where H^t is the transpose of H and I_n is the identity matrix of order n. It is well known that the order of a Hadamard matrix is 1, 2 or divisible by 4 [11]. A special kind of Hadamard matrix, called *Sylvester-Hadamard matrix* or *Walsh-Hadamard matrix*, will be relevant to this paper. A Sylvester-Hadamard matrix of order 2^n, denoted by H_n, is generated by the following recursive relation

$$H_0 = 1, H_n = \begin{bmatrix} 1 & 1 \\ 1 & -1 \end{bmatrix} \otimes H_{n-1}, n = 1, 2, \ldots$$

where $\otimes$ denotes the Kronecker product. Note that H_n can be represented as $H_n = H_s \otimes H_t$ for any s and t with $s + t = n$.

Sylvester-Hadamard matrices are closely related to linear functions, as is shown in the following lemma.

Lemma 1. *Write* $H_n = \begin{bmatrix} \ell_0 \\ \ell_1 \\ \vdots \\ \ell_{2^n-1} \end{bmatrix}$ *where* ℓ_i *is a row of* H_n. *Then* ℓ_i *is the sequence of* $h_i = \langle \alpha_i, x \rangle$, *a linear function, where* α_i *is a vector in* V_n *whose integer representation is* i *and* $x = (x_1, \ldots, x_n)$. *Conversely the sequence of any linear function on* V_n *is a row of* H_n.

From Lemma 1 the rows of H_n comprise the sequences of all linear functions on V_n. Consequently the rows of $\pm H_n$ comprise the sequences of all *affine* functions on V_n.

The following notation is very useful in obtaining the functional representation of a concatenated sequence. Let $\delta = (i_1, i_2, \ldots, i_p)$ be a vector in V_p. Then D_δ is a function on V_p defined by

$$D_\delta(y_1, y_2, \ldots, y_p) = (y_1 \oplus i_1 \oplus 1) \cdots (y_p \oplus i_p \oplus 1).$$

We now introduce the concept of bent functions.

Definition 2. A function f on V_n is called a *bent* function if

$$2^{-\frac{n}{2}} \sum_{x \in V_n} (-1)^{f(x) \oplus \langle \beta, x \rangle} = \pm 1$$

for all $\beta \in V_n$. Here $f(x) \oplus \langle \beta, x \rangle$ is regarded as a real-valued function. The sequence of a bent function is called a bent sequence.

From the definition we can see that bent functions on V_n exist only when n is even. It was Rothaus who first introduced and studied bent functions in 1960s, although his pioneering work was not published in the open literature until some ten years later [10]. Applications of bent functions to digital communications, coding theory and cryptography can be found in such as [2, 4, 7].

The following result can be found in an excellent survey of bent functions by Dillon [5].

Lemma 3. *Let f be a function on V_n, and let ξ be the sequence of f. Then the following four statements are equivalent:*

(i) f is bent.
(ii) $\langle \xi, \ell \rangle = \pm 2^{\frac{1}{2}n}$ for any affine sequence ℓ of length 2^n.
(iii) $f(x) \oplus f(x \oplus \alpha)$ is balanced for any non-zero vector $\alpha \in V_n$.
(iv) $f(x) \oplus \langle \alpha, x \rangle$ assumes the value one $2^{n-1} \pm 2^{\frac{1}{2}n-1}$ times for any $\alpha \in V_n$.

By (iv) of Lemma 3, if f is a bent function on V_n, then $f(x) \oplus h(x)$ is also a bent function for any affine function h on V_n. This property will be employed in constructing highly nonlinear balanced functions to be described in Section 4.

In this paper we are concerned with the propagation criterion whose formal definition follows (see also [1, 9]).

Definition 4. Let f be a function on V_n. We say that f satisfies

1. the ***propagation criterion with respect to a non-zero vector α in V_n*** if $f(x) \oplus f(x \oplus \alpha)$ is a balanced function.
2. the ***propagation criterion of degree k*** if it satisfies the propagation criterion with respect to all $\alpha \in V_n$ with $1 \leq W(\alpha) \leq k$.

Note that the SAC is equivalent to the propagation criterion of degree 1. Also note that the *perfect nonlinearity* studied by Meier and Staffelbach [6] is equivalent to the propagation criterion of degree n.

2 Properties of Balancedness and Nonlinearity

This section presents a number of results related to balancedness and nonlinearity. These include upper bounds for nonlinearity and properties of concatenated and split sequences. Due to the limit on space, proofs for some of the results are left to the full version of the paper [13].

The following lemma is very useful in calculating the nonlinearity of a function.

Lemma 5. *Let f and g be functions on V_n whose sequences are ξ_f and ξ_g respectively. Then the distance between f and g can be calculated by $d(f, g) = 2^{n-1} - \frac{1}{2}\langle \xi_f, \xi_g \rangle$.*

Corollary 6. *A function on V_n attains the upper bound for nonlinearities, $2^{n-1} - 2^{\frac{1}{2}n-1}$, if and only if it is bent.*

From Corollary 6, balanced functions can not attain the upper bound for nonlinearities, namely $2^{n-1} - 2^{\frac{1}{2}n-1}$. A slightly improved upper bound for the nonlinearities of balanced functions can be obtained by noting the fact that a balanced function assumes the value one an even number of times.

Corollary 7. *Let f be a balanced function on V_n $(n \geq 3)$. Then the nonlinearity N_f of f is given by*

$$N_f \leq \begin{cases} 2^{n-1} - 2^{\frac{1}{2}n-1} - 2, & n \text{ even} \\ \lfloor\lfloor 2^{n-1} - 2^{\frac{1}{2}n-1} \rfloor\rfloor, & n \text{ odd} \end{cases}$$

where $\lfloor\lfloor x \rfloor\rfloor$ denotes the maximum even integer less than or equal to x.

The following lemma, first proved in [12], gives the lower bound of the nonlinearity of a function obtained by concatenating the sequences of two functions.

Lemma 8. *Let f_1 and f_2 be functions on V_n, and let g be a function on V_{n+1} defined by*

$$g(u, x_1, \ldots, x_n) = (1 \oplus u) f_1(x_1, \ldots, x_n) \oplus u f_2(x_1, \ldots, x_n). \tag{1}$$

Suppose that ξ_1 and ξ_2, the sequences of f_1 and f_2 respectively, satisfy $\langle \xi_1, \ell \rangle \leq P_1$ and $\langle \xi_2, \ell \rangle \leq P_2$ for any affine sequence ℓ of length 2^n, where P_1 and P_2 are positive integers. Then the nonlinearity of g satisfies $N_g \geq 2^n - \frac{1}{2}(P_1 + P_2)$.

As bent functions do not exist on V_{2k+1}, an interesting question is what functions on V_{2k+1} are highly nonlinear. The following result, as a special case of Lemma 8, shows that such functions can be obtained by concatenating bent sequences. This construction has also been discovered by Meier and Staffelbach in [6].

Corollary 9. *In the construction (1), if both f_1 and f_2 are bent functions on V_{2k}, then $N_g \geq 2^{2k} - 2^k$.*

A similar result can be obtained when sequences of four functions are concatenated.

Lemma 10. *Let f_0, f_1, f_2 and f_3 be functions on V_n whose sequences are ξ_0, ξ_1, ξ_2 and ξ_3 respectively. Assume that $\langle \xi_i, \ell \rangle \leq P_i$ for each $0 \leq i \leq 3$ and for each affine sequence ℓ of length 2^n, where each P_i is a positive integer. Let g be a function on V_{n+2} defined by*

$$g(y, x) = \bigoplus_{i=0}^{3} D_{\alpha_i}(y) f_i(x) \tag{2}$$

where $y = (y_1, y_2)$, $x = (x_1, \ldots, x_n)$ and α_i is a vector in V_2 whose integer representation is i. Then $N_g \geq 2^{n+1} - \frac{1}{2}(P_0 + P_1 + P_2 + P_3)$. In particular, when n is even and f_0, f_1, f_2 and f_3 are all bent functions on V_n, $N_g \geq 2^{n+1} - 2^{\frac{1}{2}n+1}$.

We have discussed the concatenation of sequences of functions including bent functions. The following lemma deals with the other direction, namely splitting bent sequences.

Lemma 11. *Let $f(x_1, x_2, \ldots, x_{2k})$ be a bent function on V_{2k}, η_0 be the sequence of $f(0, x_2, \ldots, x_{2k})$, and η_1 be the sequence of $f(1, x_2, \ldots, x_{2k})$. Then for any affine sequence ℓ of length 2^{2k-1}, we have $-2^k \leq \langle \eta_0, \ell \rangle \leq 2^k$ and $-2^k \leq \langle \eta_1, \ell \rangle \leq 2^k$.*

A consequence of Lemma 11 is that the nonlinearity of $f(0, x_2, \ldots, x_{2k})$ and $f(1, x_2, \ldots, x_{2k})$ is at least $2^{2k-2} - 2^{k-1}$. It is interesting to note that concatenating and splitting bent sequences both achieve the same nonlinearity.

Splitting bent sequences can also result in balanced functions. Let ℓ_i be the ith row of H_k where $i = 0, 1, \ldots, 2^k - 1$. Note that ℓ_0 is an all-one sequence while ℓ_1, ℓ_2, ..., ℓ_{2^k-1} are all balanced sequences. The concatenation of the rows, $(\ell_0, \ell_1, \ldots, \ell_{2^k-1})$, is a bent sequence [1]. Denote by $f(x_1, x_2, \ldots, x_{2k})$ the function corresponding to the bent sequence. Let ξ be the second half of the bent sequence, namely, $\xi = (\ell_{2^{k-1}}, \ell_{2^{k-1}+1}, \ldots, \ell_{2^k-1})$. Then ξ is the sequence of $f(1, x_2, \ldots, x_{2k})$. Since all ℓ_i, $i = 2^{k-1}, 2^{k-1}+1, \ldots, 2^k - 1$, are balanced, $f(1, x_2, \ldots, x_{2k})$ is a balanced function. The nonlinearity of the function is at least $2^{2k-2} - 2^{k-1}$.

By permuting $\{\ell_{2^{k-1}}, \ell_{2^{k-1}+1}, \ldots, \ell_{2^k-1}\}$, we obtain a new balanced sequence $\xi' = (\ell'_{2^{k-1}}, \ell'_{2^{k-1}+1}, \ldots, \ell'_{2^k-1})$ that has the same nonlinearity as that of ξ. Now let $\xi'' = (e_{2^{k-1}}\ell'_{2^{k-1}}, e_{2^{k-1}+1}\ell'_{2^{k-1}+1}, \ldots, e_{2^k-1}\ell'_{2^k-1})$, where each e_i is independently selected from $\{1, -1\}$. ξ'' is also a balanced sequence with the same nonlinearity. The total number of balanced sequences obtained by permuting and changing signs is $2^{2^{k-1}} \cdot 2^{k-1}!$. These sequences are all different from one another but have the same nonlinearity.

3 Highly Nonlinear Balanced Functions

Note that a bent sequence on V_{2k} contains $2^{2k-1} + 2^{k-1}$ ones and $2^{2k-1} - 2^{k-1}$ zeros, or vice versa. As is observed by Meier and Staffelbach [6], changing 2^{k-1} positions in a bent sequence yields a balanced function having a nonlinearity of at least $2^{2k-1} - 2^k$. This nonlinearity is the same as that obtained by concatenating four bent sequences of length 2^{2k-2} (see Lemma 10).

It is well-known that the maximum nonlinearity of functions on V_n coincides with the covering radius of the first order binary Reed-Muller code $R(1, n)$ of length 2^n, many results on covering radius of $R(1, n)$ (see [3]) have direct implications on the nonlinearity of functions. In particular, using a result of [8], we can construct unbalanced functions on V_{2k+1}, $k \geq 7$, whose nonlinearity is at least $2^{2k} - \frac{108}{128}2^k$, a higher value than $2^{2k} - 2^k$ achieved by the construction in Corollary 9. One might tempt to think that modifying the sequences in [8] would result in balanced functions with a higher nonlinearity than that obtained by concatenating or splitting bent sequences. We find that it is not the case. We take V_{15} for an example. The Hamming weight of the sequences on V_{15}, which have the largest nonlinearity of 16276, is 16492. Changing 54 positions makes them balanced. The nonlinearity of the resulting functions is 16222, smaller than 16256 achieved by concatenating two bent sequences of length 2^{14} (see Corollary 9).

In the following we show how to modify bent sequences of length 2^{2k} constructed from Hadamard matrices in such a way that the resulting functions are balanced and have a much higher nonlinearity than that attainable by concatenating four bent sequences. This result, in conjunction with sequences in [8], allows us to construct balanced functions on V_{2k+15}, $k \geq 7$, that have a higher nonlinearity than that achieved by concatenating or splitting bent sequences.

3.1 On V_{2k}

Note that an even number $n \geq 4$ can be expressed as $n = 4t$ or $n = 4t+2$, where $t \geq 1$. As the first step towards our goal, we have the following lemma whose proof is left to the full paper [13]:

Lemma 12. *For any integer $t \geq 1$ there exists*

(i) a balanced function f on V_{4t} such that $N_f \geq 2^{4t-1} - 2^{2t-1} - 2^t$,
(ii) a balanced function f on V_{4t+2} such that $N_g \geq 2^{4t+1} - 2^{2t} - 2^t$.

With the above result as a basis, we consider an iterative procedure to further improve the nonlinearity of a function constructed. Note that an even number $n \geq 4$ can be expressed as $n = 2^m$, $m \geq 2$, or $n = 2^s(2t+1)$, $s \geq 1$ and $t \geq 1$.

Consider the case when $n = 2^m$, $m \geq 2$. We start with the bent sequence obtained by concatenating the rows of $H_{2^{m-1}}$. The sequence consists of $2^{2^{m-1}}$ sequences of length $2^{2^{m-1}}$. Now we replace the all-one leading sequence with a bent sequence of the same length, which is obtained by concatenating the rows of $H_{2^{m-2}}$. The length of the new leading sequence becomes $2^{2^{m-2}}$. It is replaced by another bent sequence of the same length. This replacing process is continued until the length of the all-one leading sequence is $2^2 = 4$. To finish the procedure, we replace the leading sequence $(1,1,1,1)$ with $(1,-1,1,-1)$. The last replacement makes the entire sequence balanced. By induction on $s = 2,3,4,\ldots$, it can be proved that the nonlinearity of the function obtained is at least

$$2^{2^m-1} - \frac{1}{2}(2^{2^{m-1}} + 2^{2^{m-2}} + \cdots + 2^{2^2} + 2 \cdot 2^2).$$

The modifying procedure for the case of $n = 2^s(2t+1)$, $s \geq 1$ and $t \geq 1$, is the same as that for the case of $n = 2^m$, $m \geq 2$, except for the last replacement. In this case, the replacing process is continued until the length of the all-one leading sequence is 2^{2t+1}. The last leading sequence is replaced by $\ell_0^* = (e_{2^t}, e_{2^t+1}, \ldots, e_{2^{t+1}-1})$, the second half of the bent sequence $(e_0, e_1, \ldots, e_{2^{t+1}-1})$, where each e_i is a row of H_{t+1}. Again by induction on $s = 1,2,3,\ldots$, it can be proved that the nonlinearity of the resulting function is at least

$$2^{2^s(2t+1)-1} - \frac{1}{2}(2^{2^{s-1}(2t+1)} + 2^{2^{s-2}(2t+1)} + \cdots + 2^{2(2t+1)} + 2^{2t+1} + 2^{t+1}).$$

We have completed the proof for the following

Theorem 13. *For any even number $n \geq 4$, there exists a balanced function f^* on V_n whose nonlinearity is*

$$N_{f^*} \geq \begin{cases} 2^{2^m-1} - \frac{1}{2}(2^{2^{m-1}} + 2^{2^{m-2}} + \cdots + 2^{2^2} + 2 \cdot 2^2), & n = 2^m, \\ 2^{2^s(2t+1)-1} - \frac{1}{2}(2^{2^{s-1}(2t+1)} + 2^{2^{s-2}(2t+1)} + \cdots + 2^{2(2t+1)} + 2^{2t+1} + 2^{t+1}), & n = 2^s(2t+1). \end{cases}$$

Let $\zeta = (\zeta_0, \zeta_1, \ldots, \zeta_{2^k-1})$ be a sequence of length 2^{2k} obtained by modifying a bent sequence. Permuting and changing signs discussed in Section 2 can also be applied to ζ. In this way we obtain in total $2^{2^k} \cdot 2^k!$ different balanced functions, all of which have the same nonlinearity. Even more functions can be obtained by observing the fact that the leading sequence ζ_0 has exactly the same structure as the large sequence ζ, and hence permuting and changing signs can also be applied to ζ_0.

3.2 On V_{2k+1}

Lemma 14. *Let f_1 be a function on V_s and f_2 be a function on V_t. Then $f_1(x_1, \ldots, x_s) \oplus f_2(y_1, \ldots, y_t)$ is a balanced function on V_{s+t} if either f_1 or f_2 is balanced.*

Let ξ_1 be the sequence of f_1 on V_s and ξ_2 be the sequence of f_2 on V_t. Then it is easy to verify that the Kronecker product $\xi_1 \otimes \xi_2$ is the sequence of $f_1(x_1, \ldots, x_s) \oplus f_2(y_1, \ldots, y_t)$.

Lemma 15. *Let f_1 be a function on V_s and f_2 be a function on V_t. Let g be a function on V_{s+t} defined by*

$$g(x_1, \ldots, x_s, y_1, \ldots, y_s) = f_1(x_1, \ldots, x_s) \oplus f_2(y_1, \ldots, y_t).$$

Suppose that ξ_1 and ξ_2, the sequences of f_1 and f_2 respectively, satisfy $\langle \xi_1, \ell \rangle \leq P_1$ and $\langle \xi_2, \ell \rangle \leq P_2$ for any affine sequence ℓ of length 2^n, where P_1 and P_2 are positive integers. Then the nonlinearity of g satisfies $N_g \geq 2^{s+t-1} - \frac{1}{2} P_1 \cdot P_2$.

Let ξ_1 be a balanced sequence of length 2^{2k} that is constructed using the method in the proof of Theorem 13, where $k \geq 2$, Let ξ_2 be a sequence of length 2^{15} obtained by the method of [8]. Note that the nonlinearity of ξ_2 is 16276, and there are 13021 such sequences. Denote by f_1 the function corresponding to ξ_1 and by f_2 the function corresponding to ξ_2. Let

$$f(x_1, \ldots, x_{2k}, x_{2k+1}, \ldots, x_{2k+15}) = f_1(x_1, \ldots, x_{2k}) \oplus f_2(x_{2k+1}, \ldots, x_{2k+15}) \quad (3)$$

Then

Theorem 16. *The function f defined by (3) is a balanced function on V_{2k+15}, $k \geq 2$, whose nonlinearity is at least*

$$N_f \geq \begin{cases} 2^{2^m+14} - 108(2^{2^{m-1}} + 2^{2^{m-2}} + \cdots + \\ \qquad 2^{2^2} + 2 \cdot 2^2), & 2k = 2^m, \\ 2^{2^s(2t+1)+14} - 108(2^{2^{s-1}(2t+1)} + 2^{2^{s-2}(2t+1)} + \cdots + \\ \qquad 2^{2(2t+1)} + 2^{2t+1} + 2^{t+1}), & 2k = 2^s(2t+1). \end{cases}$$

Proof. Let $\xi = \xi_1 \otimes \xi_2$. Then ξ is the sequence of f. Let ℓ be an arbitrary affine sequence of length 2^{2k+15}. Then $\ell = \pm\ell_1 \otimes \ell_2$, where ℓ_1 is a linear sequence of length 2^{2k} and ℓ_2 is a linear sequence of length 2^{15}. Thus

$$\langle \xi_1, \ell_1 \rangle \leq \begin{cases} 2^{2^{m-1}} + 2^{2^{m-2}} + \cdots + 2^{2^2} + 2 \cdot 2^2, & 2k = 2^m, \\ 2^{2^{s-1}(2t+1)} + 2^{2^{s-2}(2t+1)} + \cdots + 2^{2(2t+1)} + 2^{2t+1} + 2^{t+1}, & 2k = 2^s(2t+1). \end{cases}$$

and

$$\langle \xi_2, \ell_2 \rangle \leq 2 \cdot (2^{14} - 16276) = 216$$

By Lemma 15, the theorem is true. □

The nonlinearity of a function on V_{2k+15} constructed in this section is larger than that obtained by concatenating or splitting bent sequences for all $k \geq 7$.

4 Constructing Highly Nonlinear balanced Functions Satisfying High Degree Propagation Criterion

This section presents two methods for constructing highly nonlinear balanced functions satisfying the propagation criterion.

4.1 Basic Construction

On V_{2k+1} Let f be a bent function on V_{2k}, and let g be a function on V_{2k+1} defined by

$$\begin{aligned} & g(x_1, x_2, \ldots, x_{2k+1}) \\ & = (1 \oplus x_1) f(x_2, \ldots, x_{2k+1}) \oplus x_1 (1 \oplus f(x_2, \ldots, x_{2k+1})) \\ & = x_1 \oplus f(x_2, \ldots, x_{2k+1}). \end{aligned} \tag{4}$$

Lemma 17. *The function g defined in (4) satisfies the propagation criterion with respect to all non-zero vectors $\gamma \in V_{2k+1}$ with $\gamma \neq (1, 0, \ldots, 0)$.*

Proof. Let $\gamma = (a_1, a_2, \ldots, a_{2k+1}) \neq (1, 0, \ldots, 0)$ and let $x = (x_1, x_2, \ldots, x_{2k+1})$. Then $g(x) \oplus g(x \oplus \gamma) = a_1 \oplus f(x_2, \ldots, x_{2k+1}) \oplus f(x_2 \oplus a_2, \ldots, x_{2k+1} \oplus a_{2k+1})$. Since f is a bent function, $f(x_2, \ldots, x_{2k+1}) \oplus f(x_2 \oplus a_2, \ldots, x_{2k+1} \oplus a_{2k+1})$ is balanced for all $(a_2, \ldots, a_{2k+1}) \neq (0, \ldots, 0)$ (see (iii) of Lemma 3). Thus $g(x) \oplus g(x \oplus \gamma)$ is balanced for all $\gamma = (a_1, a_2, \ldots, a_{2k+1}) \neq (1, 0, \ldots, 0)$. □

From Corollary 9, the nonlinearity of the function g defined by (4) satisfies $N_g \geq 2^{2k} - 2^k$. Furthermore, by Lemma 14, g is balanced. Thus we have

Corollary 18. *The function g defined by (4) is balanced and satisfies the propagation criterion with respect to all non-zero vectors* $\gamma \in V_{2k+1}$ *with* $\gamma \neq (1, 0, \ldots, 0)$. *The nonlinearity of g satisfies* $N_g \geq 2^{2k} - 2^k$.

On V_{2k} Let f be a bent function on V_{2k-2} and let g be a function on V_{2k} obtained from f in the following way:

$$
\begin{aligned}
& g(x_1, x_2, x_3, \ldots, x_{2k}) \\
& = (1 \oplus x_1)(1 \oplus x_2) f(x_3, \ldots, x_{2k}) \oplus (1 \oplus x_1) x_2 (1 \oplus f(x_3, \ldots, x_{2k})) \\
& \quad x_1 (1 \oplus x_2)(1 \oplus f(x_3, \ldots, x_{2k})) \oplus x_1 x_2 f(x_3, \ldots, x_{2k}) \\
& = x_1 \oplus x_2 \oplus f(x_3, \ldots, x_{2k}). \qquad (5)
\end{aligned}
$$

Lemma 19. *The function g defined in (5) satisfies the propagation criterion with respect to* all but three *non-zero vectors in* V_{2k}. *The three vectors where the propagation criterion is not satisfied are* $\gamma_1 = (1, 0, 0, \ldots, 0)$, $\gamma_2 = (0, 1, 0, \ldots, 0)$, *and* $\gamma_3 = \gamma_1 \oplus \gamma_2 = (1, 1, 0, \ldots, 0)$.

Proof. Let $\gamma = (a_1, a_2, \ldots, a_{2k})$ be a non-zero vector in V_{2k} differing from γ_1, γ_2 and γ_3. Also let $x = (x_1, \ldots, x_{2k})$. Then we have $g(x) \oplus g(x \oplus \gamma) = a_1 \oplus a_2 \oplus f(x_3, \ldots, x_{2k}) \oplus f(x_3 \oplus a_3, \ldots, x_{2k} \oplus a_{2k})$. Since f is a bent function on V_{2k-2} and $(a_3, \ldots, a_{2k}) \neq (0, \ldots, 0)$, $f(x_3, \ldots, x_{2k}) \oplus f(x_3 \oplus a_3, \ldots, x_{2k} \oplus a_{2k})$ is balanced, from which it follows that $g(x) \oplus g(x \oplus \gamma)$ is balanced for any non-zero vector γ in V_{2k} differing from γ_1, γ_2 and γ_3. This proves the lemma. □

Since $x_1 \oplus x_2$ is balanced on V_2, g is balanced on V_{2k}. On the other hand, by Lemma 8, we have $N_g \geq 2^{2k-1} - 2^k$. Thus we have the following result:

Corollary 20. *The function g defined by (5) is balanced and satisfies the propagation criterion with respect to all non-zero vectors* $\gamma \in V_{2k}$ *with* $\gamma \neq (c_1, c_2, 0, \ldots, 0)$, *where* $c_1, c_2 \in GF(2)$. *The nonlinearity of g satisfies* $N_g \geq 2^{2k-1} - 2^k$.

4.2 Moving Vectors Around

Though functions constructed according to (4) or (5) satisfy the propagation criterion with respect to all but one or three non-zero vectors, they are not interesting in practical applications. We show that through linear transformation of input coordinates, the vectors where the propagation criterion is not satisfied can be transformed while the balancedness and nonlinearity of the functions are preserved. In particular, the vectors can be transformed into vectors having a high Hamming weight. In this way we obtain highly nonlinear balanced functions satisfying the high degree propagation criterion.

Let f be a function on V_n, A a nondegenerate matrix of order n with entries from $GF(2)$, and b a vector in V_n. Then $f^*(x) = f(xA \oplus b)$ defines a new function on V_n, where $x = (x_1, x_2, \ldots, x_n)$. It can be proved that the algebraic degree and the nonlinearity of f^* is the same as those of f. In addition, f^* is balanced iff f is balanced.

On V_{2k+1}

Theorem 21. *For any non-zero vector $\gamma^* \in V_{2k+1}$ $(k \geq 1)$, there exist balanced functions on V_{2k+1} satisfying the propagation criterion with respect to all non-zero vectors $\gamma \in V_{2k+1}$ with $\gamma \neq \gamma^*$. The nonlinearities of the functions are at least $2^{2k} - 2^k$.*

Proof. Let f be a bent function and let g be the function constructed by (4). From linear algebra we know that for any bases B_1 and B_2 of the vector space V_{2k+1}, where $B_1 = \{\alpha_j | j = 1, \ldots, 2k+1\}$ and $B_2 = \{\beta_j | j = 1, \ldots, 2k+1\}$, there exists a unique nondegenerate matrix A of order $2k+1$ with entries from $GF(2)$ such that $\alpha_j A = \beta_j$, $j = 1, \ldots, 2k+1$. In particular, this is true when $\alpha_1 = \gamma^*$ and $\beta_1 = (1, 0, \ldots, 0)$. Let $x = (x_1, x_2, \ldots, x_n)$ and let g^* be the function obtained from g by employing linear transformation on the input coordinates of g:

$$g^*(x) = g(xA).$$

Since A is nondegenerate, g^* is balanced and has the same nonlinearity as that of g. Now we show that g^* satisfies the propagation criterion with respect to all non-zero vectors except γ^*.

Let γ be a non-zero vector in V_{2k+1} with $\gamma \neq \gamma^*$. Consider the following function $g^*(x) \oplus g^*(x \oplus \gamma) = g(xA) \oplus g(xA \oplus \gamma A) = g(y) \oplus g(y \oplus \gamma A)$ where $y = xA$. Note that A is nondegenerate and thus y runs through V_{2k+1} while x runs through V_{2k+1}. Since $\gamma \neq \gamma^*$ we have $\gamma A \neq (1, 0, \ldots, 0)$. From (iii) of Lemma 3, $g(y) \oplus g(y \oplus \gamma A)$ is balanced and hence $g^*(x) \oplus g^*(x \oplus \gamma)$ is balanced. Consequently, g^* satisfies the propagation criterion with respect to all non-zero vectors in V_{2k+1} but γ^*. This completes the proof. □

As a consequence of Theorem 21, we obtain, by letting $\gamma^* = (1, 1, \ldots, 1)$, highly nonlinear balanced functions on V_{2k+1} satisfying the propagation criterion of degree $2k$. This is described in the following:

Corollary 22. *Let f be a bent function on V_{2k} and let $g^*(x_1, \ldots, x_{2k+1}) = x_1 \oplus f(x_1 \oplus x_2, x_1 \oplus x_3, \ldots, x_1 \oplus x_{2k+1})$. Then g^* is a balanced function on V_{2k+1} and satisfies the propagation criterion of degree $2k$. The nonlinearity of g^* satisfies $N_{g^*} \geq 2^{2k} - 2^k$.*

On V_{2k}

Theorem 23. *For any non-zero vectors $\gamma_1^*, \gamma_2^* \in V_{2k}$ $(k \geq 2)$ with $\gamma_1^* \neq \gamma_2^*$, there exist balanced functions on V_{2k} satisfying the propagation criterion with respect to* all but three *non-zero vectors in V_{2k}. The three vectors where the propagation criterion is not satisfied are γ_1^*, γ_2^* and $\gamma_1^* \oplus \gamma_2^*$. The nonlinearities of the functions are at least $2^{2k-1} - 2^k$.*

Proof. The proof is essentially the same as that for Theorem 21. The major difference lies in the selection of bases $B_1 = \{\alpha_j | j = 1, \ldots, 2k\}$ and $B_2 = \{\beta_j | j =$

$1,\ldots,2k\}$. By linear algebra, we can let $\alpha_1 = \gamma_1^*$, $\alpha_2 = \gamma_2^*$, $\beta_1 = (1,0,0,\ldots,0)$, and $\beta_2 = (0,1,0,\ldots,0)$. By the same reasoning as in the proof of Theorem 21, we can see that g^* defined by $g^*(x) = g(xA)$ satisfies the propagation criterion with respect to all but the following three non-zero vectors in V_{2k}: γ_1^*, γ_2^* and $\gamma_1^* \oplus \gamma_2^*$. Here $x = (x_1, x_2, \ldots, x_{2k})$, $g(x) = x_1 \oplus x_2 \oplus f(x_3, \ldots, x_{2k})$, and f, a bent function on V_{2k-2}, are all the same as in (5), and A is the unique nondegenerate matrix such that $\alpha_j A = \beta_j$, $j = 1, \ldots, 2k$. □

Similarly to the case on V_{2k+1}, we can obtain highly nonlinear balanced functions satisfying the high degree propagation criterion, by properly selecting vectors γ_1^* and γ_2^*. Unlike the case on V_{2k+1}, however, the degree of propagation criterion the functions can achieve is $\frac{4}{3}k$, but not $2k-1$. The construction method is described in the following corollary.

Corollary 24. *Suppose that $2k = 3t + c$ where $c = 0, 1$ or 2. Then there exist balanced functions on V_{2k} that satisfy the propagation criterion of degree $2t - 1$ (when $c = 0$ or 1), or $2t$ (when $c = 2$). The nonlinearities of the functions are at least $2^{2k-1} - 2^k$.*

Proof. Set $c_1 = 0$, $c_2 = 1$ if $c = 1$ and set $c_1 = c_2 = \frac{1}{2}c$ otherwise. Let $\gamma_1^* = (a_1, \ldots, a_{3t+c})$ and $\gamma_2^* = (b_1, \ldots, b_{3t+c})$, where

$$a_j = \begin{cases} 1 & \text{for } j = 1, \ldots, 2t + c_1, \\ 0 & \text{for } j = 2t + c_1 + 1, \ldots, 3t + c. \end{cases}$$

$$b_j = \begin{cases} 0 & \text{for } j = 1, \ldots, t + c_1, \\ 1 & \text{for } j = t + c_1 + 1, \ldots, 3t + c. \end{cases}$$

By Theorem 23 there exists a balanced function g^* on V_{2k} satisfying the propagation criterion with respect to all but three non-zero vectors in V_{2k}. The three vectors are γ_1^*, γ_2^* and $\gamma_1^* \oplus \gamma_2^*$. The nonlinearity of g^* satisfies $N_{g^*} \geq 2^{2k-1} - 2^k$.

Note that $W(\gamma_1^*) = 2t + c_1$, $W(\gamma_2^*) = 2t + c_2$, and $W(\gamma_1^* \oplus \gamma_2^*) = 2t + 2c_1 = 2t + c$. The minimum among the three weights is $2t + c_1$. Therefore, for any nonzero vector $\gamma \in V_{2k}$ with $W(\gamma) \leq 2t + c_1 - 1$, we have $\gamma \neq \gamma_1^*, \gamma_2^*$ or $\gamma_1^* \oplus \gamma_2^*$. By Theorem 23, $g^*(x) \oplus g^*(x \oplus \gamma)$ is balanced. From this we conclude that g^* satisfies the propagation criterion of order $2t + c_1 - 1$. The proof is completed by noting that $c_1 = 0$ if $c = 0$ or 1 and $c_1 = 1$ if $c = 2$. □

In the full paper [13] we shall show that functions obtained by (4) and (5) can achieve a wide range of algebraic degrees, namely 2, ... , k and 2, ... , $k - 1$ respectively. We shall also provide two concrete examples to illustrate our construction methods.

References

1. ADAMS, C. M., AND TAVARES, S. E. Generating and counting binary bent sequences. *IEEE Transactions on Information Theory IT-36 No. 5* (1990), 1170–1173.
2. ADAMS, C. M., AND TAVARES, S. E. The use of bent sequences to achieve higher-order strict avalanche criterion. Technical Report, TR 90-013, Department of Electrical Engineering, Queen's University, 1990.
3. COHEN, G. D., KARPOVSKY, M. G., H. F. MATTSON, J., AND SCHATZ, J. R. Covering radius — survey and recent results. *IEEE Transactions on Information Theory IT-31*, 3 (1985), 328–343.
4. DETOMBE, J., AND TAVARES, S. Constructing large cryptographically strong S-boxes. In *Advances in Cryptology - AUSCRYPT'92* (1993), Springer-Verlag, Berlin, Heidelberg, New York. to appear.
5. DILLON, J. F. A survey of bent functions. *The NSA Technical Journal* (1972), 191–215. (unclassified).
6. MEIER., W., AND STAFFELBACH, O. Nonlinearity criteria for cryptographic functions. In *Advances in Cryptology - EUROCRYPT'89* (1990), vol. 434, Lecture Notes in Computer Science, Springer-Verlag, Berlin, Heidelberg, New York, pp. 549–562.
7. NYBERG, K. Perfect nonlinear S-boxes. In *Advances in Cryptology - EUROCRYPT'91* (1991), vol. 547, Lecture Notes in Computer Science, Springer-Verlag, Berlin, Heidelberg, New York, pp. 378–386.
8. PATTERSON, N. J., AND WIEDEMANN, D. H. The covering radius of the $(2^{15}, 16)$ Reed-Muller code is at least 16276. *IEEE Transactions on Information Theory IT-29*, 3 (1983), 354–356.
9. PRENEEL, B., LEEKWICK, W. V., LINDEN, L. V., GOVAERTS, R., AND VANDEWALLE, J. Propagation characteristics of boolean functions. In *Advances in Cryptology - EUROCRYPT'90* (1991), vol. 437, Lecture Notes in Computer Science, Springer-Verlag, Berlin, Heidelberg, New York, pp. 155–165.
10. ROTHAUS, O. S. On "bent" functions. *Journal of Combinatorial Theory Ser. A, 20* (1976), 300–305.
11. SEBERRY, J., AND YAMADA, M. Hadamard matrices, sequences, and block designs. In *Contemporary Design Theory: A Collection of Surveys*, J. H. Dinitz and D. R. Stinson, Eds. John Wiley & Sons, Inc, 1992, ch. 11, pp. 431–559.
12. SEBERRY, J., AND ZHANG, X. M. Highly nonlinear 0-1 balanced functions satisfying strict avalanche criterion. In *Advances in Cryptology - AUSCRYPT'92* (1993), Springer-Verlag, Berlin, Heidelberg, New York. to appear.
13. SEBERRY, J., ZHANG, X. M., AND ZHENG, Y. Nonlinearity and propagation characteristics of balanced boolean functions. Submitted for Publication, 1993.

A Low Communication Competitive Interactive Proof System for Promised Quadratic Residuosity*

Toshiya Itoh[1] Masafumi Hoshi[1] Shigeo Tsujii[2]

[1] Department of Information Processing,
Interdisciplinary Graduate School of Science and Engineering,
Tokyo Institute of Technology,
4259 Nagatsuta, Midori-ku, Yokohama 227, Japan.

[2] Department of Electrical and Electronic Engineering,
Disciplinary Graduate School of Science and Engineering,
Tokyo Institute of Technology,
2-12-1 O-okayama, Meguro-ku, Tokyo 152, Japan.

Abstract. A notion of "competitive" interactive proof systems is defined by Bellare and Goldwasser as a natural extension of a problem whether computing a witness w of $x \in L$ is harder than deciding $x \in L$ for a language $L \in \mathcal{NP}$. It is widely believed that quadratic residuosity (QR) does not have a competitive interactive proof system. Bellare and Goldwasser however introduced a notion of "representative" of Z_N^* and showed that there exists a competitive interactive proof system for promised QR, i.e., the moduli N is guaranteed to be the product of $k = O(\log\log|N|)$ distinct odd primes. In this paper, we consider how to reduce the communication complexity of a competitive interactive proof system for promised QR and how to relax the constraint on k from $O(\log\log|N|)$ to $O(\log|N|)$. To do this, we introduce a notion of "dominant" of Z_N^* and show that promised QR with the constraint that $k = O(\log|N|)$ has a competitive interactive proof system with considerably low communication complexity.

1 Introduction

1.1 Background and Motivation

Is proving membership harder than deciding membership? This is one of the most basic questions in theoretical computer science. It has been known that if a language L is $\mathcal{NP}$-complete then computing a witness w for $x \in L$ is polynomially equivalent to deciding $x \in L$. How about the languages that are not known to be $\mathcal{NP}$-complete? In general, it has been widely believed that this is not the case. Recently, Bellare and Goldwasser [2], [3] showed that there exists a language $L \in \mathcal{NP} - \mathcal{P}$ for which computing a witness w for $x \in L$ is exactly harder than deciding $x \in L$ if the class of deterministic double exponential

* Sponsored by Okawa Institute of Information and Telecommunication grant 93-01.

time is not equal to the class of nondeterministic double exponential time. The language $L \in \mathcal{NP} - \mathcal{P}$ found by Bellare and Goldwasser [2], [3] satisfies the uniformly log-sparse property and thus it is somewhat unnatural. On the other hand, there exist several natural languages $L \in \mathcal{NP}$ for which computing a witness w for $x \in L$ may be harder than deciding $x \in L$, e.g., quadratic residuosity (QR), quadratic nonresiduosity (QNR), etc.

What will happen when interactions and randomization are allowed in the proving process of membership? This way of the proving process of membership is formulated by Goldwasser, Micali, and Rackoff [7] (resp. independently by Babai and Moran [4]) as interactive proof systems (resp. Arthur-Merlin games). Informally, a language L has an interactive proof system $\langle P, V \rangle$ if for the honest prover P and for any $x \in L$, the honest verifier V accepts $x \in L$ with probability at least 2/3 and for any all powerful (dishonest) prover P^* and for any $x \notin L$, the honest verifier V accepts $x \notin L$ with probability at most 1/3. Bellare and Goldwasser [2], [3] extended the problem whether computing a witness w for $x \in L$ is harder than deciding $x \in L$ for a language $L \in \mathcal{NP}$ to the case of interactive proof systems and formulated the problem to be "competitive" interactive proof systems. Informally, an interactive proof system $\langle P, V \rangle$ for a language L is *competitive* if for the (probabilistic polynomial time bounded) honest prover P with an access to the oracle L and for any $x \in L$, the honest verifier V accepts $x \in L$ with probability at least 2/3 and for any all powerful (dishonest) prover P^* and for any $x \notin L$, the honest verifier V accepts $x \notin L$ with probability at most 1/3. It should be noted that in interactive proof systems $\langle P, V \rangle$, the honest prover P is allowed to be a computationally unbounded Turing machine, while in competitive interactive proof systems $\langle P, V \rangle$, the honest prover P must be a probabilistic polynomial time bounded oracle Turing machine with an access to the underlying language as an oracle.

Then is proving membership still harder than deciding membership in competitive interactive proof systems? In some cases, the interactions and the randomization alleviate the proving task, but in another cases, they may not. To see this more precisely, let us first consider the language QNR. It has not been known that computing a witness w for $x \in$ QNR is polynomially equivalent to deciding $x \in$ QNR. Indeed it is believed that computing a witness w for $x \in$ QNR may be harder than deciding $x \in$ QNR. Goldwasser, Micali, and Rackoff [7] however showed that QNR has a competitive interactive proof system and this implies that the (honest) prover P suffices to have the computational ability of deciding $x \in$ QNR in order to prove membership of $x \in$ QNR in an interactive and a randomized manner. Next let us consider the language QR. It is also believed that computing a witness w for $x \in$ QR may be harder than deciding $x \in$ QR. Contrary to QNR, in all known interactive proof systems $\langle P, V \rangle$ for QR (see, e.g., [10], [6]), the (honest) prover P requires to have at least the computational ability of computing square roots modulo a composite number N (equivalently the computational ability of factoring a composite number N).

Bellare and Goldwasser [2], [3] observed that the interactions and the randomization do not necessarily alleviate the proving task and showed that there exists a language $L \in \mathcal{NP} - \mathcal{BPP}$ that does not have a competitive interactive proof system if the class of nondeterministic double exponential time is not included in the class of bounded probabilistic double exponential time. (Independently, Beigel and Feigenbaum [1] showed for a different purpose that there exists an *incoherent* language $L \in \mathcal{NP}$ if the class of nondeterministic triple exponential time is not included in the class of bounded probabilistic triple exponential time.) Again, the language $L \in \mathcal{NP} - \mathcal{BPP}$ shown by Bellare and Goldwasser [2], [3] satisfies the uniformly log-sparse property and thus it is also somewhat unnatural. This result only guarantees the existence of a language $L \in \mathcal{NP} - \mathcal{BPP}$ that does not have a competitive interactive proof system under the complexity assumption but does not necessarily imply that QR never has a competitive interactive proof system. Then is it possible to construct a competitive interactive proof system for QR like in the case for QNR?

This has not been solved yet but is believed that this is not the case. To affirmatively solve this open problem, Bellare and Goldwasser [3] investigated QR in a *promised* form. Intuitively, a promise problem (see, e.g., [5], [9], etc) is specified by a pair of disjoint sets A and B and for $x \in A \cup B$ we have to decide whether $x \in A$ or $x \in B$. It should be noted that the promise problem is different from the language membership problem, because the former imposes restrictions on inputs but the latter does not. In this setting, Bellare and Goldwasser [3] introduced a notion of *representative* of Z_N^* and showed that there exists a competitive interactive proof system for promised QR, i.e., the moduli N is guaranteed to be the product of $k = O(\log\log|N|)$ distinct odd primes. Informally, a vector $y = (y_1, y_2, \ldots, y_{2^k-1})$ over Z_N^* is said to be *representative* of Z_N^* if each y_i $(1 \le i \le 2^k - 1)$ belongs to a distinct residue class except for quadratic residues modulo N. The basic idea behind the result above is to use the fact that there exist $2^k = O(\log|N|)$ distinct residues classes under a relation appropriately defined on Z_N^* and to reduce a quadratic residuosity test to a collection of quadratic nonresiduosity tests. Then the protocol following this idea requires about 2^{2k} quadratic nonresiduosity tests and thus the communications complexity of the resulting protocol is comparatively large — in the protocol, the prover P sends to the verifier V about $2^k(|N| + 2^k)$ bits and the verifier V sends to the prover P about $2^{2k}\,|N|$ bits.

1.2 Results

In this paper, we consider how to reduce the communication complexity of a competitive interactive proof system for promised QR and how to relax the constraint on k from $O(\log\log|N|)$ to $O(\log|N|)$. For this purpose, we first introduce a notion of *dominant* of Z_N^*, which plays a role very similar to a basis in a linear space over $GF(2)$. Informally, a vector $y = (y_1, y_2, \ldots, y_k)$ over Z_N^* is said to be *dominant* of Z_N^* if for any vector $e = (e_1, e_2, \ldots, e_k) \in \{0,1\}^k$ such that $e \ne 0$, $z \equiv y_1^{e_1} y_2^{e_2} \cdots y_k^{e_k} \pmod N$ is not a square modulo N. Then we investigate several properties of a dominant vector of Z_N^* and show that promised QR with the constraint that $k = O(\log|N|)$ has a competitive interactive proof system in which the prover P sends to the verifier V about $k\,|N|$ bits and the verifier V sends to the prover P about $4|N|$ bits. The basic idea behind the result here is to use the fact that if the moduli N is guaranteed to be the product of $k = O(\log|N|)$ distinct odd primes then there exist sufficiently many (samplable) vectors $\boldsymbol{y} = (y_1, y_2, \ldots, y_k)$ over Z_N^* to uniquely specify 2^k residue classes under a relation appropriately defined on Z_N^*. The idea here is inspired by the one due to Bellare and Goldwasser [3] but its use enables us to avoid 2^{2k} invocations of quadratic nonresiduosity tests. Thus the resulting protocol based on this idea considerably reduces the communication complexity.

2 Preliminaries

In this section, we present definitions and notation necessary in the sequel.

Let $\langle P, V\rangle$ be an interactive protocol. Informally, an interactive protocol $\langle P, V\rangle$ is said to be an interactive proof system for a language L if for the honest prover P and for any $x \in L$, the honest verifier V accepts $x \in L$ with probability at least 2/3 and for any all powerful dishonest prover P^* and for any $x \notin L$, the honest verifier V accepts $x \notin L$ with probability at most 1/3. For further details on this, see, e.g., [7], [8], etc.

Definition 1 [2, 3]. An interactive proof system $\langle P, V\rangle$ for a language L is said to be competitive if

- **Completeness:** For any $x \in L$, $\text{Prob}\{\langle P^L, V\rangle \text{ accepts } x\} \geq 2/3$, where the prover P is a probabilistic polynomial time oracle Turing machine;
- **Soundness:** For any $x \notin L$ and any all powerful dishonest prover P^*, $\text{Prob}\{\langle P^*, V\rangle \text{ accepts } x\} \leq 1/3$,

where the probabilities are taken over all possible coin tosses of P and V.

It is already known that there exist competitive interactive proof systems for quadratic nonresiduosity [7], for graph nonisomorphism [8], and for graph isomorphism [8], [10], however, quadratic residuosity is believed not to have a competitive interactive proof system.

Let $\langle A, B\rangle$ be a pair of disjoint sets. Intuitively, the problem $\langle A, B\rangle$ is said to be *promised* if the inputs are guaranteed to be in $A \cup B$. Associated to the promise problem $\langle A, B\rangle$, we define a promise oracle that returns correct answers only when the queries are in $A \cup B$.

Definition 2 [3]. A promise problem is a pair of disjoint sets $\langle A, B\rangle$. A promise oracle for $\langle A, B\rangle$ is an oracle that given $q \in A \cup B$, returns 1 if $q \in A$ and returns 0 if $q \in B$.

Informally, a promise problem $\langle A, B\rangle$ has a a competitive interactive proof system $\langle P, V\rangle$ if for the (probabilistic polynomial time bounded) honest prover P with an access to the promise oracle for $\langle A, B\rangle$ and for any $x \in A$, the honest verifier V accepts $x \in A$ with probability at least 2/3 and for any all powerful dishonest prover P^* and for any $x \in B$, the honest verifier V accepts $x \in B$ with probability at most 1/3.

Definition 3 [3]. A promise problem $\langle A, B\rangle$ is said to have a competitive interactive proof system if

- **Completeness:** For any $x \in A$ and any promise oracle O for $\langle A, B\rangle$, $\text{Prob}\{\langle P^O, V\rangle \text{ accepts } x\} \geq 2/3$, where P is a probabilistic polynomial time oracle Turing machine;
- **Soundness:** For any $x \in B$ and any all powerful dishonest prover P^*, $\text{Prob}\{\langle P^*, V\rangle \text{ accepts } x\} \leq 1/3$,

where the probabilities are taken over all possible coin tosses of P and V.

A language quadratic residuosity (QR) is defined to be $\text{QR} = \{\langle x, N\rangle \mid x \in Z_N^* \text{ is a square modulo } N\}$ and a language quadratic nonresiduosity QNR is defined to be $\text{QNR} = \{\langle x, N\rangle \mid x \in Z_N^* \text{ is not a square modulo } N\}$. The problem that we are interested in is when the moduli N is guaranteed to be the product of $k \geq 1$ distinct odd primes. In the following, we define the problem "promised QR" that will be investigated in this paper.

Definition 4 [3]. A promised QR is a pair of disjoint sets $\langle \text{QR}_k, \text{QNR}_k\rangle$, where $\text{QR}_k = \{\langle x, N\rangle \in \text{QR} \mid N \text{ is the product of } k \text{ distinct odd primes}\}$, $\text{QNR}_k = \{\langle x, N\rangle \in \text{QNR} \mid N \text{ is the product of } k \text{ distinct odd primes}\}$, and $k \geq 1$.

3 Known Results

We overview the result by Bellare and Goldwasser [3], i.e., if $k = O(\log\log|N|)$, then the promised QR $\langle \text{QR}_k, \text{QNR}_k\rangle$ has a competitive interactive proof system.

Lemma 5 [3]. *If $k = O(\log\log|N|)$, then promised QR $\langle \text{QR}_k, \text{QNR}_k\rangle$ has a competitive interactive proof system.*

Here we overview the protocol given by Bellare and Goldwasser [3]. In the competitive interactive proof system for promised QR [3], Protocol QNR is used as a subprotocol.

Protocol QNR: A "Competitive" IP for QNR

common input: $\langle x, N\rangle$ and 1^s, where s is the security parameter.

V1: V chooses $c_i \in_R \{0,1\}$, $r_i \in_R Z_N^*$ and computes $z_i \equiv x^{c_i} r_i^2 \pmod N$ $(1 \leq i \leq s)$.
$V \rightarrow P$: $\langle z_1, z_2, \ldots, z_s\rangle$.
P1: For each i $(1 \leq i \leq s)$, if $z_i \in Z_N^*$ is a square modulo N, then P sets $d_i = 0$; otherwise P sets $d_i = 1$.
$P \rightarrow V$: $\langle d_1, d_2, \ldots, d_s\rangle$.
V2: V accepts iff $c_i = d_i$ for each i $(1 \leq i \leq s)$.

It is easy to see that the protocol above is a competitive interactive proof system for quadratic nonresiduosity (QNR).

To show the protocol by Bellare and Goldwasser [3], we present a notion of "representative vector" of Z_N^* and several technical lemmas on its properties.

Definition 6 [3]. Let N be the product of k distinct odd primes. A vector $y = (y_1, y_2, \ldots, y_{2^k-1})$ over Z_N^* is said to be representative of Z_N^* if (1) for each i ($1 \le i \le 2^k - 1$), $y_i \in Z_N^*$ is not a square modulo N and (2) for each i, j ($1 \le i < j \le 2^k - 1$), $z_{ij} \equiv y_i y_j \pmod{N}$ is not a square modulo N.

The following is the key proposition on the reduction of a quadratic residuosity test to a collection of quadratic nonresiduosity tests.

Proposition 7 [3]. *Let N be the product of k distinct odd primes and let $y = (y_1, y_2, \ldots, y_{2^k-1})$ be representative of Z_N^*. Then $\langle x, N\rangle \in \mathrm{QR}_k$ iff $w_i \equiv xy_i \pmod{N}$ is not a square modulo N for each i ($1 \le i \le 2^k - 1$).*

Bellare and Goldwasser [3] showed an efficient way to find a representative vector y of Z_N^*, i.e., if $k = O(\log\log|N|)$, then there exists a probabilistic polynomial time oracle Turing machine with an access to the promise oracle for $\langle \mathrm{QR}_k, \mathrm{QNR}_k\rangle$ that samples with probability at least 3/4 a representative vector $y = (y_1, y_2, \ldots, y_{2^k-1})$ of Z_N^*.

Proposition 8 [3]. *If $k = O(\log\log|N|)$, then exists a probabilistic polynomial time oracle Turing machine R with an access to the promise oracle for $\langle \mathrm{QR}_k, \mathrm{QNR}_k\rangle$ that on input $\langle x, N\rangle \in \mathrm{QR}_k \cup \mathrm{QNR}_k$ outputs either a representative vector $y = (y_1, y_2, \ldots, y_{2^k-1})$ of Z_N^* with probability at least 3/4 or $\perp$ with probability at most 1/4.*

The basic idea behind the result by Bellare and Goldwasser [3] is as follows: (1) The prover P generates a representative vector y of Z_N^* (see Proposition 8); (2) The prover P shows to the verifier V that the vector y is really representative of Z_N^* (see Definition 6) by the interactive proof system for QNR [7]; and (3) The prover P shows to the verifier V that $\langle x, N\rangle \in \mathrm{QR}_k$ (see Proposition 7) by the interactive proof system for QNR [7].

The following is the competitive interactive proof system for promised QR [3] under the constraint that $k = O(\log\log|N|)$.

Protocol PQR-1: A Competitive IP for Promised QR

common input: $\langle x, N\rangle \in \mathrm{QR}_k \cup \mathrm{QNR}_k$, where $k = O(\log\log|N|)$.

P1: P runs the machine R to sample a vector $y = (y_1, y_2, \ldots, y_{2^k-1})$ as a candidate of representative of Z_N^*

$P \to V$: $y = (y_1, y_2, \ldots, y_{2^k-1})$.

V1: If V receives $\perp$ from P, then V halts and rejects $\langle x, N\rangle \in \mathrm{QR}_k \cup \mathrm{QNR}_k$; otherwise V continues.

$P \leftrightarrow V$: P shows to V by Protocol QNR with $s = 2$ that y_i is not a square modulo N for each i ($1 \le i \le 2^k - 1$).

V2: If V does not accept $\langle y_i, N\rangle$ for some i ($1 \le i \le 2^k - 1$), then V halts and rejects $\langle x, N\rangle \in \mathrm{QR}_k \cup \mathrm{QNR}_k$; otherwise V continues.

$P \leftrightarrow V$: P shows to V by Protocol QNR with $s = 2$ that $z_{ij} \equiv y_i y_j \pmod{N}$ is not a square modulo N for each i, j ($1 \le i < j \le 2^k - 1$).

V3: If V does not accept $\langle z_{ij}, N\rangle$ for some i, j ($1 \le i < j \le 2^k - 1$), then V halts and rejects $\langle x, N\rangle \in \mathrm{QR}_k \cup \mathrm{QNR}_k$; otherwise V continues.

$P \leftrightarrow V$: P shows to V by Protocol QNR with $s = 2$ that $w_i \equiv xy_i \pmod N$ is not a square modulo N for each i ($1 \le i \le 2^k - 1$).

V4: If V does not accept $\langle w_i, N\rangle$ for some i ($1 \le i \le 2^k - 1$), then V halts and rejects $\langle x, N\rangle \in \mathrm{QR}_k \cup \mathrm{QNR}_k$; otherwise V halts and accepts $\langle x, N\rangle \in \mathrm{QR}_k \cup \mathrm{QNR}_k$.

The correctness of Protocol PQR-1 follows from that of Protocol QNR [3].

4 Main Results

In this section, we show that if $k = O(\log |N|)$, then promised QR $\langle \mathrm{QR}_k, \mathrm{QNR}_k\rangle$ has a competitive interactive proof system with much lower communication complexity than the one by Bellare and Goldwasser [3].

4.1 Technical Lemmas

Let $M \ge 2$ be an odd integer. For any $x \in Z_M^*$, let $Q_M(x) = 0$ if $x \in Z_M^*$ is a square modulo M and let $Q_M(x) = 1$ if $x \in Z_M^*$ is not a square modulo M. Let $N = p_1^{\alpha_1} p_2^{\alpha_2} \cdots p_k^{\alpha_k}$, where $p_1, p_2, \ldots p_k$ are distinct odd primes and $\alpha_i \ge 1$ for each i ($1 \le i \le k$). For any $x, y \in Z_N^*$, define a binary relation $\simeq$ on Z_N^* to be $x \simeq y$ iff $Q_{p_i}(x) = Q_{p_i}(y)$ for each i ($1 \le i \le k$).

It is easy to see that the relation $\simeq$ on Z_N^* is an equivalence relation on Z_N^*. The equivalence class $R_N(x)$ of $x \in Z_N^*$ under the relation $\simeq$ on Z_N^*, i.e., $R_N(x) = \{y \in Z_N^* \mid x \simeq y\}$, is called to be a residue class of $x \in Z_N^*$.

Definition 9. Let $N = p_1^{\alpha_1} p_2^{\alpha_2} \cdots p_k^{\alpha_k}$ be the product of k distinct odd primes. Then for $x \in Z_N^*$, a vector $\mathbf{c}_x = (c_1^x, c_2^x, \ldots, c_k^x) \in \{0,1\}^k$ is said to be associated with $x \in Z_N^*$ if $c_i^x = Q_{p_i}(x)$ for each i ($1 \le i \le k$).

The following lemmas show basic properties of a vector $\mathbf{c}_x \in \{0,1\}^k$ associated with $x \in Z_N^*$.

Lemma 10. *Let N be the product of k distinct odd primes. For any $x, y \in Z_N^*$, let $z \equiv xy \pmod N$ and let $\mathbf{c}_x, \mathbf{c}_y, \mathbf{c}_z \in \{0,1\}^k$ be vectors associated with $x, y, z \in Z_N^*$, respectively. Then $\mathbf{c}_z \equiv \mathbf{c}_x + \mathbf{c}_y \pmod 2$.*

Lemma 11. *Let N be the product of k distinct odd primes and let $\mathbf{c}_x \in \{0,1\}^k$ be a vector associated with $x \in Z_N^*$. For any integer $e \ge 0$, let $\mathbf{c}_y \in \{0,1\}^k$ be a vector associated with $y \equiv x^e \pmod N$. Then $\mathbf{c}_y \equiv e\mathbf{c}_x \pmod 2$.*

The following notion of "dominant" is one of the most important ones in our main result here. It plays a role similar to a basis in a linear space over $GF(2)$.

Definition 12. Let N be the product of k distinct odd primes. A vector $\mathbf{y} = (y_1, y_2, \ldots, y_k)$ is said to be dominant of Z_N^* if vectors $\mathbf{c}_{y_1}, \mathbf{c}_{y_2}, \ldots, \mathbf{c}_{y_k} \in \{0,1\}^k$, each of which is associated with $y_i \in Z_N^*$ ($1 \le i \le k$), are linearly independent over $GF(2)$.

Hereafter, we use d_i for a dominant vector $y = (y_1, y_2, \ldots, y_k)$ of Z_N^* to denote a vector associated with $y_i \in Z_N^*$ instead of c_{y_i} $(1 \le i \le k)$.

Let $y = (y_1, y_2, \ldots, y_k)$ be a vector over Z_N^* and let $e = (e_1, e_2, \ldots, e_k)$ be a vector over $GF(2)$. For simplicity, here we use $y \uparrow e$ to denote $y \uparrow e \equiv y_1^{e_1} y_2^{e_2} \cdots y_k^{e_k} \pmod N$. In the following lemma, we show that if $k = O(\log |N|)$, then a dominant vector y of Z_N^* can be efficiently sampled by a probabilistic polynomial time oracle Turing machine with an access to the promise oracle for promised QR $\langle \mathrm{QR}_k, \mathrm{QNR}_k \rangle$.

Lemma 13. *If $k = O(\log |N|)$, then there exists a probabilistic polynomial time oracle Turing machine D with an access to the promise oracle for $\langle \mathrm{QR}_k, \mathrm{QNR}_k \rangle$ that on input $\langle x, N \rangle \in \mathrm{QR}_k \cup \mathrm{QNR}_k$ outputs either a dominant vector y of Z_N^* with probability at least 3/4 or $\perp$ with probability at most 1/4.*

Proof. Let $y = (y_1, y_2, \ldots, y_k)$ be a vector over Z_N^* and let $c_i \in \{0,1\}^k$ be a vector associated with $y_i \in Z_N^*$ for each i $(1 \le i \le k)$. The probability P_{ind} that the vectors $c_1, c_2, \ldots, c_k$ are linearly independent over $GF(2)$ is bounded by

$$P_{ind} = \frac{1}{\|Z_N^*\|^k} \prod_{j=0}^{k-1} \left(\|Z_N^*\| - 2^j \cdot \frac{\|Z_N^*\|}{2^k} \right) = \prod_{i=1}^{k} (1 - 2^{-i})$$

$$\ge \prod_{i=1}^{\infty} (1 - 2^{-i}) = 1 + \sum_{i=1}^{\infty} \frac{(-1)^i}{(2-1)(2^2-1)\cdots(2^i-1)} > \frac{2}{7},$$

where $\|A\|$ denotes the cardinality of a (finite) set A. Then the machine D randomly chooses m vectors $y_j = (y_{1j}, y_{2j}, \ldots, y_{kj})$ over Z_N^* $(1 \le j \le m)$. For each y_j $(1 \le j \le m)$, the machine D computes $q_j^\ell \equiv y_j \uparrow \mathrm{bin}(\ell) \pmod N$ for each ℓ $(1 \le \ell \le 2^k - 1)$, and queries q_j^ℓ to the promise oracle for $\langle \mathrm{QR}_k, \mathrm{QNR}_k \rangle$ to get the answer $a_j^\ell \in \{0,1\}$, where $\mathrm{bin}(\ell)$ is the binary representation of an integer ℓ $(1 \le \ell \le 2^k - 1)$. If there exists an index j $(1 \le j \le 2^k - 1)$ such that $a_j^\ell = 0$ for each ℓ $(1 \le \ell \le 2^k - 1)$, then the machine D outputs $y = y_j$ as a dominant vector of Z_N^*; otherwise the machine D outputs $\perp$.

The vector y sampled by the machine D is always dominant of Z_N^*. We show this by contradiction. We assume that the vector $y = (y_1, y_2, \ldots, y_k)$ sampled by D is not dominant of Z_N^*. Then for a vector c_i associated with $y_i \in Z_N^*$ $(1 \le i \le k)$, there exists a nonzero vector $e = (e_1, e_2, \ldots, e_k)$ over $GF(2)$ such that $e_1 c_1 + e_2 c_2 + \cdots + e_k c_k \equiv 0 \pmod 2$. This implies that $z \equiv y \uparrow e \pmod N$ is a square modulo N and this contradicts the fact that $q^\ell \equiv y \uparrow \mathrm{bin}(\ell) \pmod N$ is not a square modulo N for each ℓ $(1 \le \ell \le 2^k - 1)$. The probability P_{dom} that the machine D samples a dominant vector y of Z_N^* is bounded by $P_{dom} = 1 - (1 - P_{ind})^m > 1 - (1 - 2/7)^m$. Then letting $m \ge 5$, $P_{dom} \ge 3/4$. Since the machine D queries to the promise oracle for $\langle \mathrm{QR}_k, \mathrm{QNR}_k \rangle$ at most $m2^k$ times, it runs in probabilistic polynomial (in $|N|$) time.

Thus on input $\langle x, N \rangle \in \mathrm{QR}_k \cup \mathrm{QNR}_k$, the machine D with an access to the promise oracle for promised QR $\langle \mathrm{QR}_k, \mathrm{QNR}_k \rangle$ outputs either a dominant vector y of Z_N^* with probability at least 3/4 or $\perp$ with probability at most 1/4. ∎

The lemma below is the essential to reduce the communication complexity of a competitive interactive proof system for promised QR $\langle \mathrm{QR}_k, \mathrm{QNR}_k \rangle$.

Lemma 14. *Let N be the product of k distinct odd primes and let a vector $\boldsymbol{y} = (y_1, y_2, \ldots, y_k)$ be dominant of Z_N^*, Then for any $x \in Z_N^*$, there exists a* unique *vector $\boldsymbol{e} = (e_1, e_2, \ldots, e_k)$ over $GF(2)$ such that $x \simeq \boldsymbol{y} \uparrow \boldsymbol{e}$.*

Proof. We assume that $N = p_1^{\alpha_1} p_2^{\alpha_2} \cdots p_k^{\alpha_k}$, where $p_1, p_2, \ldots, p_k$ are distinct odd primes and $\alpha_i \geq 1$ for each i $(1 \leq i \leq k)$. Let $\boldsymbol{c}_x$ be a vector associated with $x \in Z_N^*$ and let $\boldsymbol{d}_i$ be a vector associated with $y_i \in Z_N^*$ for each i $(1 \leq i \leq k)$. Since $\boldsymbol{y}$ is dominant of Z_N^*, $\boldsymbol{d}_1, \boldsymbol{d}_2, \ldots, \boldsymbol{d}_k$ are linearly independent over $GF(2)$. Then there exists a unique vector $\boldsymbol{e} = (e_1, e_2, \ldots, e_k)$ over $GF(2)$ such that $\boldsymbol{c}_x \equiv e_1 \boldsymbol{d}_1 + e_2 \boldsymbol{d}_2 + \cdots + e_k \boldsymbol{d}_k \pmod 2$. Here we define $z \in Z_N^*$ to be $z \equiv \boldsymbol{y} \uparrow \boldsymbol{e} \pmod N$. From the property of the Jacobi symbol, it follows that $Q_{p_i}(x) = Q_{p_i}(z)$ for each i $(1 \leq i \leq k)$ and thus $x \simeq z \simeq \boldsymbol{y} \uparrow \boldsymbol{e}$.

The uniqueness of a vector $\boldsymbol{e} = (e_1, e_2, \ldots, e_k)$ can be shown by contradiction. Here we assume that there exist distinct vectors $\boldsymbol{e} = (e_1, e_2, \ldots, e_k)$ and $\boldsymbol{f} = (f_1, f_2, \ldots, f_k)$ over $GF(2)$ such that $x \simeq \boldsymbol{y} \uparrow \boldsymbol{e} \simeq \boldsymbol{y} \uparrow \boldsymbol{f}$. Then for a vector $\boldsymbol{c}_x$ associated with $x \in Z_N^*$, $\boldsymbol{c}_x \equiv e_1 \boldsymbol{d}_1 + e_2 \boldsymbol{d}_2 + \cdots + e_k \boldsymbol{d}_k \equiv f_1 \boldsymbol{d}_1 + f_2 \boldsymbol{d}_2 + \cdots + f_k \boldsymbol{d}_k \pmod 2$. This implies that there exists a nonzero vector $\boldsymbol{g} = (g_1, g_2, \ldots, g_k)$ over $GF(2)$ such that $g_1 \boldsymbol{d}_1 + g_2 \boldsymbol{d}_2 + \cdots + g_k \boldsymbol{d}_k \equiv 0 \pmod 2$, where $g_i \equiv e_i + f_i \pmod 2$ for each i $(1 \leq i \leq k)$, and this contradicts the assumption that $\boldsymbol{d}_1, \boldsymbol{d}_2, \ldots, \boldsymbol{d}_k$ are linearly independent over $GF(2)$. ■

The following lemma shows that if $k = O(\log |N|)$, then there exists an efficient algorithm that for a dominant vector $\boldsymbol{y}$ of Z_N^* and any $z \in Z_N^*$, finds a (unique) vector $\boldsymbol{f} \in \{0,1\}^k$ satisfying $z \simeq \boldsymbol{y} \uparrow \boldsymbol{f}$.

Lemma 15. *Let N be the product of k distinct odd primes. Let $\boldsymbol{y}$ be dominant of Z_N^* and let $z \in Z_N^*$. If $k = O(\log |N|)$, then there exists a deterministic polynomial time algorithm* FIND *with an access to the promise oracle for $\langle \mathrm{QR}_k, \mathrm{QNR}_k \rangle$ that on input $\langle \boldsymbol{y}, z \rangle$ always outputs a (unique) vector $\boldsymbol{f} \in \{0,1\}^k$ such that $z \simeq \boldsymbol{y} \uparrow \boldsymbol{f}$.*

Proof. The following is an algorithm with an access to the promise oracle for $\langle \mathrm{QR}_k, \mathrm{QNR}_k \rangle$ that on input $\langle \boldsymbol{y}, z \rangle$ outputs $\boldsymbol{f} \in \{0,1\}^k$ such that $z \simeq \boldsymbol{y} \uparrow \boldsymbol{f}$.

Algorithm FIND:

Input: $\langle \boldsymbol{y}, z \rangle$, where $\boldsymbol{y}$ is dominant of Z_N^* and $z \in Z_N^*$.
Step 1: Compute $q_\ell \equiv (\boldsymbol{y} \uparrow \mathrm{bin}(\ell)) \times z \pmod N$ for each ℓ $(0 \leq \ell \leq 2^k - 1)$.
Step 2: Query q_ℓ to the promise oracle for $\langle \mathrm{QR}_k, \mathrm{QNR}_k \rangle$ to get the answer $a_\ell \in \{0,1\}$ for each ℓ $(0 \leq \ell \leq 2^k - 1)$.
Step 3: If $a_\ell = 1$ for some ℓ $(0 \leq \ell \leq 2^k - 1)$, then outputs $\boldsymbol{f} = \mathrm{bin}(\ell)$; otherwise output $\perp$.
Output: $\boldsymbol{f} \in \{0,1\}^k$ such that $z \simeq \boldsymbol{y} \uparrow \boldsymbol{f}$.

It follows from Lemma 14 that if $\boldsymbol{y}$ is dominant of Z_N^*, the algorithm FIND always finds a *unique* $\boldsymbol{f} \in \{0,1\}^k$ such that $z \simeq \boldsymbol{y} \uparrow \boldsymbol{f}$. Since $k = O(\log|N|)$ and the algorithm FIND queries to the promise oracle for $\langle \mathrm{QR}_k, \mathrm{QNR}_k \rangle$ at most 2^k times, the algorithm FIND runs in deterministic polynomial (in $|N|$) time. ■

4.2 A Low Communication Competitive IP for Promised QR

We now describe the whole protocol of a competitive interactive proof system for promised QR $\langle \mathrm{QR}_k, \mathrm{QNR}_k \rangle$ with considerably low communication complexity.

Protocol PQR-2: A Competitive IP for Promised QR

common input: $\langle x, N \rangle \in \mathrm{QR}_k \cup \mathrm{QNR}_k$, where $k = O(\log|N|)$.

P1: P runs the machine D to sample a vector $\boldsymbol{y} = (y_1, y_2, \ldots, y_k)$ as a candidate of dominant of Z_N^*.

$P \rightarrow V$: $\boldsymbol{y} = (y_1, y_2, \ldots, y_k)$.

V1-1: If V receives $\perp$ from P, then V halts and rejects $\langle x, N \rangle \in \mathrm{QR}_k \cup \mathrm{QNR}_k$; otherwise V continues.

V1-2: V chooses $\boldsymbol{a}_j \in_{\mathrm{R}} \{0,1\}^k$ and $r_j \in_{\mathrm{R}} Z_N^*$ and computes $z_j \equiv (\boldsymbol{y} \uparrow \boldsymbol{a}_j) \times r_j^2 \pmod N$ for each j $(0 \le j \le 1)$.

$V \rightarrow P$: $z_0, z_1 \in Z_N^*$.

P2: P computes $\boldsymbol{\alpha}_j \in \{0,1\}^k$ such that $z_j \simeq \boldsymbol{y} \uparrow \boldsymbol{\alpha}_j$ for each j $(0 \le j \le 1)$.

$P \rightarrow V$: $\boldsymbol{\alpha}_0, \boldsymbol{\alpha}_1 \in \{0,1\}^k$.

V2-1: V checks that $\boldsymbol{\alpha}_j = \boldsymbol{a}_j$ for each j $(0 \le j \le 1)$.

V2-2: If either $\boldsymbol{\alpha}_0 \ne \boldsymbol{a}_0$ or $\boldsymbol{\alpha}_1 \ne \boldsymbol{a}_1$, then V halts and rejects $\langle x, N \rangle \in \mathrm{QR}_k \cup \mathrm{QNR}_k$; otherwise V continues.

V2-3: V chooses $e_j \in_{\mathrm{R}} \{0,1\}$, $\boldsymbol{b}_j \in_{\mathrm{R}} \{0,1\}^k$, $s_j \in_{\mathrm{R}} Z_N^*$ for each j $(0 \le j \le 1)$.

V2-4: V computes $w_j \equiv x^{e_j} \times (\boldsymbol{y} \uparrow \boldsymbol{b}_j) \times s_j^2 \pmod N$ for each j $(0 \le j \le 1)$.

$V \rightarrow P$: $w_0, w_1 \in Z_N^*$.

P3: P computes $\boldsymbol{\beta}_j \in \{0,1\}^k$ such that $w_j \simeq \boldsymbol{y} \uparrow \boldsymbol{\beta}_j$ for each j $(0 \le j \le 1)$.

$P \rightarrow V$: $\boldsymbol{\beta}_0, \boldsymbol{\beta}_1 \in \{0,1\}^k$.

V3-1: V checks that $\boldsymbol{\beta}_j = \boldsymbol{b}_j$ for each j $(0 \le j \le 1)$.

V3-2: If either $\boldsymbol{\beta}_0 \ne \boldsymbol{b}_0$ or $\boldsymbol{\beta}_1 \ne \boldsymbol{b}_1$, then V halts and rejects $\langle x, N \rangle \in \mathrm{QR}_k \cup \mathrm{QNR}_k$; otherwise V halts and accepts $\langle x, N \rangle \in \mathrm{QR}_k \cup \mathrm{QNR}_k$.

Correctness of PQR-2: We show that even when $k = O(\log|N|)$, Protocol PQR-2 is a competitive interactive proof system for promised QR $\langle \mathrm{QR}_k, \mathrm{QNR}_k \rangle$.

(*Completeness*) Assume that $\langle x, N \rangle \in \mathrm{QR}_k$. It follows from Lemma 13 that in step V1-1, V receives a dominant vector $\boldsymbol{y} = (y_1, y_2, \ldots, y_k)$ of Z_N^* from P with probability at least 3/4.

Assume that $\boldsymbol{y}$ is dominant of Z_N^*. Then it follows from Lemma 14 that there exists a unique vector $\boldsymbol{\alpha}_j \in \{0,1\}^k$ such that $z_j \simeq \boldsymbol{y} \uparrow \boldsymbol{\alpha}_j$ for each j $(0 \le j \le 1)$. To find such a vector $\boldsymbol{\alpha}_j \in \{0,1\}^k$, P executes the algorithm FIND on input $\langle \boldsymbol{y}, z_j \rangle$ for each j $(0 \le j \le 1)$. Since $k = O(\log|N|)$, P runs in deterministic polynomial (in $|N|$) time in step P2 (see Lemma 15). From the assumption that

y is dominant of Z_N^*, it follows that $\alpha_j = a_j$ for each j $(0 \le j \le 1)$. This implies that if y is dominant of Z_N^*, then V never rejects $\langle x, N\rangle \in \mathrm{QR}_k$ in step V2-2.

For any $r \in Z_N^*$, let $z \equiv xr \pmod N$. From the fundamental property of the Jacobi symbol, it is easy to see that $z \simeq r$ if $\langle x, N\rangle \in \mathrm{QR}_k$. This implies that $w_j \simeq y \uparrow \beta_j$ regardless of the value of $e_j \in \{0,1\}$ for each j $(0 \le j \le 1)$. Then it follows from Lemma 14 that P can find a *unique* vector $\beta_j \in \{0,1\}^k$ by running the algorithm FIND on input $\langle y, w_j\rangle$ for each j $(0 \le j \le 1)$. Since $k = O(\log|N|)$, P runs in deterministic polynomial (in $|N|$) time in step P3 (see Lemma 15). From the assumption that y is dominant of Z_N^*, it follows that $\beta_j = b_j$ for each j $(0 \le j \le 1)$. This implies that if y is dominant of Z_N^*, then V always accepts $\langle x, N\rangle \in \mathrm{QR}_k$ in step V3-2.

Thus for any $\langle x, N\rangle \in \mathrm{QR}_k$, the (probabilistic polynomial time bounded) honest prover P with an access to the promise oracle for $\langle \mathrm{QR}_k, \mathrm{QNR}_k\rangle$ can cause the honest verifier V to accept $\langle x, N\rangle \in \mathrm{QR}_k$ with probability at least 3/4.

(*Soundness*) Assume that $\langle x, N\rangle \in \mathrm{QNR}_k$. If V receives $\perp$ from P in step V1-1, then V halts and rejects $\langle x, N\rangle \in \mathrm{QNR}_k$. Then any dishonest prover P^* needs to send to V a vector $y = (y_1, y_2, \ldots, y_k)$ over Z_N^*. Assume that y is not dominant of Z_N^*. For each $z_j \in Z_N^*$ $(0 \le j \le 1)$ in step V1-2, there are 2^t $(1 \le t \le k)$ possible $\alpha_j \in \{0,1\}^k$ such that $z_j \simeq y \uparrow \alpha_j$ for each j $(0 \le j \le 1)$. This implies that if y is not dominant of Z_N^*, then with probability at most $2^{-2t} \le 1/4$, any all powerful P^* can find a vector $\alpha_j \in \{0,1\}^k$ such that $\alpha_j = a_j$ for each j $(0 \le j \le 1)$ in step P2. Thus if y is not dominant of Z_N^*, then V halts and rejects $\langle x, N\rangle \in \mathrm{QNR}_k$ in step V2-2 with probability at least 3/4.

Assume that y is dominant of Z_N^*. Since $\langle x, N\rangle \in \mathrm{QNR}_k$, there exists a unique vector $e \in \{0,1\}^k$ such that $x \simeq y \uparrow e$ and $e \ne 0$. For each j $(0 \le j \le 1)$, there exist $\beta_j^0, \beta_j^1 \in \{0,1\}^k$ such that $w_j \simeq y \uparrow \beta_j^0$ and $w_j \simeq x \times (y \uparrow \beta_j^1)$. Indeed, for i, j $(0 \le i, j \le 1)$, $\beta_j^i \equiv b_j + \{(e_j + i) \times e\} \pmod 2$. This implies that any dishonest prover P^* cannot guess better at random the value of $e_j \in \{0,1\}$ for each j $(0 \le j \le 1)$ even if it is infinitely powerful. Thus if y is dominant of Z_N^*, then with probability at most 1/4, any all powerful P^* can find a vector $\beta_j \in \{0,1\}^k$ such that $\beta_j = b_j$ for each j $(0 \le j \le 1)$ in step P3. Then V halts and rejects $\langle x, N\rangle \in \mathrm{QNR}_k$ in step V3-2 with probability at least 3/4.

Thus for any $\langle x, N\rangle \in \mathrm{QNR}_k$, any all powerful dishonest prover P^* can cause the honest verifier V to accept $\langle x, N\rangle \in \mathrm{QNR}_k$ with probability at most 1/4. ∎

5 Discussions

Let $CC_P(\mathrm{A})$ (resp. $CC_V(\mathrm{A})$) be the total number of bits sent by the prover P (resp. the verifier V) to the verifier V (resp. the prover P) in the protocol A.

On one hand, in Protocol PQR-1 (see section 3), we have $CC_P(\text{PQR-1}) = (2^k - 1)(|N| + 2^k + 2)$ and $CC_V(\text{PQR-1}) = (2^k - 1)(2^k + 2)\,|N|$. On the other hand, in Protocol PQR-2 (see section 4), we have $CC_P(\text{PQR-2}) = k\,|N| + 2k + 2k = k(|N| + 4)$ and $CC_V(\text{PQR-2}) = 2\,|N| + 2\,|N| = 4\,|N|$. From the fact that $k \ge 1$, it immediately follows that

$$\frac{CC_P(\text{PQR-2})}{CC_P(\text{PQR-1})} = \frac{k(|N|+4)}{(2^k-1)(|N|+2^k+2)} \le \frac{k(|N|+4)}{(2^k-1)(|N|+2+2)} = \frac{k}{2^k-1};$$
$$\frac{CC_V(\text{PQR-2})}{CC_V(\text{PQR-1})} = \frac{4\,|N|}{(2^k-1)(2^k+2)\,|N|} \le \frac{4\,|N|}{(2^{2k}+2-2)\,|N|} = \frac{4}{2^{2k}},$$

and thus Protocol PQR-2 considerably reduces the communication complexity.

In Protocol PQR-1, the constraint that $k = O(\log\log|N|)$ is caused by the task in step P1. In step P1, the prover P queries to the promise oracle for $\langle \text{QR}_k, \text{QNR}_k \rangle$ at most 2^{2^k} times to sample a representative vector $\boldsymbol{y}$ of Z_N^*. Then we must assume that $k = O(\log\log|N|)$ in Protocol PQR-1 to guarantee that P runs in probabilistic polynomial (in $|N|$) time. In Protocol PQR-2, however, the prover P queries to the promise oracle for $\langle \text{QR}_k, \text{QNR}_k \rangle$ to sample a dominant vector $\boldsymbol{y}$ of Z_N^* at most 2^k times. Then we must assume that $k = O(\log|N|)$ in Protocol PQR-2 to guarantee that P runs in probabilistic polynomial (in $|N|$) time. The essential of a dominant vector $\boldsymbol{y} = (y_1, y_2, \ldots, y_k)$ of Z_N^* is that it can generate a representative vector $\boldsymbol{y}' = (y_1', y_2' \ldots, y_{2^k-1}')$ of Z_N^* by $y_\ell' \equiv \boldsymbol{y} \uparrow \text{bin}(\ell)$ (mod N) for each ℓ $(1 \le \ell \le 2^k - 1)$.

References

1. Beigel, R. and Feigenbaum, J., "On Being Coherent Without Being Very Hard," *Computational Complexity*, Vol.2, No.1, pp.1-17 (1992).
2. Bellare, M. and Goldwasser, S., "The Complexity of Decision versus Search," MIT/LCS/TM-444 (April 1991).
3. Bellare, M. and Goldwasser, S., "The Complexity of Decision versus Search," to appear in *SIAM J. on Comput.*
4. Babai, L. and Moran, S., "Arthur-Merlin Games: A Randomized Proof Systems and a Hierarchy of Complexity Classes," *JCSS*, Vol.36, pp.254-276 (1988).
5. Even, S., Selman, A., and Yacobi, Y., "The Complexity of Promise Problems with Applications to Public-Key Cryptography," *Information and Control*, Vol.61, pp.159-173 (1984).
6. Feige, U., Fiat, A., and Shamir, A., "Zero-Knowledge Proofs of Identity," *J. of Cryptology*, Vol.1, pp.77-94 (1988).
7. Goldwasser, S., Micali, S., and Rackoff, C., The Knowledge Complexity of Interactive Proof Systems," *SIAM J. on Comput.*, Vol.18, No.1, pp.186-208 (1989).
8. Goldreich, O., Micali, S., and Wigderson, A., "Proofs That Yield Nothing But Their Validity or All Languages in $\mathcal{NP}$ Have Zero-Knowledge Interactive Proof Systems," *J. of the ACM*, Vol.38, No.1, pp.691-729 (1991).
9. Grollmann, J. and Selman, A., "Complexity Measures for Public-Key Cryptosystems," *SIAM J. on Comput.*, Vol.17, No.2, pp.309-335 (1988).
10. Tompa, M. and Woll, H., "Random Self-Reducibility and Zero-Knowledge Interactive Proofs of Possession of Information," *Proc. of FOCS*, pp.472-482 (1987).

Secret Sharing and Perfect Zero Knowledge*

A. De Santis,[1] G. Di Crescenzo,[1] G. Persiano[2]

[1] Dipartimento di Informatica ed Applicazioni,
Università di Salerno, 84081 Baronissi (SA), Italy

[2] Dipartimento di Matematica,
Università di Catania, 95125 Catania, Italy

Abstract. In this work we study relations between secret sharing and perfect zero knowledge in the non-interactive model. Both secret sharing schemes and non-interactive zero knowledge are important cryptographic primitives with several applications in the management of cryptographic keys, in multi-party secure protocols, and many other areas. Secret sharing schemes are very well-studied objects while non-interactive *perfect* zero-knowledge proofs seem to be very elusive. In fact, since the introduction of the non-interactive model for zero knowledge, the only perfect zero-knowledge proof known was for quadratic non residues.
In this work, we show that a large class of languages related to quadratic residuosity admits *non-interactive perfect zero-knowledge proofs.* More precisely, we give a protocol for proving non-interactively and in perfect zero knowledge the veridicity of any "threshold" statement where atoms are statements about the quadratic character of input elements. We show that our technique is very general and extend this result to any secret sharing scheme (of which threshold schemes are just an example).

1 Introduction

Secret Sharing. The fascinating concept of *Secret Sharing* scheme has been first considered in [18] and [3]. A secret sharing scheme is a method of dividing a secret s among a set of participants in such a way that only qualified subsets of participants can reconstruct s but non-qualified subsets have absolutely no information on s. Secret sharing schemes are very useful in the management of cryptographic keys and in multi-party secure protocols (see for instance [13]). For an unified description of recent results in the area of secret sharing schemes, and for a complete bibliography, see [20] and [19].

Zero Knowledge. The seminal concept of a *Zero-Knowledge* proof has been introduced in [15] that gave zero-knowledge proofs for the number-theoretic languages of quadratic residues and quadratic non residues. A Zero-Knowledge (ZK) proof is a special kind of proof that allows an all-powerful prover to convince

* Work supported by MURST and CNR

a poly-bounded verifier that a certain statement is true without revealing any additional information.

The theory of zero knowledge has been greatly extended by the work of [13] that proved that indeed all NP languages have zero-knowledge proofs. This breakthrough work caused much excitement both for its theoretical importance and for its impact on the design of cryptographic protocols (see [14]).

The zero-knowledge proof for all NP of [13], are *computational*zero-knowledge, that is secure only against poly-bounded adversaries, whereas those in [15] are perfect, that is secure against unlimited-power adversaries. Moreover, the proofs of [13] are based on the unproven complexity assumption of the existence of one-way functions. Perfect zero knowledge is a desirable property for a proof as one can never be sure of the computational power of the person he is giving the proof to. On the other hand, it is very unlikely that perfect zero-knowledge proofs for all NP exist, as their complexity-theoretic consequences (the collapse of the polynomial hierarchy, see [7] and [11]) are considered to be false. However, perfect zero-knowledge proofs have been given for some languages in NP which are not believed to be neither NP-complete nor in BPP and are either number-theoretic or have the property of random self-reducibility [15, 13, 12, 21, 4].

The shared-string model for Non-Interactive ZK has been put forward in [6] and further elaborated by [8, 5]. In this model, prover and verifier share a random string and the communication is monodirectional. In [5] it is proved that under the quadratic residuosity assumption all NP languages have non-interactive computational zero-knowledge proofs in this model. Subsequently, in [10] it was proved that certified trapdoor permutations are sufficient for proving non-interactively and in zero knowledge membership to any language in NP ([1] relaxed the assumption by proving that trapdoor permutations are sufficient). In the non-interactive model the only perfect zero-knowledge proof has been given in [5] for the language of quadratic non residuosity modulo *Regular*(2) integers.

Because of their importance, obtaining perfect-ZK proofs for certain classes of languages still remains an important research area.

Organization of the paper and Our Results In the next section we review some number-theoretic results about quadratic residuosity and the definition of perfect zero-knowledge in the non-interactive model of [5].

In Section 3 we present two simple proof systems. The first is to prove that an integer is a Blum integer while the second is for the *logical or* of quadratic non residuosity. More precisely, for the language OR of triples (x, y_1, y_2) such that at least one of y_1, y_2 is a quadratic non residue modulo x and x is a Blum integer.

In Section 4, we present our main result: a perfect non-interactive zero-knowledge proof system for any threshold statement of quadratic residuosity of any number of inputs. That is, for all $k \leq m$, we give a proof system for the language T(k, m) of $(m+1)$-uples $(x, y_1, \cdots, y_m)$ such that less than k of the y_i's are quadratic non residues modulo x and x is a Blum integer. Our construction is based upon the properties of secret sharing schemes. We give a way of constructing a set of shares from the random string that has the following property. If less than k of the y_i's are quadratic non residues modulo x, then this set can be opened by the prover as a sharing both of the bit $b = 0$ and of the bit $b = 1$. On the other hand, if at least k of the y_i's are quadratic non residues modulo x, then this set can be opened in a unique way. Then a bit b is taken from the random string and the prover has to construct a set of shares for it. Thus, if the input pair $(x, \vec{y})$ does not belong to $T(k, m)$, the prover has probability less than 1/2 of success. By repeating m times the protocol with different pieces of the reference string, we force the probability of cheating to be negligible. The construction of the shares employs the protocol for the language OR of Section 3.

In Section 5, we briefly discuss the generalization of the tecnique of the previous section to statements based on secret sharing schemes in which the subsets of participants that recover the secret are arbitrary. That is, given a secret sharing scheme for an access structure on a set of participants, then we can construct a non-interactive perfect zero-knowledge proof system for a special formula based on the access structure given.

In all of our proof systems the prover's program can be performed in polynomial-time provided that the factorization of the modulus is given as an additional input.

2 Background and Notations

2.1 Notations

We identify a binary string σ with the integer x whose binary representation is σ. If σ and τ are binary strings, we denote their concatenation by either $\sigma \circ \tau$ or $\sigma\tau$. By the expression $\vec{w}$ we denote the $k-$uple $(w_1, \ldots, w_k)$ of numbers or bits. We say that $\vec{w} \in S$ meaning that $w_i \in S$, for $i = 1, \ldots, k$. By the expression $|x|$ we denote the length of x if x is a string. If $\vec{x}$ is a k-uple, by the expression $|\vec{x}|$ we denote the number k of components of $|\vec{x}|$.

2.2 Number Theory

We refer the reader to [17] and [5] for the definitions of quadratic residues, Jacobi symbol and Regular(s) integers. We define the ***quadratic residuosity predicate*** as $\mathcal{Q}_x(y) = 0$ if y is a quadratic residue modulo x and 1 otherwise. Moreover, we let Z_x^{+1} and Z_x^{-1} denote, respectively, the sets of elements of Z_x^* with Jacobi symbol $+1$ and -1 and $QR_x = \{y \in Z_x^{+1} | \mathcal{Q}_x(y) = 0\}$, $NQR_x = \{y \in Z_x^{+1} | \mathcal{Q}_x(y) = 1\}$.

In this paper we will be mainly concerned with the special moduli called Blum integers.

Definition 1. An integer x is a Blum integer, in symbols $x \in \text{BL}$, if and only if $x = p^{k_1} q^{k_2}$, where p and q are different primes both $\equiv 3 \bmod 4$ and k_1 and k_2 are odd integers.

From Euler's criterion it follows that, if x is a Blum integer, $-1 \bmod x$ is a quadratic non residue with Jacobi symbol $+1$. Moreover we have the following fact.

Fact 1. On input a Blum integer x, it is easy to generate a random quadratic non residue in Z_x^{+1}: randomly select $r \in Z_x^*$ and output $-r^2 \bmod x$.

The following lemmas prove that the Blum integers enjoy the elegant property that each quadratic residue has a square root which is itself a quadratic residue. Thus each quadratic residue modulo a Blum integer has also a fourth root.

Lemma 2. Let x be a Blum integer. Every quadratic residue modulo x has at least one square root which is itself a quadratic residue modulo x.

On the other hand, if x is a product of two prime powers, but not a Blum integer, then the above lemma does not hold.

Lemma 3. Let $x = p^{k_1} q^{k_2}$, where $p \equiv 1 \bmod 4$. Then, at least one half of the quadratic residues has no square root which is itself a quadratic residue modulo x.

The following characterization of Blum integers will be used to obtain a non-interactive perfect zero-knowledge proof system for the set of Blum integers.

Fact 2. An integer x is a Blum integer if and only if $x \in Regular(2)$, $-1 \bmod x \in NQR_x$, and for each $w \in QR_x$ there exists an r such that $r^4 \equiv w \bmod x$.

2.3 Non-Interactive Perfect Zero Knowledge

Let us now review the definition of Non-Interactive Perfect ZK of [5] (we refer the reader to the original paper for motivations and discussion of the definition). We denote by L the language in question and by x an instance to it. Let P a probabilistic Turing machine and V a deterministic Turing machine that runs in time polynomial in the length of its first input.

Definition 4. We say that (P, V) is a Non-Interactive Perfect Zero-Knowledge Proof System (Non-Interactive Perfect ZK Proof System) for the language L if there exists a positive constant c such that:

1. *Completeness.* $\forall x \in L$, $|x| = n$ and for all sufficiently large n,

$$\mathbf{Pr}(\sigma \leftarrow \{0,1\}^{n^c}; Proof \leftarrow P(\sigma, x) : V(\sigma, x, Proof) = 1) > 1 - 2^{-n}.$$

2. *Soundness.* For all probabilistic algorithms *Adversary* outputting pairs $(x, Proof)$, where $x \notin L$, $|x| = n$, and all sufficiently large n,

$$\mathbf{Pr}(\sigma \leftarrow \{0,1\}^{n^c}; (x, Proof) \leftarrow Adversary(\sigma) : V(\sigma, x, Proof) = 1) < 2^{-n}.$$

3. *Perfect Zero Knowledge.* There exists an efficient simulator algorithm S such that $\forall x \in L$, the two probability spaces $S(x)$ and $View_V(x)$ are equal, where by $View_V(x)$ we denote the probability space

$$View_V(x) = \{\sigma \leftarrow \{0,1\}^{|x|^c};\ Proof \leftarrow P(\sigma, x) : (\sigma, Proof)\}.$$

We notice that in soundness, we let the adversary choose the false statement he wants to prove after seeing the random string. Nonetheless, he has only negligible probability of convincing V.
We say that (P, V) is a Non-Interactive Proof System for the language L if completeness and soundness are satisfied.
We call the "common" random string σ, input to both P and V, the *reference string.* (Above, the common input is σ and x.)

2.4 Secret Sharing

Shamir [18] and Blackley [3] were the first to consider the problem of secret sharing and gave secret sharing schemes known as *threshold schemes.* We review the notion of threshold scheme as it will be instrumental for our construction of a non-interactive perfect zero-knowledge proof system for threshold statements of quadratic residuosity. A (k, m)-threshold scheme is an efficient algorithm that on input a data S outputs m *pieces* $S_1, \ldots, S_m$, such that:
- knowledge of any k or more pieces S_i makes S easily computable;
- knowledge of any $k-1$ or fewer pieces S_i leaves S completely undetermined (all its possible values are equally likely).

Shamir [18] shows how to construct such threshold schemes using interpolation of polynomials. We have the following fact:

Fact 3. The following is a (k, m)-threshold scheme. Let $(\mathcal{E}, +, \cdot)$ be a finite field with more than m elements and let S be the value to be shared. Choose at random $a_1, \ldots, a_{k-1} \in \mathcal{E}$, construct the polynomial $q(x) = S + a_1 \cdot x + \cdots + a_{k-1} \cdot x^{k-1}$ and output $S_i = q(i)$ (all operations are performed in $\mathcal{E}$).

We say that a sequence $(S_1, \ldots, S_m)$ is a (k, m)*-sequence of admissible shares* for S (we will call it *sequence of admissible shares* when k and m are clear from the context) if there exists a polynomial $q(x) = a_0 + a_1 x + \cdots + a_{k-1} x^{k-1}$ with coefficients in $\mathcal{E}$, such that $a_0 = S$ and $S_i = q(i)$ for $i = 1, \ldots, m$.

Remark. Let $I \subseteq \{1, \ldots, m\}$ and suppose $|I| < k$. Then given S and a sequence $(S_i | i \in I)$ of values, it is always possible to efficiently generate random values S_i, $i \notin I$, such that $(S_1, \ldots, S_m)$ is a sequence of admissible shares for S (for random values S_i, $i \notin I$ we mean that the S_i's for $i \notin I$ are uniformly distributed among the S_i's such that $(S_1, \ldots, S_m)$ is a sequence of admissible shares for S). Moreover, given a sequence $(S_i | i \in I)$ of values, if the values S_i for $i \notin I$ are chosen with uniform distribution among the S_i's such that $(S_1, \ldots, S_m)$ is a sequence of admissible shares for some S, then S is uniformly distributed in $\mathcal{E}$. On the other hand, if $|I| = k$, then a sequence $(S_i | i \in I)$ of values uniquely determines a value S and values S_i for $i \notin I$ such that $(S_1, \ldots, S_m)$ is a sequence of admissible shares for S.

3 Preliminary Results

In this section we discuss two simple non-interactive perfect zero-knowledge proof systems for the language BL of Blum integers and for the language OR of logical or of quadratic residuosity that we will define later. They will be useful in the construction of our main result.

3.1 The Proof System for BL

A proof system (A,B) for BL is easily obtained using the characterization of Blum integers given by Fact 2. In fact, it is sufficient for the prover to first prove that x is a *Regular*(2) integer and that -1 is a quadratic non residue modulo x using the proof system given in [5]. Then, all it is left to prove is that every quadratic residue has a fourth root modulo x. This is done by giving, for each element $y \in Z_x^{+1}$ taken from the random string, a fourth root modulo x of y or $-y$, depending on the quadratic residuosity of y. Completeness, soundness, and perfect zero knowledge are easily seen to be satisfied.

3.2 The Proof System for OR

We now describe a non-interactive perfect ZK proof system (C,D) for the language

$$\text{OR} = \{(x, y_1, y_2) \mid x \in \text{BL}, y_1, y_2 \in Z_x^{+1} \text{ and } (y_1 \in NQR_x) \vee (y_2 \in NQR_x)\}.$$

This is an extension of the proof system for quadratic non residuosity given in [5].

In our construction we will use the following

Definition 5. For any positive integer x, define the relation $\approx_x$ on $Z_x^{+1} \times Z_x^{+1}$ as follows: $(a_1, a_2) \approx_x (b_1, b_2) \iff a_1 b_1 \in QR_x$ and $a_2 b_2 \in QR_x$.

We write $(a_1, a_2) \not\approx_x (b_1, b_2)$ when (a_1, a_2) is not $\approx_x$ equivalent to (b_1, b_2). One can prove that for each integer $x \in$ *Regular*(s), $\approx_x$ is an equivalence relation on $Z_x^{+1} \times Z_x^{+1}$ and that there are $2^{2(s-1)}$ equally numerous $\approx_x$ equivalence classes.

Let us informally describe the protocol (C,D). By (A,B) we denote the non-interactive perfect zero-knowledge proof system for the language BL described above. On input (x, y_1, y_2), C and D share a reference string $\gamma = \rho \circ \sigma$, where σ is split into pairs $(\sigma_{i,1}, \sigma_{i,2})$. First C proves that $x \in$ BL by running the algorithm A on input x and using the random string ρ. C partitions the pairs $(\sigma_{i,1}, \sigma_{i,2})$ belonging to $Z_x^{+1} \times Z_x^{+1}$ according to the relation $\approx_x$ into 4 classes. It is easy for C to prove that two pairs $(\sigma_{i,1}, \sigma_{i,2})$ and $(\sigma_{j,1}, \sigma_{j,2})$ belong to the same class: C just gives a square root modulo x of the products $\sigma_{i,1}\sigma_{j,1} \bmod x$ and $\sigma_{i,2}\sigma_{j,2} \bmod x$.

Once all the pairs, including the input pair (y_1, y_2), have been assigned to an equivalence class, C uncovers the class of pairs made of two quadratic residues by giving the square root of both elements of one of its pairs. D checks first that x is a Blum integer, by running the algorithm B, and then that the pair (y_1, y_2) is in a different class from that whose pairs are both quadratic residues.

Now, suppose $(x, y_1, y_2) \notin$ OR. Then two situations may happen: (a) $x \notin$ BL or (b) $x \in$ BL and $y_1, y_2 \in QR_x$. In the first D accepts with negligible probability because of soundness of the proof system (A,B). In the second C can perform the protocol if and only if one of the three classes of pairs, for which at least one element is a quadratic non residue, does not appear in the random string. In fact the prover has to uncover the class of pairs made of two quadratic residues and thus (y_1, y_2) has to be assigned to one of the three remaining classes. However, this means that all the pairs in that class must be made of two quadratic residues and thus we would only have representatives from three classes. This happens with negligible probability.

Let us now give a sketch of the proof that (C,D) is perfect zero-knowledge. On input $(x, y_1, y_2) \in$ OR, the simulator S has to generate uniformly distributed pairs $(\sigma_{i,1}, \sigma_{i,2})$ belonging to each of the four classes of $Z_x^{+1} \times Z_x^{+1}$ determined by the relation $\approx_x$. Notice that, on input $(x, y_1, y_2) \in$ OR, it is possible to efficiently construct four pairs $(\alpha_1, \beta_1), \ldots, (\alpha_4, \beta_4)$, each belonging to a different $\approx_x$ class, in the following way. Randomly choose $r, s \in Z_x^*$, and output

$$\begin{aligned}(\alpha_1, \beta_1) &= (y_1, y_2),\\ (\alpha_2, \beta_2) &= (r^2 \bmod x, s^2 \bmod x),\\ (\alpha_3, \beta_3) &= (y_1 y_2 \bmod x, y_1),\\ (\alpha_4, \beta_4) &= (y_2, y_1 y_2 \bmod x).\end{aligned}$$

Thus it is possible to efficiently generate a uniformly distributed pair belonging to one (randomly chosen) of the four $\approx_x$ classes, in the following way. Randomly choose $j \in \{1, \ldots, 4\}$ and $u, v \in Z_x^*$, and output the pair $(\sigma_{j,1}, \sigma_{j,2})$, where $\sigma_{j,1} = \alpha_i^{-1} u^2 \bmod x$ and $\sigma_{j,2} = \beta_i^{-1} v^2 \bmod x$. Moreover, it is easy for S to give random square roots modulo x of the products $\sigma_{j,1}\alpha_i \bmod x$ and $\sigma_{j,2}\beta_i \bmod x$: he simply gives u and v.

4 Non-Interactive Perfect Zero-Knowledge for Threshold Statement

In this section we give a non-interactive perfect zero-knowledge proof system (P,V) for the language T(k, m) of pairs $(x, \vec{y})$ where less than k elements of $\vec{y} = (y_1, \ldots, y_m)$ are quadratic non residue modulo x. That is, the language

$$\mathrm{T}(k, m) = \left\{(x, \vec{y}) \,|\, x \in \mathrm{BL}, y_i \in Z_x^{+1},\ i = 1, \ldots, m \text{ and } \left|\{y_i | y_i \in NQR_x\}\right| < k\right\},$$

for $1 \leq k \leq m + 1$. For instance, T$(1, m)$ is the language of pairs $(x, \vec{y})$ that satisfy $(y_1 \in QR_x) \wedge \ldots \wedge (y_m \in QR_x)$ and T(m, m) is the language of pairs $(x, \vec{y})$ that satisfy $(y_1 \in QR_x) \vee \ldots \vee (y_m \in QR_x)$.

The prover P wants to convince the polynomial-time verifier V that less than k of the m integers $y_1, \ldots, y_m$ are quadratic non residue modulo the Blum integer x without giving away any information that V was not able to compute alone before. V cannot compute by himself whether $(x, \vec{y}) \in \mathrm{T}(k, m)$, since the fastest way known for deciding quadratic residuosity modulo a composite integer x consists of first factoring x. Thus no efficient algorithm is known to decide if $(x, \vec{y}) \in \mathrm{T}(k, m)$. Moreover, the proof is non-interactive (P sends only one message to V), and perfect zero-knowledge (V does not gain any additional information even if not restricted to run in polynomial time).
We use the proof systems (A,B) and (C,D) of previous sections as subroutines for (P,V).

4.1 The Proof System (P,V) for $\mathrm{T}(k, m)$

Let us now introduce a bit of notation that we will use in the description of our proof system. Let $x \in \mathrm{BL}$, w and $y \in Z_x^{+1}$ and $b \in \{0,1\}$. We define the predicate $\mathcal{B}(x, y, w, b)$ in the following way:

$$\mathcal{B}(x, y, w, b) = ((-1)^b w \bmod x \in QR_x) \vee (y \in QR_x).$$

We say that the prover (x, y)*-opens* w as b if he proves that $\mathcal{B}(x, y, w, b) = 1$.
If $y \in QR_x$ then $\mathcal{B}(x, y, w, 0) = \mathcal{B}(x, y, w, 1) = 1$ regardless of the quadratic residuosity of w and thus the prover can (x, y)-open w both as a 0 and as a 1. Instead, if $y \in NQR_x$ then the prover can open w in just one way determined by the quadratic residuosity of w. In fact, suppose that $w \in QR_x$. Then obviously $\mathcal{B}(x, y, w, 0) = 1$ (and thus the prover can (x, y)-open w as a 0) and $\mathcal{B}(x, y, w, 1) = 0$, as, by the fact that -1 is a quadratic non residue modulo x, $-w \bmod x$ is a quadratic non residue modulo x. Now, suppose that $w \in NQR_x$. Then $\mathcal{B}(x, y, w, 1) = 1$, as, by the fact that -1 is a quadratic non residue modulo x, $-w \bmod x$ is a quadratic residue modulo x, (and thus the prover can (x, y)-open w as a 1) and obviously $\mathcal{B}(x, y, w, 0) = 0$. In our protocol, the (x, y)-opening of w as b is done in a zero-knowledge fashion by using the proof system (C,D) of the previous section. More precisely, $\mathcal{B}(x, y, w, b)$ is proven to hold by running C on input $(x, (-1)^{1-b} w \bmod x, -y \bmod x)$.

An informal description. Let us now informally describe our proof system. Let $(x, \vec{y}) \in \mathrm{T}(k, m)$ and let $|x| = n$ and $\vec{y} = (y_1, \ldots, y_m)$. First P proves that $x \in \mathrm{BL}$ by running the algorithm A on input x and using a first part of the reference string η. Then, from the reference string η P picks $m\lceil\log(m+1)\rceil$ integers $\rho_{ij} \in Z_x^{+1}$ and a bit b and (x, y_j)-opens each ρ_{ij} as a bit s_{ij} in such a way that the following condition is satisfied: denoted by S_j the integer whose binary representation is $s_{1j} \cdots s_{\lceil\log(m+1)\rceil j}$, the m-uple $(S_1, \ldots, S_m)$ represents a (k, m)-sequence of admissible shares for b. Now, why is this a proof that less than k elements of $\vec{y}$ are quadratic non residues?
Let I be the set of i such that $y_i \in NQR_x$. Then the value of S_i is fixed for all $i \in I$. Thus, if $|I| < k$ then it is always possible to choose S_i for $i \notin I$ such that $(S_1, \ldots, S_m)$ is a sequence of admissible shares for b.

Suppose now that $|I| \geq k$. Then the values S_i for which $i \in I$ completely determine S. Moreover, the S_i's are uniformly distributed and thus the probability that $S = b$ is at most $1/|\mathcal{E}| \leq 1/m$. Thus, the probability that the prover convinces the verifier can be made negligible by repeating the protocol on different parts of the reference string.

A formal description of the proof system (P,V) can be found in Figure 1. Here (C,D) is a non-interactive perfect ZK proof system for OR. Our field $\mathcal{E}$ is the field with $2^{\lceil \log(m+1) \rceil}$ elements.

Theorem 6. (P,V) is a Non-Interactive Perfect Zero-Knowledge Proof System for the language $\mathrm{T}(k,m)$.

Proof. Omitted.

Remark. The protocol (P,V) can be easily extended to a proof system for the language of pairs $(x, \vec{y})$, where the number of quadratic non residues in $\vec{y}$ is greater or equal to k, that is:

$$\overline{\mathrm{T}}(k,m) = \{ (x, \vec{y}) \mid x \in \mathrm{BL},\ y_i \in Z_x^{+1},\ \text{for } i = 1, \ldots, m, \text{ and } |\{y_i | y_i \in NQR_x\}| \geq k \}.$$

The prover uses the algorithm P on input $(x, -y_1 \bmod x, \ldots, -y_m \bmod x)$.

5 Non-Interactive Perfect Zero-Knowledge for General Access Structures Statements

In general a secret sharing scheme is a procedure to share a secret among a certain number of participants so that only qualified subsets of participants can reconstruct the secret. Threshold schemes are particular secret sharing schemes where the set of qualified subsets consists of all the subsets with at least k participants. The set of qualified subsets is called access structure. For obvious reasons an access structure $\mathcal{A}$ has to be monotone; i.e., if $A \in \mathcal{A}$ then all A' that contain A also belong to $\mathcal{A}$.

As done for threshold scheme, to each access structure $\mathcal{A}$, we associate a language $\mathrm{AS}(\mathcal{A})$ of pairs $(x, \vec{y})$, where x is a Blum integer and $\vec{y}$ is an m-uple of elements of Z_x^{+1}, in the following way. For each access structure $\mathcal{A} = \{A_1, \cdots, A_k\}$, where $A_i = \{a_{i1}, \cdots, a_{ik_i}\} \subseteq \{0, 1, \cdots, m\}$, we can define a predicate $p_{\mathcal{A}}(x, \vec{y})$ as follows:

$$p_{\mathcal{A}}(x, \vec{y}) = \begin{cases} 1 & \text{if for each } i = 1, \ldots, k, \text{ at least one out of } \{y_{i1}, \cdots, y_{ik_i}\} \\ & \text{is a quadratic residue modulo } x \text{ and} \\ 0 & \text{otherwise.} \end{cases}$$

Then the language $\mathrm{AS}(\mathcal{A})$ is the language of pairs $(x, \vec{y})$ for which $p_{\mathcal{A}}(x, \vec{y}) = 1$.

We consider polynomial-time ideal secret sharing schemes, that is schemes in which the secret and its shares are taken from the same set (for instance $\mathrm{GF}(q)$, where q is a prime), the algorithm of the dealer is polynomial-time, and it is possible to verify in polynomial time that a given set of shares reconstructs the

Input to P and V:

- A reference string η.
- $(x, \vec{y}) \in \mathrm{T}(k, m)$, where $|x| = n$ and $\vec{y} = (y_1, \ldots, y_m)$.

(Set $\eta = \tau \circ \sigma$, where $\sigma = b \circ \rho \circ \gamma_{11} \circ \cdots \circ \gamma_{1m} \circ \cdots \circ \gamma_{\lceil\log(m+1)\rceil 1} \circ \cdots \circ \gamma_{\lceil\log(m+1)\rceil m}$, $|b| = 1$ and ρ is the concatenation of $\rho_{ij} \in Z_x^{+1}$, $1 \leq i \leq \lceil\log(m+1)\rceil$ and $1 \leq j \leq m$.)

Instructions for P.

P.1 (*Prove that x is a Blum integer.*)
Run the algorithm A on input x using the random string τ and send its output *Pf*.

P.2 (*Construct the sequence of admissible shares.*)
For j such that $y_j \in NQR_x$,
for $i = 1, \ldots, \lceil\log(m+1)\rceil$,
if $\rho_{ij} \in QR_x$ then set $s_{ij} \leftarrow 0$, else set $s_{ij} \leftarrow 1$;
let S_j be the integer whose binary representation is $s_{1j} \ldots s_{\lceil\log(m+1)\rceil j}$.
Randomly choose S_l, with l such that $y_l \in QR_x$, in such a way that $(S_1, \ldots, S_m)$ constitutes a (k, m)-sequence of admissible shares for the bit b.
For l such that $y_l \in QR_x$,
let $s_{1l} \ldots s_{\lceil\log(m+1)\rceil l}$ be the binary representation of S_l.

P.3 (*Prove that the sequence of admissible shares has been correctly constructed.*)
For $i = 1, \ldots, \lceil\log(m+1)\rceil$,
for $j = 1, \ldots, m$,
(x, y_j)-open ρ_{ij} as s_{ij} by running the algorithm C on input $(x, (-1)^{1-s_{ij}} \rho_{ij} \bmod x, -y_j \bmod x)$ using γ_{ij} as random string and obtaining as output Π_{ij}. Send s_{ij} and Π_{ij}.

Input to V:

- A proof *Pf* that $x \in BL$.
- A sequence of shares $(S_1, \ldots, S_m)$.
- A sequence of proofs Π_{ij}, for $1 \leq i \leq \lceil\log(m+1)\rceil$, $1 \leq j \leq m$.

Instructions for V.

V.1 (*Verify that x is a Blum integer.*)
Run the algorithm B on input x and τ thus verifying *Pf*.

V.2 (*Verify the admissibility of the sequence of shares.*)
Verify that the m-uple $(S_1, \ldots, S_m)$ is a (k, m)-sequence of admissible shares for the bit b.

V.3 (*Verify that the sequence of admissible shares has been correctly constructed.*)
For $i = 1, \ldots, \lceil\log(m+1)\rceil$,
for $j = 1, \ldots, m$,
verify that the proof Π_{ij} is correct by running the program of D on input $(x, (-1)^{1-s_{ij}} \rho_{ij} \bmod x, -y_j \bmod x)$ using γ_{ij} as random string.
If all verifications are successful then ACCEPT else REJECT.

Fig. 1. The proof system (P,V) for $\mathrm{T}(k, m)$.

secret. Supposing the existence of such a secret sharing scheme, a non-interactive perfect ZK proof system for AS($\mathcal{A}$) can be obtained in the following way. Let $(x, \vec{y}) \in$ AS($\mathcal{A}$) be the input to the protocol. Similarly to the protocol (P,V), the prover picks from the reference string $m\lceil\log(m+1)\rceil$ integers $\rho_{ij} \in Z_x^{+1}$ and a bit b and (x, y_j)-opens each ρ_{ij} as a bit s_{ij} in such a way that the following condition is satisfied: denoted by S_j the integer whose binary representation is $s_{1j} \cdots s_{\lceil\log(m+1)\rceil j}$, the m-uple $(S_1, \ldots, S_m)$ represents a secret sharing scheme for b for the access structure $\mathcal{A}$ on $\mathcal{P}$.

Theorem 7. Let $\mathcal{P}$ be a set of m participants and $\mathcal{A}$ a monotone access structure on it. Suppose there exist a polynomial-time secret sharing scheme for the access structure $\mathcal{A}$. Then the protocol described above is a non-interactive perfect zero-knowledge proof system for AS($\mathcal{A}$).

Proof's sketch: Using the existence of a polynomial-time ideal secret sharing scheme for the access structure $\mathcal{A}$, and ideas similar to those of protocol (P,V), one can see that the theorem holds. □

Notice that in the above theorem no restriction is imposed upon the size of $\mathcal{A} = \{A_1, \cdots, A_k\}$; e.g., k can be exponential in m (the length of the vector $\vec{y}$). This is actually what happens in threshold schemes.

References

1. M. Bellare and M. Yung, *Certifying Cryptographic Tools: The case of Trapdoor Permutations*, in CRYPTO 92.
2. M. Ben-Or, O. Goldreich, S. Goldwasser, J. Hastad, S. Micali, and P. Rogaway, *Everything Provable is Provable in Zero Knowledge*, in "Advances in Cryptology – CRYPTO 88", vol. 403 of "Lecture Notes in Computer Science", Springer Verlag, pp. 37–56.
3. G. R. Blackley, *Safeguarding Chryptographic Keys*, Proceedings AFIPS 1979 National Computer Conference, pp. 313–317, June 1979.
4. J. Boyar, K. Friedl, and C. Lund, *Practical Zero-Knowledge Proofs: Giving Hints and Using Deficiencies*, Journal of Cryptology, n. 4, pp. 185–206, 1991.
5. M. Blum, A. De Santis, S. Micali, and G. Persiano, *Non-Interactive Zero-Knowledge*, SIAM Journal of Computing, vol. 20, no. 6, Dec 1991, pp. 1084–1118.
6. M. Blum, P. Feldman, and S. Micali, *Non-Interactive Zero-Knowledge and Applications*, Proceedings of the 20th Annual ACM Symposium on Theory of Computing, 1988, pp. 103–112.
7. R. Boppana, J. Hastad, and S. Zachos, *Does co-NP has Short Interactive Proofs ?*, Inf. Proc. Lett., vol. 25, May 1987, pp. 127–132.
8. A. De Santis, S. Micali, and G. Persiano, *Non-Interactive Zero-Knowledge Proof-Systems*, in "Advances in Cryptology – CRYPTO 87", vol. 293 of "Lecture Notes in Computer Science", Springer Verlag, pp. 52–72.
9. A. De Santis, G. Persiano, and M. Yung, *Perfect Zero-Knowledge Proofs for Graph Isomorphism Languages*, manuscript.

10. U. Feige, D. Lapidot, and A. Shamir, *Multiple Non-Interactive Zero-Knowledge Proofs Based on a Single Random String,* in Proceedings of 22nd Annual Symposium on the Theory of Computing, 1990, pp. 308–317.
11. L. Fortnow, *The Complexity of Perfect Zero-Knowledge*, Proceedings of the 19th Annual ACM Symposium on Theory of Computing, 1987, pp. 204–209.
12. O. Goldreich and E. Kushilevitz, *A Perfect Zero Knowledge Proof for a Decision Problem Equivalent to Discrete Logarithm,* in "Advances in Cryptology - CRYPTO 88", Ed. S. Goldwasser, vol. 403 of "Lecture Notes in Computer Science", Springer-Verlag, pp. 57–70.
13. O. Goldreich, S. Micali, and A. Wigderson, *Proofs that Yield Nothing but their Validity and a Methodology of Cryptographic Design,* Proceedings of 27th Annual Symposium on Foundations of Computer Science, 1986, pp. 174–187.
14. O. Goldreich, S. Micali, and A. Wigderson, *How to Play Any Mental Game*, Proceedings of the 19th Annual ACM Symposium on Theory of Computing, pp. 218–229.
15. S. Goldwasser, S. Micali, and C. Rackoff, *The Knowledge Complexity of Interactive Proof-Systems*, SIAM Journal on Computing, vol. 18, n. 1, February 1989.
16. R. Impagliazzo and M. Yung, *Direct Minimum Knowledge Computations* "Advances in Cryptology – CRYPTO 87", vol. 293 of "Lecture Notes in Computer Science", Springer Verlag pp. 40–51.
17. I. Niven and H. S. Zuckerman, *An Introduction to the Theory of Numbers,* John Wiley and Sons, 1960, New York.
18. A. Shamir, *How to share a secret*, Communication of the ACM, vol. 22, n. 11, November 1979, pp. 612–613.
19. G. J. Simmons, *An Introduction to Shared Secret and/or Shared Control Schemes and Their Application*, Contemporary Cryptology, IEEE Press, pp. 441–497, 1991.
20. D. R. Stinson, *An Explication of Secret Sharing Schemes*, Design, Codes and Cryptography, Vol. 2, pp. 357–390, 1992.
21. M. Tompa and H. Woll, *Random Self-Reducibility and Zero-Knowledge Interactive Proofs of Possession of Information*, Proc. 28th Symposium on Foundations of Computer Science, 1987, pp. 472–482.

One Message Proof Systems with Known Space Verifiers

Yonatan Aumann and Uriel Feige

Dept. Computer Science, The Weizmann Institute of Science, Rehovot 76100, Israel.

Abstract. We construct a proof system for any NP statement, in which the proof is a single message sent from the prover to the verifier. No other interaction is required, neither before nor after this single message is sent. In the "envelope" model, the prover sends a sequence of envelopes to the verifier, where each envelope contains one bit of the prover's proof. It suffices for the verifier to open a constant number of envelopes in order to verify the correctness of the proof (in a probabilistic sense). Even if the verifier opens polynomially many envelopes, the proof remains perfectly zero knowledge.
We transform this proof system to the "known-space verifier" model of De-Santis *et al.* [7]. In this model it suffices for the verifier to have space S_{min} in order to verify proof, and the proof should remain statistically zero knowledge with respect to verifiers that use space at most S_{max}. We resolve an open question of [7], showing that arbitrary ratios S_{max}/S_{min} are achievable. However, we question the extent to which these proof systems (that of [7] and ours) are really zero knowledge. We do show that our proof system is witness indistinguishable, and hence has applications in cryptographic scenarios such as identification schemes.

1 Introduction

We construct a proof system for the NP-language 3-SAT. The common input is a 3-CNF formula ψ. Prover P tries to convince the verifier V that ψ is satisfiable, without revealing additional information. The prover in our proof system need not be stronger than polynomial time, provided that he is given a satisfying assignment for ψ. We place a limitation on the space S of the verifier. Namely, $S_{min} < S < S_{max}$, where S_{min} is the amount of space that suffices in order to verify the proof, and S_{max} is a bound on the space used by V, so that if $S < S_{max}$ then V does not learn "too much" (in a sense to be defined shortly). We stress that S is polynomial in the input length. The proof itself is given as one message sent from P to V. There is no interaction between P and V other than this one message sent by P.

Our work was motivated by the work of De-Santis, Persiano and Yung [7]. They construct such a proof system for an NP-complete language. An open question that they pose concerns the ratio S_{max}/S_{min}, known as the *tolerance* of the proof system. In their proof system, S_{max}/S_{min} is bounded by 2. Hence, the proof system can be employed only if the space of the verifier can be characterized within a very narrow range (not only the space of the truthful verifier, but also the space available to potential cheating verifiers). De-Santis *et al.* ask whether the tolerance of such proof systems can be improved. We answer this question in the affirmative. We construct a proof system scheme, parameterized by k, where k may be polynomially related to the length of ψ. For any desired value of k, our proof system scheme gives a proof system for 3-SAT with tolerance $S_{max}/S_{min} = k$.

Our original intention was to show that our proof system is zero knowledge (see Definition 2). However, we could not prove this. We could not prove the zero knowledge property even for the protocol in [7]. We will return to this issue in Sect. 1.4. Falling short of proving the zero knowledge property, we show that our proof system is *witness indistinguishable.* This property, which is weaker than zero knowledge, suffices for some cryptographic applications (see Sect. 1.5). It remains a major open question if a one message proof systems can be zero knowledge with respect to a known space verifier (under some reasonable definition of the concept of zero knowledge).

1.1 Definitions and Statement of our Results

The output of an algorithm A on input x is denoted by $A(x)$, which can be a random variable if A is a randomized algorithm.

Let $R = (x, w)$ be a relation testable in polynomial time in which the sizes of x and w are polynomially related, and let L_R be the NP-language associated with it. (E.g., x may be a satisfiable 3-CNF formula, w its satisfying assignment, and L_R the language 3-SAT.) In a proof system (P, V) both prover P and verifier V are probabilistic polynomial time. They both see a common input x, for which P tries to convince V that $x \in L_R$. The truthful prover is given a witness w such that $(x, w) \in R$ as auxiliary input. The size S of the work space available to the verifier is known to satisfy $S_{min} < S < S_{max}$. The verifier V may also have auxiliary input y of polynomial length (e.g., leftover information from executions of previous protocols). We do not require that $|y| < S_{max}$.

Definition 1. Language L_R has a *one message proof system* (P, V) with error ϵ if the following holds:

1. *Completeness:* the truthful prover can convince the truthful verifier to accept true statements. If $(x, w) \in R$, then $V(x, P(x, w))$ accepts.
2. *Soundness:* even a computationally unbounded cheating prover has only small probability of convincing the truthful verifier to accept false statements. If $x \notin L$, then for any message m, $Pr[V(x, m) = accept] \leq \epsilon$.

The proof system is a *proof of knowledge* if there exists a polynomial time "knowledge extractor" M such that for any x and m, if $Pr[V(x, m) = accept] > \epsilon$, then $(x, M(x, m)) \in R$.

In the following definitions of *zero knowledge* and *witness indistinguishability*, one may assume that at the end of the protocol the verifier outputs the contents of its work space (it "dumps" its memory).

Definition 2. A one message proof system for L_R is *perfectly (statistically, computationally) zero knowledge* if for any (possibly cheating) verifier V with work space $S < S_{max}$, there exists an expected polynomial time simulator M, such that for any $(x, w) \in R$, and any auxiliary input y, the distributions $V(x, y, P(x, w))$ and $M(x, y)$ are perfectly (statistically, computationally) indistinguishable.

Definition 3. A one message proof system for L_R is *perfectly (statistically, computationally) witness indistinguishable* if for any (possibly cheating) verifier V with work space $S < S_{max}$, for any $x \in L_R$, for any witnesses w_1 and w_2 that satisfy $(x, w_1) \in R$ and $(x, w_2) \in R$, and any auxiliary input y, the distributions $V(x, y, P(x, w_1))$ and $V(x, y, P(x, w_2))$ are perfectly (statistically, computationally) indistinguishable.

Theorem 4. *For any polynomial ratio $S_{max}/S_{min} = k$, there is a statistically witness indistinguishable one message proof of knowledge for 3-SAT (and hence for any NP-statement) with tolerance k.*

We remark that theorem 4 requires no cryptographic assumptions.

1.2 Related Work

Interactive proof systems and the concept of zero knowledge were introduced by Goldwasser *et al.* [14]. In the [14] model the verifier is probabilistic polynomial time, with no restrictions on its space. In Goldreich *et al.* [13] and Brassard *et al.* [4] it was shown that under certain computational complexity assumptions, all NP-languages have zero knowledge interactive proofs. Interaction seems to be an essential ingredient for zero knowledge. Goldreich and Krawczyk [15] prove that at least two rounds of messages are required for zero knowledge proofs for languages not in BPP, if the zero knowledge property is proved by *blackbox simulation*. The need for interaction can be replaced by the assumption that the prover and verifier share a common random string. In this model, non-interactive zero knowledge proof systems can be constructed under certain cryptographic assumptions [3, 12].

The study of interactive proof systems with space bounded verifiers was initiated by Condon [5]. Dwork and Stockmeyer [8] studied zero knowledge aspects when the verifier is a finite automaton. Kilian [17] constructed a proof system for any language in PSPACE, which is zero knowledge with respect to a log-space verifier. The model studied in our paper, that of one message known space verifier, was introduced by De Santis *et al.* [7]. Its goal was to achieve zero knowledge in one message, no cryptographic assumptions, and with respect to reasonably strong verifiers. It turns out that the main issue concerned is obtaining a one message proof, since Kilian's protocol uses interaction extensively, but does not require cryptographic assumptions, and can easily be adapted to the scenario of known space verifier with polynomial space bounds.

1.3 Main Ideas in our Construction

It is convenient to consider the following envelope scenario. An *envelope* is an idealized version of *bit commitment*. The prover can commit to a bit by placing it inside an envelope, and sealing the envelope. Thereafter, the prover cannot change the value of the bit. The verifier cannot see the value of the committed bit until she explicitly opens the envelope.

In the envelope scenario, in order to convince the verifier that $x \in L_R$, the prover sends his proof hidden in a sequence of ℓ sealed envelopes. If the prover is truthful, then whichever envelopes the verifier chooses to open, the verifier accepts. If $x \notin L_R$, then it suffices for the verifier to open ℓ_{min} envelopes (where $\ell_{min} < \ell$) in order to have non-negligible probability of rejecting. The proof is zero knowledge with respect to verifiers that open no more than ℓ_{max} envelopes (where $\ell_{min} < \ell_{max} < \ell$). The tolerance of the proof system is ℓ_{max}/ℓ_{min}.

We first construct a zero knowledge proof system for 3-SAT in the envelope scenario. Our construction is based on the concept of *randomized tableau*, first introduced by Kilian [16]. We modify his original constructs so as to get control of the tolerance of the proof system, and so as to improve efficiency. The details of our protocol appear in Sect. 2.

Once we have a protocol in the envelope scenario, we transform it to a protocol in the known space scenario. For this we replace the ideal bit commitments based on envelopes by a computational form of bit commitment based on *inner products*, as introduced by Kilian [17] and modified by De Santis *et al.* [7]. The idea is as follows. Let S_{min} denote the minimum space of the verifier, and let $b \simeq S_{min}/\ell_{min}$. P commits to the value of a bit z_i by selecting two vectors $v_1(z_i), v_2(z_i) \in \{0,1\}^b$ at random, subject to $v_1(z_i) \bullet v_2(z_i) = z_i$, where $\bullet$ denotes the inner product operation (the inner product of two vectors is the *sum mod 2* of the *and* of the respective bits). In order to send the sequence of ℓ bits $\{z_1, z_2, ..., z_\ell\}$, the prover first sends the sequence $v_1(z_1)$, $v_1(z_2)$, ... , $v_1(z_\ell)$, and then the sequence $v_2(z_1)$, $v_2(z_2)$, ... , $v_2(z_\ell)$. The verifier can "open" ℓ_{min} committed bits by first saving their respective v_1 vectors (this requires space roughly $b\ell_{min} \simeq S_{min}$), and then performing the inner products with the respective v_2 vectors online. Intuitively, the desired property that the verifier cannot open more that ℓ_{max} of the committed bits follows from the fact that in space $S_{max} \simeq b\ell_{max}$ the verifier cannot save more than ℓ_{max} of the v_1 vectors, and hence lacks sufficient information to recover the values of $\ell_{max} + 1$ bits at the time that the v_2 vectors arrive.

1.4 The Problems with Zero Knowledge

There is no problem in showing that the protocol in the envelope model is zero knowledge. This is true also of the protocol constructed in the [7] paper. Hence intuitively, it seems that also the protocol in the known space model is zero knowledge. This intuition is supported by the following communication complexity game. Player A receives in private a random vector $v_1 \in \{0,1\}^b$ and player B receives in private a random vector $v_2 \in \{0,1\}^b$. Player A can send a message of s bits to B. How large should s be so that the probability that player B computes $z = v_1 \bullet v_2$ is significantly greater than 1/2? Clearly, if $s = b$, then A can send v_1 to B, and B can compute z. However, if s is significantly smaller than b, then B's probability of guessing the value of z is essentially 1/2, as proved in [6]. Returning to our scenario of known space verifier, this implies that the verifier needs to store almost b bits of information regarding the vector $v_1(z_i)$, if she is later to open the committed bit z_i. Hence if the verifier's space is limited to S_{max}, it seems that she cannot store sufficient information in order to recover more than ℓ_{max} of the committed bits.

The above intuitive argument was formalized in [7] in the following way. They considered a game in which each player receives k random vectors (each of b bits), player A sends a single message of s bits, and player B has to compute the value of the k respective inner products. [7] prove that unless $s \simeq kb$, there is only negligible probability that B computes correctly all k inner products. From this [7] conclude that the verifier cannot recover more then ℓ_{max} of the committed bits, which would seem to imply that the proof system is zero knowledge in the known space model. However, this line of argument does not address the following issues:

1. It still has to be established that in order for the verifier to get meaningful information, she must explicitly open committed bits. Perhaps after seeing P's proof, the verifier can output the lexicographically first satisfying assignment for ψ, without outputting any of the committed bits z_i. (Its hard to imagine how such a thing can be done, but it was not shown that it cannot be done.)
2. There is side information available to the verifier. The committed bits z_1, ... ,z_ℓ, encode a satisfying assignment for ψ. They are not truly random and independent bits and this causes dependencies among the vectors $v_1(z_1)$, ... ,$v_1(z_\ell)$, $v_2(z_1)$, ... , $v_2(z_\ell)$.

We were unable to complete the argument to a full rigid proof that the protocol is zero-knowledge. The following are the types of problems we have encountered.

1. Communication complexity arguments alone do not seem to suffice, as they only address the space of the verifier. The verifier has polynomial space, and this suffices for her in order to *find* a satisfying assignment for ψ. Thus obviously, communication complexity arguments cannot exclude the possibility that V eventually outputs a satisfying assignment for ψ.
2. If one attempts to construct a simulator for V, then one encounters the following problem. In all previous protocols that we know of (e.g [14]), it is clear from the protocol which of the committed bits are revealed. This fact is later used in constructing the simulation. But in our model, there is no indication of which bit commitments the verifier chooses to open (if any).

3. The most successful way of proving that a protocol is zero knowledge is by *blackbox simulation*, in which the simulator M treats the possibly cheating verifier as a blackbox, and studies its input/output relations at intermediate steps of the protocol. However, a one message proof system leaves no room for blackbox simulation. In fact, it can be shown that only languages in BPP have a one message proof system that is blackbox-simulation zero knowledge. We know of only one case in which something different from blackbox simulation was used in order to prove that a protocol is zero knowledge. This was done in an interactive scenario where both verifier and simulator where restricted to logarithmic space [17], but does not seem to apply in our context.

We regard it as a highly challenging open question to prove that a one message proof system is zero knowledge, with respect to some reasonable definition of zero knowledge.

1.5 Witness Indistinguishability

We prove that our protocol is witness indistinguishable (in the known space model). Our proof is based on communication complexity arguments, and takes into account the dependencies between the committed bits. See Sect. 4 for details.

The concept of witness indistinguishability first received comprehensive treatment in [11] (though it was used implicitly also in earlier works). It turns out that if certain easy to meet restrictions are placed on the distribution of inputs to the protocol, then any witness indistinguishable protocol is also *witness hiding* - at the end of the protocol the verifier cannot compute any witness to the input statement, unless she could do so before the protocol began. Witness hiding is a natural property to consider in the context of proofs of knowledge. If a proof of knowledge is witness hiding, then the verifier of the proof cannot use it to become a prover later, since becoming a prover requires knowledge of a witness to the input statement. Witness hiding proofs of knowledge can serve as identification schemes in the spirit of [9]. For more information on the theory of witness indistinguishability and witness hiding see [11], and for applications see [10, 12].

2 The Protocol

Consider a predicate $\psi \in$ 3–CNF. Prover P claims he has a satisfying assignment w for ψ, and wishes to prove this to the verifier V. We describe the protocol for doing so in several stages (the protocol we describe is a simplified version of the protocol, a somewhat more efficient version of the protocol is noted upon in Sect. 3 and will appear in the full version of our paper). First we describe a transformation of ψ into a different predicate Ψ, where each variable is represented by K variables, where $K = \Theta(k)$ (k being the tolerance). Then we convert this predicate into a *Permutation Branching Program (PBP)* format. Next we construct a *Random Computation Tree (RCT)* for the PBP, to compute the value of the PBP, while hiding intermediate results. Next we describe the protocol in the pure envelope scenario, and show how the prover can prove his claim, with no interaction, and zero–knowledge. The final stage, the implementation of envelopes in the non–interactive bounded–space scenario, was described in Sect. 1.3.

2.1 Splitting the Variables

Let

$$\psi = \bigwedge_{i=1}^{m} C_i = \bigwedge_{i=1}^{m} (L_{i,1} \vee L_{i,2} \vee L_{i,3}) \ ,$$

where

$$L_{i,l} \in \{x_j\}_{j=1}^{T} \cup \{\bar{x}_j\}_{j=1}^{T} \ .$$

We replace each variable x_j of ψ by the *exclusive "or"* of K new variables. For each x_j of ψ, set $x_j = \bigoplus_{q=1}^{K} y_j^q$, *symbolically*, and set $\bar{x}_j = \bar{y}_j^1 \oplus (\bigoplus_{q=2}^{K} y_j^q)$. Set

$$\Psi = (\bigwedge_{i=1}^{m_1} C_i')$$

where C_i' is the symbolic representation of C_i with each literal replaced by its symbolic representation as an exclusive "or" of K variables.

The new predicate Ψ has a satisfying assignment iff ψ has one. Furthermore, any satisfying assignment for Ψ induces one for ψ, and any satisfying assignment for ψ induces several ones for Ψ, in a natural way.

2.2 Permutation Branching Program Format

Next, we describe how to transform each clause of Ψ into a *Permutation Branching Program (PBP)* format. For the general case, Barrington [2] describes such a transformation yielding polynomial size programs with permutations in S_5. For our case, however, we can achieve more efficient representation, with permutations in S_3 and linear size programs.

A 3–PBP, B, of length l, over the set of Boolean variables Y, is an ordered list of triplets

$$B = ((v^1, \sigma_0^1, \sigma_1^1), \ldots, (v^l, \sigma_0^l, \sigma_1^l)),$$

where $v_i \in Y$, and σ_0^i, σ_1^i, are permutations in S_3. For a given assignment $f : Y \to \{0,1\}$, the program B *yields* the value

$$\mathrm{val}(B, f) = \prod_{i=1}^{l} \sigma_{f(v^i)}^i .$$

We say that B *accepts* f if $\mathrm{val}(B, f) = e$, where e is the identity permutation.

Given a clause C'_i we show how to transform it into a PBP $B(C'_i)$ such that $B(C'_i)$ accepts f iff f is a satisfying assignment. Furthermore if it does not accept, then it yields some fixed permutation π.

Set

$$\pi_1 = (1\ 2)(3),\ \pi_2 = (1)(2\ 3),\ \pi_3 = (1\ 3)(2)$$

For $C'_i = (L'_{i_1} \vee L'_{i_2} \vee L'_{i_3})$, and $p = 1, 2, 3$, define B_i^p of length K

$$B_i^p = ((y_{i_p}^1, \pi_p, e), (y_{i_p}^2, e, \pi_p), \ldots, (y_{i_p}^{K-1}, e, \pi_p), (y_{i_p}^K, e, \pi_p))$$

if L'_{i_p} corresponds to a positive occurrence of variable x_{i_p}, and

$$B_i^p = ((y_{i_p}^1, e, \pi_p), (y_{i_p}^2, e, \pi_p), \ldots, (y_{i_p}^{K-1}, e, \pi_p), (y_{i_p}^K, e, \pi_p))$$

if L'_{i_p} corresponds to a negative occurrence of variable x_{i_p}.

Define

$$B(C'_i) = B_i^1 \circ B_i^2 \circ B_i^1 \circ B_i^2 \circ B_i^3 \circ B_i^2 \circ B_i^1 \circ B_i^2 \circ B_i^1 \circ B_i^3$$

where $B \circ B'$ is the concatenation of B and B'.

We claim $B(C'_i)$ is the desired PBP.

Lemma 5. *Let f be an assignment to the $y_{i_p}^q$'s, then*

$$\text{val}(B(C'_i), f) = \begin{cases} e & f \text{ satisfies } C'_i \\ \pi & \text{otherwise} \end{cases}$$

where $\pi = (1\ 3\ 2)$.

Proof: Consider B_i^1. By construction,

$$\text{val}(B_i^1, f) = \begin{cases} \pi_1 & y_{i_1}^1 \oplus y_{i_1}^2 \oplus \cdots \oplus y_{i_1}^K = 0 \\ e & y_{i_1}^1 \oplus y_{i_1}^2 \oplus \cdots \oplus y_{i_1}^K = 1 \end{cases}$$

The same holds for B_i^2 and B_i^3, with π_2 and π_3. Calculation now shows that if any of the B_i^p's yields e then so does the full $B(C'_i)$, and otherwise $B(C'_i)$ yields π. ∎

Clearly, the representation of $B(C'_i)$ is linear in that of C'_i. In fact the representation is rather concise: for a clause C'_i of length $3K$, $B(C'_i)$ has $10K$ entries.

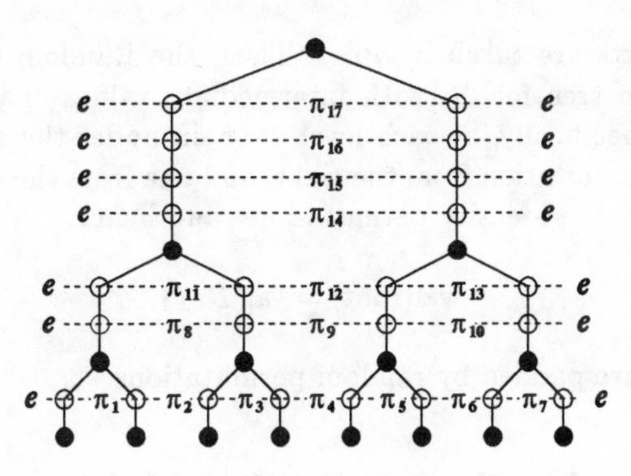

Fig. 1. An RCT for $t = 8$

2.3 Randomized Computation Trees

Consider a PBP, B. For a given assignment f, we have $\text{val}(B, f) = \prod_{j=1}^{t} \sigma_j$ (where the σ_j depends on the assignment f, and the PBP B). W.l.o.g. assume t is a power of 2. Define the following tree structure: take a full binary tree with t leaves, and replace each node of height l by a chain of length $2^l + 1$. The root node is not replaced. A picture of this structure for $t = 8$ is depicted in Fig. 1. Call the lowest node of each chain a *combining node*, and the other nodes *chain nodes*. To the right and the left of each chain node we place a permutation. Neighboring nodes share permutations (see Fig. 1). We call these permutations *hiding permutations*. For a node, w, denote by $LP(w)$ and $RP(w)$ its left and the right hiding permutations, respectively. The outer permutations, on each side, are fixed to be the identity permutation (e). Other permutation are chosen randomly.

We define the values at the nodes of the computation tree recursively. For a combining node w, denote its left and right children by w_L and w_R, respectively. For a chain node w, denote its child by w_C. The value of a node is defined:

$$\text{val}(w) = \begin{cases} \sigma_i & w \text{ is the } i\text{-}th \text{ leaf.} \\ \text{val}(w_L) \cdot \text{val}(w_R) & w \text{ is a combining node.} \\ \text{val}(v) = LP(w)^{-1} \cdot \text{val}(w_C) \cdot RP(w) & w \text{ is a chain node.} \end{cases} \quad (1)$$

For a given permutation branching program B, assignment f, and set of hiding permutations R, we denote the corresponding *Random Computation Tree (RCT)*, by $\mathcal{T} = \mathcal{T}(B, f, R)$.

For a node w, let R-path(w) be right–most path leading from the node down to a leaf, and let L-path(w) be the left–most path leading from a leaf to the node. Let Span(w) be the sequence of leaf nodes under w, in order from left to right. Then

$$\text{val}(w) = \prod_{u \in \text{L-path}(w)} LP(u)^{-1} \cdot \prod_{u \in \text{Span}(w)} \text{val}(u) \cdot \prod_{u \in \text{R-path}(w)} RP(u),$$

where all products are taken in order. Thus, the Random Computation Tree is a computation tree for B, with intermediate values "padded" by random permutations. Specifically, at each level of chain nodes the value is padded by one additional permutation from the right, and one from the left. The root node is padded only by the identity permutations, and hence,

$$\mathrm{val}(\mathrm{root}) = \mathrm{val}(B, f).$$

All other nodes are padded by random permutations.

2.4 The Protocol in The Pure Envelope Model

Let w be the witness available to P. Formally $w : \{x_j\} \to \{0,1\}$ is a truth assignment to the variables of ψ. For each variable x_j, P chooses at a random truth assignment, denoted by f, for $(y_j^1, \ldots, y_j^K)$, subject to $\bigoplus_{q=1}^{K} f(y_j^q) = w(x_j)$. Next, for each PBP $B(C_i')$, P chooses at random a set of hiding permutations, and prepares the corresponding RCT $\mathcal{T}_i$. In all appearances of a variable x_j, the *same* values for the y_j^q's are being used. Next, for all the y_j^q's, prover P puts the assignment values for these variables in separate envelopes. In addition, for each RCT, $\mathcal{T}_i$, corresponding to the PBP $B(C_i')$, P puts the value of each interior node in a separate envelope, as well as the set of all hiding random permutations, each permutation in a separate envelope.

All envelopes are now sent to the verifier. The verifier can open between four and $K-1$ envelopes. The verifier performs a *correctness test*:

Correctness test: The verifier chooses at random one RCT $\mathcal{T}$, and one non-leaf node $w \in \mathcal{T}$. For this node the verifier checks that:

1. The value $\mathrm{val}(w)$ was correctly constructed. V opens the related envelopes and checks that $\mathrm{val}(w)$ satisfies eq. (1).
2. If w is the root node then $\mathrm{val}(w) = e$ (implying that the corresponding clause is satisfied).

If any of these tests fail, the verifier rejects, otherwise, she accepts.

The above envelopes contain permutation in S_3. Each permutation can be represented by 3 bits. Confining ourselves to single-bit envelopes, we have the prover send each permutation in a sequence of 3 envelopes. We assume the verifier can open between $\ell_{min} = 12$ and $\ell_{max} = K - 1$ of these envelopes[1]. The tolerance of the system is $k = K/12$.

Lemma 6. *The above protocol is a perfectly zero knowledge proof of knowledge for 3-SAT in the envelope model.*

[1] Kilian (private communication) remarks that ℓ_{min} can be reduced to 3. However, for a given tolerance, this reduction entails a degradation in the error probability and complexity. Consequently, this does not provide a more efficient protocol in the known space scenario.

The proof of this lemma is omitted due to space limitations. It is based on the construction of a simulator M that sends "empty" envelopes to V. For any set of less than K envelopes that V chooses to open, M supplies "contents" to these envelopes, with a distribution that is perfectly indistinguishable from the distribution that would arise if a real prover was sending the envelopes.

The envelope protocol is transformed into a protocol for the bounded space scenario as described in Sect. 1.3.

3 Cheating Probability

Suppose $\psi \notin$ 3-SAT. The proof-message sent by P, induces an assignment $w : \{x_j\} \rightarrow \{0,1\}$ of the variables of ψ. Let C_i be a clause not satisfied by w (there must be at least one such clause). Suppose that for the correctness test, verifier V chooses to test $\mathcal{T}_i$ (the RCT associated with C_i). Then there is at least one node $w \in \mathcal{T}_i$ for which the test fails. The number of nodes in an RCT is $\leq 20K \log K = O(k \log k)$, where k is the tolerance. Thus, for a given clause C_i, not satisfied by w, Pr[correctness test fails] $\geq 1/O(k \log k)$. Let m be the total number of clauses in ψ, and let $\hat{m}$ be the maximum number of clauses which can be satisfied simultaneously. We obtain

$$p = \Pr[V \text{ accepts}] \leq 1 - \left(\frac{m - \hat{m}}{m} \cdot \frac{1}{O(k \log k)}\right). \tag{2}$$

For any $\psi \notin$ SAT this probability is $\leq (1 - 1/O(nk \log k))$. By [1], for $\psi \in$ MAX-SNP, the fraction $m/(m - \hat{m}) = O(1)$, and hence the probability in (2) gives $\leq (1 - 1/O(k \log k))$. The error probability can be further reduced by sequential repetition of the protocol. With t sequential repetitions the acceptance probability decreases to p^t. Clearly, the repeated protocol still remains a one message proof. Note that in the envelope model sequential repetition requires that the bounds on the number of envelopes V can open ($\ell_{min} \leq i \leq \ell_{max}$) must hold for each repetition separately. For the known space scenario, this automatically holds.

Finally we note on a method to further reduce the cheating probability (and consequently the complexity of the protocol for any fixed security parameter). Consider again the envelope model, and suppose that the RCT's are sent one by one, and that for *each* RCT, verifier V can open between 12 and $K-1$ envelopes. In this case V can test each RCT separately. Thus, for *each* non-satisfied clause, C_i, V will detect that C_i is not satisfied with probability $\geq 1/O(k \log k)$. Thus,

$$\Pr[V \text{ accepts}] \leq \left(1 - \frac{1}{O(k \log k)}\right)^{(m - \hat{m})}.$$

However, this later protocol, as described above, is not zero-knowledge, even in the pure envelope setting (a sequence x_j may be sent many times, until all its y_j^q's are revealed). In order to preserve the zero-knowledge property, further modification to the construction are introduced. We omit the details here. When proving witness indistinguishability, this modified protocol introduces extra complexities in the analysis. The proof we give in the next section is for the original protocol.

4 Proof of Witness Indistinguishability

We now sketch the proof that the proof system is statistically witness indistinguishable in the known space model.

As corollary of Lemma 6 and of the theory of witness indistinguishability [11], we have:

Corollary 7. *The proof system is perfectly witness indistinguishable in the envelope model.*

We shall use the above corollary in our analysis of the known space model. We use the following notation:

n - length of input statement.
w_1, w_2 - possible witnesses to the input statement.
ℓ - number of envelopes.
ℓ_{min} - number of envelopes that V needs to open.
ℓ_{max} - number of envelopes that V is allowed to open.
b - number of bits per envelope in the inner product representation.
S - space of verifier.
S_{min} - space guaranteed to be available to V. We assume $S_{min} > n$.
S_{max} - space guaranteed not to be available to V.
$Z = z_1, z_2, ..., z_\ell$ - message sent by P in envelope model.
R - number of possible values of Z consistent with a single witness.
$v_1(z_i)$, $v_2(z_i)$ - vectors of b bits, $v_1(z_i) \bullet v_2(z_i) = z_i$.
m - message sent by P in the known space model.
m_1 - first part of message (only the v_1 vectors).
m_2 - second part of message (only the v_2 vectors).

In the envelope model, all messages Z consistent with a specific witness are equi-probable. We require that for every z_i, $v_1(z_i) \neq 0^b$.

The protocol is repeated sequentially many times, with independent random bits for the prover, so as to decrease the error probability. It suffices to prove that a single iteration is witness indistinguishable, since in our model we allow the verifier to have auxiliary input, and hence witness indistinguishability is preserved under repetition.

Consider any two witnesses w_1 and w_2, and construct the following matrix M. The rows of M are labeled by messages m_1 and the columns by messages m_2. Hence M has $(2^b - 1)^\ell$ rows and $2^{b\ell}$ columns. For each entry M_{ij} interpret i and j as m_1 and m_2 respectively, and set $M_{ij} = 1$ if i, j is a message that corresponds to P using w_1, $M_{ij} = -1$ if i, j is a message that corresponds to P using w_2, and $M_{ij} = 0$ otherwise. Then each row of M has exactly $R2^{-\ell}2^{b\ell}$ entries that are 1, and $R2^{-\ell}2^{b\ell}$ entries that are -1. If P is using w_1 (w_2, respectively) as a witness, then in effect his message m corresponds to a random 1 entry (-1 entry, respectively) in M.

We model the scenario as a communication complexity game. $m = (m_1, m_2)$ is picked with uniform probability from the nonzero entries of M. Alice sees m_1 and Bob sees m_2. Alice sends to Bob a message of length S. Bob outputs either

1 or -1. We want to bound $Pr[B(m) = M(m)]$. The probability is taken over the random choice of $m = (m_1, m_2)$. (The optimal strategy for Alice and Bob is deterministic.)

Lemma 8. *If in the communication game* $Pr[B(m) = M(m)] = 1/2 + o(n^{\omega(1)})$, *then the protocol is statistically witness indistinguishable.*

Proof: V's behavior can be simulated in the communication complexity game. Alice simulates $V(m_1)$, then sends the contents of the workspace of V to Bob (at most S bits), and Bob completes the simulation on m_2. If V can distinguish between w_1 and w_2, then so can Alice and Bob. ∎

It remains to analyze the communication complexity game. Alice sends one of 2^S possible messages. This partitions the rows of M into 2^S horizontal blocks. Bob holds a column. We need to prove that with high probability, this column is balanced along the block (sum of entries is approximately 0).

A block is *heavy* if it contains at least $2^{b\ell - S - \alpha}$ rows. The probability that a random row is in a heavy block is at least $1 - 2^{-\alpha}$ (ignoring the negligible factor of $(\frac{2^b-1}{2^b})^\ell$ due to the fact that $v_1 \neq 0^b$). We show that for heavy blocks, most columns are balanced.

Consider an arbitrary row r, and a random column c. Then

$$Pr[M(r,c) = 1] = Pr[M(r,c) = -1] = R2^{-\ell}$$

Consider now a heavy block B. Take a random column c and consider the random variable $X_c = \sum_{r \in B} M(r,c)$. Since $Pr[M(r,c) = 1] = Pr[M(r,c) = -1]$, then the expectation satisfies $E[X_c] = 0$. To show that most columns are nearly balanced we will bound $var[X_c]$, and use Chebyshev's inequality, $Pr[|X - E[X]| > \gamma] \leq \frac{var[X]}{\gamma^2}$.

$var[X_c] = E[(X - E[X])^2] = E[X^2] = \sum_{r_1, r_2} E[M(r_1,c)M(r_2,c)]$. To bound this last sum we use the fact that the protocol is perfectly witness indistinguishable in the envelope model, and hence knowledge of the contents of ℓ_{max} envelopes gives no information on whether P is using w_1 or w_2.

Consider two rows $r_1, r_2 \in B$. Each string r_i is composed of ℓ vectors, each of b bits. It follows from the witness indistinguishability property in the envelope model that:

Lemma 9. *If r_1 and r_2 agree on no more than ℓ_{max} matching vectors, then* $E[M(r_1,c)M(r_2,c)] = 0$.

Hence

$$\sum_{r_1, r_2 \in B} E[M(r_1,c)M(r_2,c)] \leq 2R2^{-\ell}|B| \sum_{j=\ell_{max}+1}^{\ell} \left(\binom{\ell}{j} 2^{-bj} 2^{\ell b}\right)$$

$$< 4R2^{-\ell}|B| \binom{\ell}{\ell_{max}+1} 2^{-b(\ell_{max}+1)} 2^{\ell b}$$

Using $4R2^{-\ell}\binom{\ell}{\ell_{max}+1} < 1$ we obtain that $var[X_c] < |B|2^{-b\ell_{max}}2^{b\ell}2^{-b}$. By Chebyshev inequality, $Pr[X_c > \gamma|B|] \leq 2^{b\ell}/2^{b\ell_{max}}2^b\gamma^2|B|$.

Call a column *biased* if $X_c > \gamma|B|$. There are at most $2^{2b\ell}/2^{b\ell_{max}}2^{-b}\gamma^2|B|$ biased columns. The total number of nonzero entries in biased columns is at most $2^{2b\ell}/2^{b\ell_{max}}2^b\gamma^2$. The total number of nonzero entries in the block B is at least $|B|2^{b\ell}R/2^{\ell}$. Thus the probability that m is chosen in a biased column is at most $2^{b\ell}2^{\ell}/2^{b\ell_{max}}\gamma^2|B|R2^b < 2^S 2^{\alpha}2^{\ell}/\gamma^2 R2^b 2^{b\ell_{max}}$. By our choice of ℓ_{max} it will follow that there are at least $|B|2^{b\ell}R/2^{\ell+1}$ nonzero entries in non-biased columns. Hence the probability that B guesses $M(m)$ correctly when m is chosen in a non-biased column is bounded by $1/2 + \frac{\gamma|B|2^{\ell+1}}{|B|R}$.

If follows from the above that

$$Pr[B(m) = M(m)] < 1/2 + 2^{-\alpha} + \frac{2^S 2^{\alpha} 2^{\ell}}{\gamma^2 R 2^b 2^{b\ell_{max}}} + \frac{\gamma 2^{\ell+1}}{R}$$

Choosing $\alpha = n+2$, $\gamma = 2^{-(n+\ell+3)}$, $\ell_{max} = \frac{S_{max}+4n+3\ell+10}{b}$, we obtain that $Pr[B(m) = M(m)] < 1/2 + 2^{-n}$. Observe that this last choice of ℓ_{max} can be made, despite the fact that ℓ is super linear in ℓ_{max}. Recall that ℓ_{min} is some universal constant, as derived from our protocol, and that $\ell = O(n\ell_{max}\log \ell_{max})$. Recall also that $S_{min} > n$, and that $b \simeq S_{min}/\ell_{min}$. For $S_{max} = n^c$ (where c is some constant), a good choice for ℓ_{max} can always be found, provided that $S_{min} >> cn\log n$.

Acknowledgments: We thank Alfredo De-Santis for sending us the full version of the [7] paper, and Joe Kilian for sending us the full version of the [17] paper.

References

1. S. Arora, C. Lund, R. Motwani, M. Sudan, M. Szegedy, "Proof Verification and Hardness of Approximation Problems", 33rd *FOCS, 1992, 14-23.*
2. D. Barrington, "Bounded width polynomial size branching programs recognize exactly those languages in NC^1", 18th *STOC, 1986, 1-5.*
3. M. Blum, A. De Santis, S. Micali, G. Persiano, *Noninteractive Zero-Knowledge,* SIAM Jour. on Computing, Vol. 20, No. 6, December 1991, pp. 1084-1118.
4. G. Brassard, D. Chaum, C. Crepeau, "Minimum disclosure proofs of knowledge", *JCSS, 37, 1988, 156-189.*
5. A. Condon, "Computational models of games", *MIT Press.*
6. B. Chor, O. Goldreich, "Unbiased bits from sources of weak randomness and probabilistic communication complexity", *SIAM J. of Computing, 1988.*
7. A. De-Santis, G. Persiano, M. Yung, "One-message statistical zero-knowledge proofs with space-bounded verifier", 19th *ICALP, 1992.*
8. C. Dwork, L. Stockmeyer, "Finite state verifiers II: zero knowledge" *JACM, 39, 4, 1992, 829-858.*
9. U. Feige, A. Fiat, A. Shamir, "Zero Knowledge Proofs of Identity", *Journal of Cryptology, 1988, Vol. 1, pp. 77-94.*

10. U. Feige, A. Shamir, "Zero Knowledge Proofs of Knowledge in Two Rounds", *Advances in Cryptology – CRYPTO '89 Proceedings, Lecture Notes in Computer Science, Springer-Verlag 435, pp. 526-544.*
11. U. Feige, A. Shamir, *Witness Indistinguishable and Witness Hiding Protocols*, Proc. of 22nd ACM Symposium on Theory of Computing, 1990, pp. 416-426.
12. U. Feige, D. Lapidot, A. Shamir, "Multiple Non-Interactive Zero Knowledge Proofs Based on a Single Random String" *Proc. of 31st FOCS, 1990, pp. 308-317.*
13. O. Goldreich, S. Micali, A. Wigderson, *"Proofs that Yield Nothing But Their Validity or All Languages in NP Have Zero-Knowledge Proof Systems,* Journal of ACM, Vol. 38, No. 1, July 1991, pp. 691-729.
14. S. Goldwasser, S. Micali, C. Rackoff, *The Knowledge Complexity of Interactive Proof Systems,* SIAM J. Comput. Vol. 18, No. 1, pp. 186-208, February 1989.
15. O. Goldreich, H. Krawczyk, "On the composition of zero knowledge proof systems", 17th *ICALP, 1990, 268-282.*
16. J. Kilian, "Uses of randomness in algorithms and protocols", *MIT Press, 1990.*
17. J. Kilian, "Zero knowledge with log space verifiers" 29th *FOCS, 1988, 25-35.*

Interactive Hashing can Simplify Zero-Knowledge Protocol Design Without Computational Assumptions

Extended Abstract

Ivan B. Damgård

Aarhus University

Abstract. We show that any 3-round protocol (in general, any bounded round protocol) in which the verifier sends only random bits, and which is zero-knowledge against an *honest* verifier can be transformed into a protocol that is zero-knowledge *in general*. The transformation is based on the interactive hashing technique of Naor, Ostrovsky, Venkatesan and Yung. No assumption is made on the computing power of prover or verifier, and the transformation therefore is valid in both the proof and argument model, and does not rely on any computational assumptions such as the existence of one-way permutations. The technique is also applicable to proofs of knowledge. The transformation preserves perfect and statistical zero-knowledge. As corollaries, we show first a generalization of a result by Damgård on construction of bit-commitments from zero-knowledge proofs. Other corollaries give results on non-interactive zero-knowledge, one-sided proof systems, and black-box simulation.

1 Introduction

In this paper, we consider protocols in which a prover tries to convince a verifier that some claim is true. Some protocols can be shown to not reveal anything to the verifier, other than the fact that the claim is indeed true, by demonstrating that the verifier could have simulated the protocol himself. Such a protocol is said to be zero-knowledge [8]. Such protocols can be considered in the proof-model, where the prover is unbounded while the verifier is polynomial time restricted; or in the argument model, where the prover is poly-time bounded, while the verifier may (in most cases) be unbounded [1].

It is well known that the design of zero-knowledge proof or arguments is a difficult task. A main complicating factor is the demand that the protocol must be secure against dishonest behavior by both the prover and the verifier. For example, by allowing the prover too much control over the conversation in an effort to protect her privacy, we risk also allowing a dishonest prover to cheat. It would be much easier if we could assume that the verifier was honest.

[1] This does not mean that an honest verfier needs infinite computing power to execute the protocol, only that the protocol is secure, even against a cheating verifier with unbounded resources

In this paper, we consider this problem for protocols in which the verifier sends only random bits, sometimes called public-coin protocols. This is quite a large class of protocols, for example the general zero-knowledge proof systems and arguments for any NP-problem in [7], [2] are of this type. where the private are

For any bounded round public-coin protocol secure (zero-knowledge) against an *honest* verifier, we present a generic method transforming it into a protocol that is zero-knowledge against *any* verifier. This is based on the interactive hashing technique of [10]. The transformation preserves the proving capabilities of the original protocol, i.e. if it was a proof system, resp. an argument, resp. a proof of knowledge, the resulting protocol will be a proof/argument/proof of knowledge for the same problem. Also, if the original protocol was perfect or statistical zero-knowledge, so is the transformed protocol. In this paper, we show explicitly only the case of 3-round protocols. The generalization to any bounded number of rounds is in principle simple, but the proof is technically cumbersome. It is not clear how to generalize to any polynomial number of rounds. We will return to this problem later.

As a corollary of the transformation, we obtain a generalization of the result from [5], on constructing bit commitments from a 3-move public coin zero-knowledge proof of knowledge. The result from [5] needed a restriction on the number of random bits sent by the verifier, or on the error probability of the protocol; we show that this restriction is unnecessary.

Another corollary is that any non-interactive zero-knowledge proof can be transformed into an ordinary interactive zero-knowledge proof for the same problem. Also this transformation preserves perfect and statistical zero-knowledge.

Finally, we show that for bounded round, public coin, statistical zero knowledge proofs, requiring one-sidedness (i.e. completeness with probability 1) or black-box simulation is not a restriction on the set of statements that can be proved.

2 Related Work

The idea of transforming honest-verifier zero-knowledge into zero-knowledge in general was first studied by Bellare, Micali and Ostrovsky [4]. Their transformation needed a computational assumption of a specific algebraic type.

Since then several constructions have reduced the computational assumptions needed. The latest in this line of work is by Ostrovsky, Venkatesan and Yung [11], who give a transformation which is based on interactive hashing and preserves statistical zero-knowlegde. This transformation works for any protocol, but only in the zero-knowledge proof model, in which the verifier is assumed to be polynomial time bounded: the transformation relies on the existence of a one-way permutation which the verifier cannot invert.

Thus compared to [11], the contribution of this paper is a new technique for using interactive hashing, showing that if we restrict to bounded round public-coin protocols, we can get a transformation that does not need any computational assumptions and therefore works even if the verifier (and/or the prover) is unbounded. Moreover, our transformation preserves both perfect and statistical zero-knowledge.

Finally some recent independent work should be mentioned, in which Ostrovsky and Wigderson [12] show that the existence of honest verifier zero-knowledge proofs for non-trivial (i.e. non-BPP) problems impliy the existence of certain kinds of one-way functions. It appears that this could be immediately combined with the result in [3] that everything provable is provable in computational zero-knowledge assuming that one-way functions exist. This would give a transformation without assumptions (but one that would not preserve statistical or perfect zero-knowledge). However, the notion of one-way functions used in [12] is technically different from the standard one needed in [3], and the results therefore cannot be combined without solving some technical problems.

3 Notation and Definitions

In this section, the technical definitions and notation for zero-knowledge and probabilistic algorithms are given.

As the results below are only presented informally in this extended abstract, we omit technicalities here and present a minimum of notation.

We will restrict ourselves to 3-move public coin protocols for simplicity.

Thus we can describe the protocol we start with as follows: common input to the prover and verifier is a word x of length n bits in a language L. The prover P sends a message m_1, receives a random bit string c from the verifier V. Finally P sends a message m_2 to V, who then outputs accept or reject. We let $t = t(n)$ be the length of c.

Any function of n converging to 0 faster than any polynomial fraction will be called negligible.

We assume the following about the protocol: if $x \in L$, and P, V are honest, the probability that V accepts is at least $1 - \epsilon(n)$, where ϵ is negligible. If $x \notin L$, then the probability that V accepts, when talking to an arbitrary prover P^* is negligible. For arguments, we need to replace "arbitrary prover" by "any polynomial time prover". We do not make this explicit here, however, because the results below do not depend on any such restriction.

Finally, we assume that the protocol is zero-knowledge against the honest verifier: there is an expected polynomial time probabilistic machine S, which on input x produces a simulated conversation which is indistinguishable from the real conversation between P and the honest V on input x. We do not make explicit here, which flavor of zero-knowledge we talk about (perfect, statistical or computational) because the construction to follow will work for all flavors. In accordance with the usual models, we do assume, however, that cheating verifiers are polynomial time bounded in the case of computational zero-knowledge, but may be unbounded in case of statistical or perfect zero-knowledge.

4 The Transformation

For the transformation of the protocol, we need the technique of interactive hashing [10], which we repeat here with some changes to match our context: we work with the vector space $GF(2)^t$, and we assume that P is capable of computing some function g on values in this space.

1. P selects a random t-bit vector c, which is kept secret from V.
2. V selects at random $t-1$ vectors in $GF(2)^t$, $h_1, ..., h_{t-1}$, such that the h_i's are linearly independent over $GF(2)$.
3. For $j = 1$ to $t-1$:
   - V sends h_j to P.
   - P sends $b_j := h_j \cdot c$ (the inner product) to V.
4. Both parties compute the two vectors c_1, c_2, with the property that $c_i \cdot h_j = b_j$ for $j = 1..t-1$, where of course c is one of c_1, c_2. We say that the hashing isolates the values c_1, c_2 and that P answers consistently with c.
5. V sends $v = 1$ or 2 to P.
6. P returns $g(c_v)$ to V.

In [10], the following is proved about this procedure (some of the wording has been changed):

Lemma 1 At the end of Step 4, an arbitrary cheating V^* playing the role of V has no Shannon information about which of c_1, c_2 P has answered consistently with.

Lemma 2 Let P^* be any machine which plays the role of P and is capable of returning both $g(c_1)$ and $g(c_2)$ with probability ϵ. Then there is a probabilistic polynomial time algorithm M using P^* as an oracle, which can, on input a random c, compute $g(c)$ with probability $T(t, \epsilon)$, where T is a function polynomial in $1/t$ and ϵ. This probability is over the choice of c and internal coin tosses of M.

While the proof of Lemma 1 is easy, the proof of Lemma 2 is very technical and complicated. We refer to [10] for details.

The reader may notice that the counterpart of P^* in [10] (called S' there) was assumed to be polynomial time bounded. But this was only necessary there because g was the inverse of a one-way permutation, and the purpose was to show in a proof by contradiction that S' could compute this inverse. However, the analysis in [10] essentially shows that *for any* strategy used by P^* to choose the b_j's, at least one of c_1, c_2 will be almost uniformly chosen. Their construction of M will therefore work in our case for any P^*.

We are now ready to show our transformed protocol for proving that $x \in L$. The prover and verifier from the original protocol will be called P_0, V_0.

Intuitively, the reason why the original protocol may not be secure against an arbitrary verifier V^*, is that even though we can, using S, generate valid looking transcripts m_1, c, m_2 of the protocol where c is uniform, this distribution of c may not be the one that V^* would generate. In particular, there could be some dependency between c and m_1 that a random c-value has only negligible probability of satisfying. Hence trying to use the honest-verifier simulator directly might take expected exponential time. We therefore apply in the transformation the interactive hashing to cut down the number of possible c's to a small number. This increases our chance of hitting the right c-value in the simulation enough to make the expected time polynomial. We cut down to 2 possibilities for simplicity because this allows us to use the results of [10] without modifications. In practice, one could gain efficiency by stopping the hashing earlier, leaving more possibilities for c. But note that if the transformation is to remain provably secure, the number of possible values must be kept polynomial.
The transformed protocol consists of repeating the following $n = |x|$ times:

1. P starts running P_0 on input x. She sends the m_1 generated by P_0 to V.
2. P and V go through the interactive hashing process described above. The value $g(c)$ is defined to be the m_2 P_0 would return given that the initial message was m_1 and the challenge from V_0 was c. Therefore, to compute $g(c_v)$, P passes c_v on to P_0, gets m_2 back and sends it to V.
3. V uses V_0 to decide if the conversation m_1, c_v, m_2 would lead to accept by V_0. If so, he outputs accept, otherwise he rejects.

Remark: to improve readability, we have simplified things a little by defining g as a simple function of c: in general there may be more than one valid answer from P given that the first part of the conversation was m_1, c, so that $g(c)$ would in fact be a random variable. This makes no difference in the following, however.

Lemma 3 If (P_0, V_0) is perfect resp. statistical resp. computational honest verifier zero-knowledge, then (P, V) is perfect resp. statistical resp. computational zero-knowledge.

Proof Sketch We show a simulator S^*, working with any verifier V^*. One iteration of steps 1-3 above will be called a round. It is sufficient to show how to simulate 1 round:

1. Run the honest-verifier simulator S on input x. This results in a conversation m_1, c, m_2. Send m_1 to V^*.
2. Go through the interactive hashing with V^* while answering consistently with c.
3. Receive v from V^*. If $c = c_v$, send m_2 to V^*, stop. Else rewind V^* to the start of Step 1. Go to 1.

It is clear that, by Lemma 1, the probability that $c = c_v$ would be exactly 1/2 if the c we use in Step 2 had been uniformly chosen. The c we actually use is produced by S. In the perfect/statistical zero-knowledge case c cannot be distinguished from a uniform choice by any unbounded verifier. In the computational case c is only computationally indistinguishable from a uniform choice, but the difference cannot be told by a polynomially bounded verifier. Hence in any case the probability that $c = c_v$ is away from 1/2 by at most a negligible amount, and therefore simulation of one round needs an expected number of rewinds that is $O(1)$. Thus the complete simulation takes expected linear time.

To show correctness of the output distribution, first note that it is sufficient for all flavors of indistinguishability to show that the simulation of 1 round is indistinguishable from a real execution, since we have used a back-box simulation, which is closed under serial composition.

Then observe the following: the messages sent by the verifier V^* in real conversations are computed from m_1 chosen by P_0 and answers from P consistent with a random c-value. Our simulator uses m_1, c as produced by S, but otherwise uses the same algorithm as P to compute the answers in the interactive hashing. The final message m_2 is a random sample of P_0's final message, given that the first part of the conversation was m_1 and c_v as chosen by V^*. In the simulation we have a random sample of what S would produce, given that the first part of the conversation was m_1, c_v.

This immediately implies that if the output of S is perfectly indistinguishable from real conversations between P_0 and V_0, then the output of S^* is perfectly indistinguishable from conversations between P and V^*. Statistical indistinguishability introduces at most a negligible deviation from real conversations.

For the computational case, we give a proof by contradiction: assume that we have a successful distinguisher D for S^*. Then we can turn D into a distinguisher for S: it is well known that indistinguishablity does not depend on whether the distinguisher gets 1 or any polynomial number of samples of the distributions in question. So assume we are given a list L containing a linear number of samples produced by either S or P_0, V_0. We then use the algorithm of S^* to make from this a conversation C, seemingly between P and V^*. This is done by changing Step 1 such that we take the next sample in the input list, in stead of running S. The only problem is if many rewinds are necessary, so that L is exhausted before we finish. But this only happens with negligible probability, as we have a linear number of samples. We give the conversation produced to D and output its result.

It is now clear that if L was produced by S, C will be statistically indistinguishable from the output of S^*, while if it was produced by P_0, V_0, C will be statistically indistinguishable from conversations between P and V. Therefore, if D is a successful distinguisher for S^*, we get a successful distinguisher for S.

Lemma 4 For any prover P^* that convinces the verifier in step 1-3 of the transformed protocol with probability $1/2 + \epsilon$, there is a prover P_0^* in the original protocol which convinces V_0 with probability $T(t(n), \epsilon)$, where T is a function polynomial in $1/t(n)$ and ϵ. P_0^* is polynomial time, using P^* as an oracle.

Proof We show the algorithm of P_0^*:

1. Start running P^*. Get m_1, and send it to V_0.
2. Receive c from V_0.
3. Use the procedure of Lemma 2 to get an valid $m_2 = g(c)$ from P^*. Send it to V_0

To see that this works, observe that the success probability of $1/2 + \epsilon$ of P^* implies that for at least a fraction ϵ of the possible choices of the h_j's, we can, by rewinding P^* get correct replies to both $v = 1$ and 2. Therefore P^* satisfies the requirements of Lemma 2, and we get that we convince V_0 with probability $T(t(n), \epsilon)$.

Theorem 1 Assume that a language L has an honest verifier zero-knowlegde proof system, resp. argument that is a 3-round public coin protocol. Any such protocol can be transformed into an ordinary zero-knowledge proof, resp. argument for L.

Similarly, any honest verifier zero-knowledge proof of knowledge for a predicate Pre that is a 3-round public coin protocol can be transformed into an ordinary zero-knowledge proof of knowledge for Pre.

Both types of transformations preserve perfect and statistical zero-knowledge. The transformations are efficient in the sense that provers and verifiers in the transformed protocols are polynomial time machines that use the original provers and verifiers as oracles.

Proof In all cases, the transformation given above will work.

The statements on zero-knowledge are clear from Lemma 3.

If the original protocol is a proof system (an argument), it is clear from Lemma 4 that the transformed protocol is also a proof system (an argument).

Lemma 4 also shows that if the original protocol was a proof of knowledge, so is the transformed one: we can construct a knowledge extractor for P^*, by first using Lemma 4 to get a prover P_0^* in the original protocol and then use the knowledge extractor we know exists for the original protocol.

Remark on generalizations of Theorem 1 To generalize Theorem 1 to any bounded number of rounds, we make the transformation by simply doing one interactive hashing process for each random string sent by the verifier. Zero-knowledge is proved essentially as before. Showing soundness of the transformed proof can be done by proving a multi-round version of Lemma 4 by essentially applying the technique of Lemma 2 once for each interactive hashing done. It appears, however, that the overall success probability of the resulting reduction will tend to 0 exponentially in the number of rounds, and therefore it is not clear how to generalize Theorem 1 to any polynomial number of rounds.

5 Consequences of the main result

5.1 Bit Commitments from Zero-Knowledge Proofs

In [5], it was shown how to construct bit commitments from 3-move public-coin zero-knowledge proofs of knowledge. These commitments will hide the bits committed to statistically, resp. perfectly if the protocol used is statistical, resp. perfect zero-knowledge.

The technique used in [5], however, only works if either the number of bits sent by the verifier is constant, or the error probability decreases sufficiently fast as a function of the security parameter.

What we have shown in the previous section is how to transform any 3-move public coin protocol into one that is essentially equivalent to a protocol where the verifier sends only 1 bit. Hence the technique from [5] can be used on the transformed protocol, and we get directly from Theorem 1:

Corollary 1 Assume there exists any 3-move public coin zero-knowledge proof of knowledge for a problem of which hard instances can be sampled efficiently. Then bit commitments exist. The commitments will hide the bits committed to statistically, resp. perfectly if the protocol used is statistical, resp. perfect zero-knowledge. The commitments are computationally binding.

The prime interest of this result is its ability to produce perfectly or statistically hiding bit commitments. Such commitments otherwise seem to require one-way permutations or collision intractable hash functions, and neither assumption seems to follow from the assumption of Corollary 1.

5.2 Non-Interactive Zero-Knowledge

In the non-interactive zero-knowledge model [1], prover and verifier both have access to a uniformly chosen, random bit string. Based on this string, the prover can produce a proof consisting of just 1 message, that can be checked by the verifier. A simulator in this model produces both a simulated shared random string and a proof. Security from both the verifier's and prover's point of view is based on trust in the randomness of the shared string.

Thus the non-interactive model is less powerful than the ordinary one because interaction is not allowed, but more powerful because a shared random string is assumed as a part of the model. It is therefore not immediately clear, if a language that has a non-interactive (perfect) zero-knowledge proof also has an ordinary (perfect) zero-knowledge proof.

From Theorem 1, however, we can get the following:

Corollary 2 If a language L has a non-interactive perfect/statistical/computational zero-knowledge proof, then L has an ordinary perfect/statistical/computational zero-knowledge proof.

Proof Just observe that from the non-interactive proof we can trivially get an interactive honest-verifier zero-knowledge proof, by just letting the verifier choose the "shared" random string, and then let the prover respond with the proof. This is of course a 3-round public coin protocol (where the prover's first message is empty), so we can use Theorem 1.

5.3 One-sidedness and Black-box Simulation

In [11], Ostrovsky, Venkatesan and Yung show a transformation from honest verifier zero-knowledge to ordinary zero-knowledge. As mentioned in the introduction, this transformation works for both public and secret coin protocols, but needs to assume existence of one-way permutations. They show show two implications, which also rely on one-way permutations. These results were first shown in [4], under a stronger computational assumption. With our result, we get those two corollaries for bounded round public coin protocols without computational assumptions.

In [6], it is shown that any language which has an Arthur-Merlin proof (public coin protocol), also has a *one-sided* Arthur-Merlin proof, i.e. one in which the verfier always accepts if the common input is in L. Their proof is a transformation that builds a one-sided proof system in which the number of rounds in the original proof system is preserved. The transformed protocol is not necessarily zero-knowledge, even if the original protocol was. But it can (with minor modifications) be shown to be honest verifier statistical zero-knowledge, if the original protocol was honest verifier statistical zero-knowledge. This means that we get the following:

Corollary 3 If a language L has a bounded round public coin statistical zero-knowledge proof system, it also has a one-sided public coin statistical zero-knowledge proof system.

A final implication concerns black-box simulation, which is a special case of zero-knowledge, in which the simulator is only allowed to use the verifier as an oracle. It is unknown in general whether black-box simulation is more restrictive than the most liberal definition, where the simulator is allowed to depend on the verifier. But for public coin bounded round protocols, we can show that black box simulation is not a restriction for any flavor of zero-knowledge, simply from the fact that the simulation guaranteed by Theorem 1 and its generalization is black-box:

Corollary 4 If a language L has a bounded round public coin perfect, statistical or computational zero-knowledge proof system, it also has a public coin perfect, statistical or computational *black-box* zero-knowledge proof system.

6 Open Problem

The main open problem arising from the results in this paper is of course whether a *general* transformation from honest verifier zero-knowledge can be found that preserves perfect and statistical zero-knowledge. It is worth noting that this problem would be immediately solved if perfectly hiding bit commitments could be implemented based on any one-way function. But also this problem is open so far.

References

1. Blum, De Santis, Micali and Persiano: *Non-Interactive Zero-Knowledge*, SIAM Journal of Computing, vol.20, no.6, 1991.
2. G.Brassard, D.Chaum and C.Crépeau: *Minimum Disclosure Proofs of Knowledge*, JCSS.
3. Ben-Or, Goldreich, Goldwasser, Håstad, Killian, Micali and Rogaway: *Everything Provable is Provable in Zero-Kowledge*, Proc. of Crypto 88.
4. Bellare, Micali and Ostrovsky: *The (true) Complexity of Statsitical Zero-Knowledge*, STOC 90.
5. I.Damgård: *On the Existence of Bit Commitment Schemes and Zero-Knowledge Proofs*, Proc. of Crypto 89, Springer Verlag LNCS series.
6. Goldreich, Mansour and Sipser: *Poofs that Never Fail and Random Selection*, FOCS 87.
7. O.Goldreich, S.Micali and A.Wigderson: *Proof that Yield Nothing but their Validity and a Methodology of Cryptographic Protocol Design*, Proc. of FOCS 86.
8. S.Goldwasser, S.Micali and C.Rackoff: *The Knowledge Complexity of Interactive Proof Systems*, SIAM J.Computing, Vol.18, pp.186-208, 1989.
9. Impagliazzo and Yung: *Direct Minimum-Knowledge Computations*, Crypto 87.
10. Naor, Ostrovsky, Venkatesan and Yung: *Zero-Knowledge Arguments for NP can be based on General Complexity Assumptions*, Proc. of Crypto 92, Springer Verlag LNCS series.
11. Ostrovsky, Venkatesan and Yung: *Interactive Hashing Simplifies Zero-Knowledge Protocol Design*, to appear in Proc. of EuroCrypt 93, Springer Verlag LNCS Series.
12. Ostrovsky and Wigderson: *One-way functions are essntial for non-trivial zero-knowledge proofs*, preliminary manuscript.

Fully Dynamic Secret Sharing Schemes *

C. Blundo,[1] A. Cresti,[2] A. De Santis,[1] and U. Vaccaro[1]

[1] Dipartimento di Informatica ed Applicazioni, Università di Salerno, 84081 Baronissi (SA), Italy

[2] Dipartimento di Scienze dell' Informazione, Università di Roma "La Sapienza", 00198 Roma, Italy

Abstract. We consider secret sharing schemes in which the dealer has the feature of being able (after a preprocessing stage) to activate a particular access structure out of a given set and/or to allow the participants to reconstruct different secrets (in different time instants) by sending to all participants the same broadcast message. In this paper we establish a formal setting to study such secret sharing schemes. The security of the schemes presented is unconditional, since they are not based on any computational assumption. We give bounds on the size of the shares held by participants and on the size of the broadcast message in such schemes.

1 Introduction

A secret sharing scheme is a method of dividing a secret s among a set $\mathcal{P}$ of participants in such a way that: if the participants in $A \subseteq \mathcal{P}$ are qualified to know the secret then by pooling together their information they can reconstruct the secret s; but any set A of participants not qualified to know s has absolutely no information on the secret. The collection of subsets of participants qualified to reconstruct the secret is usually referred to as the *access structure* of the secret sharing scheme.

Secret sharing schemes are useful in any important action that requires the concurrence of several designed people to be initiated, as launching a missile, opening a bank vault or even opening a safety deposit box. Secret sharing schemes are also used in management of cryptographic keys and multi-party secure protocols (see [7] for example). We refer the reader to the excellent surveys papers [13] and [16] for a detailed discussion of secret sharing schemes and for a complete bibliography on the argument.

Simmons [13] pointed out the practical relevance of secret sharing schemes having the feature of being able (after some preprocessing stage) to activate a particular access structure out of a given set and/or to allow the participants to

---

* Partially supported by Italian Ministry of University and Research (M.U.R.S.T.) and by National Council for Research (C.N.R.).

reconstruct different secrets (in different time instants) simply by sending to all participants the same broadcast message. Harn, Hwang, Laih, and Lee [8] gave an algorithm to set up threshold secret sharing schemes (i.e., characterized by an access structure consisting of all subsets of participants of cardinality not less than some integer k), in which participants could be qualified to recover different secrets in different time instants simply by receiving a broadcast message. However, they assumed that the access structure remained the same in each time instant. Martin [11] presented a technique to realize secret sharing schemes for general access structures in which by sending a broadcast message to all participants, at each time instant a new secret is activated and a participant is disenrolled from the scheme. Blakley, Blakley, Chan, and Massey [1] considered the problem of constructing threshold secret sharing schemes with disenrollment capability. The threshold of the secret sharing schemes is not changed at each disenrollment. They gave a lower bound on the size of the shares held by each participant in such schemes.

In this paper we establish a formal setting to study secret sharing schemes in which different access structures and/or different secrets can be activated in subsequent time instants simply by sending the same broadcast message to all participants. Our approach is information–theoretic based. The security of the schemes presented in this paper is unconditional, since they are not based on any computational assumption. We first study the case in which we have different access structures and we want to enable one of them to reconstruct a predefined secret. In this model we show that the size of shares held by any participant and the size of the broadcast message are bounded from below by the size of the secret. We show that these bounds are optimal if one considers either the share of the participant or the broadcast message (see Theorem 6 and Theorem 7). Motivated by this result we define *Ideal Secret Sharing Schemes with Broadcast* as schemes for which the size of the shares held by participants and the size of the broadcast messages are the same as the size of the secret. We analyze ideal secret sharing schemes with broadcast messages when the family of the access structures that can be activated contains threshold access structures only. In Section 6 we consider the general case in which one wants to activate different access structures to recover possibly different secrets at subsequent time instants. We give sufficient conditions for the existence of a participant whose share size is lower bounded by the sum of the sizes of the secrets. This result generalizes the result of [1].

2 Secret Sharing Schemes with Broadcast Message

In this section we define secret sharing schemes with broadcast message. Let $\mathcal{P}$ be the set of participants. Denote by $\mathcal{A}$ the family of subsets of participants which we desire to be able to recover the secret; hence $\mathcal{A} \subseteq 2^{\mathcal{P}}$. $\mathcal{A}$ is called the *access structure* of the secret sharing scheme. In this section we consider the situation in which we have *more* than one access structure and we want to enable only one of them to be active to recover a predefined secret. In this scheme there is a special participant called the *dealer*. The dealer is denoted by D and we assume $D \notin \mathcal{P}$. Let $\mathsf{A} = \{\mathcal{A}_1, \ldots, \mathcal{A}_m\}$ be a family of access structures on the set of participants $\mathcal{P}$ and let $\{p_S(s)\}_{s \in S}$ be a probability distribution on the set of secrets S. The dealer in the preprocessing phase, knowing $\{p_S(s)\}_{s \in S}$ (but not knowing the value of the secret) and A, generates and distributes shares to participants in $\mathcal{P}$. The dealer, in the message-generation phase, on input a secret s randomly chosen accordingly to $\{p_S(s)\}_{s \in S}$, the access structures $\mathcal{A}_1, \ldots, \mathcal{A}_m$, the shares of participants $P_1, \ldots, P_n$, and an index $i \in \{1, 2, \ldots, m\}$ (arbitrarily chosen) computes a message b_i and broadcasts it to all participants in $\mathcal{P}$. At the end of the message-generation phase, only the subsets of participants in $\mathcal{A}_i$ are able to recover s. These phases are described in the following algorithms.

Preprocessing-Algorithm

Input: $\{p_S(s)\}_{s \in S}$, $\mathcal{P} = \{P_1, \ldots, P_n\}$, and $\mathcal{A}_1, \ldots, \mathcal{A}_m$.

Output: The shares $a_1, \ldots, a_n$ for participants $P_1, \ldots, P_n$, respectively.

Message-Generation

Input: $s \in S$, $\mathcal{A}_1, \ldots, \mathcal{A}_m$, $a_1, \ldots, a_n$, and $i \in \{1, 2, \ldots, m\}$.

Output: The broadcast message b_i that enables the access structure $\mathcal{A}_i$.

In this section we consider the case in which we want to enable only one access structure among the family A, the case in which we want to enable different access structures at different times will be analyzed in Section 6.

Let $\mathcal{P} = \{P_1, \ldots, P_n\}$ be the set of participants and let $\mathcal{A}$ be an access structure on $\mathcal{P}$. It is reasonable to require that $\mathcal{A}$ be *monotone*, that is if $A \in \mathcal{A}$ and $A \subseteq A' \subseteq \mathcal{P}$, then $A' \in \mathcal{A}$. In this paper, we assume that the access structures are not trivial, that is, there is always at least a subset of participants who can reconstruct the secret, i.e., $\mathcal{A} \neq \emptyset$, and that not all possible subsets of participants are able to recover the secret, i.e., $\mathcal{A} \neq 2^{\mathcal{P}}$.

If $\mathcal{A}$ is an access structure on $\mathcal{P}$, then $B \in \mathcal{A}$ is a *minimal* authorized subset if $A \notin \mathcal{A}$ whenever $A \subset B$. The set of minimal authorized subsets of $\mathcal{A}$ is denoted $\mathcal{A}^0$ and is called the *basis* of $\mathcal{A}$. $\mathcal{A}$ is uniquely determined as a function of $\mathcal{A}^0$, as we have $\mathcal{A} = \{B \subseteq \mathcal{P} : A \subseteq B,\ A \in \mathcal{A}^0\}$. We say that $\mathcal{A}$ is the *closure* of $\mathcal{A}^0$ and write $\mathcal{A} = cl(\mathcal{A}^0)$.

Let $\mathsf{A} = \{\mathcal{A}_1, \ldots, \mathcal{A}_m \mid \mathcal{A}_i \subseteq 2^{\mathcal{P}}, 1 \leq i \leq m\}$ be a family of monotone access structures on $\mathcal{P}$. For $1 \leq j \leq m$, define $\mathcal{P}_j = \bigcup_{X \in \mathcal{A}_j} X$; $\mathcal{P}_j$ denotes the set of participants in the scheme with access structure $\mathcal{A}_j$. Let S be the set of secrets, $\{p_S(s)\}_{s \in S}$ be a probability distribution on S, and let a secret sharing scheme with broadcast message for secrets in S be fixed. For any participant $P \in \mathcal{P}$, let us denote by $K(P)$ the set of all possible shares given to participant P. Given a set of participants $A = \{P_{i_1}, \ldots, P_{i_r}\} \subseteq \mathcal{P}$, where $i_1 < i_2 < \ldots < i_r$, denote by $K(A)$ the set $K(P_{i_1}) \times \cdots \times K(P_{i_r})$. A secret sharing scheme with broadcast message for secrets in S and a probability distribution $\{p_S(s)\}_{s\in S}$ naturally induce a probability distribution on $K(A)$, for any $A \subseteq \mathcal{P}$. Denote such probability distribution by $\{p_{K(A)}(a)\}_{a\in K(A)}$. Finally, denote by $H(S)$ the entropy of $\{p_S(s)\}_{s\in S}$ and by $H(A)$ the entropy of $\{p_{K(A)}(a)\}_{a\in K(A)}$, for any $A \in 2^{\mathcal{P}}$.

For any access structure $\mathcal{A}_i \in \mathsf{A}$, let us denote by b_i a generic broadcast message that enables the access structure $\mathcal{A}_i$ and by B_i the set of all possible broadcast messages enabling $\mathcal{A}_i$. A secret sharing scheme with broadcast message for $\mathsf{A} = \{\mathcal{A}_1, \ldots, \mathcal{A}_m\}$ and a probability distribution $\{p_S(s)\}_{s\in S}$ induce, through the two probabilistic algorithms above, a probability distribution on each B_i. Denote such probability distribution by $\{p_{B_i}(b)\}_{b\in B_i}$. Finally, for all $1 \le i \le m$, denote by $H(B_i)$ the entropy of $\{p_{B_i}(b)\}_{b\in B_i}$.

By using the entropy approach, as done in [9] and [6] for usual secret sharing schemes, we define a secret sharing scheme with broadcast message as follows.

Definition 1. Let $\mathsf{A} = \{\mathcal{A}_1, \ldots, \mathcal{A}_m\}$ be a family of monotone, non trivial access structures on $\mathcal{P}$. A *secret sharing scheme with broadcast message* is a sharing of secrets in S among participants in $\mathcal{P}$ such that

1. Before knowing the broadcast message any subset of participants has no information about the value of the secret:
   Formally, for any $X \in 2^{\mathcal{P}}$, *it holds* $H(S|X) = H(S)$.
2. After seeing the broadcast message, we have a perfect secret sharing scheme:
   Formally, for any $\mathcal{A}_i \in \mathsf{A}$ *and for any* $X \in 2^{\mathcal{P}}$, *it holds*

$$H(S|XB_i) = \begin{cases} H(S) & \text{if } X \notin \mathcal{A}_i \\ 0 & \text{if } X \in \mathcal{A}_i \end{cases}$$

Notice that $H(S|X) = H(S)$ is equivalent to state that S and X are statistically independent, i.e., for all $x \in K(X)$ and for all $s \in S$, $p(s|x) = p_S(s)$ and therefore the knowledge of x gives no information about the secret. Equivalently, $H(S|XB_i) = H(S)$ means that S and XB_i are statistically independent. Moreover, $H(S|XB_i) = 0$ means that each set of values of the shares and broadcast message in $K(X) \times K(B_i)$ corresponds to a unique value of the secret. In fact, by definition, $H(S|XB_i) = 0$ is equivalent to the fact that for all $x \in K(X)$ and for all $b \in K(B_i)$ with $p(x, b) > 0$ a unique $s \in S$ exists such that $p(s|x\ b) = 1$.

For any access structure $\mathcal{A}_i \in \mathsf{A}$, let $\mathcal{A}_i^B = \{X \cup \{B_i\} | X \in \mathcal{A}_i\}$, that is, $\mathcal{A}_i^B$ contains all the sets that can reconstruct the secret in the access structure $\mathcal{A}_i$ together with the broadcast message that enables this access structure. Intuitively, in $\mathcal{A}_i^B$ the broadcast message B_i "plays" the role of a participant.

As an example let us consider the following situation. Let $\mathcal{P} = \{P_1, P_2, ..., P_6\}$ be the set of participants. The family A, depicted in Figure 1, contains three access structures $\mathcal{A}_1 = \{P_1P_2, P_2P_3\}$, $\mathcal{A}_2 = \{P_3P_4\}$, and $\mathcal{A}_3 = \{P_4P_5, P_5P_6\}$. The following algorithms realize a secret sharing scheme with broadcast for A when the secret is uniformly chosen in $GF(q)$, where q is a prime power.

Preprocessing-Algorithm

Input: a prime power q, $\mathsf{A} = \{\mathcal{A}_1, \mathcal{A}_2, \mathcal{A}_3\}$, and $\mathcal{P} = \{P_1, \ldots, P_6\}$.

Randomly select $r_1, r_2, r_3, r_4, r_5, r_6 \in GF(q)$.

Let $a_1 = r_1$ be the share of participant P_1, $a_2 = r_2$ be the share of participant P_2, $a_3 = r_1, r_3$ be the share of participant P_3, $a_4 = r_4, r_5$ be the share of participant P_4, $a_5 = r_6$ be the share of participant P_5, and $a_6 = r_5$ be the share of participant P_6.

Output: The shares $a_1, \ldots, a_6$ for participants $P_1, \ldots, P_6$, respectively.

Message-Generation

Input: $s \in GF(q)$, $\mathcal{A}_1, \mathcal{A}_2, \mathcal{A}_3$, $a_1, \ldots, a_6$, and $i \in \{1, 2, 3\}$.

Compute $x_1 = r_1 + r_2 \bmod q$, $x_2 = r_3 + r_4 \bmod q$, and $x_3 = r_5 + r_6 \bmod q$.

Output: The broadcast message $b_i = s + x_i \bmod q$ that enables $\mathcal{A}_i$.

It is easy to see that previous algorithms realize a secret sharing scheme with broadcast for A.

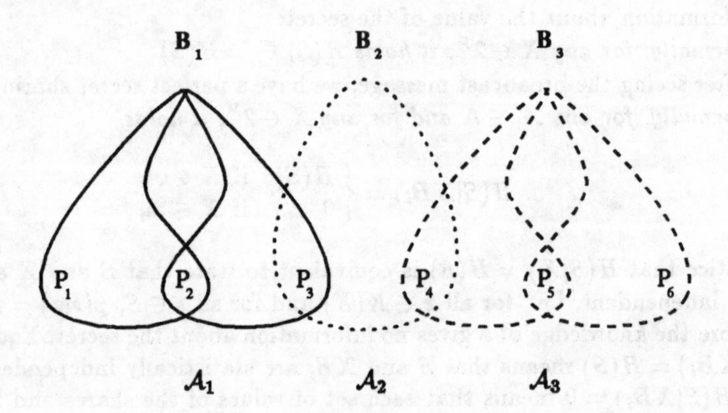

Figure 1.

3 The Size of Shares

An important issue in the implementation of secret sharing schemes is the size of the shares, since the security of a system degrades as the amount of secret information increases. Thus, one of the basic problems is to analyze the amount of information that must be kept secret. Unfortunately, in all secret sharing schemes with broadcast message the size of the shares, as well as the size of the broadcast message, cannot be less than the size of the secret as we will see in the next lemma. Moreover, there are families of access structures for which any corresponding secret sharing scheme with broadcast message must either give to some participant a share of size strictly bigger than the secret size, or the broadcast message has to have size strictly bigger than the secret size, as we will see in Section 5.

The following lemmas are a generalization to secret sharing schemes with broadcast message of the results proved in [6] for secret sharing schemes with no broadcast message.

Lemma 2. *Let* $\mathsf{A} = \{\mathcal{A}_1, \ldots, \mathcal{A}_m\}$ *be a family of monotone access structures on a set* $\mathcal{P}$ *of participants. Let* $\mathcal{A}_i \in \mathsf{A}$*, if* $Y \in 2^{\mathcal{P} \cup \{B_i\}} \backslash \mathcal{A}_i^B$ *and* $X \cup Y \in \mathcal{A}_i^B$*, then* $H(X|Y) = H(S) + H(X|YS)$.

Proof: If $Y \in 2^{\mathcal{P} \cup \{B_i\}} \backslash \mathcal{A}_i^B$ we distinguish two cases: $B_i \notin Y$ and $B_i \in Y$. If $B_i \notin Y$, then $H(S|Y) = H(S)$ by property 1 of Definition 1. If $B_i \in Y$, then $H(S|Y) = H(S)$ because of property 2 of Definition 1 since $Y \backslash \{B_i\} \notin \mathcal{A}_i$. Now, consider the conditional mutual information $I(X; S|Y)$, it can be written either as $H(X|Y) - H(X|YS)$ or as $H(S|Y) - H(S|XY)$. Hence, $H(X|Y) = H(X|YS) + H(S|Y) - H(S|XY)$. Because of $H(S|XY) = 0$ for $X \cup Y \in \mathcal{A}_i^B$ and $H(S|Y) = H(S)$, we have $H(X|Y) = H(S) + H(X|YS)$. □

As immediate consequence of the previous lemma we get the following theorem.

Theorem 3. *Let* $\mathsf{A} = \{\mathcal{A}_1, \ldots, \mathcal{A}_m\}$ *be a family of monotone access structures on a set* $\mathcal{P}$ *of participants. For any secret sharing scheme with broadcast message for* A *the following properties hold:*

1. *For any* $P \in \mathcal{P}$*, it holds* $H(P) \geq H(S)$.
2. *For* $i = 1, 2, \ldots, m$*, it holds* $H(B_i) \geq H(S)$.

If the secrets are uniformly chosen in S, that is $H(S) = \log |S|$, then we can bound both the size of the shares distributed to participants and the size of the broadcast messages.

Theorem 4. *Let* $\mathsf{A} = \{\mathcal{A}_1, \ldots, \mathcal{A}_m\}$ *be a family of monotone access structures on a set* $\mathcal{P}$ *of participants. If the secrets are uniformly chosen in* S*, then for any secret sharing scheme with broadcast message for* A *the following properties hold:*

1. *For any* $P \in \mathcal{P}$*, it holds* $\log |K(P)| \geq \log |S|$.
2. *For* $i = 1, 2, \ldots, m$*, it holds* $\log |B_i| \geq \log |S|$.

Next lemma implies that the uncertainty on shares of participants, who cannot recover the secret, it cannot be decreased by the knowledge of the secret.

Lemma 5. *Let* $\mathsf{A} = \{\mathcal{A}_1, \ldots, \mathcal{A}_m\}$ *be a family of monotone access structures on a set* $\mathcal{P}$ *of participants. Let* $\mathcal{A}_i \in \mathsf{A}$*, if* $X \cup Y \in 2^{\mathcal{P} \cup \{B_i\}} \backslash \mathcal{A}_i^B$*, then* $H(Y|X) = H(Y|XS)$.

Proof: The conditional mutual information $I(Y, S|X)$ can be written either as $H(Y|X) - H(Y|XS)$ or as $H(S|X) - H(S|XY)$. Hence, $H(Y|X) = H(Y|XS) + H(S|X) - H(S|XY)$. Because of $H(S|XY) = H(S|X) = H(S)$, for $X \cup Y \notin \mathcal{A}_i^B$, we have $H(Y|X) = H(Y|XS)$. □

Next theorems prove that for any family of monotone access structures there are secret sharing schemes with broadcast message such that the size of the shares given to a predefined participant or the size of the broadcast messages is the same than that of the secret.

The secret sharing schemes with broadcast message presented in this paper are all realized by considering uniform distributions on S. In this case, we suppose that $S = GF(q)$, where q is a prime power.

Theorem 6. *Let* $\mathsf{A} = \{\mathcal{A}_1, \ldots, \mathcal{A}_m\}$ *be a family of monotone access structures on a set* $\mathcal{P}$ *of participants and let* $P \in \mathcal{P}$ *be a fixed participant. If the secret is uniformly chosen then there exists a secret sharing scheme with broadcast message such that*

$$H(P) = H(S).$$

In Section 2 we presented a scheme for the family of monotone access structures $\mathsf{A} = \{\{P_1P_2, P_2P_3\}, \{P_3P_4\}, \{P_4P_5, P_5P_6\}\}$ on the set of participants $\mathcal{P} = \{P_1, P_2, \ldots, P_6\}$. In such scheme participants P_3 and P_4 get a share whose size is twice the size of the secret. By previous theorem there exists a scheme where either P_3 or P_4 can have a share of the same size than that of the secret. A possible scheme, in which P_3 gets a shares whose size is equal to the size of the secret, is the following. In this sheme the secret is uniformly chosen in $GF(q)$, where q is a prime power.

Preprocessing-Algorithm

Input: a prime power q, $\mathsf{A} = \{\mathcal{A}_1, \mathcal{A}_2, \mathcal{A}_3\}$, and $\mathcal{P} = \{P_1, \ldots, P_6\}$.

Randomly select $r_1, r_2, \ldots, r_9 \in GF(q)$.

Let $a_1 = r_1$ be the share of participant P_1, $a_2 = r_2, r_3$ be the share of participant P_2, $a_3 = r_4$ be the share of participant P_3, $a_4 = r_5, r_6$ be the share of participant P_4, $a_5 = r_7, r_8$ be the share of participant P_5, and $a_6 = r_9$ be the share of participant P_6.

Output: The shares $a_1, \ldots, a_6$ for participants $P_1, \ldots, P_6$, respectively.

Message-Generation

Input: $s \in S = GF(q)$, $\mathcal{A}_1, \mathcal{A}_2, \mathcal{A}_3, a_1, \ldots, a_6$, and $i \in \{1, 2, 3\}$.

Compute $b_1 = s + r_1 + r_2 \bmod q$, $s + r_3 + r_4 \bmod q$, $b_2 = s + r_4 + r_5 \bmod q$, and $b_3 = r_6 + r_7 \bmod q$, $s + r_8 + r_9 \bmod q$.

Output: The broadcast message b_i that enables the access structure $\mathcal{A}_i$.

It is easy to see that previous algorithms realize a secret sharing scheme with broadcast for A, in which the participant P_3 gets a shares whose size is equal to the size of the secret.

Next theorem states that for any family of monotone access structures there are secret sharing schemes with broadcast message such that the size of the broadcast messages is the same than that of the secret.

Theorem 7. *Let* $\mathsf{A} = \{\mathcal{A}_1, \ldots, \mathcal{A}_m\}$ *be a family of monotone access structures on a set* $\mathcal{P}$ *of participants. If the secret is uniformly chosen then there exists a secret sharing scheme with broadcast message such that, for all* $i \in \{1, 2, \ldots, m\}$, *it holds*

$$H(B_i) = H(S).$$

4 Ideal Schemes

In the previous section we have seen that for any family of access structures A either the shares given to a participant in $\mathcal{P}$, or the broadcast messages can be of the same dimension than that of the secret. In this section we give a sufficient condition for which there exists a secret sharing scheme with broadcast message for a family of access structures $\mathsf{A} = \{\mathcal{A}_1, \ldots, \mathcal{A}_m\}$ such that for any $P \in \mathcal{P}$ and for any $i \in \{1, 2, \ldots, m\}$, it holds $H(P) = H(B_i) = H(S)$. That is, we consider schemes in which both the broadcast messages and the shares of participants have the same dimension than the secret. We will use the following lemma that is an extension of a lemma proved in [6].

Lemma 8. *Let be* A, B, C, D, F, S *six random variables such that*

1. $H(S|ABF) = H(S|BCF) = H(S|ACDF) = 0,$
2. $H(S|BF) = H(S|ACF) = H(S|ADF) = H(S|F).$

Then $H(BC|F) \geq 3H(S|F)$.

An ideal secret sharing scheme with broadcast message is defined as follows.

Definition 9. Let $\mathsf{A} = \{\mathcal{A}_1, \ldots, \mathcal{A}_m\}$ be a family of monotone access structures on a set $\mathcal{P}$ of participants. A secret sharing scheme with broadcast message for A is said *ideal* if for any $P \in \mathcal{P}$ and for any $i \in \{1, 2, \ldots, m\}$, we have $H(P) = H(B_i) = H(S)$.

We first consider the simple case in which $\mathsf{A} = \{\mathcal{A}_1\}$.

Theorem 10. *Assume that the secret is uniformly chosen. An ideal secret sharing scheme with broadcast message for* $\mathsf{A} = \{\mathcal{A}_1\}$ *exists if and only if there exists an ideal secret sharing scheme for the access structure* $\mathcal{A}_1$.

Therefore, in such a case the classification of ideal secret sharing schemes with no broadcast messages given in [5] applies.

Definition 11. Two access structures $\mathcal{A}_1$ and $\mathcal{A}_2$, on the sets of participants $\mathcal{P}_1$ and $\mathcal{P}_2$ respectively, are *compatible* if and only if

$$P \in \mathcal{P}_1 \cap \mathcal{P}_2 \Rightarrow P \in \bigcap_{X \in \mathcal{A}_1^0} X \cap \bigcap_{Y \in \mathcal{A}_2^0} Y.$$

Let $\mathcal{A}_1$ and $\mathcal{A}_2$ be two access structures on the sets of participants $\mathcal{P}_1$ and $\mathcal{P}_2$, respectively. If $\mathcal{P}_1 \cap \mathcal{P}_2 \neq \emptyset$ we say that the two access structures are *connected.* If $\mathcal{A}_1$ and $\mathcal{A}_2$ are not connected, then for $\mathsf{A} = \{\mathcal{A}_1, \mathcal{A}_2\}$ there exists an ideal secret sharing scheme with broadcast message if and only if it exists an ideal secret sharing scheme with broadcast message for both $\mathsf{A}_1 = \{\mathcal{A}_1\}$ and $\mathsf{A}_2 = \{\mathcal{A}_2\}$. We say that m access structures $\mathcal{A}_1, \ldots, \mathcal{A}_m$ are *connected* if the set $\cup_{i=1}^{m} \mathcal{P}_i$ cannot be partitioned into two nonempty sets X and Y such that each $\mathcal{P}_i$, for $i = 1, \ldots, m$, is all contained either in X or in Y. When $\mathcal{A}_1, \ldots, \mathcal{A}_m$ are not connected, we can study separately each connected part.

Theorem 12. *Let $\mathsf{A} = \{\mathcal{A}_1, \ldots, \mathcal{A}_m\}$ be a family of m connected access structures pairwise compatible such that there exists an ideal secret sharing scheme with broadcast message for each $\mathcal{A}_i$, $i = 1, \ldots, m$. If the secret is uniformly chosen, then there exists an ideal secret sharing scheme with broadcast message for* A.

5 Threshold Schemes with Broadcast Message

In this section we analyze the case in which all access structures in A are distinct threshold structures. That is, $\mathsf{A} = \{\mathcal{A}_{(k_1,\mathcal{P}_1)}, \mathcal{A}_{(k_2,\mathcal{P}_2)}, \ldots, \mathcal{A}_{(k_t,\mathcal{P}_t)}\}$, where $\mathcal{A}_{(k_i,\mathcal{P}_i)}$ is the set of all subsets consisting of at least k_i participants in $\mathcal{P}_i$. In previous section we gave a sufficient condition for which ideal secret sharing schemes with broadcast message exist. Each access structure in the scheme must admit an ideal secret sharing scheme. This condition is necessary but not sufficient. In fact, threshold schemes that admit ideal secret sharing schemes not always have ideal secret sharing schemes with broadcast message as we will see in the following. If $t = 1$, then by Theorem 10 and [12] there exists an ideal secret sharing scheme with broadcast message for A. We observe that for a threshold structure $\mathcal{A}_{(k_i,\mathcal{P}_i)}$ a participant P belongs to $\bigcap_{X \in \mathcal{A}^0_{(k_i,\mathcal{P}_i)}} X$ if and only if $k_i = |\mathcal{P}_i|$, that is $|\mathcal{A}^0_{(k_i,\mathcal{P}_i)}| = 1$. Thus, two connected access structures $\mathcal{A}_{(k_1,\mathcal{P}_1)}$ and $\mathcal{A}_{(k_2,\mathcal{P}_2)}$ are compatible if and only if $k_1 = |\mathcal{P}_1|$ and $k_2 = |\mathcal{P}_2|$.

Theorem 13. *Let $\mathsf{A} = \{\mathcal{A}_{(k_1,\mathcal{P}_1)}, \mathcal{A}_{(k_2,\mathcal{P}_2)}, \ldots, \mathcal{A}_{(k_t,\mathcal{P}_t)}\}$ be a family of $t \geq 2$ (distinct) connected access structures. There exists an ideal secret sharing scheme with broadcast message for A if and only if the access structures composing A are pairwise compatible.*
If an ideal secret sharing scheme with broadcast message does not exist then, for each participant $P \in \mathcal{P}_{i_1} \cap \mathcal{P}_{i_2}$ there is an index $j \in \{1, 2\}$, such that for any secret sharing scheme with broadcast message it holds $H(P) + H(B_{i_j}) \geq 3H(S)$.

The previous theorem proves a gap for the dimension of the shares of participants and of the broadcast message. Either there is an ideal scheme (and thus they all have the same size than the secret) or the size of at least one of them is 50% bigger the the secret size. Thus, we have proved that there are families of access structures for which any corresponding secret sharing scheme with broadcast message must either give to some participant a share of size strictly bigger than the secret size, or the broadcast message has to have size strictly bigger than the secret size even though each access structure belonging to these families admits an ideal secret sharing scheme.

Next corollary is a consequence of Theorem 13.

Corollary 14. *Let* $\mathsf{A} = \{\mathcal{A}_{(1,\mathcal{P}_1)}, \cdots, \mathcal{A}_{(1,\mathcal{P}_t)}\}$ *be a family of* $t \geq 2$ *(distinct) access structures. There exists an ideal secret sharing scheme with broadcast message for* A *if and only if the access structures composing* A *are not connected, that is,* $\mathcal{P}_i \cap \mathcal{P}_j = \emptyset$*, for all* $i \neq j$*.*

In some cases a better bound on the size of the shares distributed to participants holds. Consider the set of participants $\mathcal{P} = \{X_0, X_1, X_2, \cdots, X_n\}$ and the $\mathcal{M}_n$ be the closure of $\{X_1X_2 \ldots X_n\} \cup \{X_0X_1, X_0X_2, \ldots, X_0X_{n-1}\}$. In a similar way of Theorem 4.1 in [3] one can easily prove that for any $n-2$ indices $i_1, i_2, \ldots, i_{n-2} \in \{1, 2, \ldots, n-1\}$, it holds

$$H(X_0) + H(X_{i_1}) + \ldots + H(X_{i_{n-2}}) \geq (2n-3)H(S). \qquad (1)$$

The following theorem holds.

Theorem 15. *Let* $\mathsf{A} = \{\mathcal{A}_{(k_1,\mathcal{P}_1)}, \mathcal{A}_{(k_2,\mathcal{P}_2)}\}$*, with* $k_1 \leq k_2$*, be a family of two distinct connected access structures. Let* $r = |\mathcal{P}_1 \cap \mathcal{P}_2|$*.*
If $k_1 < r$ *then*

1. *If* $k_1 < k_2$*, then for any* $P_{l_1}, \ldots, P_{l_{t-k_1}} \in \mathcal{P}_1 \cap \mathcal{P}_2$ *where* $t = \min\{k_2, r\}$*, it holds*
   $H(B_1) + \sum_{j=1}^{t-k_1} H(P_{l_j}) \geq (2(t-k_1)+1)H(S)$.
2. *If* $k_1 = k_2 = k$ *and* $r = t_2$*, then for any* $P_{l_1}, \ldots, P_{l_\ell} \in \mathcal{P}_1 \cap \mathcal{P}_2$*, where* $\ell = \min\{k-1, t_1 - r\}$*, it holds* $H(B_1) + \sum_{j=1}^{\ell} H(P_{l_j}) \geq (2\ell+1)H(S)$*.*
3. *If* $k_1 = k_2 = k$ *and* $r < t_2$*, then for any* $P_{l_1}, \ldots, P_{l_\ell} \in \mathcal{P}_1 \cap \mathcal{P}_2$*, where* $\ell = \min\{k-1, t_2 - r\}$*, it holds* $H(B_2) + \sum_{j=1}^{\ell} H(P_{l_j}) \geq (2\ell+1)H(S)$*.*

If $r \leq k_1$ *then*

1. *If* $k_2 < t_2$*, then for any* $P_{l_1}, \ldots, P_{l_{t-1}} \in \mathcal{P}_1 \cap \mathcal{P}_2$*, where* $t = \min\{r, t_2 - k_2 + 1\}$*, it holds*
   $H(B_2) + \sum_{j=1}^{t-1} H(P_{l_j}) \geq (2(t-1)+1)H(S)$.
2. *If* $k_2 = t_2$*, then for any* $P_{l_1}, \ldots, P_{l_{t-1}} \in \mathcal{P}_1 \cap \mathcal{P}_2$*, where* $t = \min\{r, t_1 - k_1 + 1\}$*, it holds*
   $H(B_1) + \sum_{j=1}^{t-1} H(P_{l_j}) \geq (2(t-1)+1)H(S)$.

We now analyze the case in which the access structures in A consist of all possible distinct threshold structures on $\mathcal{P}$, that is, $\mathsf{A} = \{\mathcal{A}_{(k,\mathcal{P}')} \mid 1 \leq k \leq |\mathcal{P}'| \leq n \text{ and } \mathcal{P}' \subseteq \mathcal{P}\} \setminus \{\mathcal{A}_{(1,\mathcal{P})}\}$. From Theorem 13 there is no ideal secret sharing scheme with broadcast message for A. A scheme based on a geometric construction (for an overview of geometric constructions for secret sharing schemes, the reader is advised to consult [13], see also [15] and [14]) is the following: Let q be a prime power, consider the $(n+1)$-dimensional vector space over $GF(q)$. Consider the $(n+1)$-dimensional affine geometry $AG(n+1,q)$. Let V_D be a fixed line in $AG(n+1,q)$ and let V_I be a hyperplane such that $|V_D \cap V_I| = 1$. The secret will be the point $s \in V_D \cap V_I$. Choose $2n$ points $y_1, y_2, \ldots, y_{2n} \in V_I$ such that no $n+1$ of the $2n+1$ points $y_1, y_2, \ldots, y_{2n}, s$ are collinear. For $i = 1, 2, \ldots, n$, give the point y_i to the participant P_i. The broadcast message $b_{k,\mathcal{P}'}$ that enables the access structure $\mathcal{A}_{(k,\mathcal{P}')}$ will be equal to

$$b_{k,\mathcal{P}'} = \Big(\bigcup_{1 \le i \le |\mathcal{P}'|-k+1} \{y_{n+i}\} \Big) \cup \Big(\bigcup_{P_i \notin \mathcal{P}'} \{y_i\} \Big).$$

It is easy to see that in the previous scheme for any P in $\mathcal{P}$ we have, $H(P) = (n+1)H(S)$. Moreover, the broadcast message $b_{k,\mathcal{P}'}$ that enables the access structure $\mathcal{A}_{(k,\mathcal{P}')}$ has entropy equal to $H(B_{k,\mathcal{P}'}) = (n-k+1)(n+1)H(S)$. With a slight modification of the previous scheme (using tecniques described in [16] and [10]), we can obtain a geometric scheme in which $H(P) = H(S)$.
The following algorithms describe a secret sharing scheme with broadcast message such that for all P in $\mathcal{P}$, $H(P) = H(S)$. We suppose that $S = GF(q)$, where $q \ge \max\{2n, m\} + 1$ is a prime power.

Threshold Preprocessing-Algorithm

Input: a prime power q and $\mathcal{P} = \{P_1, \ldots, P_n\}$.

For all $P_i \in \mathcal{P}$, randomly select $r_i \in GF(q)$ and set $a_i = r_i$ to be the share of $P_i \in \mathcal{P}$.

Output: The shares $a_1, \ldots, a_n$ for participants $P_1, \ldots, P_n$, respectively.

Threshold Message-Generation

Input: $s \in S = GF(q)$, $a_1, \ldots, a_n$, k, and $\mathcal{P}'$, such that $1 \le k \le |\mathcal{P}'| \le n$

Use a threshold scheme $(n+1, 2n)$ for the secret s to generate the shares $y_1, \ldots, y_{2n}$ in such a way that $y_i = a_i$, for $i = 1, \ldots, n$.

Compute

$$b_{k,\mathcal{P}'} = \Big(\bigcup_{1 \le i \le |\mathcal{P}'|-k+1} \{y_{n+i}\} \Big) \cup \Big(\bigcup_{P_i \notin \mathcal{P}'} \{a_i\} \Big).$$

Output: The broadcast message $b_{k,\mathcal{P}'}$ that enables the access structure $\mathcal{A}_{(k,\mathcal{P}')}$.

Notice that we can always construct the threshold scheme $(n+1, 2n)$ used in the *Message-Generation* algorithm. Indeed, we can use the threshold scheme proposed by Shamir [12]. We have to construct a polynomial $f(x)$ over $GF(q)$ of degree n such that $f(i) = y_i$, $i = 1, 2, \ldots, n$, and $f(0) = s$. This can be done by using the Lagrange interpolation. Thus, $f(i) = y_i$, $i = n+1, \ldots, 2n$. The broadcast message $b_{k,\mathcal{P}'}$ that enables the access structure $\mathcal{A}_{(k,\mathcal{P}')}$ has entropy

equal to $H(B_{k,\mathcal{P}'}) = (n-k+1)H(S)$. Moreover, the entropy of the shares of each participant $P \in \mathcal{P}$ is equal to $H(P) = H(S)$. Since each broadcast message $b_{k,\mathcal{P}'}$ consists of $n-k+1$ shares, every k participants in the threshold structure $\mathcal{A}_{(k,\mathcal{P}')}$ know $n+1$ shares and can reconstruct the secret s. But $k-1$, or less, participants are not able to recover the secret.
It is clear that the previous algorithm can be easily adapted to handle the case in which only a subset of all threshold structures can be activated by the broadcast message.

6 Fully Dynamic Secret Sharing Schemes

In previous sections we have analyzed the situation in which we have various access structures and by using a public message we enable one of them to recover the secret. A more interesting situation arises when we want to activate different access structures at subsequent times. At time i we want to enable an access structure $\mathcal{A}_{j_i}^{(i)}$, chosen in a fixed family $\mathsf{A}^{(i)}$, to recover the i-*th* secret s_i. The family $\mathsf{A}^{(i)}$ of access structures that can be enabled at time i may depend on the access structures activated at previous times. If $b_{j_1}^{(1)} \ldots b_{j_{i-1}}^{(i-1)}$ are the broadcast messages sent by the dealer from time 1 up to time $i-1$, then we should denote the family of access structures that can be enabled at time i by $\mathsf{A}_{j_1,\ldots,j_{i-1}}^{(i)}$ but to avoid overburdening the notation we will denote this family by $\mathsf{A}^{(i)}$.

Suppose that at time i the dealer enables the access structure $\mathcal{A}_{j_i}^{(i)}$. Thus, after the publication of all $i-1$ previous broadcast messages, the subsets of participants in $\mathcal{A}_{j_i}^{(i)}$ will recover the i-*th* secret after seeing the i-*th* broadcast message. Moreover, at time i each subset of participants knowing only the $i-1$ previous broadcast messages have no information on the secret s_i.

Suppose that we want to enable different access structures to reconstruct a secret a number of times, say T. Let $S^{(i)}$ be the set from which we choose the i-*th* secret, and $\mathsf{A}^{(i)} = \{\mathcal{A}_1^{(i)}, \ldots, \mathcal{A}_{m_i}^{(i)}\}$ be the family of possible access structures at time i, for $i = 1, 2, \ldots, T$. Denote by $\mathcal{P}^{(i)}$ the set of participants involved at time i and let $\mathcal{P} = \bigcup_{i=1}^{T} \mathcal{P}^{(i)}$. Denote by $\mathsf{B}^{(i)} = \{B_1^{(i)}, \ldots, B_{m_i}^{(i)}\}$ the family of all sets of broadcast messages for all possible access structures at time i. A fully dynamic secret sharing scheme is defined as follows.

Definition 16. Let $\mathsf{A}^{(1)}, \ldots, \mathsf{A}^{(T)}$ be families of monotone, non trivial access structures on $\mathcal{P}$. A *fully dynamic secret sharing scheme* is a sharing of secrets in $S^{(1)}, \ldots, S^{(T)}$ among participants in $\mathcal{P}$ such that

1. Before knowing the new broadcast message any subset of participants has no information about the new secret:
   Formally, for all $X \in 2^{\mathcal{P}}$, for all $i = 1, \ldots, T$, and for all $j_1, \ldots, j_{i-1}$, where $1 \le j_\ell \le m_\ell$, it holds $H(S^{(i)}|XB_{j_1}^{(1)} \ldots B_{j_{i-1}}^{(i-1)}) = H(S^{(i)})$.

2. **After seeing the new broadcast message, we have a new perfect secret sharing scheme:**
   Formally, for all $i = 1, \ldots, T$, for all $X \in 2^{\mathcal{P}}$, and for all $j_1, \ldots, j_i$, where $1 \leq j_\ell \leq m_\ell$, it holds

$$H(S^{(i)}|XB_{j_1}^{(1)} \ldots B_{j_i}^{(i)}) = \begin{cases} H(S^{(i)}) & \text{if } X \notin \mathcal{A}_{j_i}^{(i)} \\ 0 & \text{if } X \in \mathcal{A}_{j_i}^{(i)} \end{cases}$$

The following theorem is a generalization to fully dynamic secret sharing schemes of Theorem 3

Theorem 17. *Let $\mathsf{A}^{(1)}, \ldots, \mathsf{A}^{(T)}$ be families of monotone access structures on a set $\mathcal{P}$ of participants. In any fully dynamic secret sharing scheme for $i = 1, 2, \ldots, T$ the following properties hold:*

1. *For any $P \in \mathcal{P}^{(i)}$, it holds $H(P) \geq H(S^{(i)})$.*
2. *For $j = 1, 2, \ldots, m_i$, it holds $H(B_j^{(i)}) \geq H(S^{(i)})$.*

Definition 16 says nothing on the sets X of participants such that $X \notin \mathcal{A}_{j_i}^{(i)}$ and that know all secrets $s_1, \ldots, s_{i-1}$ previously recovered. A natural requirement is that the information of these sets of participants have on the i-*th* secret s_i given the secrets $s_1, \ldots, s_{i-1}$, is equal to zero. That is, the knowledge of previous secrets does not give information about the i-*th* secret to all sets of participants not in $\mathcal{A}_j^{(i)}$. Next we define a *strong fully dynamic secret sharing scheme*, that is a fully dynamic secret sharing scheme with an additional property.

Definition 18. Let $\mathsf{A}^{(1)}, \ldots, \mathsf{A}^{(T)}$ be families of monotone, non trivial access structures on $\mathcal{P}$. A ***strong fully dynamic secret sharing scheme*** is a fully dynamic secret sharing scheme such that after seeing the new broadcast message, any subset of participants that is not in the new access structure, even knowing all the previous secrets, has no information about new secret:
Formally, for all $i = 1, \ldots, T$, for all $j_1, \ldots, j_i$, where $1 \leq j_\ell \leq m_\ell$, and for all $X \notin \mathcal{A}_{j_i}^{(i)}$, it holds

$$H(S^{(i)}|XB_{j_1}^{(1)} \ldots B_{j_i}^{(i)} S^{(1)} \ldots S^{(i-1)}) = H(S^{(i)}).$$

Notice that the property $H(S^{(i)}|XB_{j_1}^{(1)} \ldots B_{j_i}^{(i)} S^{(1)} \ldots S^{(i-1)}) = H(S^{(i)})$ in the above definition implies that $H(S^{(i)}|XB_{j_1}^{(1)} \ldots B_{j_i}^{(i)}) = H(S^{(i)})$ if $X \notin \mathcal{A}_{j_i}^{(i)}$ in Definition 16. In fact, for all $i = 1, \ldots, T$, for all $j_1, \ldots, j_i$, where $1 \leq j_\ell \leq m_\ell$, and for all $X \notin \mathcal{A}_{j_i}^{(i)}$ it holds

$$\begin{aligned} H(S^{(i)}) &\geq H(S^{(i)}|XB_{j_1}^{(1)} \ldots B_{j_i}^{(i)}) \\ &\geq H(S^{(i)}|XB_{j_1}^{(1)} \ldots B_{j_i}^{(i)} S^{(1)} \ldots S^{(i-1)}) \\ &= H(S^{(i)}). \end{aligned}$$

If the families of monotone access structures $\mathsf{A}^{(1)},\ldots,\mathsf{A}^{(T)}$ satisfy some condition, then we can prove a lower bound on the size of shares held by a fixed participant. Next theorem holds.

Theorem 19. *Let $\mathcal{P}$ be a set of participants, $\mathsf{A}^{(1)},\ldots,\mathsf{A}^{(T)}$ be families of monotone, non trivial access structures on $\mathcal{P}$. If there exist T indices $j_1,\ldots,j_T$, a participant $P\in\mathcal{P}$, and subsets of participants $X_i\subseteq\mathcal{P}$, $i=1,\ldots,T$, such that*

- $X_i\notin\mathcal{A}^{(i)}_{j_i}$, *but* $X_i\cup\{P\}\in\mathcal{A}^{(i)}_{j_i}$, *for* $i=1,\ldots,T$,
- $X_i\subseteq X_{i+1}$, *for* $i=1,\ldots,T-1$,

then, in any strong fully dynamic secret sharing scheme for $\mathsf{A}^{(1)},\ldots,\mathsf{A}^{(T)}$ the entropy of P satisfies

$$H(P)\geq\sum_{i=1}^{T}H(S^{(i)}).$$

We point out that Theorem 19 does not hold if we assume fully dynamic secret sharing schemes instead of strong fully dynamic secret sharing schemes.
As an example, consider the following situation. Let $\mathsf{A}^{(1)}=\{\mathcal{A}^{(1)}_1\}$ and $\mathsf{A}^{(2)}=\{\mathcal{A}^{(2)}_1\}$, be two families of monotone access structures on the set of participants $\mathcal{P}=\{P_1,P_2,P_3\}$, where $\mathcal{A}^{(1)}_1=\{\{P_1P_2\}\}$ and $\mathcal{A}^{(2)}_1=\{\{P_2P_3\}\}$. Suppose that at time 1 the dealer enables $\mathcal{A}^{(1)}_1$ to reconstruct the secret $s^{(1)}$ and at time 2 the dealer enables $\mathcal{A}^{(2)}_1$ to reconstruct the secret $s^{(2)}$. The following algorithms describe a fully dynamic secret sharing scheme for $\mathsf{A}^{(1)}$ and $\mathsf{A}^{(2)}$.

Preprocessing-Algorithm

Input: a prime power q, $\mathsf{A}^{(1)}$, $\mathsf{A}^{(2)}$, and $\mathcal{P}=\{P_1,P_2,P_3\}$.
For $i=1,2,3$, randomly select $r_i\in GF(q)$ and set $a_i=r_i$ to be the share of $P_i\in\mathcal{P}$.
Output: The shares a_1,a_2,a_3 for participants P_1,P_2,P_3, respectively.

Message-Generation

Input: $s^{(1)},s^{(2)}\in GF(q)$, $\mathsf{A}^{(1)}$, $\mathsf{A}^{(2)}$, and a_1,a_2,a_3.
Compute

$$b^{(1)}_1=a_1+a_2+s^{(1)}\bmod q$$

and

$$b^{(2)}_1=a_2+a_3+s^{(2)}\bmod q,$$

that are the broadcast messages for the two access structures $\mathcal{A}^{(1)}_1$ and $\mathcal{A}^{(2)}_1$.
Output: The broadcast messages $b^{(1)}_1$ and $b^{(2)}_1$.

The scheme above is a fully dynamic secret sharing scheme, but it is not a strong one. In fact, it is easy to see that

$$H(S^{(2)}|P_1P_3B^{(1)}_1B^{(2)}_1S^{(1)})=0,\text{ but }\{P_1P_3\}\notin\mathcal{A}^{(2)}_1.$$

The scheme above satisfies the remaining hypothesis of Theorem 19 by setting $P = P_2$, $X_1 = \{P_1\}$ and $X_2 = \{P_1, P_3\}$. On the other hand, we have $H(P_2) = H(S^{(1)}) = H(S^{(2)})$, thus

$$H(P_2) < H(S^{(1)}) + H(S^{(2)}).$$

The following corollaries hold.

Corollary 20. *Let $\mathcal{P}$ be a set of participants and let $\mathsf{A}^{(1)}, \ldots, \mathsf{A}^{(T)}$ be families of monotone, non trivial access structures on $\mathcal{P}$ such that $\mathcal{A}_{(k_i,\mathcal{P}_i)} \in \mathsf{A}^{(i)}$, for $i = 1, 2, \ldots, T$. If $k_1 \leq k_2 \leq \cdots \leq k_T$ and $\mathcal{P}_1 \subseteq \mathcal{P}_2 \subseteq \cdots \subseteq \mathcal{P}_T$ then the entropy of any participant $P \in \mathcal{P}_1$ satisfies*

$$H(P) \geq \sum_{i=1}^{T} H(S^{(i)}).$$

Corollary 21. *Let $\mathcal{P}$ be a set of n participants and k an integer, $1 \leq k \leq n$. Let $\mathsf{A}^{(0)}, \ldots, \mathsf{A}^{(T)}$, $1 \leq T \leq n - k$ be families of monotone, non trivial access structures on $\mathcal{P}$ such that $\mathcal{A}_{(k,\mathcal{P}_\ell)} \in \mathsf{A}^{(\ell)}$, for $\ell = 0, 1, \ldots, T$, where $\mathcal{P} = \mathcal{P}_0 \supseteq \mathcal{P}_1 \supseteq \cdots \supseteq \mathcal{P}_T$, and $|\mathcal{P}_\ell| = |\mathcal{P}_{\ell-1}| - 1$ for $\ell = 1, \ldots, T$. Then, in any strong fully dynamic secret sharing scheme for $\mathsf{A}^{(0)}, \ldots, \mathsf{A}^{(T)}$ the entropy of any participant $P \in \mathcal{P}_T$ satisfies*

$$H(P) \geq \sum_{i=0}^{T} H(S^{(i)}).$$

A particular class of strong fully dynamic secret sharing scheme which satisfies the hypothesis of Corollary 21 are (k, n) threshold schemes with disenrollment [1]. At each subsequent time instant we disenroll a participant from the scheme, but the threshold of the new scheme remains unchanged. Thus, in any (k, n) threshold scheme with L-fold disenrollment capability (as defined in [1]), with $0 \leq L \leq n - k$, for any participant $P \in \mathcal{P}$, it holds $H(P) \geq \sum_{i=0}^{L} H(S^{(i)})$.

References

1. B. Blakley, G. R. Blakley, A. H. Chan, and J. Massey, *Threshold Schemes with Disenrollment*, in "Advances in Cryptology - CRYPTO '92", Ed. E. Brickell, "Lecture Notes in Computer Science", Springer-Verlag.
2. G. R. Blakley, *Safeguarding Cryptographic Keys*, Proceedings AFIPS 1979 National Computer Conference, pp.313–317, June 1979.
3. C. Blundo, A. De Santis, L. Gargano, and U. Vaccaro, *On the Information Rate of Secret Sharing Schemes*, in "Advances in Cryptology - CRYPTO '92", Ed. E. Brickell, "Lecture Notes in Computer Science", Springer-Verlag.

4. C. Blundo, A. De Santis, D. R. Stinson, and U. Vaccaro, *Graph Decomposition and Secret Sharing Schemes*, in "Advances in Cryptology – Eurocrypt '92", Lecture Notes in Computer Science, Vol. 658, R. Rueppel Ed., Springer-Verlag, pp. 1–24, 1993.
5. E. F. Brickell and D. M. Davenport, *On the Classification of Ideal Secret Sharing Schemes*, J. Cryptology, Vol. 4, No. 2, pp. 123–134, 1991.
6. R. M. Capocelli, A. De Santis, L. Gargano, and U. Vaccaro, *On the Size of Shares for Secret Sharing Schemes*, Journal of Cryptology, Vol. 6, No. 3, pp. 157-169, 1993.
7. O. Goldreich, S. Micali, and A. Wigderson, *How to Play any Mental Game*, Proceedings of 19th ACM Symp. on Theory of Computing, pp. 218–229, 1987.
8. L. Harn, T. Hwang, C. Laih, and J. Lee, *Dynamic Threshold Scheme based on the definition of Cross-Product in a N-dimensional Linear Space* in "Advances in Cryptology - Eurocrypt '89", Lecture Notes in Computer Science, Vol. 435, J. Brassard Ed., Springer-Verlag, pp. 286–298.
9. E. D. Karnin, J. W. Greene, and M. E. Hellman, *On Secret Sharing Systems*, IEEE Trans. on Inform. Theory, Vol. IT-29, no. 1, pp. 35–41, Jan. 1983.
10. K. Martin, *Discrete Structures in the Theory of Secret Sharing*, PhD Thesis, University of London, 1991.
11. K. Martin, *Untrustworthy Participants in Perfect Secret Sharing Schemes*, Proceedings of the 3rd IMA Conference on Coding and Cryptology, 1992.
12. A. Shamir, *How to Share a Secret,* Communications of the ACM, Vol. 22, n. 11, pp. 612–613, Nov. 1979.
13. G. J. Simmons, *An Introduction to Shared Secret and/or Shared Control Schemes and Their Application*, Contemporary Cryptology, IEEE Press, pp. 441–497, 1991.
14. G. J. Simmons, *How to (Really) Share a Secret*, in "Advances in Cryptology - CRYPTO 88", Ed. S. Goldwasser, "Lecture Notes in Computer Science", Springer-Verlag.
15. G. J. Simmons, W. Jackson, and K. Martin, *The Geometry of Shared Secret Schemes*, Bulletin of the ICA, Vol. 1, pp. 71–88, 1991.
16. D. R. Stinson, *An Explication of Secret Sharing Schemes*, Design, Codes and Cryptography, Vol. 2, pp. 357–390, 1992.
17. D. R. Stinson, *Decomposition Constructions for Secret Sharing Schemes*, Technical Report UNL-CSE-92-020, Department of Computer Science and Engineering, University of Nebraska, September 1992.

Multisecret Threshold Schemes

Wen-Ai Jackson, Keith M. Martin* and Christine M. O'Keefe*

Department of Pure Mathematics, The University of Adelaide, Adelaide SA 5005, Australia

Abstract. A threshold scheme is a system that protects a secret (key) among a group of participants in such a way that it can only be reconstructed from the joint information held by some predetermined number of these participants. In this paper we extend this problem to one where there is more than one secret that participants can reconstruct using the information that they hold. In particular we consider the situation where there is a secret s_K associated with each k–subset K of participants and s_K can be reconstructed by any group of t participants in K $(t \leq k)$. We establish bounds on the minimum amount of information that participants must hold in order to ensure that up to w participants $(0 \leq w \leq n-k+t-1)$ cannot obtain any information about a secret with which they are not associated. We also discuss examples of systems that satisfy this bound.

1 Introduction

Secret sharing schemes have received much attention in the recent literature. The basic problem is to protect a *secret* by distributing information (*shares*) relating to it among a group of n *participants* in such a way that only certain pre-specified groups of participants can reconstruct the secret from their pooled shares. The collection of sets of participants that can reconstruct the secret in this way is called the *access structure*. If there is an integer t such that the access structure consists of all the subsets of participants of size at least t then the access structure is called a (t, n)-*threshold* access structure, and the corresponding scheme is called a (t, n)-*threshold scheme*. The collection of sets of participants that are desired not to obtain any information about the secret is called the *prohibited structure*. The access and prohibited structures of the scheme together form the *structure* of the scheme.

Threshold schemes were proposed and constructed in the first papers on this subject (see Blakley [2], Shamir [15]). Since then many authors have considered more general structures (see Ito et al [9] and Benaloh and Leichter [1]).

Secret sharing schemes (and in particular threshold schemes) have many potential uses in the area of information security (see Simmons [16]). In particular, such a scheme can be used to ensure the secure implementation of a cryptographic key in a multi-user network. It is a natural generalisation to extend

* This work was supported by the Australian Research Council

this concept to a multi-user network in which many different keys need to be protected among different sets of participants in the network. The problem of determining the minimum amount of information that each user must hold in order to be able to reconstruct the appropriate keys (the minimum size of the user's share) becomes increasingly important as the complexity of the network increases.

We generalise the concept of a secret sharing scheme to allow a number of different secrets to be reconstructed by the participants. We then consider the special case of this generalisation in which each subset of k participants is associated with a secret which is protected by a (t, k)-threshold access structure. This paper is concerned with finding lower bounds on the size of a participant's share in this generalisation of a threshold scheme. This is also a generalisation of a problem previously considered by Blom [4], Matsumoto [13] and Blundo et al [5]. We call these schemes *multisecret threshold schemes* (or *multithreshold schemes* for short).

The paper is structured as follows. In Section 2 we give a formal definition of a multithreshold scheme. In Section 3 we prove lower bounds on the size of each share of a participant (or a group of participants) in a multithreshold scheme and in Section 4 we discuss some schemes that achieve these bounds.

2 Multithreshold Schemes

We first define a multisecret sharing scheme. Let $\mathcal{P}$ be a set of n *participants* and $\mathcal{K} = \{s_1, \ldots, s_r\}$ be a set of *secrets*. For each i $(1 \le i \le r)$ denote the access structure of secret s_i by Γ_i and the prohibited structure of s_i by Δ_i. Then we call $\boldsymbol{\Gamma} = (\Gamma_1, \ldots, \Gamma_r)$ the *access structure* of the multisecret sharing scheme and $\boldsymbol{\Delta} = (\Delta_1, \ldots, \Delta_r)$ the *prohibited structure* of the multisecret sharing scheme. We make the natural restriction that for each i $(1 \le i \le r)$ Γ_i is *monotone increasing* and Δ_i is *monotone decreasing.* In other words, for $A \in \Gamma_i$ we have that $A' \subseteq \mathcal{P}$, $A \subseteq A'$ implies $A' \in \Gamma_i$, and for $A \in \Delta_i$ we have that $A' \subseteq \mathcal{P}$, $A \supseteq A'$ implies $A' \in \Delta_i$. The *structure* of the multisecret sharing scheme is the pair $(\boldsymbol{\Gamma}, \boldsymbol{\Delta})$. If $\Delta_i = 2^{\mathcal{P}} \backslash \Gamma_i$ (for each $1 \le i \le r$) then we say that the structure $(\boldsymbol{\Gamma}, \boldsymbol{\Delta})$ is *complete.*

The model we now present for a perfect multisecret sharing scheme is an extension of the model for secret sharing first proposed by Brickell and Davenport [7] and used in [8, 10, 12, 17].

Let the share held by participant p $(p \in \mathcal{P})$ come from the set $\mathcal{S}_p$ and let the value of secret s $(s \in \mathcal{K})$ come from the set $\mathcal{S}_s$ of size q. We call the sets $\mathcal{S}_p$ *share spaces* and the sets $\mathcal{S}_s$ *secret spaces.* We refer to $|\mathcal{S}_p|$ as the *size* of the share held by p and refer to q as the *size* of the secrets. A *perfect multisecret sharing scheme* with structure $(\boldsymbol{\Gamma}, \boldsymbol{\Delta})$ (denoted $\mathrm{PS}(\boldsymbol{\Gamma}, \boldsymbol{\Delta}, q)$) is a collection $\mathcal{F}$ of (publically known) distribution rules where each $f \in \mathcal{F}$ is a one to one mapping from $\mathcal{P} \cup \mathcal{K}$ to $\left(\bigcup_{p \in \mathcal{P}} \mathcal{S}_p\right) \cup \left(\bigcup_{s \in \mathcal{K}} \mathcal{S}_s\right)$ with

$$f(p) \in \mathcal{S}_p \ \text{ (for all } p \in \mathcal{P}) \text{ and } f(s) \in \mathcal{S}_s \ \text{ (for all } s \in \mathcal{K}),$$

and such that for each i $(1 \le i \le r)$,

1. if $A \in \Gamma_i$ and $f, g \in \mathcal{F}$ are such that $f(x) = g(x)$ (for all $x \in A$) then $f(s_i) = g(s_i)$;
2. if $A \in \Delta_i$, and $f \in \mathcal{F}$ then there exists some integer λ such that for each $k \in \mathcal{S}_{s_i}$ there are precisely λ distribution rules $g \in \mathcal{F}$ such that $f(x) = g(x)$ (for all $x \in A$) and $f(s_i) = k$.

This is best represented by a matrix M whose rows are indexed by members of $\mathcal{F}$ and whose columns are indexed by members of $\mathcal{P} \cup \mathcal{K}$. The entry in row f and column x $(x \in \mathcal{P} \cup \mathcal{K})$ is $f(x)$. We assume that each distribution rule is equiprobable and implement the scheme by choosing a rule f at random and distributing share $f(p)$ to participant p (for all $p \in \mathcal{P}$). The secrets that these shares are protecting are the values $f(s)$ (for all $s \in \mathcal{K}$) (see [7] for more details).

We assume that a set of participants will attempt to determine a secret s by looking at their collective shares and considering only the set $\mathcal{G}$ of distribution rules f under which they could have received these shares. If the set of participants is in Γ_i then every $f \in \mathcal{G}$ will have the same value at the secret s_i. Otherwise, the structure of a perfect multisecret sharing scheme ensures that each value for the secret s_i will occur equally often among the distribution rules in $\mathcal{G}$. Note that the security offered by this model is *unconditional* in the sense that it is independent of the amount of computing time and resources that are available in any attempt to obtain a secret by some unauthorised means.

We note that the term "multisecret sharing scheme" has also been used in [6] to refer to a special class of complete multisecret sharing schemes with the same access structure for each secret.

We now introduce notation which we use for the remainder of the paper. Let $1 \le t \le k \le n$ and $r = \binom{n}{k}$. Let the collection of k-subsets of $\mathcal{P}$ be denoted by $\{X_1, \ldots, X_r\}$. Let $0 \le w \le n - k + t - 1$. Then for each i, $(1 \le i \le r)$ let

$$\Gamma_i = \{A \subseteq \mathcal{P} \mid |A \cap X_i| \ge t\},$$

and

$$\Delta_i = \{A \subseteq \mathcal{P} \mid |A| \le w\} \backslash \Gamma_i \quad .$$

Let $\boldsymbol{\Gamma} = (\Gamma_1, \ldots, \Gamma_r)$ and let $\boldsymbol{\Delta} = (\Delta_1, \ldots, \Delta_r)$. Then for $q > 1$ we refer to a $\mathrm{PS}(\boldsymbol{\Gamma}, \boldsymbol{\Delta}, q)$ as a *w-secure (t, k, n)-multithreshold scheme* with *secret size q*.

Thus a w-secure (t, k, n)-multithreshold scheme has one secret s_i for each k-subset X_i of the n participants in $\mathcal{P}$. Any set of t or more of the k participants in X_i is able to reconstruct s_i. Further, a set A of w or fewer participants in $\mathcal{P}$ is unable to obtain any information about s_i unless A contains at least t members of X_i. (If $A = \emptyset$ then this is equivalent to $|\{f \in \mathcal{F} \mid f(s_i) = k\}|$ being independent of $k \in \mathcal{S}_{s_i}$.) Note that if $0 \le w < t - 1$ then it is possible that a subset of X_i of size μ, $(w < \mu < t)$ is able to obtain some information about the value of the secret.

It follows from the definition that a w-secure (t, k, n)-multithreshold scheme is also a w'-secure (t', k, n)-multithreshold scheme for all $0 \le w' \le w$ and $t \le t' \le k$.

If $w = n-k+t-1$ then the multithreshold scheme will be complete. It makes no sense to have $w > n - k + t - 1$ since any set of this size can by definition reconstruct all of the secrets in $\mathcal{K}$.

Example 1. Let $\mathcal{P} = \{a, b, c\}$, $\mathcal{S} = \{s_1, s_2, s_3\}$ and $X_1 = \{a, b\}$, $X_2 = \{a, c\}$ and $X_3 = \{b, c\}$. The matrix in below represents a (complete) 1-secure $(1, 2, 3)$-multithreshold scheme with secret size 3.

$$
\begin{array}{cccccc}
a & b & c & s_1 & s_2 & s_3
\end{array}
$$

$$
\begin{pmatrix}
0 & 0 & 0 & 0 & 0 & 0 \\
0 & 1 & 2 & 0 & 0 & 2 \\
0 & 2 & 1 & 0 & 0 & 1 \\
1 & 4 & 7 & 1 & 2 & 0 \\
1 & 5 & 6 & 1 & 2 & 2 \\
1 & 3 & 8 & 1 & 2 & 1 \\
2 & 8 & 5 & 2 & 1 & 0 \\
2 & 6 & 4 & 2 & 1 & 2 \\
2 & 7 & 3 & 2 & 1 & 1 \\
3 & 3 & 3 & 1 & 1 & 1 \\
3 & 4 & 5 & 1 & 1 & 0 \\
3 & 5 & 4 & 1 & 1 & 2 \\
4 & 7 & 1 & 2 & 0 & 1 \\
4 & 8 & 0 & 2 & 0 & 0 \\
4 & 6 & 2 & 2 & 0 & 2 \\
5 & 2 & 8 & 0 & 2 & 1 \\
5 & 0 & 7 & 0 & 2 & 0 \\
5 & 1 & 6 & 0 & 2 & 2 \\
6 & 6 & 6 & 2 & 2 & 2 \\
6 & 7 & 8 & 2 & 2 & 1 \\
6 & 8 & 7 & 2 & 2 & 0 \\
7 & 1 & 4 & 0 & 1 & 2 \\
7 & 2 & 3 & 0 & 1 & 1 \\
7 & 0 & 5 & 0 & 1 & 0 \\
8 & 5 & 2 & 1 & 0 & 2 \\
8 & 3 & 1 & 1 & 0 & 1 \\
8 & 4 & 0 & 1 & 0 & 0
\end{pmatrix}.
$$

Fig. 1. Matrix for Example 1.

Complete (t, k, k)-multithreshold schemes are just perfect threshold schemes as first studied in [2] and [15] (see Section 4). They can be used to show the existence of complete (t, k, n)-multithreshold schemes (and thus w-secure (t, k, n)-multithreshold schemes for $0 \leq w \leq n - k + t - 1$) through the following result:

Theorem 1. *Let $1 \leq t \leq k \leq n$. If there exists a complete (t,k,k)-multithreshold scheme with secret size q then there exists a complete (t,k,n)-multithreshold scheme with secret size q.*

Proof. Let r and the sets X_i be as defined earlier. For each i $(1 \leq i \leq r)$ let $\mathcal{F}_i$ be the distribution rules of a complete (t,k,k)-multithreshold scheme defined on participant set X_i with secret s_i. For each $(f_1,\ldots,f_r)$ $(f_i \in \mathcal{F}_i)$, define a new rule f as follows: for each $p \in \mathcal{P}$, $f(p)$ is the $\binom{n-1}{k-1}$–tuple with entries $f_i(p)$ for each i with $p \in X_i$, and for each s_i, $f(s_i) = f_i(s_i)$. The collection of these rules f form a complete (t,k,n)-multithreshold scheme with secret size q. □

Although Theorem 1 guarantees the existence of multithreshold schemes it produces schemes with the relatively large share size of $q^{\binom{n-1}{k-1}}$. We address this problem in the next section.

The special case of w-secure $(1,2,n)$-multithreshold schemes was first discussed in [4] and the generalisation to $(1,k,n)$-multithreshold schemes was studied in [13] where the use of symmetric functions was considered. The special case of symmetric polynomials was analysed in [5] for w-secure $(1,k,n)$ schemes. The later paper also produced a lower bound on the size of share that each participant holds, and an example of a scheme that achieves this bound. We will generalise this bound to the case of w-secure (t,k,n)-multithreshold schemes and discuss examples of schemes that achieve this bound.

We note that the application of a w-secure (t,k,n)-multithreshold scheme with $t > 1$ to a multi-user network as mentioned in the introduction will in general be different from the scenario for $t = 1$ discussed in [4, 5, 13]. The case $t = 1$ deals with the situation where any of the k participants can at any time *non-interactively* establish a common key. If it is required that members of a k-set of participants should reach some threshold of consensus before establishing their common key, then it is necessary that $t > 1$.

3 Bounds on Share Size

In the last section we showed that it is possible to find a w-secure (t,k,n)-multithreshold scheme with $1 \leq t \leq k \leq n$ and $0 \leq w \leq n-k+t-1$. Note however, that the construction in the proof of Theorem 1 is equivalent to simply giving each participant one share from each of the $\binom{n-1}{k-1}$ complete (t,k,k)-multithreshold schemes in which the participant is involved. We would like to be able to construct multithreshold schemes in which each participant has a share which is much smaller than that which arises from the construction of Theorem 1.

We will denote by $\mathcal{M}(t,k,n,w,q)$ the minimum size of share that a participant holds in any w-secure (t,k,n)-multithreshold scheme with secret size q taken over all schemes with these parameters.

The following well known result can be found in [17].

Result 2. *Let $1 \leq t \leq k$. Then $\mathcal{M}(t,k,k,t-1,q) \geq q$.*

Since complete (t,k,k)-multithreshold schemes with participant share size q can be found for all prime powers $q \geq k$ (so $\mathcal{M}(t,k,k,t-1,q) = q$ in this case, see Section 4.4) we can combine Result 2 and Theorem 1 to obtain the following:

Corollary 3. *Suppose* $1 \leq t \leq k \leq n$ *and* $0 \leq w \leq n-k+t-1$. *Let* q *be a prime power* $(q \geq k)$. *Then* $\mathcal{M}(t,k,n,w,q) \leq q^{\binom{n-1}{k-1}}$.

We now recall a result from [5].

Result 4. *Let* $0 \leq w \leq n-1$ *and* $k \leq n$. *Then*

$$\mathcal{M}(1,k,n,w,q) \geq q^{\binom{w+k-1}{k-1}}.$$

The main result of this paper is a generalisation of Result 4 to w-secure (t,k,n)-multithreshold schemes for any t $(1 \leq t \leq k)$ and $w \geq t-1$ (see Theorem 5). For $w < t-1$ the situation is slightly different; here we obtain a bound for the share size of a *group* of $t-w$ participants (see Corollary 8).

We first discuss two operations that can be performed on a multithreshold scheme.

Let $\mathcal{F}$ be the set of distribution rules of a w-secure (t,k,n)-multithreshold scheme with secret size q. Let M be the representation of $\mathcal{F}$ as a matrix and let $X \subseteq \mathcal{P} \cup \mathcal{K}$. The *restriction* of M at X is the matrix that is obtained from M by deleting the columns in X. Now let $f \in \mathcal{F}$. The *contraction* of M at X with respect to f is the matrix obtained from M by selecting only the rows of M that agree with f on the columns of X and then taking the restriction of the resulting matrix at X. Note that these are extensions of the definitions of restrictions and contractions of a perfect secret sharing scheme that were given in [12].

Theorem 5. *Let* $1 \leq t \leq k$ *and* $t-1 \leq w \leq n-k+t-1$. *Then*

$$\mathcal{M}(t,k,n,w,q) \geq q^{\binom{w+k-2t+1}{k-t}}.$$

Proof. If $t=1$ then the theorem follows from Result 4. Let $t \geq 2$ and let M be a w-secure (t,k,n)-multithreshold scheme with secret size q with a set $\mathcal{F}$ of distribution rules. Let X be a subset of $t-1$ participants and let $W_X = \{s_i \mid X \not\subseteq X_i,\ 1 \leq i \leq r\}$ (with r, X_i as described earlier). Let $f \in \mathcal{F}$. Construct the matrix M' formed from M by taking the contraction of M at X with respect to f and then taking the restriction of the resulting matrix at W_X. The rows of M' correspond to the distribution rules of a $(w-t+1)$-secure $(1,k-t+1,n-t+1)$-multithreshold scheme defined on the participants of $\mathcal{P}\backslash X$. M' also has secret size q since X lies in the prohibited structure of each secret in M. Applying Result 4 we see that

$$\mathcal{M}(t,k,n,w,q) \geq q^{\binom{(w-t+1)+(k-t+1)-1}{(k-t+1)-1}} = q^{\binom{w+k-2t+1}{k-t}},$$

as required. $\square$

Note that putting $w = n-k+t-1$ in Theorem 5 gives us the bound on each participant's share for the case of complete multithreshold schemes.

Corollary 6. *Let M be a complete (t,k,n)-multithreshold scheme with secret size q. Then each participant must have a share of size at least $q^{\binom{n-t}{k-t}}$.*

We finish this section by considering the remaining case of w-secure (t,k,n)-multithreshold schemes with $w \leq t-1$. As mentioned earlier, in this case we do not get a bound on an *individual* participant's share; instead we get a bound for a group of $t-w$ participants.

Let $\mathcal{F}$ be a set of distribution rules for a multisecret sharing scheme defined on participant set $\mathcal{P}$ and let $X = \{x_1, \ldots, x_u\} \subseteq \mathcal{P} \cup \mathcal{K}$. Let $\mathcal{S}_X = \{(f(x_1), \ldots, f(x_u)) \mid f \in \mathcal{F}\}$ and let $\sharp(x_1 \ldots x_u) = |\mathcal{S}_X|$.

Result 7 [6]. *Let $1 \leq t \leq k$, $0 \leq w \leq t-1$ and let M be a matrix for a w-secure (t,k,k)-multithreshold scheme with secret s of size q and participant set $\mathcal{P}$. Let $X \subseteq \mathcal{P}$ such that $|X| = t-w$. Then $\prod_{p \in X} |\mathcal{S}_p| \geq q$.*

Corollary 8. *Let $1 \leq t \leq k \leq n$ and $0 \leq w \leq t-1$. Let M be a matrix for a w-secure (t,k,n)-multithreshold scheme with secret size q and participant set $\mathcal{P}$. Let $X \subseteq \mathcal{P}$ such that $|X| = t-w$. Then $\prod_{p \in X} |\mathcal{S}_p| \geq q$.*

Proof. Let $X \subseteq \mathcal{P}$ with $|X| = t-w$. Let X_i be such that $X \subseteq X_i$. The restriction of M at $(\mathcal{P} \backslash X_i) \cup (\mathcal{K} \backslash s_i)$ is a w-secure (t,k,k) multithreshold scheme and the corollary now follows from Result 7. □

Note that when $w = t-1$, Theorem 5 and Corollary 8 both give the same lower bound q for the participant share size.

4 Optimal Multithreshold Scheme Constructions

In this section we discuss some constructions for multithreshold schemes. Let $1 \leq t \leq k \leq n$ and let $0 \leq w \leq n-k+t-1$. We will call a w-secure (t,k,n)-multithreshold scheme with secret size q *optimal* if one of the following holds:

1. The size of each participant's share meets the bound for $\mathcal{M}(t,k,n,w,q)$ given in Theorem 5 (for $t-1 \leq w \leq n-k+t-1$);
2. The share size of each set of $t-w$ participants meets the bound given in Corollary 8 (for $0 \leq w < t-1$).

4.1 Case $t = 1$

Optimal w-secure $(1,k,n)$-multithreshold schemes with secret size q were constructed in [5] using symmetric polynomials ($0 \leq w \leq n-k$). Such a scheme can be found for each prime power q ($q \geq n$). In fact, Example 1 was constructed using this method.

4.2 Case $t = 2$

We have the following result for the case $t = 2$:

Theorem 9 [11]. *Let $2 \leq k \leq n$ and q be a prime power such that $q > \binom{n-1}{k-2}+1$. Then there exists an optimal complete $(2, k, n)$-multithreshold scheme with secret size q.*

4.3 Case $t = k$

From Theorem 5 it follows that in any optimal w-secure (k, k, n)-multithreshold scheme with secret size q each participant holds a share of size q. Since the share size is independent of w it is of greatest interest to construct optimal complete (k, k, n)-multithreshold schemes. This can be done as follows:

Theorem 10. *Let $1 \leq k \leq n$ and $q \geq k$. Then there exists an optimal complete (k, k, n)-multithreshold scheme with secret size q.*

Proof. To construct an optimal complete (k, k, n)-multithreshold scheme with secret size q ($q \geq k$) proceed as follows. Let $\mathcal{P} = \{p_1, \ldots, p_n\}$. Let the q^n distribution rules $\mathcal{F}$ of the scheme be such that the set $\mathcal{F}(\mathcal{P}) = \{(f(p_1), \ldots, f(p_n)) \mid f \in \mathcal{F}\}$ is equal to the set $\{(x_1, \ldots, x_n) \mid x_i \in \boldsymbol{Z}_q\}$. Label the k-subsets of $\mathcal{P}$ by X_i ($1 \leq i \leq \binom{n}{k}$) and let X_i be associated with secret s_i ($1 \leq i \leq \binom{n}{k}$). Then for any $f \in \mathcal{F}$ and i ($1 \leq i \leq \binom{n}{k}$) let $f(s_i) = \sum_{p \in X_i} f(p) \pmod q$. The rules in $\mathcal{F}$ form an optimal complete (k, k, n)-multithreshold scheme with secret size q. □

4.4 Case $k = n$

Optimal complete (t, k, k)-multithreshold schemes with secret size q have been studied extensively in the literature (originally in [2, 15]). In such a scheme there is only one secret and every participant receives a share of size q. These schemes are normally referred to as *ideal* (t, k)-threshold schemes (see [7] or [10]). In [10] they were shown to be equivalent to a certain class of transversal designs. It is known that ideal (t, k)-threshold schemes with secret size q can be constructed for all prime powers $q \geq k$ (see for example [2, 15]). The case $w \leq t - 1$ is considered in the next subsection.

4.5 Case $w \leq t-1$

Optimal w-secure (t, k, k) multithreshold schemes with $1 \leq w \leq t - 1$ have been given in [3, 14]. These schemes are examples of *linear ramp schemes.*

In an optimal w-secure (t, k, n)-multithreshold scheme with $k \leq n$ and $w \leq t - 1$, it is not inconsistent with the definition that every set X of t participants can reconstruct not just the secrets that correspond to sets Y of size k such that $X \subseteq Y$, but in fact *all* of the $\binom{n}{k}$ secrets in $\mathcal{K}$. (However, from a more practical point of view it would seem unlikely that such a property would be desirable.)

Hence, a w-secure (t, n, n)-multithreshold scheme with secret s can be thought of as a w-secure (t, k, n)-multithreshold scheme where s is the secret associated with every k-set of participants. Thus we can obtain optimal w-secure (t, k, n)-multithreshold schemes from optimal w-secure (t, n, n)-multithreshold schemes. (In fact, a reverse correspondence can also be shown.)

5 Conclusions

We have introduced the general concept of a w-secure (t, k, n)-multithreshold scheme and given a lower bound on the size of share that a participant (or a group of participants) in such a scheme must hold. We have exhibited examples of *optimal* multithreshold schemes, that is, schemes which meet this lower bound. We note that some authors (for example [5, 6]) prefer to define secret sharing schemes in information theoretic terms. We have chosen to use a combinatorial model but we remark that all the results in this paper can be translated into equivalent information theoretic statements.

The authors acknowledge their useful discussions with Peter Wild regarding some of the ideas in this paper.

References

1. J. Benaloh and J. Leichter: Generalized Secret Sharing and Monotone Functions. Advances in Cryptology – Crypto '88, Lecture Notes in Comput. Sci. **403** (1990) 27–35
2. G. R. Blakley: Safeguarding cryptographic keys. Proceedings of AFIPS 1979 National Computer Conference **48** (1979) 313–317
3. G. R. Blakley and C. Meadows: Security of Ramp Schemes. Advances in Cryptology – Crypto '84, Lecture Notes in Comput. Sci. **196** (1985) 411–431
4. R. Blom: An Optimal Class of Symmetric Key Generation Systems. Advances in Cryptology – Eurocrypt'84, Lecture Notes in Comput. Sci. **209** (1984) 335–338
5. C. Blundo, A. De Santis, A. Herzberg, S. Kutten, U. Vaccaro and M. Yung: Perfectly-Secure Key Distribution for Dynamic Conferences. Presented at Crypto'92
6. C. Blundo, A. De Santis and U. Vaccaro: Efficient Sharing of Many Secrets. Proceedings of STACS '93, Lec. Notes in Comput. Sci. **665** (1993) 692–703
7. E. F. Brickell and D. M. Davenport: On the Classification of Ideal Secret Sharing Schemes. J. Cryptology **2** (1991) 123–124
8. E. F. Brickell and D. R. Stinson: Some Improved Bounds on the Information Rate of Perfect Secret Sharing Schemes. J. Cryptology **2** (1992) 153–166
9. M. Ito, A. Saito and T. Nishizeki: Secret Sharing Scheme Realizing General Access Structure. Proceedings IEEE Global Telecom. Conf., Globecom '87, IEEE Comm. Soc. Press (1987) 99–102
10. W.-A. Jackson and K. M. Martin: On Ideal Secret Sharing Schemes. Preprint
11. W.-A. Jackson, K. M. Martin and C. M. O'Keefe: A construction for multisecret threshold schemes. Preprint

12. K. M. Martin: New Secret Sharing Schemes from Old. To appear in J. Combin. Math. Combin. Comput.
13. T. Matsumoto and H. Imai: On the KEY PREDISTRIBUTION SYSTEM: a Practical Solution to the Key Predistribution Problem. Advances in Cryptology: Crypto '87, Lecture Notes in Comput. Sci. **293** (1987) 185–193
14. R. J. McEliece and D. V. Sarwate: On Sharing Secrets and Reed Solomon Codes. Comm. ACM **24** (1981) 583–584
15. A. Shamir: How to Share a Secret. Comm. ACM Vol 22 **11** (1979) 612–613
16. G. J. Simmons: An Introduction to Shared Secret and/or Shared Control Schemes and their Application. Contemporary Cryptology: The Science of Information Integrity, IEEE Press (1992)
17. D. R. Stinson: An Explication of Secret Sharing Schemes. Des. Codes Cryptogr. **2** (1992) 357–390

Secret Sharing Made Short

Hugo Krawczyk

IBM T.J. Watson Research Center
Yorktown Heights, NY 10598

Abstract. A well-known fact in the theory of secret sharing schemes is that shares must be of length *at least* as the secret itself. However, the proof of this lower bound uses the notion of information theoretic secrecy. A natural (and very practical) question is whether one can do better for secret sharing if the notion of secrecy is *computational*, namely, against resource bounded adversaries. In this note we observe that, indeed, one can do much better in the computational model (which is the one used in most applications).
We present an m-threshold scheme, where m shares recover the secret but $m-1$ shares give no (computational) information on the secret, in which shares corresponding to a secret S are of size $\frac{|S|}{m}$ plus a short piece of information whose length does not depend on the secret size but just in the security parameter. (The bound of $\frac{|S|}{m}$ is clearly optimal if the secret is to be recovered from m shares). Therefore, for moderately large secrets (a confidential file, a long message, a large data base) the savings in space and communication over traditional schemes is remarkable.
The scheme is very simple and combines in a natural way traditional (perfect) secret sharing schemes, encryption, and information dispersal. It is provable secure given a secure (e.g., private key) encryption function.

1 Introduction

Since their invention 15 years ago, secret sharing schemes [16, 2] have been extensively investigated. In particular, much work was done on the required length of the shares relative to the secret size. It is a well known basic fact that shares of a secret have to be at least of the size of the secret itself, and most of the work on share sizes investigates when this lower bound can be achieved or must be exceeded for different kind of schemes. Having shares of the size of the secret is not a serious problem as long as these secrets are short, e.g. a short secret key, as most traditional applications require. However, this effect of information replication among the participants of a distributed environment can be very space and communication inefficient if the secret is a large confidential file, a long message to be transmitted over unreliable links, or a secret data base shared by several servers. Applications like these are becoming more and more necessary.

The mentioned lower bound on the share size is related to the treatment of these secret sharing schemes in the sense of perfect (information theoretic) secrecy. A natural (and practical!) question is what can be done if the secrecy be not perfect but in a computational sense (i.e. against resource-bounded adversaries). [1] Could the shares in this case be made significantly shorter relative to the secret size?

In this note we present a solution to the above question which is surprising in two senses.

- The resultant scheme is extremely space (and communication) efficient. It realizes an m-threshold scheme, where m shares recover the secret but $m-1$ shares give no (computational) information on the secret, in which shares corresponding to a secret S are of size $\frac{|S|}{m}$ plus a short piece of information whose length does not depend on the secret size but just in the security parameter. The bound of $\frac{|S|}{m}$ is clearly optimal if the secret is to be recovered from m shares.
- The resultant solution is strikingly simple. It just combines in a natural way traditional secret sharing schemes, with encryption and information dispersal techniques.

Our scheme is simple, practical and provable secure given a secure private key encryption system (in particular, it just requires the existence of a one-way function). In addition, we present a secret sharing scheme with the same properties as above which is also *robust*, namely, malicious participants cannot prevent the reconstruction of the secret by a legal coalition, even if they return modified shares. (Clearly, the total number of malicious participants must be under some bound). The latter can be achieved by using public-key signatures or, in a much more efficient way, by using the recently introduced *distributed fingerprints* [9], together with the above mentioned techniques.

These constructions have many applications, especially in distributed scenarios where secrecy and integrity of information are to be protected. For example, consider a group of five servers sharing a data base of confidential information such that no pair of servers is allowed to learn about the information without the collaboration of a third one. (That is, the system tolerates up to two corrupted servers without compromising the information). Using a regular secret sharing scheme the amount of information stored in each server is equivalent to the whole data base size; in contrast, using our scheme each server keeps one third of the data base size (in other words, the total amount of information in the system has a 66% increase over the data base size in our solution compared to a 400% increase using the regular schemes). Same savings correspond to the amount of communication involved between the servers. Moreover, both storage and communications in this case are secret, and therefore the savings are even more significant.

A particularly interesting application of these space-efficient and robust secret sharing schemes is to the problem of *secure message transmission* defined in

[1] Computational secrecy is in no way a practical limitation. In fact, most implementations of theoretically perfect secret sharing schemes result in actual computational secrecy. This is the case, for example, when shares are encrypted for distribution or when the shares are produced with a pseudorandom generator.

[6]. There, two parties in an (incomplete) network try to communicate a confidential message. Part of the nodes of the network are controlled by an adversary that may want to derive information about the message as well as to corrupt it. The underlying idea in [6] (which investigates this problem in the information theoretic model) is that if n disjoint paths exist between the two parties, such that at most m of them are controlled by an adversary, then a solution to the problem is to decompose the message into n secret shares corresponding to an m-threshold scheme and transmit these shares over the n paths. In this solution, the complexity is given by the cost of computing, transmitting and storing these shares. Clearly, applying our solution significantly reduces this complexity relative to traditional secret sharing schemes.

2 Computational Secret Sharing

An (n, m)-secret sharing scheme is a randomized protocol for the distribution of a secret S among n parties such that the recovery of the secret is possible out of m shares for a fixed value $m, 1 \leq m \leq n$, while $m-1$ shares give no information on the secret S.

In this paper we deal with two different notions of secrecy according to the meaning of "no information" in the above formulation. One is *perfect* secrecy where "no information" is in the information theoretic sense; the other is *computational* secrecy where "no information" means no information that can be efficiently computed. We extend on these notions below.

Therefore, an (n, m)-secret sharing scheme consists of two processes; one the *distribution* process, the other the *reconstruction* process. The distribution process gets as input the secret S (and the values of n and m) and generates n shares $S_1, S_2, \ldots, S_n$, which are *privately* delivered to the system participants. The reconstruction process reconstructs the correct secret when input with any subset of m shares. Given a particular secret sharing algorithm A we denote by $A(S)$ the n resultant shares $S_1, S_2, \ldots, S_n$. Note that $A(S)$ is a random variable depending on the internal random coins of A.
For simplicity we do not formalize the exact domain of secrets and shares; in general this domain will depend on the specific scheme being used.

The notion of perfect secret sharing is well known and formalized in the literature. The notion of computational secret sharing, although very natural and widely used, is usually implicitly understood and less explicitly formalized. Although such a formalization is not the goal of this note, we outline here the basis for a formal definition. This is required in order to be able to prove that the particular construction we present is *secure*.

We start by outlining the definition of a secure encryption system. This has two reasons, one is that it is in such a secure function that the security of our construction relies; the other is that the definition of computational secret sharing schemes closely follows the definition of secure encryption. Rigorous definitions for secure encryption were first given in [8]. The reader is also referred to [7] for a detailed presentation of these notions, and for the treatment of the security of private key systems (as used in our paper).

The definitions presented here rely on the notion of *polynomial indistinguishability*.

Roughly speaking, two probability distributions are polynomially indistinguishable if any probabilistic polynomial-time algorithm behaves essentially the same when its input is selected from either of the two distributions. This notion is formalized by means of probabilistic polynomial time *tests* that output 0 or 1 as their guesses for whether the input comes from the first distribution or from the second. The condition for indistinguishability is that any such test will succeed in guessing the correct distribution with probability at most $\frac{1}{2}$ plus a negligible fraction (this probability depends on an equiprobable selection of any of the two distributions, the choice of the input according to the selected distribution and the internal coins of the test).

The formal notion of indistinguishability is an asymptotical one and has to be stated in term of collections of probability distributions indexed, in our case, by the lengths of messages or secrets. Under such formalism, a distinguishing algorithm is one that succeeds in guessing the correct distribution with probability $\frac{1}{2} + l^{-c}$, where l is the distribution index and c a positive constant.

Definition 1. (sketch)

1. Let ENC be an encryption function and M a message in the domain of ENC. Let $\{ENC_K(M)\}_K$ be the space of encryptions of the message M under all possible keys. By $\mathcal{D}_{ENC}(M)$ we denote the probability distribution on $\{ENC_K(M)\}_K$ as induced by the distribution under which keys are selected.
2. A (private-key) *encryption function ENC is secure* if for any pair of messages M' and M'' of the same length, the distributions $\mathcal{D}_{ENC}(M')$ and $\mathcal{D}_{ENC}(M'')$ are polynomially indistinguishable.

This definition, based on the notion of indistinguishability, is equivalent to the (possibly more natural) definition of *semantic security* that states that no information on a message can be derived, in polynomial-time, from seeing its encryption if the key is not known (except for a-priori knowledge on the message). Notice that the above is a weak definition in the sense, for example, that it does not contemplate security against an adversary with additional known-plaintext information. It turns out from our results that even this weak definition suffices for constructing secure and space efficient secret sharing schemes (basically, our scheme uses a "one-time key" for each secret).

Accurate definitions, and a proof of equivalence of the indistinguishability and semantic notions can be found in [7]. (As well as a precise distinction between uniform and non-uniform definitions, an important issue that we overlook here).

We now proceed to define a *computationally secure secret sharing scheme.*

Definition 2. (sketch)

1. Let CSS be an (n,m)-secret sharing scheme ('C' is for computational). For any secret S and for any set of indices $1 \leq i_1 \leq \cdots \leq i_r \leq n$, $1 \leq r \leq n$, let $\mathcal{D}_{CSS}(S, i_1, i_2, \ldots, i_r)$ denote the probability distribution on the set of shares $S_{i_1}, S_{i_2}, \ldots, S_{i_r}$ induced by the output of $CSS(S)$.
2. An (n,m)-secret sharing scheme is *computationally secure* if for any pair of secrets S' and S'' of the same length, and for any set of indices $i_1, i_2, \ldots, i_r$, $r < m$, the distributions $\mathcal{D}_{CSS}(S', i_1, i_2, \ldots, i_r)$ and $\mathcal{D}_{CSS}(S'', i_1, i_2, \ldots, i_r)$ are polynomially indistinguishable.

As in the case of encryption, also computational secret sharing schemes can be defined in the sense of semantic security. The equivalence with the above formulation is proved in a similar way to the encryption case.

A stronger definition can be stated in terms of a dynamic and adaptive adversary that progressively chooses the $m-1$ shares to be revealed to him depending on previously opened shares. Our construction satisfies also such a stronger definition.

Finally, notice that the traditional notion of perfect secret sharing can be defined in an analogous way to Definition 2 by replacing 'polynomially indistinguishable' with 'identical' (or equivalently, by replacing polynomial-time distinguishability tests with computationally unlimited tests).

3 Secret Sharing with Short Shares

In order to achieve a space efficient secret sharing scheme we combine an information dispersal scheme with a secure encryption scheme and a perfect (e.g. Shamir's) secret sharing scheme.

Information Dispersal was introduced by Rabin [15]. It is a scheme intended for the distribution of a piece of information among n active processors, in such a way that the recovery of the information is possible in the presence of m active processors (i.e. out of m fragments), where m and n are parameters satisfying $1 \leq m \leq n$. The scheme assumes that active processors behave honestly, i.e. returned fragments are unmodified. The basic idea is to add to the information, say a file F, some amount of redundancy and then to partition it into n fragments, each transmitted to one of the parties. Reconstruction of F is possible out of m (legitimate) fragments. Remarkably, each distributed fragment is of length $\frac{|F|}{m}$ which is clearly space optimal. Information dispersal schemes can be implemented in a variety of ways, all corresponding to the notion of erasure codes in the theory of error correcting codes (see [15, 13, 9]). We note that basic information dispersal schemes do not deal with malicious parties or with secrecy of information.

Remark: For completeness, we outline the following simple information dispersal scheme based on Reed Solomon erasure codes. The information to be shared is partitioned into m equal parts where each part is viewed as an element over a finite field (e.g. $GF(p)$, for a large enough prime p). These m elements are then viewed as coefficients of a polynomial of degree $m-1$, and the n fragments for distribution are obtained by evaluating this polynomial in n different points. Clearly the whole information can be reconstructed (by interpolation) from any m fragments. We stress that for large files, it is not necessary to work on a huge field but just to view the information as the concatenation of different polynomials. Notice the difference between such a scheme and Shamir's secret sharing in which the information is represented by the free coefficient and not by the whole polynomial (this is essential in Shamir's scheme to provide perfect secrecy but not for information dispersal where secrecy is not a concern).

We proceed to show how to build a space efficient secret sharing scheme. Let n denote the number of parties among which the secret is to be shared. Let m denote the threshold for our scheme, namely, m shares suffice to construct the secret but $m-1$ give no (computational) information on the secret. Let S denote the secret being shared.

We assume a generic information dispersal algorithm that we denote by IDA, and which works for parameters n (number of file fragments) and m (number of required fragments to reconstruct the file). We also assume a secure (length preserving) private key encryption function, denoted ENC, and a perfect (n, m)-secret sharing scheme (e.g. Shamir's) which we denote PSS. The space of secrets in our scheme is the same as the space of messages for the encryption function ENC.

Distribution Scheme:

1. Choose a random encryption key K. Encrypt the secret S using the encryption function ENC under the key K, let $E = ENC_K(S)$.
2. Using IDA partition the encrypted file E into n fragments, $E_1, E_2, \dots, E_n$.
3. Using PSS generate n shares for the key K, denoted $K_1, K_2, \dots, K_n$.
4. Send to each participant P_i, $i = 1, 2, \dots, n$ the share $S_i = (E_i, K_i)$. The portion K_i is privately transmitted to P_i (e.g. using encryption or any other secure way).

Reconstruction Scheme:

1. Collect from m participants P_{i_j}, $j = 1, 2, \dots, m$ their shares $S_{i_j} = (E_{i_j}, K_{i_j})$.
2. Using IDA reconstruct E out of the collected values E_{i_j}, $j = 1, 2, \dots, m$.
3. Using PSS recover the key K out of K_{i_j}, $j = 1, 2, \dots, m$.
4. Decrypt E using K to recover the secret S.

Theorem 3. *The above scheme constitutes a computationally secure (n, m)-secret sharing scheme provided that ENC is a secure encryption function and PSS a perfect secret sharing scheme. Each share S_i is of length $\frac{|S|}{m} + |K|$.*

Proof. The feasibility to reconstruct the encrypted secret E out of the m fragments E_{i_j} is inherited from the properties of the algorithm IDA. Also the reconstruction of the key K out of K_{i_j} is guaranteed by the secret sharing scheme PSS. Knowledge of E and K permits deriving S using the decryption function. The lengths of the shares comes from the lengths of the fragments and shares, respectively, in these schemes. (We assume the shares corresponding to the key K are of the same size as K. This is always possible given that $\log n < |K|$, a very reasonable assumption).

As for the secrecy against a coalition of $m-1$ shares, the intuitive idea is clear. The $m-1$ fragments corresponding to E give no more information on S than E itself. On the other hand, the $m-1$ key-shares give no information at all on K, therefore knowing E cannot help to learn about S.

A formal proof of the secrecy uses the following simulation argument.

Assume there exists a pair of secrets S' and S'' and an algorithm A that distinguishes (with significant probability) between the space of shares corresponding to S' and the space corresponding to S''. We construct an algorithm B to break the encryption function ENC in the sense that B can distinguish between the space of encryptions of S' and the space of encryptions of S''. When B is given an encrypted version of S' or of S'', call it E, it applies IDA on E to generate n fragments $E_1, E_2, \ldots, E_n$, produces at random $m-1$ shares according to the distribution of shares[2] (the key sharing is perfect!), and gives the fragments and "shares" as input to A. Now A outputs its guess for whether the secret corresponds to S' or S''; B outputs the same guess. Since A guesses correctly with significant probability over $1/2$ then B succeeds in its distinction with same probability. Therefore, B breaks the encryption function in the sense of indistinguishability.

□

We stress that the simplicity of the above proof has an important "practical" aspect. It permits a clear evaluation of the security of the scheme relative to the underlying encryption function even when this encryption function is "practically secure" (e.g. DES), rather than formally secure. Finally, we note that the issue of implementation of the "private channel" necessary between the dealer of the secret and its recipients is orthogonal to the aspect treated here. On the other hand, an insecure implementation of that channel compromises the security of the whole scheme. Using a secure encryption function (private or public key) for these channels the security of the whole system is easily provable.

Remark: Secret sharing schemes with shares shorter than the secret itself were also investigated under the information theoretic model. See, for example, [3] for a description of the so called *ramp schemes.* These schemes give up the perfect uncertainty on the secret provided by perfect secret sharing schemes in order to reduce the length of shares. Unfortunately, the security of these schemes is questionable (and insufficient) in many applications. Although one can show that the exact value of the secret cannot be learned as long as the number of shares revealed is under some threshold, there is a leakage of information with each opened share (below that threshold). The amount of leaked information is easy to measure but not the *significance* of this information, or the hardness of learning it. In other words, the approach is a quantitative one and not semantic as required in cryptography. As a simple example, an attacker to these schemes

[2] This distribution is polynomial-time samplable, e.g. use the PSS algorithm on a randomly chosen secret.

can *efficiently* discard a particular value (or even a set of values) for the secret just after seeing a number of shares much below the reconstruction threshold. In our scheme, on the contrary, if one uses a semantically secure encryption function this (or any other useful) information cannot be efficiently learned even from $m-1$ shares.

4 Robust Secret Sharing

The basic secret sharing scheme as introduced in section 3 assumes that share holders return correct shares. In many applications this assumption is too strong. This scenario, in which some shares can be (maliciously) corrupted, was investigated in many works (e.g. [10, 14, 17, 4]). A ***robust secret sharing scheme*** is a secret sharing scheme that can correctly recover the secret even in the presence of a (bounded) number of corrupted shares, while keeping the secrecy requirement.

In this case, in addition to the threshold parameter m, a bound t on the number of malicious parties in the system is to be specified. It is necessary that $t < m$ (a coalition of only malicious parties cannot reconstruct the secret), and $m \leq n - t$ (there are enough good parties to reconstruct the secret). This two relations imply that $2t < n$, i.e. a majority of honest parties is required. Clearly, it is not possible anymore to require that *any* subset of m parties can recover the secret (since part of them can be faulty). Instead we require that any coalition *containing* m honest parties (i.e. from any subset of shares containing at least m correct shares) can reconstruct the secret.

Our goal in this section is to present a robust secret sharing scheme which preserves the space efficiency of the construction described in section 3.

The first solution to the problem of designing robust secret sharing schemes was presented by McElice and Sarwate [10] where error correcting code techniques are used to enhance the original Shamir's scheme against share corruption. It tolerates up to $n/3$ cheaters and the security is unconditional. That is, secrecy is perfect, recovery is guaranteed (with probability 1), and no computational assumptions are done. Their solution cannot be applied to our needs since it requires shares (at least) as long as the secret itself. The same (space) drawback exists in all other solutions designed against computationally unlimited adversaries. A different approach is used by Rabin in [14], where shares are fingerprinted in order to detect possible alterations. Since fingerprints intended to work against resource bounded adversaries can be small and unrelated to the information length, this approach can be used to keep the space efficiency of our construction. Rabin's solution uses public key signatures for fingerprinting. (Notice that fingerprints based on private keys are unsuitable since different – mutually suspicious – parties need to verify the shares).

Therefore, our construction of Section 3 is modified such that at time of share distribution both the fragments E_i corresponding to the encrypted secret E as well as the key shares K_i are signed (using the private signing key of the dealer). When the secret is reconstructed, the public verification key for that dealer is used to verify the correctness of the shares. The total amount of information added to each share depends only on the security parameter and not on the shares or secret themselves. Therefore, the efficiency of our scheme is preserved.

This solution, although space efficient, requires the implementation of a public key system to support public signatures. This has a significant cost in administration, key management and computation. In addition, it requires to know the identity of the dealer that generated the secret (which is not always desirable) and a time-stamp mechanism to avoid replay attacks. (Notice that in our basic solution no need for a public-key system exists).

The above drawbacks related to the use of public-key signatures are overcome using the recently introduced *distributed fingerprints* [9] that permit fingerprinting the information through a method that requires no public key system, uses no secret keys at all, is time and space efficient, does not require the signer's identity or time-stamp techniques. It just requires a global public one-way hash function and the existence of a majority of honest parties. The latter condition is also part of the requirements for any robust secret sharing scheme, and therefore does not constitute a limitation here. Moreover, the distributed fingerprint scheme uses distribution among the parties in the system for fingerprint protection, which fits naturally in the secret sharing model. We refer to [9] for details on these fingerprints.

5 Further Work

Our results have many immediate applications (examples appear in the introduction). We expect also less immediate applications to emerge in the future.

In addition, many of the questions investigated for traditional secret sharing schemes are relevant to space-efficient computational secret sharing schemes. We mention here two questions which seem particularly attractive.

In this paper we have dealt with space efficient threshold schemes. More general schemes classified according to their *access structures* (cf. [1]) are investigated in the literature, mostly in the context of perfect secrecy. A natural question is whether the space efficiency can be carried over more general access structures than just threshold schemes. Our scheme can be easily extended to deal with some of these structures but it is not clear how general these structures can be. Since in our approach we apply regular secret sharing schemes to the sharing of the encryption key, then this part of the protocol can be treated as for traditional secret sharing. In this sense the question reduces to deal with access structures for information dispersal. The interested reader is referred to [12] that deals with information dispersal over arbitrary graphs.

Another question is whether *verifiable secret sharing* can be done in a space-efficient way. While robust secret sharing deals with potentially corrupted share holders, verifiable secret sharing deals also with corrupted dealers of the secret that can distribute inconsistent shares in order to prevent legal coalitions of reconstructing the secret (cf. [5]). We mention here, very briefly, the strong relation between space efficient verifiable secret sharing and fair cryptography [11]. A more desirable approach than sharing the keys used to encrypt information with escrow agencies, is to share with them each individual message. (Here 'to share' is in a sense analogous to that of 'secret sharing', namely, the information can be acceded only if a number of escrow agencies collaborate to do that). The drawback with sharing the key is that once a key is "opened" all messages (past and future) encrypted with that key can be open. Sharing individual messages solves this problem, but it is impractical to realize it through regular secret sharing schemes which require the replication of the amount of information for encryption, transmission and storage. In this sense the need for short shares is clear. On the other hand, verifiability is also necessary, otherwise the escrow agencies can be given shares that do not reconstruct the message. Further elaboration of this application is beyond the scope of this paper.

Acknowledgement

I thank Mihir Bellare, Oded Goldreich and Moti Yung for very helpful discussions.

References

1. Benaloh, J. and Leichter J., "Generalized secret sharing and monotone functions", Proc. Crypto '88, pp. 27-35.
2. Blakley, G.R., "Safeguarding Cryptographic Keys", *Proc. AFIPS 1979 National Computer Conference*, New York, Vol. 48, 1979, pp. 313-317.
3. Blakley, G.R., and Meadows C. "Security of Ramp Schemes", in *Lecture Notes in Computer Science 196; Advances in Cryptology: Proc. Crypto '84*, Springer-Verlag, 1985, pp.242-268.
4. Brickel, E.F., and Stinson, D.R., "The Detection of Cheaters in Threshold Schemes", in *Lecture Notes in Computer Science 403; Advances in Cryptology: Proc. Crypto '88*, Springer-Verlag, 1990, pp.564-577.
5. Chor, B., S. Goldwasser, S. Micali, and B. Awerbuch, "Verifiable Secret Sharing and Achieving Simultaneity in the Presence of Faults", *Proc. 26th FOCS*, 1985, pp. 383-395.
6. Dolev, D., Dwork, C., Waarts, O., and Yung, M., "Perfectly Secure Message Transmission", *Proc. 31st IEEE Symp. on Foundations of Computer Science*, 1990, pp. 36-45.
7. Goldreich, O., "A Uniform-Complexity Treatment of Encryption and Zero-Knowlege", *Jour. of Cryptology*, Vol. 6, No. 1, 1993, pp.21-53.
8. Goldwasser, S., and S. Micali, "Probabilistic Encryption", *JCSS*, Vol. 28, No. 2, 1984, pp. 270-299.
9. Krawczyk, H., "Distributed Fingerprints and Secure Information Dispersal", *Proc. of 12th. PODC*, pp. 207-218, 1993.
10. McElice R.J., and Sarwate, D.V., "On Sharing Secrets and Reed-Solomon Codes", *Comm. ACM*, Vol. 24, No. 9, 1978, pp. 583-584.
11. Micali, S., "Fair Public-Key Cryptosystems", Crypto '92.
12. Naor, M., and Roth, R.M., "Optimal File Sharing in Distributed Networks", *Proc. 32nd IEEE Symp. on Foundations of Computer Science*, 1991, pp. 515-525.

13. Preparata, F.P., "Holographic Dispersal and Recovery of Information", *IEEE Trans. on Information Theory*, IT-35, No. 5, 1989, pp. 1123-1124.
14. Rabin, M.O., "Randomized Byzantine Agreement", *24th FOCS*, pp. 403-409, 1983.
15. Rabin, M.O., "Efficient Dispersal of Information for Security, Load Balancing, and Fault Tolerance", *Jour. of ACM*, Vol. 36, No. 2, 1989, pp. 335-348.
16. Shamir, A., "How to Share a Secret", *Comm. ACM*, Vol. 22, No. 11, 1979, pp. 612-613.
17. Tompa, M. and H. Woll, "How to share a secret with cheaters", *Journal of Cryptology*, Vol 1, 1988, pp 133-138.

A Subexponential Algorithm for Discrete Logarithms over All Finite Fields

Leonard M. Adleman and Jonathan DeMarrais

Department of Computer Science, University of Southern California, Los Angeles CA 90089

Abstract. There are numerous subexponential algorithms for computing discrete logarithms over certain classes of finite fields. However, there appears to be no published subexponential algorithm for computing discrete logarithms over all finite fields. We present such an algorithm and a heuristic argument that there exists a $c \in \Re_{>0}$ such that for all sufficiently large prime powers p^n, the algorithm computes discrete logarithms over GF(p^n) within expected time:

$$e^{c(\log(p^n)\log\log(p^n))^{1/2}}$$

1 Introduction

Given α, β in a finite field, the discrete logarithm problem is to calculate an $x \in Z_{\geq 0}$ (if such exists) such that:

$$\alpha^x = \beta$$

Interest in the discrete logarithm problem first arose when Diffie and Hellman proposed a public key cryptographic system based on the complexity of this problem[DH]. Additional systems using discrete logarithms have since been proposed, including ElGamal's crypto-system[El1]. Recently the goverment has proposed using a system of this type as a standard. These present systems are based on finite fields of special form for which subexponential algorithms already exist. However, it is likely that these systems can be generalized to work with arbitrary finite fields. Previously, no subexponential algorithms existed for **all** such fields. We present such an algorithm along with a heuristic argument that there exists a $c \in \Re_{>0}$ such that for all sufficiently large prime powers p^n, the algorithm computes discrete logarithms over GF(p^n) within expected time:

$$e^{c(\log(p^n)\log\log(p^n))^{1/2}}$$

There exist several algorithms which for all primes $p \in Z_{>0}$ compute discrete logarithms over GF(p) in time subexponential in p (e.g. [Ad1, Go1]). Further, for all primes $p \in Z_{>0}$, there exists algorithms which for all $n \in Z_{>0}$ computes discrete logarithms over GF(p^n) in time subexponential in p^n (for $p = 2$, this was first shown by Hellman and Reyneri [HR] and improved by Coppersmith [Co]; however, these approaches appear to generalize to an arbitrary prime p).

ElGamal [El2] has given an algorithm which for all primes $p \in Z_{>0}$ compute discrete logarithms over GF(p^2) in time subexponential in p^2. Previously, the most general subexponential algorithm appears to be that of Lovorn [Lo] which computes discrete logarithms in GF(p^n) for $\log(p) \leq n^{0.98}$.

Our subexponential method for all finite fields actually consists of two algorithms. They both may be described as 'index calculus' methods [WM, Od]. The first algorithm is for the case $n < p$. Here, GF(p^n) is represented by $O/(p)$ where O is a number ring and (p) is the prime ideal generated by p. An element of $O/(p)$ is considered 'smooth' iff when considered as an element of O, the ideal it generates factors into prime ideals of small norm. The second algorithm is for the case $n \geq p$. Here, GF(p^n) is represented by $(Z/pZ[x])/(f)$ where $f \in Z/pZ[x]$ is irreducible. An element of $(Z/pZ[x])/(f)$ is considered 'smooth' iff when considered as an element of $Z/pZ[x]$ it factors into irreducible polynomials of small degree. The second algorithm is rather 'routine'. The first algorithm makes use of the notions of singular integers and character signatures which were introduced in the context of integer factoring [Ad2]. The first algorithm can be thought of as reducing the computation of discrete logarithms in GF(p^n) to the computation of discrete logarithms in several fields of the form GF(p') where $p' \in Z_{>0}$ is prime.

2 Preliminaries

In this section some basic facts are presented.

2.1 Singular Integers and Character Signatures

Here, some notions about presented in [Ad2] about integer factoring are generalized.

Definition: For all number fields K with ring of integers O, for all $s \in Z_{>0}$, and for all $\sigma \in O$, σ is an s-singular integer (with respect to O) iff there exists an ideal $I \subseteq O$ such that $(\sigma) = I^s$.

Let K be a number field with ring of integers O, unit group E and ideal class group C. Let $s \in Z_{>0}$ and let σ, τ be s-singular integers. Define $\sigma \approx \tau$ iff there exists $\alpha, \beta \in O$ such that $\alpha^s\sigma = \beta^s\tau$. "$\approx$" is an equivalence relation on s-singular integers, and the set of equivalence classes form a group $G(s)$ of exponent dividing s with identity $I(s) = \{\alpha^s | \alpha \in O\}$ under the operation:

$$[\alpha][\beta] \mapsto [\alpha\beta]$$

There is a homomorphism ψ from $G(s)$ onto the group $C(s) = \{c | c \in C \;\&\; c^s = [(1)]\}$.

$$[\alpha] \overset{\psi}{\mapsto} [I]$$

where $(\alpha) = I^s$.

The kernel of ψ, $\text{Ker}(\psi) = \{[u] | u \in E\}$ and consequently $\text{Ker}(\psi) \cong E/E^s$. Hence

$$(*) \quad G(s) \cong E/E^s \bigoplus C(s)$$

Definition: For all number fields K with ring of integers O, for all $s \in Z_{>0}$, for all prime ideals $P_1, P_2, \ldots, P_z \subset O$, for all $l_1, l_2, \ldots, l_z \in O$ and for all $\sigma \in O$: if for $i = 1, 2, \ldots, z$, $(\sigma) + P_i = (1), s | (N(P_i) - 1)$ and $l_i + P_i$ is a primitive s^{th} root of unity in O/P_i^*, then the s-character signature of σ with respect to $< P_1, l_1 >, < P_2, l_2 >, \ldots, < P_z, l_z >$ is: $< e_1, e_2, \ldots, e_z >$ where for $i = 1, 2, \ldots, z$, $\sigma^{(N(P_i)-1)/s} \equiv l_i^{e_i} \bmod P_i$ and $e_i \in Z_{\geq 0}^{<s}$.

Now assume that K is Abelian over Q, then it follows from the Čebotarev density theorem that for all $s \in Z_{>0}$, for all prime ideals $P_1, P_2, \ldots, P_z \subset O$, and for all $c \in G(s)$, there exists a $\sigma \in O$ such that $[\sigma] = c$ and for $i = 1, 2, \ldots, z$, $(\sigma) + P_i = (1)$. For $< P_1, l_1 >, < P_2, l_2 >, \ldots, < P_z, l_z >$ as above, let the map θ take c to the s-character signature of σ with respect to $< P_1, l_1 >, < P_2, l_2 >, \ldots, < P_z, l_z >$. θ is well defined on $G(s)$ and is a group homomorphism into $\bigoplus_{i=1}^{z} Z_s$.

2.2 Subfields of cyclotomic fields

Let $q \in Z_{>0}$ be prime and let $n | q - 1$, then there exists a unique field $K_{q,n} \subseteq Q(\zeta_q)$, the qth cyclotomic field, such that $[K_{q,n} : Q] = n$. The following are well known [Ed]:

1. The ring of integers of $K_{q,n}$, $O_{q,n} = Z[\eta_0, \eta_1, \ldots, \eta_{n-1}]$, where for $i = 0, 1, \ldots, n-1$, $\eta_i = \eta_{q,n,i} = \sum \zeta_q^a$, where the sum is taken over the set of $a \in Z_{>0}^{\leq q-1}$ such that $ind(a) \equiv i \bmod n$ where $ind(a)$ denotes the index of a in Z/qZ^* with respect to a fixed generator.
2. $K_{q,n} = Q(\eta_0)$ (however, there exist q, n such that $O_{q,n} \neq Z[\eta_0]$).
3. The minimum polynomial for η_0 over Q is $f = f_{q,n} = \prod_{i=0}^{n-1} (x - \eta_i)$
4. If $p \in Z_{>0}$ is prime and p is inert in $K_{q,n}$, then $O_{q,n}/(p)$ is a finite field with p^n elements and

$$R = R_{q,n,p} = \{\sum_{i=0}^{n-1} a_i \eta_i | a_i \in Z_{\geq 0}^{<p}, i = 0, 1, \ldots, n-1\}$$

is a complete set of representatives.

Arithmetic in $K_{q,n}$ may be done as follows (our description is essentially that of Edwards [Ed] which in turn is derived from Kummer).

Elements in $O_{q,n}$ will be represented in terms of the integer basis η_0, η_1, ..., η_{n-1}.

First, for $i, j, k \in Z_{\geq 0}^{\leq n-1}$ calculate $c_{i,j,k} \in Z$ such that:

$$\eta_i\eta_j = \sum_{k=0}^{n-1} c_{i,j,k}\eta_k$$

then multiplication in $O_{q,n}$ is straightforward.

Prime ideals of $O_{q,n}$ will be represented as follows. Let $s \neq q$ be a rational prime and let f be the order of s in Z/qZ^*. Let $e = (q-1)/f$, then the splitting field of s is $K_{q,e}$. Let $g = (e,n)$, then s splits into g distinct prime ideals of residue class degree n/g in $O_{q,n}$.

Let $h \in Z/sZ[x]$ be an irreducible factor of $f_{q,q-1} = x^{q-1} + \ldots + x + 1$ (the qth cyclotomic polynomial) and let σ be a generator for $\mathrm{GAL}(Q(\zeta_q)/Q)$ (the construction which follows produced the correct outcome for all choices of σ).

For $i = 1, 2, \ldots, g$, let $\tilde{S}_i \subseteq O_{q,q-1}$ be the prime ideal generated by s and $(h(\zeta_q))^{\sigma^i}$, and let $S_i = \tilde{S}_i \cap O_{q,n}$. Then $(s) = \prod_{i=1}^{g} S_i$ is the prime decomposition of s in $O_{q,n}$.

For $i = 1, 2, \ldots, g$, $j = 0, 1, \ldots, e-1$, calculate $u_{i,j} \in Z_{\geq 0}^{<s}$ such that

$$u_{i,j} \equiv \eta_{q,e,j} \bmod \tilde{S}_i$$

(such $u_{i,j}$ always exist [Ed]). Let $U = \{u_{i,j} | j = 0, 1, \ldots, e-1\}$ (U is the set of roots of $f_{q,e}$ mod s and is independent of i). Let $\psi_i = \prod_{j=0}^{e-1} \prod_{u \in U, u \neq u_{i,j}} (u - \eta_{i,j})$. For $i = 1, 2, \ldots, g$, $< s, \psi_i >$ will represent the prime ideal S_i of $O_{q,n}$ lying above s.

Let $\alpha \in O_{q,n}$, and let $a \in Z_{\geq 0}$. Then:

$$S_i^a | (\alpha)$$
iff
$$S_i^a O_{q,q-1} | \alpha O_{q,q-1}$$
iff
$$\tilde{S}_i^a | \alpha O_{q,q-1}$$
iff
$$p^a | \psi_i^a \alpha$$

The penultimate statement follows from Galois theory by noting that $\alpha \in K_{q,n}$. The last statement is essentially the first proposition of section 4.10 [Ed]. Hence there is a computationally efficient method for determining the power of S_i which divides (α).

Next consider singular integers and character signatures in $K_{q,n}$. Let $s \in Z_{>0}$. By Dirichlet's unit theorem, E/E^s can be written as the direct sum of at most n cyclic groups. Because the class number of $K_{q,n}$ is less than or equal to the class number of $Q(\zeta_q)$ ([Wa], Thrm 10.1), which is less than or equal to q^{q^3} [Ne], it follows that $C(s)$ can be written as the direct sum of at most $q^3 \log_2(q)$ cyclic groups. By (*) above, $G(s)$ can be written as the direct sum of at most $n + q^3 \log_2(q)$ cyclic groups. Let $H = n + q^3 \log_2(q) + 1$. If $\sigma_1, \sigma_2, \ldots, \sigma_H$ are s-singular integers then there exist $\delta \in O_{q,n}$ and $b_1, b_2, \ldots, b_H \in Z_{\geq 0}^{<s}$ such

that $\mathrm{GCD}(b_1, b_2, \ldots, b_H) = 1$ and $\prod_{j=1}^{H} \sigma_j^{b_j} = \delta^s$. Further, if $\theta_1 = \theta(\sigma_1), \theta_2 = \theta(\sigma_2), \ldots, \theta_H = \theta(\sigma_H)$ are the s-signatures of $\sigma_1, \sigma_2, \ldots, \sigma_H$ with respect to some $< P_1, l_1 >, < P_2, l_2 >, \ldots, < P_z, l_z >$ then $\sum_{j=1}^{H} b_j\theta_j = 0$. Finally, given the prime factorization of s and given the s-signatures $\theta_1, \theta_2, \ldots, \theta_H$, it can be shown that there exists an algorithm to calculate a sequence of b_j's, such that $\sum_{j=1}^{H} b_j\theta_j = 0$. This algorithm requires time at most $O(H^2 z \log^3(s))$.

2.3 Smooth numbers

For all $\gamma \in \Re_{>0}^{\leq 1}$ and $\delta \in \Re_{>0}$, $L_x[\gamma, \delta]$ denotes the set of functions from $\Re$ to $\Re$ of the form [CEP]:

$$e^{(\delta+o(1))(\log(x))^{\gamma}(\log\log(x))^{1-\gamma}} \quad x \to \infty$$

It will be helpful in the running time analyses which follow to note that for all $\gamma \in \Re_{>0}^{\leq 1}$, $\delta \in \Re_{>0}$, $L \in L_x[\gamma, \delta]$ and $c \in Z_{>0}$:

$$(\log(x)^c)L \in L_x[\gamma, \delta]$$

For all $\alpha, \gamma \in \Re_{>0}^{\leq 1}$ with $\alpha < \gamma$, for all $\beta, \delta \in \Re_{>0}$, $L_0 \in L_x[\gamma, \delta]$ and $L_1 \in L_x[\alpha, \beta]$, there exists an $L_2 \in L_x[\gamma-\alpha, (\gamma-\alpha)\delta/\beta]$ such that for all $N \in \Re_{>0}$, the probability that a positive integer less than or equal to $L_0(N)$ is $L_1(N)$-smooth is at least $1/L_2(N)$ ($L_1(N)$-smooth means all positive prime divisors are less than or equal to $L_1(N)$).

Notation

For all $p, n \in Z_{>0}$ with p prime, if we write $f \in Z/pZ[x]$ then it will be assumed that $f = \sum_{i=0}^{n} a_i x^i$ where for $i = 1, 2, \ldots, n$, $a_i \in Z_{\geq 0}^{<p}$.

3 Algorithm I

This algorithm will be used for discrete logarithms over $\mathrm{GF}(p^n)$ when $p > n$.

First, the discrete logarithm problem over $\mathrm{GF}(p^n)$ will be reduced to the discrete logarithm problem over special finite fields of the form $O_{q,n}/(p)$ (see Preliminaries section).

Let $p \in Z_{>0}$ be prime and $f_1 \in Z/pZ[x]$ irreducible, monic of degree n. Then $(Z/pZ[x])/(f_1)$ is a finite field with p^n elements. Let $\alpha_1, \beta_1 \in Z/pZ[x]$ of degree less than n such that $[\alpha_1]$ generates $(Z/pZ[x])/(f_1)^*$ and $\beta_1 \not\equiv 0 \bmod f_1$. (If α_1 is not a generator, randomly choose one, and solve for both α_1 and β_1.) Hence there exists an x such that $0 \leq x \leq p^n - 1$ and $\alpha_1^x \equiv \beta_1 \bmod f_1$. Assume that $p, f_1, \alpha_1, \beta_1$ are given and x is sought. Then one may proceed as follows:

Using the construction in [AL] find an $f \in Z/pZ[x]$ irreducible of degree n in random time polynomial in $\log(p)$ and n (assuming ERH). Hence $(Z/pZ[x])/(f) \cong (Z/pZ[x])/(f_1)$. Using [Le] calculate α_2 and $\beta_2 \in Z/pZ[x]$ of degree less than n such that $[\alpha_2]$ is the image of $[\alpha_1]$ and $[\beta_2]$ is the image of $[\beta_1]$ under this isomorphism. Hence our original problem is reduced to the problem: given p, f, α_2, β_2 with $[\alpha_2]$ generating $(Z/pZ[x])/(f)^*$ and $\beta_2 \not\equiv 0 \bmod f$, calculate x such that $0 \leq x \leq p^n - 1$ and $\alpha_2^x \equiv \beta_2 \bmod f$.

By the construction in [AL] (also see [BS]), there exists a $\tilde{c} \in Z_{>0}$ such that $f = f_{q,n}$ for some prime $q \in Z_{>0}$ with $q \leq \tilde{c}n^4(log(np))^2$ (assuming ERH). Since f is irreducible in $Z/pZ[x]$, it follows that p is inert in $K_{q,n}$. There exists the following isomorphism from $(Z/pZ[x])/(f)$ to $O_{q,n}/(p)$:

$$[\sum_{i=0}^{n-1} g_i x^i] \mapsto [\sum_{i=0}^{n-1} g_i (\sum_{j=0}^{n-1} d_{i,j}\eta_{q,n,j})]$$

where for $i = 0, 1, \ldots, n-1$, $\eta_{q,n,0}^i = \sum_{j=0}^{n-1} d_{i,j}\eta_{q,n,j}$, where $d_{i,j} \in Z$.

Calculate $\alpha_3, \beta_3 \in O$ such that $[\alpha_3]$ is the image of $[\alpha_2]$ and $[\beta_3]$ is the image of $[\beta_2]$ under this isomorphism. By reducing coefficients modulo p find $\alpha, \beta \in R_{q,n,p}$ such that $\alpha \equiv \alpha_3 \bmod p$ and $\beta \equiv \beta_3 \bmod p$. Hence the original problem becomes that of calculating x such that $0 \leq x \leq p^n - 1$ and $\alpha^x \equiv \beta \bmod p$.

Below, a family of algorithms $\{A_y\}_{y \in Z_{>0}}$ is presented. It will be argued that for sufficiently large y, A_y on all inputs q, n, p, α, β such that $p, q \in Z_{>0}$ are prime, $n < p$, $n | q-1$, $q \leq \tilde{c}n^4(log(np))^2$, p inert in $K_{q,n}$, and α, $\beta \in R_{q,n,p}$ with $[\alpha]$ generating $O_{q,n}/(p)^*$ and $\beta \not\equiv 0 \bmod p$, outputs x such that $0 \leq x \leq p^n - 1$ and $\alpha^x \equiv \beta \bmod p$.

Let $L_0 \in L_x[1/2, \sqrt{1/2}]$.

Algorithm A_y

Stage 0 input q, n, p, α, β

Stage 1 Set $N = p^{yn}$. Set (the 'smoothness bound') $B = L_0(N)$. Set $H = n + q^3 \log_2(q) + 1$.

Stage 2 Calculate $T = \{I | I$ is a prime ideal of O, $q \notin I$ and I lies over a rational prime $< B\}$. Let $w = \#T$ and let $< I_1, I_2, \ldots, I_w >$ be an ordering of T.

Stage 3 Set $j = 1$. While $j \leq H$:

Stage 3(a) Set $z = 1$. While $z \leq w + 1$: Choose random r, s with $0 \leq r, s \leq p^n - 1$ and calculate $\gamma \in R_{q,n,p}$ such that $\gamma \equiv \alpha^r \beta^s \bmod (p)$. If $(\gamma) = \prod_{i=1}^{w} I_i^{e_i}$ (i.e. if the ideal generated by γ is B-smooth) then set $\gamma_{j,z} = \gamma$, $r_{j,z} = r$, $s_{j,z} = s$, $v_{j,z} =< e_1, e_2, \ldots, e_w >$ and $z = z + 1$.

Stage 3(b) Calculate $a_1, a_2, \ldots, a_{w+1} \in Z_{\geq 0}^{<p^n-1}$ such that GCD(a_1, a_2, $\ldots$, a_{w+1}) = 1 and $\sum_{i=1}^{w+1} a_i v_{j,i} \equiv < 0, 0, \ldots, 0 > \bmod p^n - 1$. Calculate $\sigma_j = \prod_{i=1}^{w+1} \gamma_{j,i}^{a_i}$. Set $j = j + 1$.

Stage 4 For $j = 1, 2, \ldots, H$, calculate θ_j the $(p^n - 1)$-signature of σ_j with respect to $< S_1, m_1 >, < S_2, m_2 >, \ldots, < S_{2H}, m_{2H} >$. For $j = 1, 2, \ldots, H$, $k = 1, 2, \ldots, 2H$, $S_k \subset O_{q,n}$ is a prime ideal such that $(\sigma_j) + S_k = (1)$, $(p^n - 1) | N(S_k) - 1$ and m_k is a primitive $(p^n - 1)^{th}$ root of unity in O/S_k.

Stage 5 Calculate $b_1, b_2, \ldots, b_H \in Z_{\geq 0}^{<p^n-1}$ such that $GCD(b_1, b_2, \ldots, b_H) = 1$ and $\sum_{j=1}^{H} b_j \theta_j \equiv < 0, 0, \ldots, 0 > \bmod (p^n - 1)$.

Stage 6 Calculate $k = \sum_{j=1}^{H} \sum_{i=1}^{w+1} (r_{j,i} a_i b_j)$ and $l = \sum_{j=1}^{H} \sum_{i=1}^{w+1} (s_{j,i} a_i b_j)$. If $\alpha^k \beta^l \not\equiv 1 \bmod (p)$ then go to stage 3.

Stage 7 If $(l, p^n - 1) \neq 1$ then go to stage 3, Else calculate and output $x \equiv -k/l \bmod p^n - 1$ and halt.

4 Analysis of Algorithm I

In this section computational details of Algorithm I will be described and there will be an analysis of the expected number of steps required by the algorithm on all inputs q, n, p, α, β such that $p, q \in Z_{>0}$ are prime with $n < p$, $n | q-1$, $q \leq \tilde{c} n^4 (log(np))^2$, p inert in $K_{q,n}$, and $\alpha, \beta \in R_{q,n,p}$ with $[\alpha]$ generating $O_{q,n}/(p)^*$ and $\beta \not\equiv 0 \bmod p$. For convenience the argument will be for p^n sufficiently large.

To begin, consider the expected number of steps required by a single pass through each of the stages of the algorithm.

The time required for stages 0,1,6 and 7 are dominated by the time required by other stages.

Stage 2: Test all numbers less than or equal to B for primality. For each prime $s \neq q$ found, calculate the representatives $< s, \psi_i >$ of the prime ideals of $O_{q,n}$ lying above s and add them to T (see Preliminaries section).

Using random polynomial time primality testing [SS, AH] and random polynomial time finite field polynomial factorization [Be] and observing that because of the size constraints on q, orders can be computed naively, it follows that there exists an $L_1 \in L_x[1/2, \sqrt{1/2}]$ such that the expected number of steps for a pass through stage 2 is at most $L_1(N)$.

Further, since each rational prime has at most n primes lying over it in $O_{q,n}$, it follows that there exists an $L_2 \in L_x[1/2, \sqrt{1/2}]$ such that $w = \#T \leq L_2(N)$.

Stage 3(a): A γ will be tested for B-smoothness by the following method: First the norm of γ will be calculated and tested for B-smoothness. Those γ which have B-smooth norms will then be factored as ideals (see Preliminaries section).

A bound on the norm of γ will be needed.

$$\gamma = \sum_{i=0}^{n-1} g_i \eta_i,$$

where $0 \leq g_i \leq p-1$ for $i = 0, 1, \ldots, n-1$. Hence γ is the sum of $q-1$ terms each of the form $g\zeta_q^c$ where $0 \leq g \leq p-1$ and $c \in Z_{\geq 0}^{<q}$. This is also the form of the n conjugates of γ. Hence the norm of $\gamma = \prod_{\sigma \in Gal(K_{q,n}/Q)} \gamma^\sigma$ is the sum of $(q-1)^n$ terms, the largest of which has absolute value p^n. By the constraints on q and n, it follows that there exists a $y_0 \in Z_{>0}$ such that $N(\gamma) \leq p^{y_0 n} \leq N$ for all algorithms A_y with $y \geq y_0$. Henceforth assume that $y \geq y_0$.

Making the usual assumption [LLMP] that the probability that $N(\gamma)$ is B-smooth (the exception of the prime q is inconsequential) is equal to the probability that a random positive integer less than N is B-smooth, (see Preliminaries section) there exists an $L_3 \in L_x[1/2, \sqrt{1/2}]$ such that the probability that γ is

B-smooth is at least $1/L_3(N)$ (B-smooth means that all prime ideals dividing (γ) have norm less than or equal to B). Since w B-smooth γ's are needed, it follows that there exists an $L_4 \in L_x[1/2, \sqrt{2}]$ such that the expected number of γ's which must be generated and tested for B-smoothness is at most $L_4(N)$.

The norm of each γ may be tested for B-smoothness naively. Hence there exists an $L_5 \in L_x[1/2, 3/\sqrt{2}]$ such that the expected number of steps required for a single pass through stage 3(a) will be at most $L_5(N)$.

Stage 3(b) There must exist $a_1, a_2, \ldots, a_{w+1} \in Z_{\geq 0}^{<p^n-1}$ such that GCD(a_1, a_2, ..., a_{w+1}) = 1 and $\sum_{i=1}^{w+1} a_i v_{j,i} \equiv < 0, 0, \ldots, 0 > \bmod (p^n - 1)$. Further, there exists an algorithm which will find $a_1, a_2, \ldots, a_{w+1}$ in $O(w^3 \log^2(p^n))$ steps. Hence there exists an $L_6 \in L_x[1/2, 3/\sqrt{2}]$ such that the expected time for a single pass through stage 3(b) is at most $L_6(N)$.

Stage 4: Check numbers of the form $1 + a(q(p^n - 1))$ until primes s_1, s_2, ..., $s_{2H/n}$ are found. For $k = 1, 2, \ldots, 2H/n$, let $g_k \in Z_{\geq 0}^{<s_k}$ generate Z/s_kZ^* and let $g \in Z_{\geq 0}^{<q}$ generate Z/qZ^*. For $k = 1, 2, \ldots, 2H/n$, $l = 1, 2, \ldots, n$: Let $\tilde{S}_{k,l} \subseteq O_{q,q-1}$ be the prime ideal generated by s and $\zeta_q^{d_l} - c_k$, where $c_k \equiv g_k^{a(p^n-1))} \bmod s$ and $d_l \equiv g^l \bmod q$. Let $S_{k,l} = \tilde{S}_{k,l} \cap O_{q,n}$. $S_{k,1}, S_{k,2}, \ldots, S_{k,n}$ are the (distinct, residue class degree 1) prime ideals of $O_{q,n}$ lying above s_k. Since $s_k \equiv 1 \bmod q(p^n - 1)$, it follows that $(p^n - 1) | (N(S_{k,l}) - 1)$ and $N(S_{k,l}) > B$. Since for $j = 1, 2, \ldots, H$, (σ_j) is B-smooth, it follows that $(\sigma_j) + S_{k,l} = (1)$. Let $m_k = g_k^{aq} \bmod s_k$. Then the $2H$ pairs $< S_{k,l}, m_k >$ will be as required for stage 4.

Assume that approximately the 'expected' number of primes will be found in an arithmetic progression: assume that for all $m, b \in Z_{>0}$, with $b > m \log(m)^3$: $\#\{a | 1 + am < b \ \& \ 1 + am \text{ prime}\} > b/m \log(b)^2$. Letting $v = 2H/n$ and $m = q(p^n - 1)$, then all of the v primes needed above can be found by checking less than $v \log(v)^3 \log(m)^3$ a's and each prime s found will be less than $mv \log(v)^3 \log(m)^3$. The constraints on n and q imply that there exists a $c_1, c_2 \in Z_{>0}$ such that $v \log(v)^3 \log(m)^3 < (n \log(p))^{c_1}$ and $mv \log(v)^3 \log(m)^3 < p^n (n \log(p))^{c_2}$. Hence the required primes can be found and tested for primality [AH, SS] in a negligible number of steps.

Generators for Z/s_kZ^* are abundant ([AH], Lemma 4). Checking a candidate g to determine whether it is a generator will be done by factoring $s-1$ and testing that for all primes $t | s - 1$, $g^{(s-1)/t} \not\equiv 1 \bmod s$. The factorization can be done using an '$L[1/2, 1]$' factoring method (e.g. [Le2]). A similar argument shows that a generator for Z/qZ^* can be found in a negligible number of steps.

$O_{q,n}/S_{k,l} \cong Z/s_kZ$ where the isomorphism is induced by $\zeta_q^{d_l} \mapsto c_k$. Hence the calculations of the (p^n-1)-signatures of the σ_j's is a set of discrete logarithm problem over Z/s_kZ. Using the bounds on $2H$ and the primes s together with an '$L[1/2, 1]$' discrete logarithm algorithm for finite prime fields (e.g. [Po]), it follows that there exists an $L_7 \in L_x[1/2, 1]$ such that the expected number of steps required for a single pass through stage 4 is at most $L_7(N)$.

Stage 5: The required $b_1, b_2, \ldots, b_H$ can be shown to always exist and can be found in time $O(H^3 \log^3(p^n - 1))$. Using the bounds on q, it follows that the number of steps required for a single pass through stage 5 is negligible.

It will next be shown that the expected number of passes through stages of the algorithm is negligible. Stages will be repeated only if required in stage 6 or stage 7.

Stage 6 will cause stages of the algorithm to be repeated only if $\alpha^k\beta^l \not\equiv 1 \bmod (p)$. One has:

$$\begin{gathered}\alpha^k\beta^l = \\ \prod_{i,j} \alpha^{r_{j,i}a_ib_j}\beta^{s_{j,i}a_ib_j} = \\ \prod_j(\prod_i(\alpha^{r_{j,i}}\beta^{s_{j,i}})^{a_i}))^{b_j} \equiv \\ \prod_j(\prod_i \gamma_{j,i}^{a_i})^{b_j} = \\ \prod_j \sigma_j^{b_j}\end{gathered}$$

By the construction, the σ_j's are (p^n-1)-singular integers. By the arguments in the Preliminaries section there exists a $\delta \in O_{q,n}$ and $b_1, b_2, \ldots, b_H \in Z_{\geq 0}^{<p^n-1}$ such that $\mathrm{GCD}(b_1, b_2, \ldots, b_H) = 1$ and $\prod_{j=1}^{H} \sigma_j^{b_j} = \delta^{p^n-1}$. $G(p^n-1)$ is a group of index dividing p^n-1 which is the direct product of at most $H-1$ cyclic groups (see Preliminaries section). The signature homomorphism θ maps $G(p^n-1)$ into a group which is the direct product of $2H$ cyclic groups of order p^n-1. It is reasonable to assume that this map is an embedding and hence that these $b_1, b_2, \ldots, b_H$ are the ones found in stage 5. It follows that:

$$\begin{gathered}\alpha^k\beta^l \equiv \\ \prod_j \sigma^{b_j} = \\ \delta^{p^n-1} \equiv \\ 1\end{gathered}$$

Stage 7 will cause stages of the algorithm to be repeated only if $(l, p^n-1) \neq 1$. However, $(l, p^n-1) = 1$ with probability $\phi(p^n-1)/(p^n-1) \geq 1/c\log p^n$ where $c \in \Re_{>0}$ is independent of p and n ([AH], Lemma 4). Briefly, this can be argued as follows: Since from stage 3(b) $\mathrm{GCD}(a_1, a_2, \ldots, a_{w+1}) = 1$ and from stage 5 $GCD(b_1, b_2, \ldots, b_H) = 1$ it follows that for all primes t dividing p^n-1, there exist $i \in Z_{>0}^{\leq w+1}$ and $j \in Z_{>0}^{\leq H}$ such that a_ib_j is relatively prime to t. Consider $\gamma_{j,i} \equiv \alpha^{r_{j,i}}\beta^{s_{j,i}}$, and observe that for all $s \in Z_{\geq 0}^{<p^n-1}$, there exists a unique $r \in Z_{\geq 0}^{<p^n-1}$ such that $\gamma_{j,i} \equiv \alpha^r\beta^s$. Hence $s_{j,i}$ is 'random' mod t and consequently $l = \sum_{j=1}^{H}\sum_{i=1}^{w+1}(s_{j,i}a_ib_j)$ is also 'random' mod t.

Recalling that in algorithm A_y, $N = p^{yn}$, it follows that there exists a $c_I \in \Re_{>0}$ and an $L_I \in L_x[1/2, c_I]$ such that for all sufficiently large y, the expected number of steps required by Algorithm A_y is $L_I(p^n)$ on all inputs q, n, p, α, β such that $p, q \in Z_{>0}$ are prime, $n < p$, $n|q-1$, $q \leq \tilde{c}n^4(log(np))^2$, p inert in $K_{q,n}$, and α, $\beta \in R_{q,n,p}$ with $[\alpha]$ generating $O_{q,n}/(p)^*$ and $\beta \not\equiv 0 \bmod p$. Hence there exists a $c_I \in \Re_{>0}$ such that the expected number of steps required by Algorithm I (when $n < p$) is:

$$e^{c_I(\log(p^n)\log\log(p^n))^{1/2}}$$

Finally, it is clear from stages 6 and 7 that the output of the algorithm is x such that $\alpha^x \equiv \beta \bmod p$.

5 Algorithm II

This algorithm will be used for discrete logarithms over $GF(p^n)$ when $p \leq n$.

Algorithm II is a generalization of the algorithm for $GF(2^n)$ by Hellman and Reyneri discussed in Coppersmith [HR, Co].

It is assumed that the inputs to the algorithm are p, f, α, β such that $p \in Z_{>0}$ is prime, $f \in Z/pZ[x]$ is monic, irreducible of degree $n \geq p$, and $\alpha, \beta \in Z/pZ[x]$ of degree less than n with $[\alpha] \in (Z/pZ[x])/(f)$, and $[\alpha]$ a generator of the multiplicative group and $\beta \not\equiv 0 \bmod f$.

For purposes of brevity, we have not included analysis of Algorithm II. Lovorn [Lo] gives detailed analysis of a similar algorithm.

Algorithm II

Stage 0 input f, p, α, β

Stage 1 Set n =degree of f, $m = \lfloor\sqrt{n}\rfloor$

Stage 2 Calculate $T = \{f_i | f_i \in Z/pZ[x],\ deg(f_i) \leq m,\ f_i$ irreducible and monic$\}$. Let $w = \#T$ and let $< f_1, f_2, \ldots, f_w >$ be an ordering of T.

Stage 3 Set $z = 1$, While $z \leq w+1$: Choose random r, s with $0 \leq r, s \leq p^n - 1$ and calculate $\gamma \in Z/pZ[x]$, of degree less than n such that $\gamma \equiv \alpha^r\beta^s \bmod f$. If $\gamma = \widetilde{\gamma}\prod_{i=1}^{w} f_i^{e_i}$ where $\widetilde{\gamma}$ is the leading coefficient of γ (i.e. if γ is m-smooth) then set $\gamma_z = \gamma$, $r_z = r$, $s_z = s$, $v_z =< e_1, e_2, \ldots, e_w >$ and $z = z + 1$.

Stage 4 Calculate $a_1, a_2, \ldots, a_{w+1} \in Z_{\geq 0}^{<p^n-1}$ such that GCD(a_1, a_2f, ..., a_{w+1}) $= 1$ and $\sum_{i=1}^{w+1} a_i v_i \equiv < 0, 0, \ldots, 0 > \bmod (p^n - 1)$.

Stage 5 Calculate $k = \sum_{i=1}^{w+1}(r_i a_i)$ and $l = \sum_{i=1}^{w+1}(s_i a_i)$. Calculate $s \in Z_{>0}^{<p}$ such that $s \equiv \alpha^k\beta^l \bmod f$.

Stage 6 Calculate $y \in Z_{\geq 0}^{<p-1}$ such that $\alpha^{y((p^n-1)/(p-1))} \equiv s \bmod f$.

Stage 7 If $(l, p^n - 1) \neq 1$ then go to Stage 3, Else calculate and output $x \equiv (y((p^n-1)/(p-1)) - k)/l \bmod p^n - 1$ and halt.

Acknowledgments

We would like to thank Dennis Estes and Bob Guralnick for their help.

Research supported by NSF CCR-9214671.

Discussion

Little effort was made to 'optimize' the algorithm presented here. It is possible to improve the running time in several ways. Sparse matrix methods can be used to find some dependencies[Wi]. A better bound on q in Algorithm I can be argued heuristically. Smoothness of norms can be tested using the 'elliptic curve methods' [Le]. The integer factoring done in various parts can probably be avoided if necessary or 'L[1/3]' methods can be used (e.g. [AH, LLMP]). The use of Algorithm II can perhaps be avoided altogether by adopting Algorithm I to a more general setting. Alternatively the 'L[1/3,c]' method of Coppersmith [Co] might be adapted for the case $n \geq p$.

It terms of running time there appear to be several natural open problems:

- Do there exist a $c \in Z_{>0}$ and an algorithm for discrete logarithms over GF(p^n) with provable expected running time in $L_x[1/2, c]$?
- Does there exist an algorithm for discrete logarithms over GF(p^n) with heuristic expected running time in $L_x[1/2, 1]$?
- Does there exist an algorithm for discrete logarithms over GF(p^n) with provable expected running time in $L_x[1/2, 1]$?
- Do there exist a $c \in Z_{>0}$ and an algorithm for discrete logarithms over GF(p^n) with heuristic expected running time in $L_x[1/3, c]$?

References

[Ad1] Adleman L.M., A subexponential algorithm for discrete logarithms with applications to cryptography. *Proc. 20th IEEE Found. Comp. Sci. Symp.* 1979, pp. 55-60.

[Ad2] Adleman L.M., Factoring numbers using singular integers, *Proc. 23rd Annual ACM Symposium on Theory of Computing*, 1991, pp. 64-71.

[AH] Adleman L.M. and Huang M., *Primality Testing and Abelian Varieties Over Finite Fields*, Lecture Notes In Mathematics 1512, Springer-Verlag, 1992.

[AL] Adleman L.M. and Lenstra H.W. Jr., Finding irreducible polynomials over finite fields. *Proc. 18th Annual ACM Symposium on Theory of Computing*, 1986, pp. 350-355.

[Be] Berlekamp E., Factoring polynomials over large finite fields. *Math. Comp.* 24, 1970. pp. 713-735.

[BS] Bach E. and Shallit J., Factoring with cyclotomic polynomials. *Proc. 26th IEEE Found. Comp. Sci. Symp.* 1985, pp. 443-450.

[CEP] Canfield E.R., Erdös P. and Pomerance C., On a problem of Oppenhiem concerning "Factorisatio Nemerorum". *J. Number Theory*, 17, 1983 pp. 1-28.

[Co] Coppersmith D., Fast Evaluation of Logarithms in Fields of Characteristic Two. *IEEE Trans on Information Theory*, vol IT-30, No 4, July 1984, pp. 587-594.

[COS] Coppersmith D., Odlyzko A.M. and Schroeppel R., Discrete logarithms in *GF(p)*, *Algorithmica*, v. 1, 1986, pp 1-15.

[DH] Diffie W. and Hellman M.E., New Directions in Cryptography, *IEEE Trans. Inform Theory*, vol IT-22, pp 644-654, 1976

[Ed] Edwards H.M., *Fermat's Last Theorem*, Graduate Texts in Mathematics 50, Springer-Verlag, 1977.

[El1] ElGamal T., A public key cryptosystem and a signature scheme based on discrete logarithms, *IEEE Trans. Info. Theory*, vol IT-31 pp. 469-472, 1985

[El2] ElGamal T., A subexponential-time algorithm for computing discrete logarithms over GF(p^2),*IEEE Trans. Info. Theory*, vol IT-31 pp. 473-481, 1985

[Ga] Gauss K.F., *Disquisitiones Arithmeticae*, translation A.C. Clarke, S.J., Yale University Press, 1966.

[Go1] Gordon D.M., Discrete logarithms in GF(p) using the number field sieve, manuscript, April 4, 1990.

[HR] Hellman M. E., Reyneri J. M. Fast computation of discrete logarithms in GF(q). *Advances in Cryptography: Proceedings of CRYPTO '82*, pp. 3-13

[Le] Lenstra H.W. Jr., Finding isomorphisms between finite fields. *Math Comp* 56, 1991, pp. 329-347.

[Le2] Lenstra H.W. Jr., Factoring integers with elliptic curves. *Ann. of Math.* 126, 1987, pp. 649-673.

[LLMP] Lenstra A.K., Lenstra H.W., Jr., Manasse M.S. and Pollard J.M. The number field sieve. *Proc. 22nd STOC*, 1990, pp. 564-572.

[Lo] Lovorn R., Rigorous, subexponenial algorithms for discrete logarithms over finite fields, PhD Thesis, University of Georgia, May 1992

[Ne] Newman M., Bounds for class numbers, *Proc. Sympos. Pure Math.* American Mathematics Society, Vol. VIII, 1965, pp 70-77.

[Od] Odlyzko A. M., Discrete Logarithms in Finite Fields and their Cryptographic Significance, *Proceedings of Eurocrypt '84*, Lecture Notes in Computer Science, Springer-Verlag. 1985. pp. 224-314.

[Po] Pomerance C. Fast, rigorous factorization and discrete logarithms, *Discrete Algorithms and Complexity.* ED. Johnson D.S., Nishizeki T., Nozaki A. and Wilf H.S. Academic Press, 1987. pp. 119-144.

[RA] Rabin M. O., Probabilistic Algorithms in Finite Fields. *SIAM Journal of Computing*, Vol 9, No 2, May 1980, pp. 273-280

[SS] Solovay R. and Strassen V., A fast Monte-Carlo test for primality. *Siam Journal of Computing* 6, 1977. pp. 84-85.

[Wa] Washington L.C., *Introduction to Cyclotomic Fields*, Graduate Texts in Mathematics 83, Springer-Verlag, 1982.

[Wi] Wiedermann D. Solving sparse linear equations over finite fields. *IEEE Trans. Inform. Theory.* IT-32, pp. 54-62

[WM] Western A.E. and Miller J.C.P., *Tables of Indices and Primitive Roots*, Royal Society Mathematical Tables, vol. 9., Cambridge University Press, 1968.

An implementation of the general number field sieve

J. Buchmann J. Loho J. Zayer
Extended abstract

Fachbereich Informatik
Universität des Saarlandes
66041 Saarbrücken
Germany

Abstract. It was shown in [2] that under reasonable assumptions *the general number field sieve* (GNFS) is the asymptotically fastest known factoring algorithm. It is, however, not known how this algorithm behaves in practice. In this report we describe practical experience with our implementation of the GNFS whose first version was completed in January 1993 at the Department of Computer Science at the Universität des Saarlandes.

1 Introduction

Factoring rational integers into primes is one of the most important and most difficult problems of computational number theory. It was shown in [2] that under reasonable assumptions *the general number field sieve* (GNFS) is the asymptotically fastest known factoring algorithm. It is, however, not known how this algorithm behaves in practice. In this report we describe practical experience with the first version of our implementation of the GNFS. For our implementation we used the methods described in [2], [3], and [7]. In the course of the implementation we have found several improvements which we will describe in the full version of this paper. In this extendend abstract we restrict ourselves to the presentation of a brief sketch of the algorithm and the numerical results.

2 The GNFS

Let $n \in \mathbb{N}$. If one can find two integers x and y with

$$x^2 \equiv y^2 \text{ modulo } n \tag{1}$$

and $x \not\equiv \pm y$ modulo n, then $\gcd(x-y, n)$ is a non trivial divisor of n. Like many other factoring algorithms the GNFS factors n by producing such a pair x, y. This is done in the following way: Let $f(x) = f_0 + f_1 \cdot x + \ldots + f_{d-1} \cdot x^{d-1} + x^d \in \mathbb{Z}[x]$ be an irreducible polynomial for which there exits $m \in \mathbb{Z}$ with $f(m) \equiv 0$ modulo

n. Let ρ be a zero of $f(x)$. The algorithm determines a non-empty set S of pairs (a, b) of relatively prime integers with the following properties

$$X = \prod_{(a,b)\in S} (a + bm) = x^2 \text{ with } x \in \mathbb{Z} \tag{2}$$

$$\gamma = \prod_{(a,b)\in S} (a + b\rho) = \delta^2 \text{ with } \delta \in \mathbb{Z}[\rho] \tag{3}$$

The map $\varphi : \mathbb{Z}[\rho] \to \mathbb{Z}/n\mathbb{Z}$, $\rho \mapsto m \bmod n$ is a ring homomorphism. Therefore we have $x^2 \equiv \varphi(\delta^2) \equiv \varphi(\delta)^2 \bmod n$. If we set $y = \varphi(\delta)$ then we have found a congruence of the form (1) which with high probability yields a factorization of n.

The algorithm can thus be divided into three parts: determining the polynomial, finding the squares and extracting the square roots. In the remaining sections we describe our implementation of those parts and we give numerical examples. For background and details we refer to [2], [3] and [7].

3 Determining the polynomial

The first step of GNFS is to find an irreducible polynomial $f(x) \in \mathbb{Z}[x]$ of degree d and a rational integer m, such that $f(m) \equiv 0 \bmod n$. For $n \leq 10^{60}$ we use $d = 3$ and for $10^{60} < n < 10^{180}$ we use $d = 5$. We choose $i \in \mathbb{Z}$ such that for $m = \lfloor n^{\frac{1}{d}} \rfloor + i$ there is an expansion $n = m^d + f_{d-1}m^{d-1} + \ldots + f_1 m + f_0$ with $-m/2 \leq f_j < m/2$. We determine that expansion and we set $f(x) = x^d + f_{d-1}x^{d-1} + \ldots + f_1 x + f_0$. There are various ways of modifying f. We can, for example, replace f by $f + \sum_{j=1}^{d-1} c_j(x^j - mx^{j-1})$. It is still an open question how an optimal polynomial f can be found. We intend to use our implementation of the GNFS to study this question in detail. A few remarkable experimental results can be found in section 6.

4 Finding the squares

To find the set S of coprime pairs $(a, b) \in \mathbb{Z}^2$ satisfying (2) and (3) we use the *standard sieve* which is described in [2] or the lattice sieve which was suggested in [7].

In both algorithms we must choose two factor bases. The *rational factor base* F_R is the set of all rational primes below some bound $s_R \in \mathbb{R}_{>0}$. The *algebraic factor base* is the set F_A of all degree one prime ideals of $\mathbb{Z}[\rho]$ of norm below $s_A \in \mathbb{R}_{>0}$. The values for s_R and s_A are chosen according to experimental experience. Each prime in F_A is represented by a pair (p, c_p) where c_p is a zero of f modulo p. We also need large prime bounds L_R and L_A which are roughly $100 \cdot s_R$ or $100 \cdot s_A$, respectively.

To apply the standard sieve, we fix bounds $A, B \in \mathbb{Z}_{>0}$ on a and b, respectively. Again those values are chosen according to experimental experience. For each $b \in \{1, 2, \ldots, B\}$ we determine all a with $-A < a < A$ such that $\gcd(a, b) = 1$, all of the prime factors of $a + bm$ except for at most one factor $l_R(a, b)$ belong to F_R and all of the prime ideal factors of $(a + b\rho)\mathbb{Z}[\rho]$ except for at most two factors $l_{A,1}(a, b)$ and $l_{A,2}(a, b)$ belong to F_A. Also, the extra rational prime factors are called *large rational primes* and they must be below L_R. Analogously, the extra algebraic prime factors are called *large algebraic primes* and their norms must be below L_A. Any such pair (a, b) is called a *good pair*. We say that a good pair without large primes is of *type fff*, if there is a large rational prime it is of *type pff*. The definition of the *types fpf, fpp, ppf* and *ppp* is analogous. For a more detailed description of the sieve algorithm see [2].

To use the lattice sieve we divide the factor bases into two parts. The set $F_{r,s}$ of *small rational primes* contains all elements of F_R no larger than s_R/t where t may be chosen between 2 and 10. The set $F_{r,m}$ of *medium primes* is the complement of $F_{r,s}$ in F_R. For $q \in F_{r,m}$ the set $LR_q = \{(a, b) : q | a + bm\}$ is a two dimensional lattice in $\mathbb{Z}^2$. If $(\underline{u}, \underline{v})$ is a basis of LR_q then one can find good pairs (a, b) whose small primes are bounded by q by inspecting the vectors $c\underline{u} + d\underline{v}$ for $c, d \in \mathbb{Z}$, $-C < c < C$, $0 < d < D$ where $C \in \mathbb{R}_{>0}$ and $D \in \mathbb{R}_{>0}$ are chosen according to experimental experience. For any fixed d this can be done by a sieving procedure which is described in [7]. In this procedure we take advantage of the following fact: For $p \geq 2C$ and $d \in \{1, \ldots, D\}$ there is exactly one c_d such that p is a divisor of $a + bm$ for $(a, b) = c_d\underline{u} + d\underline{v}$ and $-p/2 \leq c_d < p/2$. Since $c_d = c_{d-1} + c_1 \bmod p$ those numbers can be very easily computed. It is even possible to determine the interesting values of c_d for which $-C < c_d < C$ immediately. This leads to a significant speed up of the lattice sieve. A similiar trick can be applied to find $a + b\rho$ which factors up to large primes over F_A.

Once sufficiently many good pairs are found, we determine for each good pair (a, b) the decompositions $a + bm = l_R(a, b) \cdot \prod_{p \in F_R} p^{e_p(a,b)}$ and $(a + b\rho)\mathbb{Z}[\rho] = l_{A,1}(a, b) \cdot l_{A,2}(a, b) \cdot \prod_{P \in F_A} P^{e_P(a,b)}$, where $l_R(a, b), l_{A,1}(a, b)$ and $l_{A,2}(a, b)$ also may be 1. We also determine a small set F_Q of degree one prime ideals of $\mathbb{Z}[\rho]$ of norms bigger than L_A and for each $Q \in F_Q$ we set $e_Q(a, b) = 0$ if $a + b\rho$ is a square in $\mathbb{Z}[\rho]/Q$ and $e_Q(a, b) = 1$ otherwise. The large primes are handled by constructing cycles as discribed in [1] and [6]. By calculating a non trivial linear dependency among the vectors $((e_p(a, b))_{p \in F_R}(e_P(a, b))_{P \in F_A}(e_Q(a, b))_{Q \in F_Q})$ over $\mathbb{F}_2$ we determine the subset S of the set of all pairs (a, b) that we are looking for. As noted in [2] it may be necessary to replace γ in (3) by $(f'(\rho))^2\gamma$ to guarantee that the square belongs to $\mathbb{Z}[\rho]$ rather than to the maximal order of the field $\mathbb{Q}[\rho]$.

5 Finding the square roots

Suppose we have found the set S of coprime pairs $(a, b) \in \mathbb{Z}^2$ satisfying (2) and (3). Let $X = \prod_{(a,b) \in S} a + bm$ and let $\gamma = (f'(\rho))^2 \prod_{(a,b) \in S} a + b\rho$. Extracting the

square root x of X is very simple since we know the prime factorization of X. Computing the square root δ of γ is, however, quite difficult since the coefficients in the representation $\delta = \delta_0 + \delta_1 \cdot \rho + \ldots + \delta_{d-1} \cdot \rho^{d-1}$ may be very large. In our implementation we use the method of Couveignes [3]. He suggests to determine a set I of prime numbers which are inert in $\mathbb{Z}[\rho]$ and for each $p \in I$ to compute δ_p such that $\delta_p^2 \equiv \gamma \bmod p$. This can easily be effected by applying a variant of Shanks' RESSOL algorithm [8]. Since we want to apply Chinese remaindering we must determine the image of the same square root for every $p \in I$. Using Newton iteration one can lift any δ_p to a number $\delta_{p^{2^k}}$ such that $\gamma \equiv \delta^2_{p^{2^k}} \bmod p^{2^k}$ where the exponent k is chosen according to experimental experience. Chinese remaindering yields $y = \varphi(\delta)$.
The square γ can be reduced in size by dividing it by some $(a+b\rho)^2$, where (a, b) is a good pair without large primes on the algebraic side. Whether γ is divisible by such a square can be easily checked by inspecting the vectors $((e_P(a,b))_{P \in F_A})$ and $((e_P(\gamma))_{P \in F_a})$. The following table shows the effect of this reduction when used in the factorization of the third number number in section 7.

$\lvert S\rvert$	$\#((a+b\rho)^2)$ reduced	$\lvert I\rvert = \#$ of inert primes *	max. exp. 2^k	maximal # of digits of δ_j	running time in mips h
25022	0	115	256	133777	62.82
25022	7398	60	256	69120	41.01

6 Quality of the polynomials

The least well understood part in the GNFS is how to find the best polynomial f. In this section we illustrate that the algorithm behaves quite differently for different choices of polynomials. Let n = 6809 47738 35969 19453 31142 12277. Except for m all the parameters were chosen identically as described in the next section. The next two tables show how different polynomials yield a different number of good pairs. For the first table we used the m-adic expansion as described in section 3 to find the polynomial, where $m = \lfloor n^{1/3} \rfloor + i$. From a bigger experiment we present the most interesting results.

* all inert primes about $3 \cdot 10^4$

i	F_A = size of the algebraic factor base	# good pairs of type fff	# good pairs of types $pff\ldots ppp$	# cycles among large primes
−27137	2537	6049	32154	18573
−27139	2532	5019	26906	13801
+23	2524	4811	27812	14665
+13	2492	4365	24790	12139
0	2493	4390	24533	11931
−50467	2498	3552	21016	8985
+27140	2484	3354	19689	7843
−43467	2514	3240	18966	7499
−27142	2454	3181	18552	6998
+27138	2533	2797	16307	5407

For the second table we modified the polynomial $f(x)$ obtained with $m = \lfloor n^{1/3} \rfloor$ by adding g(x).

$g(x)$	$\lvert F_A \rvert$	# good pairs of type fff	# good pairs of types $pff\ldots ppp$	# cycles among large primes
$-x^2 + mx$	2535	6014	33224	20213
0	2493	4390	24533	11931
$-x^2 + (m+1)x - m$	2522	3245	18657	7204
$x - m$	2533	3080	16856	5620
$-2(x^2 - (m-1)x - m)$	2348	1780	11312	2339

7 Some full factorizations

The first numbers we factored with GNFS were

1. $n = 6809\,47738\,35969\,19453\,31142\,12277$
   using $f(x) = x^3 + x^2 - 5524\,50799x + 2195\,69758$, $m = 40835\,50467$
2. $n = 82935\,75851\,23433\,22909\,99689\,74960\,03250\,42327$
   using $f(x) = x^3 + 301\,13501\,57913x + 594\,61180\,91613$, $m = 2024\,17135\,03301$
3. $n = 3488\,17079\,74401\,66635\,06963\,23211\,22160\,51028\,26088\,93989$
   using $f(x) = x^3 + 2x^2 + 5\,13769\,39621\,45733x + 2\,78963\,78107\,83197$,
   $m = 15\,16582\,05880\,38497$
4. $n = 9 \cdot 436\,22325\,30202\,01660\,81169\,50834\,54211\,20979\,47919\,09269\,39307$
   $24927\,93753\,70109\,41445\,21495\,39140\,12056\,52499\,95711\,63723\,68586$
   $19995\,36219\,76543\,09529\,71290$
   using $f(x) = x^5 + 9$, $m = 3^{56}$

All relations were found by Pollard's lattice sieve algorithm [7]. The most important datas of these factorizations are summarized in the following table.

# digits of n	29	40	49	134
factor bases				
biggest prime of the rational factor base	5279	22307	30559	951161
size of the rational factor base	700	2500	3300	75000
bound for the large primes on the rational side	10^5	$6 \cdot 10^5$	10^6	10^8
biggest prime p of the pairs (p,cp) of the algebraic factor base	22291	104729	224737	951109
size of the algebraic factor base	2493	9794	19944	74952
bound for the large prime on the algebraic side	10^5	$1.5 \cdot 10^6$	10^7	10^8
# additional pairs (p,cp) with p bigger than large prime bound	10	10	20	25
finding the squares with the lattice sieve				
sieving bound C	500	500	5000	10000
sieving bound D	50	200	1000	5250
# good pairs of type fff	4390	9133	8010	73798
# good pairs of type pff	4733	13020	16906	184864
# good pairs of type fpf	6515	30937	46531	344560
# good pairs of type ppf	8214	42681	109389	1031253
# good pairs of type fpp	2272	17849	69304	0*
# good pairs of type ppp	2799	22862	153719	0*
cycles among large primes	11931	23386	19371	69103
sieving time in mips days	1.75	118	717	41010
extracting the square root				
# inert primes	150	175	240	105
size of inert primes about	$3 \cdot 10^4$	$3 \cdot 10^4$	$3 \cdot 10^4$	$1.1 \cdot 10^5$
max. exponent for lifting	16	64	64	256
# digits of coefficients of the root	~ 8000	~ 51000	~ 69120 **	~ 135500 **
running time in mips hours	7.5	36	41	484.5

* only one large prime on each side

** with square reduction

The factorizations are

1. 6809 47738 35969 19453 31142 12277
   = 1785 89908 07069 · 3 81291 27547 91033
2. 82935 75851 23433 22909 99689 74960 03250 42327
   = 1301 67526 01273 98757 · 63 71463 07169 01048 08011
3. 3488 17079 74401 66635 06963 23211 22160 51028 26088 93989
   = 22036 72182 80384 74120 85111 · 15828 90061 71597 82957 88099
4. 436 22325 30202 01660 81169 50834 54211 20979 47919 09269 39307 24927 93753 70109 41445 21495 39140 12056 52499 95711 63723 68586 19995 36219 76543 09529 71290
   = 2 · 5 · 557 · 11 07553 · 8 20739 81221 45081·
   1 38579 05391 45329 24856 06236 63377 62045 74597·
   62 17073 56762 16461 88942 98788 28272 87720 85730 54231 32773 87634 13782 17457

References

1. J. Buchmann, J. Loho, J. Zayer, ***An implementation of the general number field sieve, full version***, to appear 1993
2. J. P. Buhler, H. W. Lenstra, C. Pomerance, ***Factoring integers with the number field sieve***, Lecture Notes in Mathematics 1554, pp. 50 - 94, Springer Verlag, 1993
3. J. - M. Couveignes, ***Computing a square root for the number field sieve***, Lecture Notes in Mathematics 1554, pp. 95 - 102, Springer Verlag, 1993
4. D. E. Knuth, *The Art of Computer Programming, vol. 2*, Second Edition, Addison Wesley, 1981
5. A. K. Lenstra, H. W. Lenstra, M. S. Manasse, J. M. Pollard, *The number field sieve*, Abstract: Proc. 22nd Ann. ACM Symp. on Theory of Computing (STOC)(1990),564-572
6. A. K. Lenstra, M. Manasse, ***Factoring with two large primes***, preprint 1992
7. J. M. Pollard, *The Lattice Sieve*, Lecture Notes in Mathematics 1554, pp. 43 - 49, Springer Verlag, 1993
8. D. Shanks *Five Number-Theoretic Algorithms* , Proc. Second Manitoba Conference On Numerical Math., 1972, pp. 51-70

On the factorization of RSA-120

T. Denny[1], B. Dodson[2], A. K. Lenstra[3], M. S. Manasse[4]

[1] Lehrstuhl Prof. Buchmann, Fachbereich Informatik , Universität des Saarlandes, Postfach 1150, 66041 Saarbrücken, Germany
E-mail: denny@cs.uni-sb.de

[2] Department of Mathematics, Lehigh University, Bethlehem, PA 18015-3174, U. S. A
E-mail: bad0@Lehigh.EDU

[3] MRE-2Q334, Bellcore, 445 South Street, Morristown, NJ 07960, U. S. A
E-mail: lenstra@bellcore.com

[4] DEC SRC, 130 Lytton Avenue, Palo Alto, CA 94301, U. S. A
E-mail: msm@src.dec.com

Abstract. We present data concerning the factorization of the 120-digit number RSA-120, which we factored on July 9, 1993, using the quadratic sieve method. The factorization took approximately 825 MIPS years and was completed within three months real time. At the time of writing RSA-120 is the largest integer ever factored by a general purpose factoring algorithm. We also present some conservative extrapolations to estimate the difficulty of factoring even larger numbers, using either the quadratic sieve method or the number field sieve, and discuss the issue of the crossover point between these two methods.

On the factorization of RSA-120

Evaluation of integer factoring algorithms, both from a theoretical and practical point of view, is of great importance for anyone interested in the security of factoring-based public key cryptosystems. In this paper we concentrate on the practical aspects of factoring. Furthermore, we restrict ourselves to *general purpose* factoring algorithms, i.e., algorithms that do not rely on special properties the numbers to be factored or their factors might have. These are the algorithms that are most relevant for cryptanalysis.

Currently the two leading general purpose factoring algorithms are the *quadratic sieve* (QS) and the *number field sieve* (NFS), cf. [12] and [2]. Throughout this paper, NFS is the generalized version (from [2]) of the algorithm from [8]; the latter algorithm is much faster, but can only be applied to composites of a very special form, cf. [9]. Let

$$L_x[a, b] = \exp\left((b + o(1))(\log x)^a(\log\log x)^{1-a}\right)$$

for real a, b, x, and $x \to \infty$. To factor an odd integer $n > 1$ which is not a prime power, QS runs in time

$$(1) \qquad L_n[1/2, 1],$$

and NFS in

(2) $$L_n[1/3, 1.923].$$

It follows that NFS is asymptotically superior to QS, but for numbers in the 100 to 150-digit range it is not immediately obvious which of the two methods is faster. If the $o(1)$'s in both runtimes are set to zero, then the crossover occurs for n around 124 digits. This computation neglects many aspects affecting the practical runtimes of both methods, and is thus oversimplistic. Similarly, it does not make much sense to evaluate any of these expressions for some particular n (again with $o(1) = 0$), and to interpret the result as 'the number of instructions' needed to factor that n, as sometimes happens in the literature or on sci.crypt. They can be used, however, to predict how hard factoring m will be, when it is known how hard n is, and if n and m differ by not more than, say, 15 digits. Although such a prediction can be helpful, it is of only limited practical value, in particular if m is considerably larger than n or if n was already testing the limits of our capabilities. This is caused by the fact that the $o(1)$ is *not* a constant and by a variety of other issues that will be discussed below.

The QS-factorization of a 116-digit factor of $10^{142}+1$ was the largest general purpose factorization reported in the literature [11]. As mentioned in [11] various parameters for that factorization were deliberately chosen suboptimally, in order to keep memory requirements acceptable to the contributors of cycles (the authors of [11] use their 'electronic mail' approach (cf. [10]) to get the cycles necessary for this factorization, so they had every reason to avoid complaints from prospective contributors). The factorization was completed in approximately 400 MIPS years, which corresponds to roughly $400 \cdot 10^6 \cdot 365 \cdot 24 \cdot 3600 \approx 1.3 \cdot 10^{16}$ instructions. Application of the ill-advised practice described above to this 116-digit n would lead to an estimate of $L_n[1/2, 1] \approx 5.7 \cdot 10^{16}$ instructions, which is off by only a factor 4.4, and thereby unusually accurate.

In the present paper we consider the 120-digit number RSA-120. Until June 9, 1993, RSA-120 was the smallest unfactored number on the 'RSA challenge list,' which is a list of composite numbers of d digits, for $d = 100, 110, 120, \ldots, 490, 500$. This list was compiled by RSA Data Security Corporation Inc. in the following manner (cf. [14]):

> Each RSA number is the product of two randomly chosen primes of approximately the same length. These primes were both chosen to be congruent to 2, modulo 3, so that the product could be used in an RSA public-key cryptosystem with public exponent 3. The primes were tested for primality using a probabilistic primality testing routine. After each product was computed, the primes were discarded, so no one—not even the employees of RSA Data Security—knows any product's factors.

RSA-100 was factored in April 1991 by the third and fourth author into two 50-digit primes, and RSA-110 was factored in April 1992 by the third author into two 55-digit primes [4]. Here we discuss some of the data that we gathered during our QS-factorization of RSA-120, and we present its two 60-digit prime factors. This factorization is a new general purpose factoring record, breaking the old record by 4 digits. A reader who wished to argue that the 116-digit record has not been broken would need to assert that there is no reason to believe the integrity of the above statement of RSA Data Security Corporation Inc., of its employees, or of the authors of this paper.

There are several points where our effort differs from the old 116-digit record. In the first place we did not employ the approach from [10] where an unknown number of anonymous volunteers on the internet contribute virtually all necessary cycles, all using the same program. Instead, we used *three* independently coded programs that ran at only *four* different sites, unevenly spread over two continents: the first author used his QS-implementation on workstations at the university of Saarbrücken [3], the other authors used the program from [10; 11] on workstations at Lehigh University, Bellcore, and DEC SRC, and the third author used his SIMD QS-implementation on Bellcore's massively parallel machine (MasPar), as described in [4].[5]

Secondly, we did not impose artificial restrictions on any of the parameters that have to be chosen, as the authors of [11] readily admit to have done (for the reasons mentioned above). As a consequence, the memory demands of our programs, as well as the further storage requirements, vastly exceeded those of any previous factoring efforts of which we are aware. This includes factorizations using the number field sieve applied to composites of a very special form, a detailed account of which can be found in [1; 8; 9]. While we have made every attempt to minimize the total effort to factor RSA-120 we also gathered experience with the much larger programs and files that one would have to deal with when attempting to factor the 512-bit moduli used for security in cryptosystems such as RSA. We note that, as a practical matter, many people would regard 1024-bit moduli as the minimum necessary for longer term security; and that our result provides a benchmark that indicates how far our best current efforts are from being able to factor such 308-digit numbers.

As will be shown in the next section our factorization of RSA-120 took at most 825 MIPS years. As mentioned above, we tried to minimize this runtime, so we do not expect that RSA-120 could have been factored by QS in substantially less time. Extrapolation of the 400 MIPS years for the 116-digit number from [11] to RSA-120, using (1), gives approximately 950 MIPS years; here we use the fact that the multiplier for the 116-digit number was 1, but that we used 7 for RSA-120, i.e., we factored 7·RSA-120. (The use of multipliers is one of the reasons why extrapolation of QS-runtimes may be unreliable.) This shows that extrapolation based on (1) can give reasonably accurate results. When we use (1) to extrapolate the 825 MIPS years for RSA-120 to the runtime that would be needed for the QS-factorization of a 129-digit number, we would get approximately 5000 MIPS years. At the time of writing a group of people on the internet is actually working

[5] When we were almost finished we were joined by Walter Lioen and Herman te Riele from the Centre for Mathematics and Computer Science (CWI) in Amsterdam, The Netherlands, who were using their QS-implementation on workstations at the CWI. We gratefully acknowledge their assistance.

on the QS-factorization of a 129-digit number,[6] following the approach and using an adapted version of the code of [10]. Current predictions indicate that it will take between 5000 and 6000 MIPS years to complete this factorization, which is close to our prediction based on extrapolation.

An important issue in the study of the practical behavior of factoring algorithm is the crossover point between QS and NFS. Experiments indicate that the NFS-implementation from [1] would need at least 300 MIPS years for a general 110-digit number. Optimistic extrapolation to RSA-120, using (2), suggests that it would take at least 1300 MIPS years using NFS; similarly, a 129-digit number would take at least 5000 MIPS years. These figures suggest that the crossover point between QS and NFS lies beyond 130 digits.

Unlike QS, however, NFS is a very new algorithm: hardly anybody has much practical experience using it on numbers with more than 110 digits, and most implementations are still in an experimental stage. At the time of writing the third author is working on various improvements of his NFS-implementation that increase the sieving and trial division speed, and that at the same time improve the yield by including relations with triple and quadruple large primes (cf. [8: 7.3]). While it is still too early to present runtime estimates for this new NFS implementation, experiments suggest that the expected NFS runtimes given above can be substantially improved and that the crossover point lies closer to 125 than to 130 digits.

In the remainder of this paper we present some of the data and statistics that we have gathered during our factorization of RSA-120, aimed at a reader with a reasonable background in current factoring terminology. We also present the factorization of RSA-120.

QS-data for RSA-120

We present QS-data related to the following number:

RSA-120 = 2270104812954373633342599609474936688958753364660847800
38173258247009162675779735389791151574049166747880487470296548479,

where the first line consists of the first 55 digits, and the second gives the last 65 digits (cf. [14]). We used 7 as multiplier, and a factor base size of 245810. These choices are based on several experimental runs with these and other choices. For the 116-digit factorization of $n = (10^{142}+1)/(101 \cdot 569 \cdot 7669 \cdot 380623849488714809)$ reported in [11] multiplier 1 and a factor base size of 120000 were used, but the authors of [11] remark that 160000 would have been a better choice. Because the factor base size is supposed to grow with the square root of the runtime, this remark would indicate that

[6] The number they are trying to factor is the 129-digit RSA challenge number that was published in Martin Gardner's column in the August 1977 issue of Scientific American.

$$160000 \cdot \sqrt{\frac{L_m[1/2,1]}{L_n[1/2,1]}} \approx 246264$$

(with $m = 7 \cdot$ RSA-120) would be a good choice for RSA-120, and this is indeed close to our choice. The memory requirements of our programs varied between 4 and 7 megabytes. As far as we know this is considerably more than was ever used before in a large factoring project using workstations; the reason that we could afford this 'luxury' is that most new workstations, at least most of the machines that we were using, are normally equipped with 16 or more megabytes of main memory. On such machines our program fitted comfortably alongside space consuming standard software packages (windows, etc.), something which is much harder on machines with only 8 megabytes or less.

Sieving stage. We used the 'double large prime variation' of QS from [11], but we allowed large primes up to 2^{30} which is an order of magnitude larger than 10^8—more or less the customary bound.[7] As a result we collected far more data than would have followed from a straightforward extrapolation of the data from the 116-digit factorization. For that factorization 1.25 million relations involving one or two large primes were sufficient to generate 120000 combinations. For RSA-120 we collected 5105500 relations (1175252338 bytes of data), 48665 of which were full relations, 884323 were partial relations involving a single large prime, and 4172512 were partial relations involving two large primes.[8] The 5056835 partial relations generated 203557 combinations, where only 653899 of the partials actually occurred in at least one combination. This led to a total of $48665 + 203557 = 252222$ relations, which is considerably more than the $\approx 245810 + 10$ relations that we would need to factor RSA-120. The lengths of the relations in terms of the original relations are given in the table.

length	number of combinations
1	48665
2	41493
3	42958
4	37018
5	29910
6	20753
7	13821
8	7934
9	4587
10	2484
11	1305

length	number of combinations
12	646
13	303
14	173
15	86
16	48
17	22
18	6
19	5
20	1
21	3
22	1

[7] Among the 1149564 relations contributed by the first author there were 16203 single large prime relations with large prime between 2^{30} and 2^{31}.

[8] The second author contributed 30.3%, Bellcore's MasPar 26.7%, the first author 22.5%, and workstations at Bellcore, DEC SRC, and the CWI the remaining 20.5%.

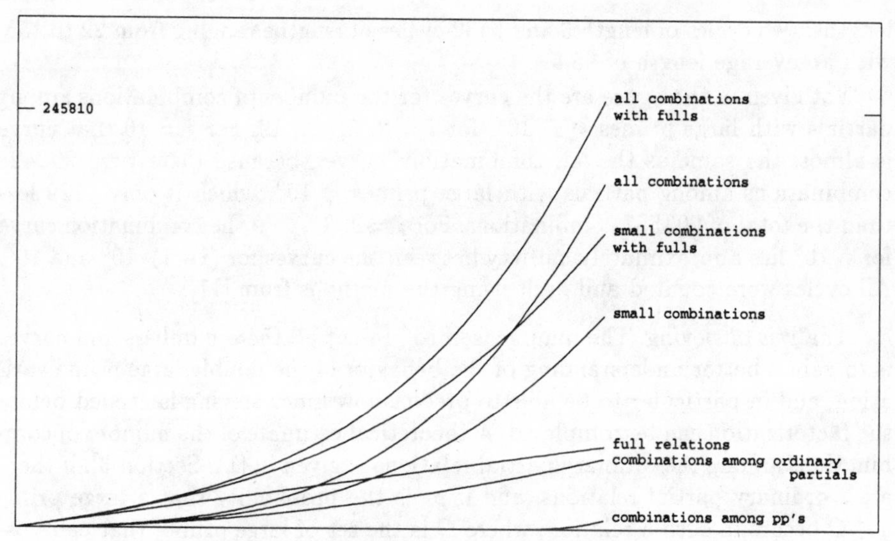

Fig. 1. Progress in the sieving stage

The total number of polynomials generated during the sieving stage was approximately 13 million, and the total number of sieve reports (i.e., actual trial divisions carried out) was approximately 100 million.

The figure presents the progress (as a function of computing time) of the relation collecting stage. The full and the partial relations behaved as usual almost as linear functions of time; the partials are not given in the figure. The upper curve, the sum of the full relations and the independent combinations among all partials, hit the 245810 mark on June 7, after 82 days of sieving. The 'small combinations' gives the growth of the number of combinations among all partials with large prime(s) less than 10^8. From this curve and the curve giving its sum with the fulls, it follows that using large primes larger than 10^8 saved us approximately two weeks of sieving; although considerably more disk space is needed to store the huge amounts of data, it follows that it is indeed a good idea to relax the usual bound of 10^8 on the large primes.

The 'combinations among ordinary partials' gives the growth of the number of combinations in the ordinary single large prime variation. These combinations of ordinary partials are known to accumulate according to a quadratic curve, and we determine the leading term of this quadratic below. The reader might observe that this leading term is sufficiently small that the factorization was completed before the quadratic term becomes visible in the picture. The lowest curve, the 'combination among pp's', is merely a curiosity, as it gives the number of cycles among the partial relations involving two large primes, excluding the partials with a single large prime. These cycles were never actually constructed, because there is no reason at all to exclude the ordinary partials, but we counted their lengths: two cycles of length 3 and 5432 cycles of lengths ranging from 22 to 155, with an average length of 95.5.

Not given in the figure are the curves for the number of combinations among partials with large primes $< i \cdot 10^8$, for $i = 2, 3, \ldots, 10$. For $i = 10$ that curve is almost the same as the 'all combinations curve', because there were 202428 combinations among partials with large primes $< 10^9$, which is only 1129 less than the total of 203557 combinations. For $i = 2, 3, \ldots, 9$ the combination curve for $i \cdot 10^8$ lies approximately halfway between the curves for $(i-1) \cdot 10^8$ and 10^9. All cycles were counted and built using the methods from [11].

Analysis of sieving. The main reason to present all these numbers and curves is to gain a better understanding of the behavior of the double large prime variation, and in particular to be able to predict how much sieving is needed before the factorization can be completed. A theoretical estimate of the number of combinations among the ordinary partial relations is given in [11: Section 3]: if there are t ordinary partial relations, and if p_q is the probability that a large prime $q \in Q$ occurs in such a relation, where Q is the set of large primes that can possibly occur in these relations, then the number of combinations is approximately $c \cdot t^2$, for $c = (\sum_{q \in Q} p_q^2)/2$.

To apply this estimate, let $Q = \{q : q \text{ prime}, 7216241 < q < 2^{30}, (\frac{m}{q}) = 1\}$, where 7216241 is the largest element in our factor base, and $m = 7 \cdot$ RSA-120. For an initial estimate of p_q one might assume that a particular large prime q occurs with probability inversely proportional to q, since $1/q$ of all numbers are multiples of q. This assumption neglects the fact that after the large prime q is removed from the number, the resulting co-factor is (almost) smooth, i.e., factors over the factor base. Since smaller numbers are more likely to be smooth than larger numbers, this may make the occurrence of larger large primes more likely. On the other hand, due to the way the sieving process works, partial relations with smaller large primes are easier to find than those with larger large primes. To take these considerations into account, we take as our model that q occurs with probability proportional to $1/q^\alpha$, for some positive $\alpha < 1$. For the 868120 ordinary partials with large prime q in Q (where $884323 - 868120 = 16203$ had large prime $> 2^{30}$), we found that $\alpha = 0.79$. This leads to $c = 5.832 \cdot 10^{-8}$, and an estimate of $c \cdot 867268^2 = 43866$ combinations. This is reasonably close to the 41490 combinations that we actually got ($41493 - 41490 = 3$ combinations of length two were among the 16203 ordinary partials with large prime $> 2^{30}$ referred to above), and we may conclude that the number of combinations of length two can fairly well be predicted, as soon as α can be estimated reliably enough given some initial collection of ordinary partials. Although other QS-factorizations lead to very similar curves, even two factorizations with the same factor base size and $\#Q$ may have entirely different α's. The curves for one of them may therefore lie much 'higher' or 'lower' than the curves for the other. This is another reason why extrapolation of QS-runtimes may be unreliable.

A theoretical analysis of the expected number of combinations of length more than two is much harder and has to our knowledge not yet been given. Consequently, we do not know of a better way to predict the yield than to plot the initial parts of the curves and extrapolate them in some reasonable way. The third author has all large prime data (138995818 bytes) available for anyone who wants to analyse these data further, compute the corresponding α (we only computed the α corresponding to the single large prime relations), or attack this problem in any other way that might take the guess-work out of future QS-runtime predictions.

We used two methods to estimate the runtime of the sieving stage. From a combination of all log-files of all our runs on DEC5000/240 workstations, we found that 5105500 relations could have been found on a single DEC5000/240 workstation in 33.02 years. Because a DEC5000/240 is a 25 MIPS machine, it follows that sieving took approximately 825 MIPS years. Another estimate is based on the MasPar runtime. Using only 3/4 of the full memory, the MasPar produced on average 480 full relations per CPU-day. This implies that 101.4 CPU-days would produce 48665 full relations, and since the MasPar is rated at approximately 3000 MIPS, we derive an estimate of $(3000 \cdot 101.4)/365 \approx 830$ MIPS years. This is too pessimistic because using the full MasPar would produce more fulls per day, and because the partials produced by the MasPar had on average smaller large primes than the partials produced by the other machines, and are therefore more likely to be combined with other partials.

We have carried out extensive tests with other choices of the factor base size, both before we started and after we had finished the sieving step. According to our computations a factor base size of 240000 would have required one percent more time, 230000 four percent, and 220000 more than seven percent. We did not attempt to analyse the effects of a factor base that is even larger than 245810.

Matrix reduction. As a result of the sieving stage we got a 252222×245810 bit-matrix. To find dependencies modulo two among its rows the third author used the technique described in [9] with the extension from [1]: structured Gaussian elimination (cf. [5; 13]) followed by the incremental version from [1] of the MasPar dense matrix eliminator described in [6]. Structured Gauss managed to reduce the size of the matrix to 89304×89088. This took 15 hours on a Sparc10 workstation, 10 of which were needed to build the 994489344 byte dense matrix, in 47 separate files of more than 21MB each. This dense matrix was smaller than expected (and smaller than one of the dense matrices from [1]) and just fitted in core on the MasPar. Reading it into core took 13 minutes, finding dependencies took 4 CPU-hours. The second dependency produced the two 60-digit prime factors of RSA-120:

$$327414555693498015751146303749141488063642403240171463406883$$

and

$$693342667110830181197325401899700641361965863127336680673013.$$

Acknowledgments. Acknowledgments are due to Paul Leyland for his helpful comments.

References

1. Bernstein, D. J., Lenstra, A. K.: A general number field sieve implementation. 103–126 in [7]
2. Buhler, J. P., Lenstra, Jr., H. W., Pomerance, C.: Factoring integers with the number field sieve. 50–94 in [7]
3. Buchmann, J. A., Denny, T. F.: An implementation of the multiple polynomial quadratic sieve, University of Saarbrücken (to appear)
4. Dixon, B., Lenstra, A. K.: Factoring integers using SIMD sieves. Advances in Cryptology, Eurocrypt '93, Lecture Notes in Comput. Sci. (to appear)
5. LaMacchia, B. A., Odlyzko, A. M.: Computation of discrete logarithms in prime fields. Designs, Codes and Cryptography **1** (1991) 47–62
6. Lenstra, A. K.: Massively parallel computing and factoring. Proceedings Latin'92, Lecture Notes in Comput. Sci. **583** (1992) 344–355
7. Lenstra, A. K., Lenstra, Jr. H. W. (eds): The development of the number field sieve. Lecture Notes in Math. **1554**, Springer-Verlag, Berlin, 1993
8. Lenstra, A. K., Lenstra, Jr., H. W., Manasse, M. S., Pollard, J. M. : The number field sieve. 11–42 in [7]
9. Lenstra, A. K., Lenstra, Jr., H. W., Manasse, M. S., Pollard, J. M.: The factorization of the ninth Fermat number. Math. Comp. **61** (1993) 319–349
10. Lenstra, A. K., Manasse, M. S.: Factoring by electronic mail. Advances in Cryptology, Eurocrypt '89, Lecture Notes in Comput. Sci. **434** (1990) 355–371
11. Lenstra, A. K., Manasse, M. S.: Factoring with two large primes. Math. Comp. (to appear); extended abstract in: Advances in Cryptology, Eurocrypt '90, Lecture Notes in Comput. Sci. **473** (1991) 72–82
12. Pomerance, C.: The quadratic sieve factoring algorithm. Lecture Notes in Comput. Sci. **209** (1985) 169–182
13. Pomerance, C., Smith, J. W.: Reduction of huge, sparse matrices over finite fields via created catastrophes. Experiment. Math. **1** (1992) 89–94
14. RSA Data Security Corporation Inc., sci.crypt, May 18, 1991; information available by sending electronic mail to challenge-rsa-list@rsa.com

Comparison of three modular reduction functions

Antoon Bosselaers, René Govaerts and Joos Vandewalle

Katholieke Universiteit Leuven, Laboratorium ESAT,
Kardinaal Mercierlaan 94, B-3001 Heverlee, Belgium.
antoon.bosselaers@esat.kuleuven.ac.be

Abstract. Three modular reduction algorithms for large integers are compared with respect to their performance in portable software: the classical algorithm, Barrett's algorithm and Montgomery's algorithm. These algorithms are a time critical step in the implementation of the modular exponentiation operation. For each of these algorithms their application in the modular exponentiation operation is considered. Modular exponentiation constitutes the basis of many well known and widely used public key cryptosystems. A fast and portable modular exponentiation will considerably enhance the speed and applicability of these systems.

1 Introduction

The widely claimed poor performance of public key cryptosystems in portable software usually results in faster, but non-portable assembly language implementations. Although they always will remain faster than their portable counterparts, their major drawback is the fact that their applicability is restricted to a limited number of computers. This means that the development effort has to be repeated for a different processor. A way out is to develop portable software that approaches the speed of an assembly language implementation as closely as possible. A primary candidate for the high level language is the versatile and standardized C language.

A basic operation in public key cryptosystems is the modular reduction of large numbers. An efficient implementation of this operation is the key to high performance. Three well known algorithms are considered and evaluated with respect to their software performance. It will be shown that they all have their specific behavior resulting in a specific field of application. No single algorithm is able to meet all demands. However a good implementation will leave minor differences in performance between the three algorithms.

In Section 2 the representation of large numbers in our implementation is discussed. The three reduction algorithms are described and evaluated in Section 3 and their behavior with respect to their argument is considered in Section 4. Section 5 looks at their use in the modular exponentiation operation. Finally, the conclusion is formulated in Section 6.

2 Representation of numbers

The three algorithms for modular reduction are described for use with large nonnegative integers expressed in radix b notation, where b can be any integer ≥ 2. Although the descriptions are quite general and unrelated to any particular computer, the best choice for b will of course be determined by the computer and the programming language used for the implementation of these algorithms. In particular, b should be chosen such that multiplications with, divisions by, and reductions modulo b^k ($k > 0$) are easy. The most obvious choice for b will therefore be one of the programming language's available integer types, in which case these three operations are reduced to respectively shifting to the left over k digits, shifting to the right over k digits (i.e., discarding the least significant k digits) and discarding all but the least significant k digits. Moreover the larger b is, the smaller the number of radix b operations to perform the same operation, and hence the faster it will be. On the other hand all multiprecision operations are performed using a number of primitive single precision operations, one of which is the multiplication of two one-digit integers giving a two-digit answer. This means that besides a basic integer type that can represent the values 0 through $b-1$, we need an integer type that is able to represent the values 0 through $(b-1)^2$. Since we normally want the ability to add and multiply concurrently [5, Algorithm 4.3.1M], we need an integer type that is able to represent the values 0 through b^2-1, i.e., a type which is at least twice as long as the basic type.

In the sequel let m be the modulus

$$m = \sum_{i=0}^{k-1} m_i b^i\,, \quad 0 < m_{k-1} < b \text{ and } 0 \leq m_i < b\,, \text{ for } i = 0, 1, \ldots, k-2\,,$$

and $x \geq m$ be the number to be reduced modulo m

$$x = \sum_{i=0}^{l-1} x_i b^i\,, \quad 0 < x_{l-1} < b \text{ and } 0 \leq x_i < b\,, \text{ for } i = 0, 1, \ldots, l-2\,,$$

both expressed in radix b notation.

3 Comparative Descriptions and Evaluation

The three algorithms to compute $x \bmod m$ are stated in terms of addition, subtraction and multiplication of both single and multiple precision integers, as well as single precision division, division by a power of b and reduction modulo a power of b. All algorithms require a precalculation, that depends only on the modulus, and hence has to be performed once for a given modulus m. Barrett's and Montgomery's methods require that the argument x is smaller than respectively b^{2k} and mb^k, where $k = \lfloor \log_b m \rfloor + 1$. If, as is mostly the case, these algorithms are used to reduce the product of two integers smaller than the modulus, this restriction will have no impact on their applicability, for then $x < m^2 < mb^k < b^{2k}$. The classical algorithm on the other hand imposes no restriction on the size of x and can easily be adapted to a general purpose division algorithm giving both quotient and remainder.

The *classical algorithm* is a formalization of the ordinary $l-k$ step pencil-and-paper method, each step of which is the division of a $(k+1)$-digit number z by the k-digit divisor m, yielding the one-digit quotient q and the k-digit remainder r. Each remainder r is less than m, so that it can be combined with the next digit of the dividend into the $(k+1)$-digit number $rb+$ (next digit of dividend) to be used as the new z in the next step.

The formalization by *D. Knuth* [5, Algorithm 4.3.1A] consists in estimating the quotient digit q as accurately as possible. Dividing the two most significant digits of z by m_{k-1} will result in an estimate that is never too small and, if $m_{k-1} \geq \lfloor \frac{b}{2} \rfloor$, at most two in error. Using an additional digit of both z and m (i.e., using the three most significant digits of z and the two most significant digits of m) this estimate can be made almost always correct, and at most one in error (an event occurring with probability $\approx 2/b$). The pseudocode of this algorithm is given in Algorithm 1.

```
if (x > mb^(l-k)) then
    x = x - mb^(l-k);
for (i = l - 1; i > k - 1; i--) do {
    if (x_i == m_(k-1)) then
        q = b - 1;
    else
        q = (x_i b + x_(i-1)) div m_(k-1);
    while (q(m_(k-1) b + m_(k-2)) > x_i b^2 + x_(i-1) b + x_(i-2)) do
        q = q - 1;
    x = x - q m b^(i-k);
    if (x < 0) then
        x = x + m b^(i-k);
}
```

Algorithm 1. Classical Algorithm ($m_{k-1} \geq \lfloor \frac{b}{2} \rfloor$)

In general the normalization $m^* = \lfloor \frac{b}{m_{k-1}} \rfloor m$ will ensure that $m^*_{k-1} \geq \lfloor \frac{b}{2} \rfloor$. On a binary computer b will be a power of 2, and hence the normalization process can be implemented more efficiently as a shift over so many bits to the left as is necessary to make the most significant bit of the most significant digit of m equal to 1. At the end the correct remainder r is obtained by applying to it the inverse of the normalization on m, i.e., by dividing it by $\lfloor \frac{b}{m_{k-1}} \rfloor$ or by shifting it to the right over the same number of bits as m was shifted over to the left during normalization.

A slightly more involved kind of normalization [7, 10] fixes one or more of the modulus' most significant digits in such a way that the most significant digit of x can be used as a first estimate for q, resulting in a faster reduction. However this normalization will increase the length of a general modulus by at least one digit, and hence all intermediate results of a modular exponentiation as well. First experiments seem to indicate that what is saved during a modular exponentiation in the modular reductions, is lost again in additional multiplications. It is as yet unclear whether further optimalization will result in a faster modular exponentiation.

P. Barrett [1] introduced the idea of estimating the quotient x div m with operations that either are less expensive in time than a multiprecision division by m (viz., 2 divisions by a power of b and a partial multiprecision multiplication), or can be done as a precalculation for a given m (viz., $\mu = b^{2k}$ div m, i.e., μ is a scaled estimate of the modulus' reciprocal). The estimate $\hat{q}$ of x div m is obtained by replacing the floating point divisions in $q = \lfloor (x/b^{2k-t})(b^{2k}/m)/b^t \rfloor$ by integer divisions:

$$\hat{q} = \left((x \text{ div } b^{2k-t})\mu\right) \text{ div } b^t .$$

This estimate will never be too large and, if $k < t \leq 2k$, the error is at most two:

$$x \text{ div } m - 2 \leq \hat{q} \leq x \text{ div } m , \quad \text{for } k < t \leq 2k .$$

It can be shown that for about 90% of the values of $x < m^2$ and m the initial value of $\hat{q}$ will be equal to x div m and only in 1% of cases $\hat{q}$ will be two in error. The only influence of the t least significant digits of the product $(x \text{ div } b^{2k-t})\mu$ on the most significant part of this product is the carry from position t to position $t+1$. This carry can be accurately estimated by only calculating the digits at position $t-1$ and t, which has the advantage that the calculation of the $t-2$ least significant digits of the product is avoided. The resulting quotient is never too large and almost always the same as $\hat{q}$, and, if $b > l-k$, at most one in error. Moreover the number of single precision multiplications and the resulting error are more or less independent of t. The best choice for t, resulting in the least single precision multiplications and the smallest maximal error, is $k+1$, which also was Barrett's original choice. The calculation of $\hat{q}$ can be speeded up even slightly more by normalizing m, such that $m_{k-1} \geq \lfloor \frac{b}{2} \rfloor$. This way $l-k+1$ single precision multiplications can be transformed into as many additions.

An estimate $\hat{r}$ for x mod m is then given by $\hat{r} = x - \hat{q}m$, or, as $\hat{r} < b^{k+1}$ (if $b > 2$), by

$$\hat{r} = (x \bmod b^{k+1} - (\hat{q}m) \bmod b^{k+1}) \bmod b^{k+1} ,$$

which means that once again only a partial multiprecision multiplication is needed. At most two further subtractions of m are required to obtain the correct remainder. Barrett's algorithm can therefore be implemented according to the pseudocode of Algorithm 2.

```
q = ((x div b^{k-1})μ) div b^{k+1};
x = x mod b^{k+1} - (qm) mod b^{k+1};
if (x < 0) then
    x = x + b^{k+1};
while (x ≥ m) do
    x = x - m;
```

Algorithm 2. Barrett's Algorithm ($\mu = b^{2k}$ div m)

By representing the residue classes modulo m in a nonstandard way, *Montgomery's method* [6] replaces a division by m with a multiplication followed by a division by a power of b. This operation will be called Montgomery reduction.

Let $R > m$ be an integer relatively prime to m such that computations modulo R are easy to process: $R = b^k$. Notice that the condition $\gcd(m, b) = 1$ means that this method can not be used for all moduli. In case b is a power of 2, it simply means that m should be odd. The m-residue with respect to R of an integer $x < m$ is defined as $xR \bmod m$. The set $\{xR \bmod m \,|\, 0 \leq x < m\}$ clearly forms a complete residue system. The Montgomery reduction of x is defined as $xR^{-1} \bmod m$, where R^{-1} is the inverse of R modulo m, and is the inverse operation of the m-residue transformation. It can be shown that the multiplication of two m-residues followed by Montgomery reduction is isomorphic to the ordinary modular multiplication.

The rationale behind the m-residue transformation is the ability to perform a Montgomery reduction $xR^{-1} \bmod m$ for $0 \leq x < Rm$ in almost the same time as a multiplication. This is based on the following theorem:

Theorem 1 P. Montgomery. *Let $m' = -m^{-1} \bmod R$. If $\gcd(m, R) = 1$, then for all integers x, $(x + tm)/R$ is an integer satisfying*

$$\frac{x + tm}{R} \equiv xR^{-1} \pmod{m}$$

where $t = xm' \bmod R$.

It can easily be verified that the estimate $\hat{x} = (x + tm)/R$ for $xR^{-1} \bmod m$ is never too small and the error is at most one. This means that a Montgomery reduction is not more expensive than two multiplications, and one can do even better: almost twice as fast. Hereto, it is sufficient to observe [2] that the basic idea of Montgomery's Theorem is to make x a multiple of R by adding multiples of m. Instead of computing all of t at once, one can compute one digit t_i at a time, add $t_i m b^i$ to x, and repeat. This change allows to compute $m_0' = -m_0^{-1} \bmod b$ instead of m'. It turns out to be a generalization of Hensel's odd division for computing inverses of "2-adic" numbers (introduced by *K. Hensel* around the turn of the century, see e.g., [3]) to a representation using b-ary numbers that have $\gcd(m_0, b) = 1$ [9].

A Montgomery modular reduction can be implemented according to the pseudocode of Algorithm 3. If x is the product of two m-residues, the result is the m-residue of the remainder, and the remainder itself is obtained by applying one additional Montgomery reduction. However both the initial m-residue transformation of the argument(s) and the final inverse transformation (Montgomery reduction) are only necessary at the beginning, respectively the end of an operation using Montgomery reduction (e.g., a modular exponentiation).

```
for (i = 0; i < k; i++) do {
    t_i = (x_i · m'_0) mod b;
    x = x + t_i m b^i;
}
x = x div b^k;
if (x >= m) then
    x = x - m;
```

Algorithm 3. Montgomery's Algorithm ($m'_0 = -m_0^{-1} \bmod b$, Hensel's b-ary division)

An indication of the attainable performance of the different algorithms will be given by the number of single precision multiplications and divisions necessary to reduce an argument twice as long as the modulus ($l = 2k$). This approach is justified by the fact that a multiplication and a division are the most time consuming operations in the inner loops of all three algorithms, with respect to which the others are negligible. The number of multiplications and divisions in Table 1 are only for the reduction operation, i.e., they do not include the multiplications and divisions of the precalculation, the argument transformation, and the postcalculation. Our reference operation is the multiplication of two k-digit numbers.

Table 1 indicates that if only the reduction operation is considered (i.e., without the precalculations, argument transformations, and postcalculations) and for arguments twice the length of the modulus, Montgomery's algorithm (only for moduli m for which $\gcd(m_0, b) = 1$) is clearly faster than both Barrett's and the classical one and almost as fast as a multiplication. Barrett's and the classical algorithm will be almost equally fast, with a slight advantage for Barrett.

These observations are confirmed by a software implementation of these algorithms, see Table 2. The implementation is written in ANSI C [4] and hence should be portable to any computer for which an implementation of the ANSI C standard exists. All figures in this article are obtained on a 33 MHz 80386 based

Table 1. Complexity of the three reduction algorithms in reducing a $2k$-digit number x modulo a k-digit modulus m.

Algorithm	Classical	Barrett	Montgomery	Multiplication
Multiplications	$k(k+2.5)$	$k(k+4)$	$k(k+1)$	k^2
Divisions	k	0	0	0
Precalculation	Normalization	b^{2k} div m	$-m_0^{-1}$ mod b	None
Arg. transformation	None	None	m-residue	None
Postcalculation	Unnormalization	None	Reduction	None
Restrictions	None	$x < b^{2k}$	$x < mb^k$	None

Table 2. Execution times for the reduction of a $2k$-digit number modulo a k-digit modulus m for the three reduction algorithms compared to the execution time of a $k \times k$-digit multiplication ($b = 2^{16}$, on a 33 MHz 80386 based PC with WATCOM C/386 9.0).

	Length of	Times in mseconds			
k	m in bits	Classical	Barrett	Montgomery	Multiplication
8	128	0.278	0.312	0.205	0.182
16	256	0.870	0.871	0.668	0.632
32	512	3.05	2.84	2.43	2.36
48	768	6.56	5.96	5.33	5.19
64	1024	11.39	10.23	9.33	9.12

PC using the 32-bit compiler WATCOM C/386 9.0. The radix b is equal to 2^{16}, which means that Montgomery's algorithm is only applicable to odd moduli.

However an operation using Barrett's or Montgomery's modular reduction methods will only be faster than the same operation using the classical modular reduction if the pre- and postcalculations and the m-residue transformation (only for Montgomery) are subsequently compensated for by enough (faster) modular reductions. An example of such an operation is modular exponentiation. This also means that for a single modular reduction the classical algorithm is the obvious choice, as the pre- and postcalculation only involve a very fast and straightforward normalization process.

4 Behavior w.r.t. argument

The execution time for the three reduction functions depends in a different way on the length of the argument. The time for a reduction using the classical algorithm or Barrett's method will vary linearly between their maximum value (for an argument twice as long as the modulus) and almost zero (for an argument as long as the modulus). For arguments smaller than the modulus no reduction takes place, as they are already reduced. On the other hand, the time for a reduction using Montgomery's method will be independent of the length of the argument. This is a consequence of the fact that in all cases, whatever the value of the argument, a modular multiplication by R^{-1} takes place. This means that both the classical algorithm and Barrett's method will be faster than Montgomery's method below a certain length of the argument. This is illustrated in Figure 1 for a 512-bit modulus. However in most cases the argument will be close to twice the length of the modulus, as it normally is the product of two values close in length to that of the modulus.

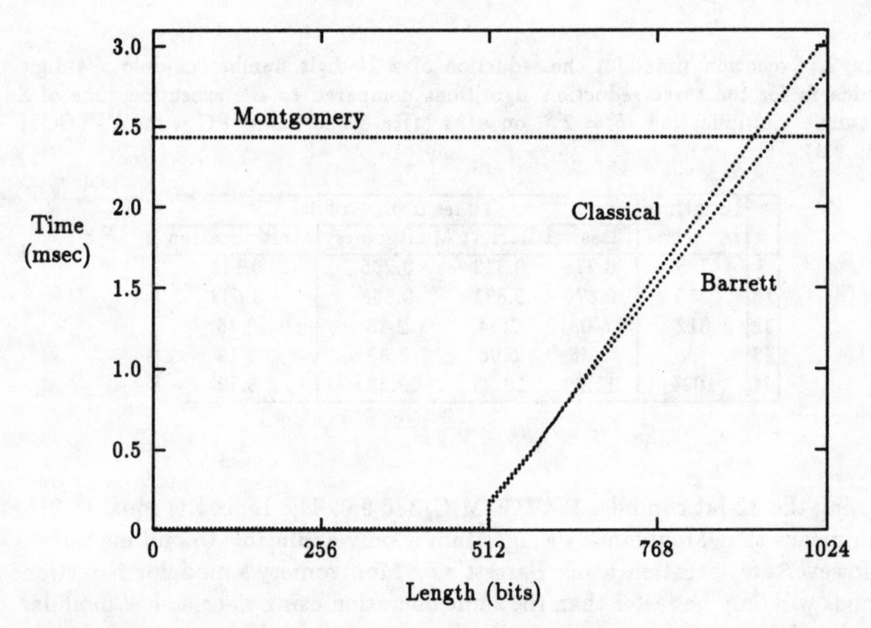

Fig. 1. Typical behavior of the three reduction functions in reducing a number up to twice the length of the modulus ($b = 2^{16}$, length of the modulus = 512 bits, on a 33 MHz 80386 based PC with WATCOM C/386 9.0).

In addition, all the modular reduction functions have, for a given length, input values for which they perform faster than average in reducing them. For some of these inputs the gain in speed can be quite substantial. Since these input values are different for each of the reduction functions, none of the functions is the fastest for all inputs of a given length.

Montgomery's method will be faster than average in reducing m-residues with consecutive zeroes in its *least* significant digit positions. The gain in speed will be directly proportional to the number of zero digits. The same applies to arguments that produce, after n steps ($0 < n < k$) in Montgomery's algorithm, a number of consecutive zero digits in the intermediate value x. For example, the argument

$$x = hb^k + b^k - (\sum_{i=0}^{n-1} t_i m b^i) \bmod b^k ,$$

where

$$0 < h < b^{l-k}$$
$$t_i = (y_i(-m_0^{-1} \bmod b)) \bmod b , \quad 0 \le y_i < b ,$$

produces after n steps $k-n$ consecutive zeroes, with once again a speed gain directly proportional to the number of consecutive zero digits.

Barrett's method will be faster than average, and possibly faster than Montgomery's method, for an argument x with zero digits among its $k+1$ *most* significant digits or that produces an approximation $\hat{q}$ of x div m containing zero digits. An example of the latter will be encountered in the next paragraph.

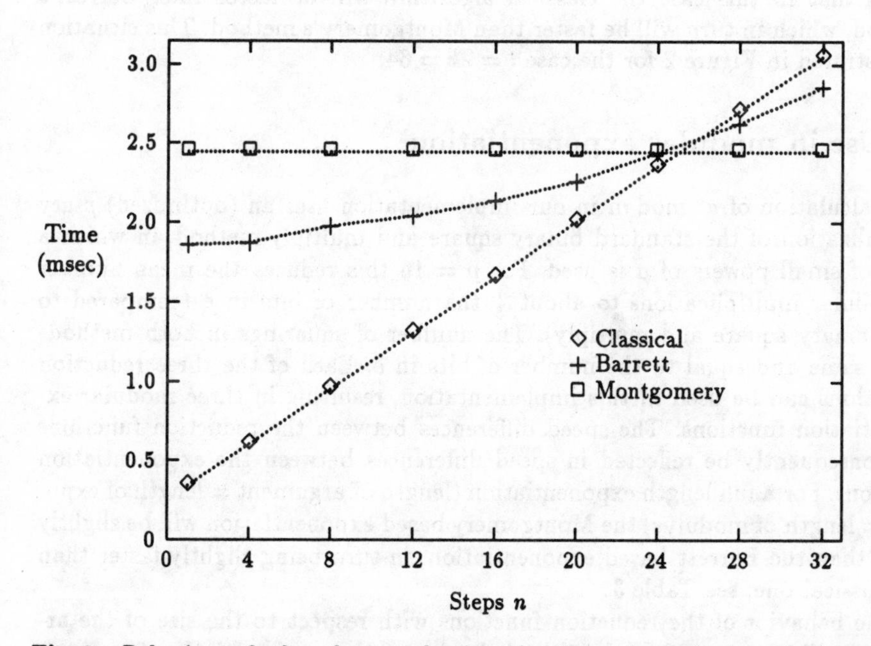

Fig. 2. Behavior of the three reduction functions in reducing the argument $x = gmb^{k-n} + h$, where $0 < n \le k$, $0 < g < b^n$ and $0 \le h < m$ for the case $k = 32$ ($b = 2^{16}$, length of the modulus = 512 bits, on a 33 MHz 80386 based PC with WATCOM C/386 9.0).

The central part of the classical algorithm is the $(l-k)$-fold loop, in each iteration of which a digit of the quotient x div m is determined. Therefore the classical algorithm will be faster than average, and possibly faster than Montgomery's and Barrett's method, for an argument that produces a quotient with a number of zero digits. For example, the argument

$$x = gmb^{l-k-n} + h, \quad \begin{array}{l} k < l \leq 2k \\ 0 < n \leq l-k \\ 0 < g < b^n \\ 0 \leq h < m, \end{array}$$

produces a quotient $q = gb^{l-k-n}$ containing $l-k-n$ zero digits in its *least* significant positions, and hence only n steps of the central loop will be executed. As the time for a reduction using the classical algorithm is clearly directly proportional to the number of non-void steps in the central loop, the reduction of the above argument will be considerably faster than average. Moreover, since the actual quotient contains $l-k-n$ zero digits, the reduction of this argument using Barrett's method will be faster than average as well: in 90% of the cases the approximation $\hat{q}$ will be equal to q, and hence the multiplication $\hat{q}m \bmod b^{k+1}$ will consist of n steps only instead of the $l-k$ steps in the average case. This means that in this case the classical algorithm will be faster than Barrett's method, which in turn will be faster than Montgomery's method. This situation is illustrated in Figure 2 for the case $l = 2k = 64$.

5 Use in modular exponentiation

The calculation of $a^e \bmod m$ in our implementation uses an (optimized) p-ary generalization of the standard binary square and multiply method, in which a table of small powers of a is used. For $p = 16$ this reduces the mean number of modular multiplications to about $\frac{1}{5}$ the number of bits in e (compared to $\frac{1}{2}$ for binary square and multiply). The number of squarings in both methods is the same and equal to the number of bits in e. Each of the three reduction algorithms can be used in this implementation, resulting in three modular exponentiation functions. The speed differences between the reduction functions will consequently be reflected in speed differences between the exponentiation functions. For a full length exponentiation (length of argument = length of exponent = length of modulus) the Montgomery based exponentiation will be slightly faster than the Barrett based exponentiation, in turn being slightly faster than the classical one, see Table 3

The behavior of the reduction functions with respect to the size of the argument will also be reflected in the behavior of the exponentiation functions. The exponentiation of an argument a smaller in length than the modulus will for the classical and Barrett's algorithm result in a table of small powers of a containing values which are still smaller in length than the modulus. Hence each multiplication by an entry of this table will yield a product that is shorter than twice the length of the modulus. The subsequent reduction will be faster than average, as the execution time of the classical and Barrett's algorithm depends linearly on the length of its argument. For these two algorithms the exponentiation of an argument smaller in length than the modulus will thus be faster than an exponentiation of a full length argument. Moreover for small enough

Table 3. Execution times for a full length modular exponentiation (length of argument = length of exponent = length of modulus, $b = 2^{16}$, on a 33 MHz 80386 based PC with WATCOM C/386 9.0).

Length of m in bits	Times in seconds Classical	Barrett	Montgomery
128	0.072	0.078	0.062
256	0.430	0.430	0.366
512	2.95	2.83	2.55
768	9.46	8.90	8.28
1024	21.74	20.30	19.14

arguments the exponentiation using these algorithms will be even faster than the exponentiation using Montgomery's modular reduction, which is explained by the fact that for these arguments not only the products but also some squares will be shorter than twice the length of the modulus. This is illustrated in Figure 3. Barrett based exponentiation is therefore the best choice to perform Rabin primality tests [8] with small bases.

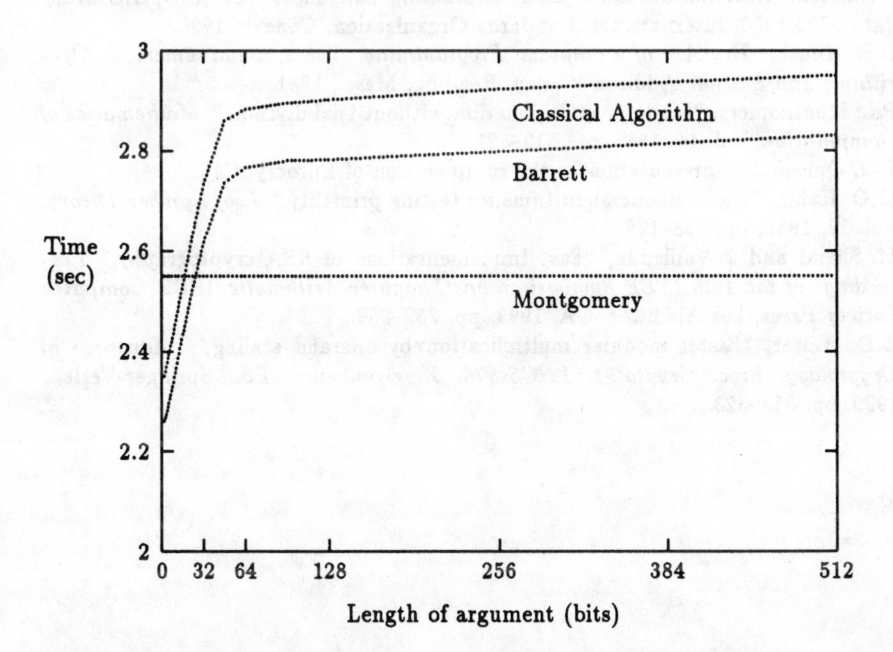

Fig. 3. Typical behavior of the exponentiation functions based on the three reduction functions in exponentiating a number up to the length of the modulus ($b = 2^{16}$, length of modulus and exponent = 512 bits, on a 33 MHz 80386 based PC with WATCOM C/386 9.0).

6 Conclusion

A theoretical and practical comparison has been made of three algorithms for the reduction of large numbers. It has been shown that in a good portable implementation the three algorithms are quite close to each other in performance. The classical algorithm is the best choice for single modular reductions. Modular exponentiation based on Barrett's algorithm is superior to the others for small arguments. For general modular exponentiations the exponentiation based on Montgomery's algorithm has the best performance.

References

1. P.D. Barrett, "Implementing the Rivest Shamir and Adleman public key encryption algorithm on a standard digital signal processor," *Advances in Cryptology, Proc. Crypto'86, LNCS 263*, A.M. Odlyzko, Ed., Springer-Verlag, 1987, pp. 311–323.
2. S.R. Dussé and B.S. Kaliski, "A cryptographic library for the Motorola DSP56000," *Advances in Cryptology, Proc. Eurocrypt'90, LNCS 473*, I.B. Damgård, Ed., Springer-Verlag, 1991, pp. 230–244.
3. K. Hensel, *Theorie der algebraischen Zahlen*, Leipzig, 1908.
4. *"American National Standard for Programming Languages—C,"* ISO/IEC Standard 9899:1990, International Standards Organization, Geneva, 1990.
5. D.E. Knuth, *The Art of Computer Programming, Vol. 2, Seminumerical Algorithms, 2nd Edition*, Addison-Wesley, Reading, Mass., 1981.
6. P.L. Montgomery, "Modular multiplication without trial division," *Mathematics of Computation*, Vol. 44, 1985, pp. 519–521.
7. J.-J. Quisquater, presentation at the rump session of Eurocrypt'90.
8. M.O. Rabin, "Probabilistic algorithms for testing primality," *J. of Number Theory*, Vol. 12, 1980, pp. 128–128.
9. M. Shand and J. Vuillemin, "Fast Implementations of RSA cryptography," *Proceedings of the 11th IEEE Symposium on Computer Arithmetic*, IEEE Computer Society Press, Los Alamitos, CA, 1993, pp. 252–259.
10. C.D. Walter, "Faster modular multiplication by operand scaling," *Advances in Cryptology, Proc. Crypto'91, LNCS 576*, J. Feigenbaum, Ed., Springer-Verlag, 1992, pp. 313–323.

Differential Cryptanalysis of Lucifer

Ishai Ben-Aroya Eli Biham
Computer Science Department
Technion - Israel Institute of Technology
Haifa 32000, Israel

Abstract

Differential cryptanalysis was introduced as an approach to analyze the security of DES-like cryptosystems. The first example of a DES-like cryptosystem was Lucifer, the direct predecessor of DES, which is still believed by many people to be much more secure than DES, since it has 128 key bits, and since no attacks against (the full variant of) Lucifer were ever reported in the cryptographic literature. In this paper we introduce a new extension of differential cryptanalysis, devised to extend the class of vulnerable cryptosystems. This new extension suggests key-dependent characteristics, called *conditional characteristics*, selected to enlarge the characteristics' probabilities for keys in subsets of the key space. The application of conditional characteristics to Lucifer shows that more than half of the keys of Lucifer are insecure, and the attack requires about 2^{36} complexity and chosen plaintexts to find these keys. The same extension can also be used to attack a new variant of DES, called RDES, which was designed to be immune against differential cryptanalysis. These new attacks flash new light on the design of DES, and show that the transition of Lucifer to DES strengthened the later cryptosystem.

1 Introduction

Differential cryptanalysis was introduced in [3,2] as an approach to analyze the security of DES-like cryptosystems. In a series of papers[3,4,5,6] this approach was used to attack the blockciphers DES[17], Feal[21,16], Khafre[14], REDOC-II[23], LOKI[8] and one variant of Lucifer[10], along with the hash functions N-Hash[15], and Snefru[13]. Lai et al[12] viewed a variant of this approach as a Markov chain and applied this approach to the PES cipher. Other researchers studied how to immune cryptosystems against differential cryptanalysis (some of which are [1,7,9,18,19,20]).

In this paper we extend differential cryptanalysis in several directions: The main extension of this paper lets differential cryptanalysis to analyze a wider set of cryptosystems. We define *conditional characteristics* as key-dependent characteristics selected to maximize the characteristic's probability (the fraction of right pairs) for only a specific subset of the key space. The required coverage of (almost) all the key space is done via selection of several conditional characteristics designed for different fractions of the key space.

In the attack on the full 16-round DES[6], structures which allow to gain one additional round for free with no additional cost are used. We extend this idea and show an implementation in which we gain two additional rounds for free, using the observations that the blocksize of Lucifer is larger than the one of DES and that the avalanche is slower. We also show two additional tools: a tool that gains a free additional round in Lucifer (described in the attack on the eight-round variant), and a tool that can enlarge the fraction of keys covered by differential cryptanalytic attacks when conditional characteristics are used. We suggest to use sets of characteristics whose Ω_P are the same, but which differ in their Ω_T. Since the same plaintexts can be shared for all these characteristics, the efficiency of the attacks is enlarged.

Many people still believe that Lucifer[22], the direct predecessor of DES, is stronger than DES, since it has 128 key bits rather than the 56 key bits of DES, and since they believe that the strength of DES was intentionally reduced by its designers. In this paper we study the strength of the variant of Lucifer described in [22] (the final variant of the Lucifer project, rather than the variant described in [10]). We apply our new techniques to this variant, and show an attack which can find the key with complexity about 2^{36}, if only the key resides within a particular subset of the key space containing about 55% of the keys. It is of interest to note that if the order of the two S boxes of Lucifer was reversed, a similar attack could cover more than 90% of the keys, but their replacement by S boxes satisfying the design rules of DES would invalidate the conditional characteristics used in this attack.

Several researchers studied how to make cryptosystems immune against differential cryptanalysis, but till now, this effort was not very successful. Many of them[1,9,18] suggested the use of S boxes whose difference distribution tables are uniform, and in particular they suggested the use of bent functions. However, the application of this suggestion to DES was studied in [2,7], and it was shown that the resultant cryptosystems become much weaker than DES.

Recently, Koyama and Terada[11] suggested to replace the deterministic swapping of the halves of the data between rounds in DES by a conditional swapping, which swap the halves only if a particular key bit (different for each round) has the value 1. They claim that the resultant cryptosystem, called RDES, is about 2^{15} times stronger than DES, although a small fraction of the keys, which do not swap the data even once, are bad. Our new extension developed in this paper can be applied to RDES, and shows that RDES is weaker than DES for almost all keys in the key space, leaving only a relatively small number of "good" keys, whose trial complexity is much smaller than exhaustive search of the whole key space.

2 Description of Lucifer

Lucifer[22] is the cryptosystem from which DES[17] was developed by IBM in the 1970's. Like DES, Lucifer has 16 rounds, but it has no initial and final permutations, and the sizes of its blocks and keys are 128 bits. The F function of Lucifer operates on the 64-bit right half of the data, 64-bit subkey and eight interchange control bits (ICBs). The F function uses only two four-bit to four-bit S boxes, called *S0* and *S1*. It swaps the two nibbles (four bits) of its input bytes whose corresponding interchange control bit (ICB) is zero. Then, the S box S0 operates on the least significant nibble, and S1 on the most significant nibble of every byte. The output of the S boxes is XORed with the subkey, in an operation called *key interruption*. The last stage of the F function permutes the output bits. Sorkin[22] describes

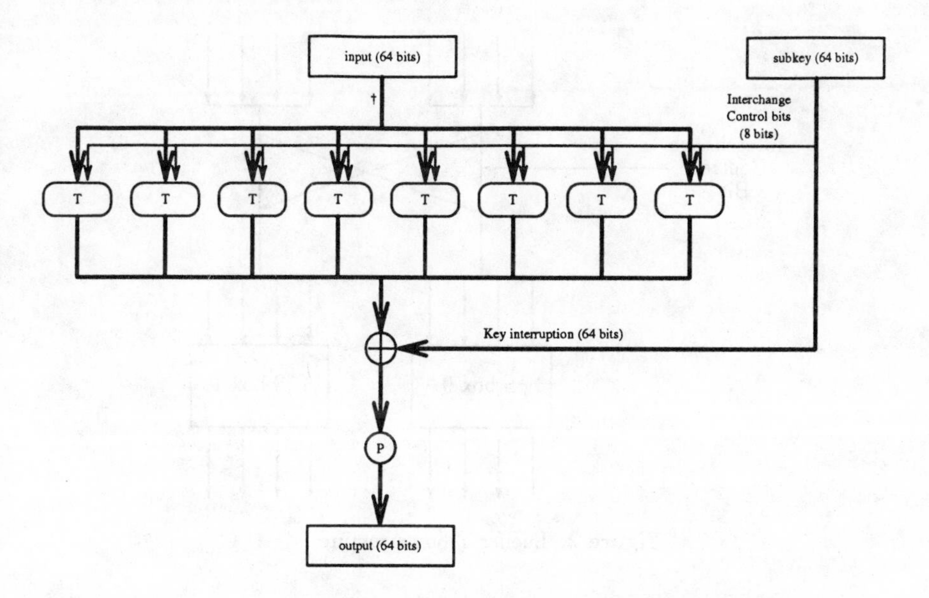

Figure 1. The F function of Lucifer.

the final bit permutation in two steps: each byte undergoes a fixed permutation (denoted P in [22]), and then the bits are mixed between the bytes – every bit enters a different byte in the same position in which it was in the original byte. This later step is called *diffusion*. We denote the product of these two permutations by P.

Figure 1 describes the F function of Lucifer. The pairs of adjacent S boxes are viewed as single combined boxes, to which we call *T boxes* (Transposition boxes). The T boxes are functions from nine bits to eight bits, whose one input bit is an ICB, and the eight others are data. The T boxes are defined by

$$\begin{aligned} T_0[XY] &= S0[X]\,S1[Y] \\ T_1[XY] &= S0[Y]\,S1[X]. \end{aligned}$$

and are described in Figure 2.

The key scheduling algorithm of Lucifer is much simpler than the one of DES. The key is assigned into a 128-bit shift register. Every round the subkey is chosen as the leftmost 64 bits of the register, the interchange control bits are chosen as the leftmost eight bits of the register, and after each round the shift register is rotated 56 bits to the left.

For the analysis it is convenient to use the following equivalent description: The key interruption is moved from after the S boxes to become the first operation in the F function (where a † is marked in Figure 1), and an initial XOR of the plaintext with a 128-bit subkey is added before the first round. The subkeys of this form are called *actual subkeys*, and are denoted by AKi. The actual subkey of the last round ($AK16$) is zero. $AK15$ is just the permuted value of the subkey of the last round ($AK15 = P(K16)$). The other actual subkeys $AK1,\ldots,AK14$ are $AK_i = AK(i+2) \oplus P(K(i+1))$, and the initial subkey is $(AK2 \oplus P(K1), AK1)$. In this description the last round becomes very simple, with a zero

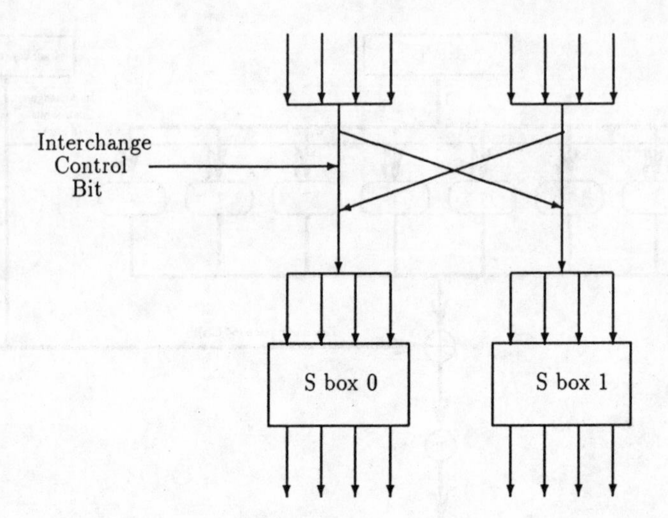

Figure 2. Lucifer T box structure.

actual subkey and the actual subkey of the first round is cancelled by the initial subkey. Thus effectively both the first and the last rounds have no key interruption.

We also denote the ciphertext by T, and its left and right halves by T_L and T_R respectively.

3 Conditional Characteristics

Differential cryptanalysis requires one to find good characteristics, i.e., to find pairs of messages, such that the difference of the output of the nth round during encryption of these messages is predictable with a relatively high probability. The key-dependent swaps make it quite difficult to find such characteristics, especially since characteristics which can predict the output for all the keys have a very low probability – thus making an attack infeasible. In order to solve this difficulty we define key-dependent characteristics that depends on the value of some ICBs. In [3,2] the characteristic's probability is defined as the probability that a random pair (whose plaintext difference is Ω_P) is a right pair with respect to a random key, and it is shown that the probability that a random pair is a right pair with respect to a fixed key may depend on the choice of the key. In this paper we are interested in characteristics for which the probability that a random pair is a right pair vary between different keys. We call these characteristics *conditional characteristics.*

Definition 1 The *probability of a characteristic Ω with respect to a fixed key K* is the probability that a random pair (whose plaintext difference is Ω_P) is a right pair with respect to the fixed key K.

Definition 2 The *probability of a characteristic Ω with respect to a set of keys U* is the minimal probability of the characteristic Ω with respect to a key K in U.

Definition 3 A *conditional characteristic* is a tuple (Ω, U, p_U^Ω) where Ω is a characteristic, U is a subset of the key space, and p_U^Ω is the probability of the characteristic Ω with respect to the subset U.

Definition 4 The *key fraction* of a conditional characteristic (Ω, U, p_U^Ω) is the ratio $|U|/|K|$ between the size of the set U and the size of the key space.

These definitions suggest a tradeoff between the probability of a conditional characteristic and its key fraction. By reducing the size of U we can enlarge the probability of the conditional characteristic, but the key fraction is reduced. By enlarging the size of U we enlarge the key fraction, but the probability may be reduced.

Whenever a conditional characteristic (Ω, U, p_U^Ω) improves the probability over the best probability of a non-conditional characteristic by a factor higher that the inverse of the key fraction ($|K|/|U|$), the usage of the conditional characteristic is advisable. There are several additional cases in which the usage of conditional characteristics is advisable as well, especially if several such characteristics can efficiently share the same structure of chosen plaintexts.

We found four six-round iterative conditional characteristics of Lucifer. One of them is (only three rounds are described; the other three rounds are symmetric):

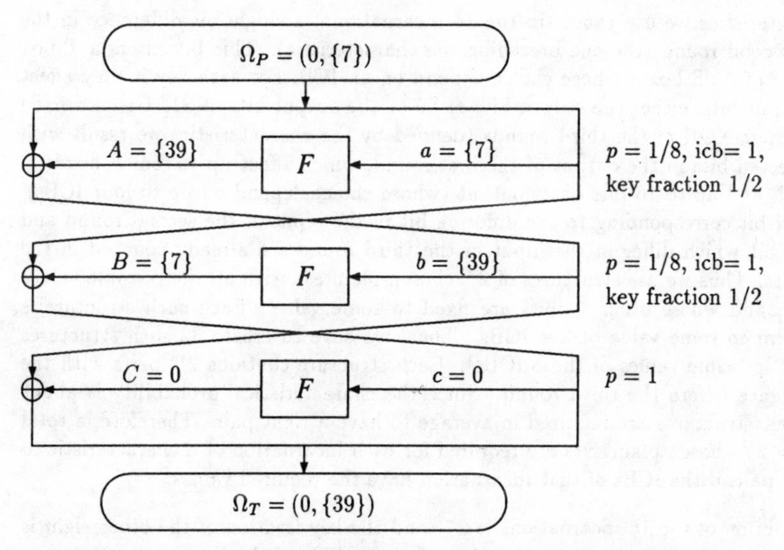

where $\{n\}$ denotes a 64-bit value whose nth bit ($n \in \{0, \ldots, 63\}$) is one and all the others are zero. The other three iterative conditional characteristics are similar with the replacement of the constants $\{7\}$ and $\{39\}$ by the constants (1) $\{15\}$ and $\{47\}$, (2) $\{23\}$ and $\{55\}$, and (3) $\{31\}$ and $\{63\}$. Each of these characteristics has six incarnations, starting from the six possible rounds.

4 The Attack on Lucifer

The differential cryptanalysis of Lucifer is slightly different than the cryptanalysis of DES. We describe the differential cryptanalysis of Lucifer in the following subsections. In the first subsection we describe the required structures and chosen plaintexts, then we describe the cryptanalysis, and finally we study modified variants and strength factors.

4.1 The Data

In order to pack all the required pairs into as few chosen plaintexts as possible, we use structures similar to the ones used in [6]. In [6] an additional "free" first round is gained and the characteristic starts only at the second round. Due to the larger blocksize of Lucifer, and to the slower avalanche, we can use two such "free" rounds in our attack on Lucifer. We use 3R-attacks, and thus, 11-round characteristics are required. The above conditional characteristics, iterated to 11 rounds, have probability 2^{-21} and a key fraction 2^{-7} in 16 of the incarnations, and probability 2^{-24} and a key fraction 2^{-8} in eight of the incarnations. In the rest of this section we ignore the details of the required data and the analysis of the eight incarnations, since (paradoxically) they require fewer chosen plaintexts and simpler structures than the other 16 incarnations.

The characteristics we use cause (in the 16 incarnations) a single bit difference in the input to the second round (the one preceding the characteristic). This bit enters a T box and affects one of its S boxes whose choice depend on an ICB. For each key it may affect up to four output bits, either the output bits of S0 or the output bits of S1. Given a fixed value of the input XOR of the third rounds (defined by the characteristic) we result with up to four affected bits in the output of the first round, which affect up to four S boxes in the first round, and up to 16 bits of its output (whose choice depend on up to four ICBs). The additional bit corresponding to the differing bit in the input of the second round and the (possible) bit which differ in the input of the third round are already counted in the $4+16=20$ bits. Thus we use structures of 2^{20} chosen plaintext with all the possible values of the 20 bits, and whose other 44 bits are fixed to some value. Each such structure is built to conform to some value of five ICBs. Thus, we have to create 32 such structures with all the 32 possible values of these ICBs[1]. Each structure contains 2^{19} pairs with the required difference before the third round. Since the characteristics' probability is about 2^{-21}, about four structures are required in average to have a right pair. Therefore, a total of $2^{20} \cdot 4 \cdot 32 = 2^{27}$ chosen plaintexts are required for each incarnation of a characteristic to have one right pair, if the ICBs of that incarnation have the required values.

The key fraction of the 16 incarnations is 2^{-7} and the key fraction of the other eight is 2^{-8}. The 24 incarnations cover a total fraction of about 15% of the key space. However, when we use some duplication techniques, which duplicate either the required data or the analysis for the two possible values of the extreme ICBs of the characteristics, we can enrich the set of covered keys and cover a fraction of about 25% of the key space. For this fraction, about $2^{27} \cdot 24 \cdot 16 \cdot 2 \approx 2^{36}$ chosen plaintexts are required (24 incarnations, 16 right pairs, 2 is the maximal duplication of the data).

[1] If the number of chosen plaintexts required was much larger, we could build huge more efficient structures for which such duplication is not required.

We can enlarge the fraction of covered keys further using the observation that there are several conditional characteristics with the same Ω_P as of the characteristics we use, but with different Ω_T's and different key subsets U. Each Ω_P we use have about 9–10 such characteristics whose total key fraction is about three times the original key fraction, and their probabilities are about the same as of the original characteristics. In the final version of this paper we will show these additional characteristics. Due to the almost perfect identification of wrong pairs this attack have, we can analyze these characteristics with a negligible additional cost with the same data. Thus, this attack covers a fraction of about 55% of the keyspace[2]. We can still enlarge this fraction slightly using characteristics whose key fraction is slightly smaller than of the ones described, but whose Ω_P's have many additional characteristics with different Ω_T's.

4.2 The Analysis

For the analysis we use the notation h to be the input of the F function of the last round in the equivalent description of Lucifer, g and f are the inputs to the two preceding rounds, and H, G and F are the outputs of the F function in these rounds.

The first step of the analysis discards as many wrong pairs as possible. The value of f' contains at most one non-zero bit, thus, many bits in F' are zero, and at most four bits of F' are non-zero, the exact four bits are ICB dependent. The value of g' may contain at most five non-zero bits (these four bits plus one bit from e'), which may affect the output of at most five S boxes in G', and thus, $h' = T'_R$ may have at most $5 \cdot 4 + 1 = 21$ non-zero bits in positions depending on at most five ICBs. Thus the probability that a random T'_R is zero at all the 43 bits suggested by one of the 2^5 choices of the five ICBs is about $2^{-43} \cdot 2^5 = 2^{-38}$. Therefore, the identification of wrong pairs in a structure can be done efficiently by sorting (or hashing) by these bits, and choosing only pairs with common values. Each structure contains up to $(2^{21})^2/2 = 2^{41}$ potential pairs, and thus the average number of remaining (wrong) pairs per structure is expected to be less than eight for each characteristic.

Since effectively there is no key interruption in the last round, and since $h = T_R$, we can calculate for any ciphertext the 256 possible outputs H of the F function of the last round using the 256 possible choices of the interchange control bits, and get 256 possible values for H' for any pair. Independently, we can calculate 56 bits of H' for any pair, using the facts that $H' = T'_L \oplus F' \oplus e'$ and that 56 particular bits of F' are zero. This value should match one of the 256 possible values calculated directly. If it does not match, the pair is clearly a wrong pair, and should be discarded. The probability of a random pair to pass this test is about $2^8 \cdot 2^{-56} = 2^{-48}$. Thus, the average number of wrong pairs in a structure which pass both the previous test and this test is $8 \cdot 2^{-48} = 2^{-45}$ for each characteristic. In practice only right pairs are expected to pass both tests. From these right pairs we can easily derive the values of seven ICBs of the last round, the seven ICBs controlling the conditional characteristic, the five ICBs affecting rounds 14 and 15 during the analysis, and the five ICBs affecting the choice of the chosen plaintexts in the first two rounds. All these ICBs are different (since each key bit is used only once as an ICB) and thus we get a total of $7 + 7 + 5 + 5 = 24$ bits of the key.

Now we can calculate the output of the F function of the last round for any given

[2]It can be verified easily (but inaccurately) by $1 - (1 - 0.25)^3 = 0.58$. The exact calculation results with a value slightly higher than 0.55.

ciphertext, and find the value of g, effectively reducing the cryptosystem to 15 rounds. The value of G' can be calculated from the characteristic and the ciphertexts by $G' = T'_R \oplus f'$, where f' is the value suggested from the characteristic. Thus, we can mount a simple counting scheme to find many additional bits of the actual subkey $AK15$, and then use other standard differential cryptanalytic techniques to complete the rest of the key.

4.3 Modified Variants and Weaknesses

As in DES, the order of the S boxes is important. If we only replace the S boxes S0 and S1 by each other, the number of (iterative) conditional characteristics grows to 20 (rather than four) and the fraction of the keys vulnerable to these attacks grows to more than 60% using about 2^{38} chosen plaintexts (rather than 25%). When using several characteristics with the same Ω_P's, the fraction of keys vulnerable to the attack grows to more than 90%.

On the other hand, replacement of the S boxes by single lines of the S boxes of DES (or by S boxes satisfying the design rules of DES) would invalidate the kind of characteristics used in the above attacks, in which a difference of one input bit of an S box may cause a difference of only one output bit. However, in order to strengthen the cryptosystem, we should make sure that no other kinds of high probability characteristics exist.

In order to disable conditional characteristics, we may choose the interchange control bits as combinations of key bits and data bits, rather than of key bits alone. This is really done in DES.

The key interruption in the F function is done in Lucifer after the S boxes. This order effectively eliminates the key interruption in the first round and in the last round and allows the analyst to analyze an equivalent description with one or two fewer rounds. The replacement of the order of the key interruption and the S boxes, as was done in DES, solves this weakness (but enables complementation properties).

The F function of Lucifer has a rotational symmetry, in which rotations by multiples of eight bits of the input half, the subkey, and concurrent rotation by the same multiple of one bit of the interchange control bits cause rotation of the same multiple of eight bits in the output. Therefore, characteristics can be rotated by multiples of eight bits as well, causing each characteristic to have seven rotated counterparts (when a characteristic is a rotation of itself we get less counterparts; the four characteristics used to attack Lucifer are such an example). In order to disable this property we have to use different S boxes in different entries, as was done in DES.

The eight-round reduced variant of Lucifer is very weak. Using the same conditional characteristics, with a new first round technique, we result with an attack requiring 256 chosen plaintexts, which cover about 90% of the keys. This attack places four-round characteristics built from the iterative characteristics described for the full variant, such that the $0 \rightarrow 0$ rounds are set in rounds 2 and 5, and such that rounds 3 and 4 have probabilities 1/8 and key fraction 1/2. There are eight such possible four-round characteristics, which cover together 90% of the key space. In order to get the first round for free, we can simply choose the right half of the plaintexts in any way (with the required input difference) and to calculate the output of the F function of the first round in the equivalent description (which can be done since $AK1$ equals the right half of the initial subkey). Then we have only to choose the left halves in such a way that cancels the difference received from the output of the first round. Two structures of all the eight characteristics are used, one assumes that the affecting ICBs in the first round are zero, and the other assumes they are one. These structures contain 128 pairs for each characteristic. Since the characteristics' probability is 1/64, we get in average two right pairs which can be used to find directly many key bits. Additional standard techniques using the same structure can complete the key.

The fact that the fraction of right pairs may depend on the choice of the key was already noted in [3]. It was shown that the conditional characteristics of DES can enrich the fraction of right pairs by a medium factor, but the key fraction of these characteristics is too small to make an attack feasible. It was concluded that the use of these characteristics does not help to attack DES.

5 RDES

RDES[11] (Randomized DES) is an attempt to strengthen DES against differential cryptanalysis. In order to reduce the probability of characteristics, the designers suggested to replace the deterministic swaps of the halves of the data between rounds by key dependent swaps. They claim that since the 15 key dependent swaps occur with 2^{15} possible instances, the probability of the characteristics used against DES, is reduced by a similar factor. As a result, they claim that RDES is much stronger than DES, and that the differential cryptanalytic technique of the full 16-round DES[6] is not applicable to RDES.

The new conditional technique suggested in this paper reverts the cryptanalytic effect of the key dependent swaps, and shows that RDES is weaker than DES.

The simplest weakness of RDES (already noted by the designers) is that one of every 2^{15} keys does not swap the data even once. Thus, half of the ciphertext bits (corresponding to the right half of the data during the various rounds) are the same in both the plaintext and the ciphertext. If this property is found under an attack, the attacker can immediately conclude the value of the 15 key bits affecting the swaps, and thus, an exhaustive search for the remaining key bits would require only about 2^{41} steps. Such property should be avoided in cryptosystems, and thus keys leading to this property are weak, and should not be used.

The next simplest weakness of RDES is that one of every 2^{15} keys swaps the data just once before the last round. In this case, if the attacker can easily derive the output of the F function of the last round, along with its input, and can find all the 48 bits of the subkey $K16$, resulting with at most 256 possibilities for the key.

These two examples show that many keys are quite weak, thus it is interesting to ask whether elimination of these weak keys would make RDES more secure. Using the conditional differential cryptanalytic technique we can show that almost any key of RDES is weaker than the corresponding key of DES, and thus that RDES should not be used.

In DES the following two-round iterative characteristic is used:

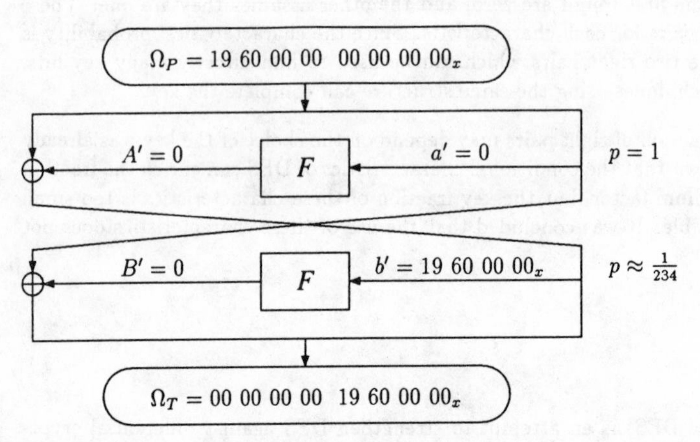

This characteristic can be iterated any number of times, since there is a deterministic swap between any two consecutive rounds. In RDES many swaps are cancelled due to the key dependent swapping policy. Thus, this characteristic cannot be iterated, and cannot be used (as is) against RDES.

However, when we look carefully, we see that whatever is the choice of the swaps, these two one-round characteristics (the two rounds of this two-round characteristic) can be combined to longer characteristics in two ways: In the first, choose the first one-round characteristic ($0 \to 0$) to appear in the first round, and the second to appear in the round after the first swap. The rest of the rounds can be completed uniquely using these two one-round characteristics. In the second way we replace all the occurrences of the one-round characteristics by each other. These two combined characteristics are duals: when one one-round characteristic occurs in a round of one combination, then the other one-round characteristic occurs in the same round of the other combination. As a results, such two r-round combined characteristics have probabilities $(\frac{1}{234})^q$ and $(\frac{1}{234})^{r-q}$, when q is the number of occurrences of $19\ 60\ 00\ 00_x \to 0$ in the first combined characteristic, and $r - q$ is the number of occurrences of $19\ 60\ 00\ 00_x \to 0$ in the second combined characteristic. Thus, for any choice of the key dependent swapping, we can easily find at least one r-round characteristic with probability $p \geq (\frac{1}{234})^{\lfloor r/2 \rfloor}$ (the r-round iterative characteristics of DES have probability exactly $p = (\frac{1}{234})^{\lfloor r/2 \rfloor}$).

It only remains to prepare characteristics for all the 2^{15} possible swap choices and choose sufficient number of plaintexts for all these choices. Fortunately, all these characteristics have only two possible values for Ω_P, and the same two possible values for Ω_T: $19\ 60\ 00\ 00\ \ 00\ 00\ 00\ 00_x$ and $00\ 00\ 00\ 00\ \ 19\ 60\ 00\ 00_x$. Therefore, the number of chosen plaintexts required for this attack is only up to twice the number required for the attack on DES, if characteristics with the same probability are used. However, for most keys these characteristics have probabilities smaller than $(\frac{1}{234})^{\lfloor r/2 \rfloor}$. The swap choice of many keys have q much smaller than $r/2$. Even when $q \approx r/2$ and the characteristics have two (or more) consecutive rounds of $19\ 60\ 00\ 00_x \to 0$, the probability is larger than $(\frac{1}{234})^{\lfloor r/2 \rfloor}$ since the probability that the exclusive-or of two (or more) output XORs (in which only in three particular S boxes the output XOR can be non-zero) is at least 2^{-12}, and not $(\frac{1}{234})^2 \approx 2^{-16}$, 2^{-24}, etc. We can conclude that the probability of one of the two dual characteristics must be

$$p \geq 2^{(-8(s+1)-4(r-s-1))/2} = 2^{-2s-2r-2},$$

where s is the number of swaps during the r rounds (we approximate $\frac{1}{234}$ by 2^{-8}). The application of this formula to the attack on the full 16-round DES, which require a 13-round characteristic, shows that any choice of up to nine swaps during these 13 rounds would result with characteristic probabilities greater than $2^{-2\cdot 9-2\cdot 13-2} = 2^{-46}$. Therefore the attacks on these cases are faster than the attacks on DES and require less chosen plaintexts. Note that these attacks usually find the subkey of the last round, but if there is no single swap in the final few rounds, they identify this fact along with the number s of swaps (estimated from the probability). Using auxiliary techniques the full key can later be completed in both cases.

The fraction of keys which cause up to nine swaps during the 13 rounds is

$$\frac{\sum_{s=0}^{9} \binom{12}{s}}{2^{12}} \approx 0.98$$

so, at most one of every 50 keys may be strong against this attack. Even if only such "strong" keys are used, an exhaustive search of all the possibilities of these keys takes only about 2^{50} steps. Therefore, RDES is not more secure than DES, and for most keys it is even much weaker.

Unlike in Lucifer, even if we replace the swap control bits by combinations of key bits and data bits, the cryptosystem does not become more secure, since then for any key, a fraction of 2^{-15} of the plaintexts would be encrypted to ciphertexts whose right halves are just the same as those of the plaintexts.

References

[1] Carlisle M. Adams, *On Immunity against Biham and Shamir's "Differential Cryptanalysis"*, Information Processing Letters, Vol. 41, No. 2, pp. 77–80, 1992.

[2] Eli Biham, Adi Shamir, *Differential Cryptanalysis of the Data Encryption Standard*, Springer-Verlag, 1993.

[3] Eli Biham, Adi Shamir, *Differential Cryptanalysis of DES-like Cryptosystems*, Journal of Cryptology, Vol. 4, No. 1, pp. 3–72, 1991.

[4] Eli Biham, Adi Shamir, *Differential Cryptanalysis of FEAL and N-Hash*, technical report CS91-17, Department of Applied Mathematics and Computer Science, The Weizmann Institute of Science, 1991. The extended abstract appears in Lecture Notes in Computer Science, Advances in Cryptology, proceedings of EUROCRYPT'91, pp. 1–16, 1991.

[5] Eli Biham, Adi Shamir, *Differential Cryptanalysis of Snefru, Khafre, REDOC-II, LOKI and Lucifer*, technical report CS91-18, Department of Applied Mathematics and Computer Science, The Weizmann Institute of Science, 1991. The extended abstract appears in Lecture Notes in Computer Science, Advances in Cryptology, proceedings of CRYPTO'91, pp. 156–171, 1991.

[6] Eli Biham, Adi Shamir, *Differential Cryptanalysis of the full 16-round DES*, Lecture Notes in Computer Science, Advances in Cryptology, proceedings of CRYPTO'92, to appear.

[7] Lawrence Brown, Matthew Kwan, Josef Pieprzyk, Jennifer Seberry, *Improving Resistance to Differential Cryptanalysis and the Redesign of LOKI*, Lecture Notes in Computer Science, Advances in Cryptology, proceedings of ASIACRYPT'91, to appear.

[8] Lawrence Brown, Josef Pieprzyk, Jennifer Seberry, *LOKI - A Cryptographic Primitive for Authentication and Secrecy Applications*, Lecture Notes in Computer Science, Advances in Cryptology, proceedings of AUSCRYPT'90, pp. 229–236, 1990.

[9] M. H. Dawson, S. E. Tavares, *An Expanded Set of S-box Design Criteria Based On Information Theory and its Relation to Differential-Like Attacks*, Lecture Notes in Computer Science, Advances in Cryptology, proceedings of EUROCRYPT'91, pp. 352–367, 1991.

[10] H. Feistel, *Cryptography and Data Security*, Scientific American, Vol. 228, No. 5, pp. 15–23, May 1973.

[11] Kenji Koyama, Routo Terada, *How to Strengthen DES-like Cryptosystems against Differential Cryptanalysis*, IEICE Transactions on Fundumentals of Electronics, Communications and Computer Science, Vol. E76-A, No. 1, pp. 63–69, January 1993.

[12] Xuejia Lai, James L. Massey, Sean Murphy, *Markov Ciphers and Differential Cryptanalysis*, Lecture Notes in Computer Science, Advances in Cryptology, proceedings of EUROCRYPT'91, pp. 17–38, 1991.

[13] Ralph C. Merkle, *A Fast Software One-Way Hash Function*, Journal of Cryptology, Vol. 3, No. 1, pp. 43-58, 1990.

[14] Ralph C. Merkle, *Fast Software Encryption Functions*, Lecture Notes in Computer Science, Advances in Cryptology, proceedings of CRYPTO'90, pp. 476–501, 1990.

[15] S. Miyaguchi, K. Ohta, M. Iwata, *128-bit hash function (N-Hash)*, proceedings of SECURICOM'90, pp. 123–137, March 1990.

[16] Shoji Miyaguchi, Akira Shiraishi, Akihiro Shimizu, *Fast Data Encryption Algorithm FEAL-8*, Review of electrical communications laboratories, Vol. 36, No. 4, pp. 433–437, 1988.

[17] National Bureau of Standards, *Data Encryption Standard*, U.S. Department of Commerce, FIPS pub. 46, January 1977.

[18] Kaisa Nyberg, *Perfect nonlinear S-boxes*, Lecture Notes in Computer Science, Advances in Cryptology, proceedings of EUROCRYPT'91, pp. 378–386, 1991.

[19] Luke O'Connor, *On the Distribution of Characteristics in Bijective Mappings*, Lecture Notes in Computer Science, Advances in Cryptology, proceedings of EUROCRYPT'93, to appear.

[20] Luke O'Connor, *On the Distribution of Characteristics in Composite Permutations*, Lecture Notes in Computer Science, Advances in Cryptology, proceedings of CRYPTO'93, to appear.

[21] Akihiro Shimizu, Shoji Miyaguchi, *Fast Data Encryption Algorithm FEAL*, Lecture Notes in Computer Science, Advances in Cryptology, proceedings of EUROCRYPT'87. pp. 267–278, 1987.

[22] Arthur Sorkin, *Lucifer, a Cryptographic Algorithm*, Cryptologia, Vol. 8, No. 1, pp. 22–41, January 1984.

[23] Michael C. Wood, technical report, Cryptech Inc., Jamestown, NY, July 1990.

Differential Attack on Message Authentication Codes

Kazuo Ohta[1] and Mitsuru Matsui[2]

[1] NTT Network Information Systems Laboratories
Nippon Telegraph and Telephone Corporation
1-2356, Take, Yokosuka-shi, Kanagawa-ken, 238-03 Japan
[2] Computer & Information Systems Laboratory
Mitsubishi Electric Corporation
5-1-1 Ofuna, Kamakura-shi, Kanagawa-ken, 247 Japan

Abstract. We discuss the security of Message Authentication Code (MAC) schemes from the viewpoint of differential attack, and propose an attack that is effective against DES-MAC and FEAL-MAC. The attack derives the secret authentication key in the chosen plaintext scenario. For example, DES(8-round)-MAC can be broken with 2^{34} pairs of plaintext, while FEAL8-MAC can be broken with 2^{22} pairs. The proposed attack is applicable to any MAC scheme, even if the 32-bits are randomly selected from among the 64-bits of ciphertext generated by a cryptosystem vulnerable to differential attack in the chosen plaintext scenario.

1 Introduction

Authentication, which certifies data integrity and data origin, is becoming an important technique because the transfer of valuable information needed for electronic funds transfer, business contracts, etc. must be made across computer networks. Data integrity ensures that the data has not been modified or destroyed. Data origin authentication is the verification that the source of data received is as claimed.

There are two frameworks of authentication: digital signature schemes using public key cryptosystems and message authentication code (MAC) schemes based on secret key cryptosystems.

MAC schemes have been standardized and are being discussed by ISO/TC68/SC2 [ISO8731/2] for banking services. MAC is produced by the sender using a secret authentication key, which is known only to the sender and receiver of the authenticated message, and appended to the original message. Upon reception, the receiver uses the authentication key to check whether the received message-MAC pair is valid. The MAC of message m is the left half 32-bits of the last ciphertext block of m in CBC-mode. The Data Encryption Standard(DES) [FIPS77, ISO8731/1] and the Fast Data Encryption Algorithm(FEAL) [MSS88, MKOM90] cryptosystems are used to generate the ciphertext. Hereafter, we denote the MAC scheme based on DES by DES-MAC, and the MAC scheme based on FEAL by FEAL-MAC. We describe their specific iteration number versions as DES(N-round)-MAC and FEAL N-MAC where N is the iteration number.

It is already known that differential attack is effective against DES [BS90, BS92], FEAL [BS91-1] , and other iterated cryptosystems [BS91-2]. Differential attack requires many ciphertext pairs corresponding to plaintext pairs where the plaintext pairs are different in a significant way. Thus the differential attack is a type of chosen plaintext attack, although the value of plaintexts are not used in the attack procedure directly. Differential attack can derive the encryption/decryption key with less computational time than required by exhaustive attack using an appropriate number of ciphertext pairs in the chosen plaintext scenario. DES(8-round) can be broken with 2^{15} plaintext pairs, and FEAL8 can be broken with 2^{10} plaintext pairs. Some known plaintext attacks on these iterated cryptosystems have also been proposed [CG91, K91, MY92, M93].

In this paper, we analyze the security of MAC schemes from the viewpoint of differential attack. Since the left half 32-bits of ciphertext in CBC-mode is used as MAC, the following question is important to establishing the security of MAC schemes: Is differential attack effective against the MAC schemes? That is, how many plaintext pairs with the left-half (partial information) of their ciphertexts are necessary to derive the secret authentication key? We introduced an attack that is effective against both DES-MAC and FEAL-MAC, and estimate the number of appropriate plaintext pairs needed for the attack to derive the secret authentication key.

It is interesting that our procedure is also effective against any MAC scheme, where the 32-bits are randomly selected among the 64-bits of ciphertext generated by a cryptosystem vulnerable to differential attack in the chosen plaintext scenario. We will also discuss the influence of the bit length of ciphertext available as the MAC.

2 Related Works

2.1 Differential Attack against Ciphertext Case

Biham and Shamir proposed a differential attack against various iterated cryptosystems [BS90, BS91-1, BS91-2, BS92]. Each iteration is a function usually based on S boxes, bit permutations, arithmetic operations, and exclusive-or operations (denoted by XOR). The S boxes are known to be nonlinear. Since they are usually the only part of the cryptosystem that is not linear, the security of the cryptosystem depends on which type of S box is selected.

Differential attack depends on the following fact: Even though we cannot determine the XOR value of the S-box output from its input XOR value, some specific input XOR values yield specific output XOR value with high probability.

To find the subkeys entering each iteration function, differential attack proceeds as follows (see p.16 of [BS90]):

Step 1: Choose an appropriate plaintext XOR.

Step 2: Create an appropriate number of plaintext pairs with the plaintext XOR chosen at **Step 1**, encrypt them and keep only the resultant ciphertext pairs.

Step 3: For each pair, derive the expected output XOR of as many S boxes in the last round as possible from the plaintext XOR and the ciphertext pair.

Step 4: For each possible subkey value, count the number of pairs that result with the expected output XOR using this subkey value in the last round, and choose the value that is counted most often as the subkey candidate.

Note that the input pair of the last round is known, since it appears here as part of the ciphertext pair.

The pushing mechanism, with which the knowledge of the XORs of the plaintext pairs is passed through as many rounds as possible without making them zero, is termed the *statistical characteristic* of the cryptosystem in [BS90]. A pair whose intermediate XORs equal the values specified by the characteristic is called a *right pair* with respect to the characteristic. Any other pair is called a *wrong pair* with respect to the characteristic.

[BS90] showed many characteristics for several variants of DES with different round number. The two characteristics listed below will be referred to in a later section. These characteristics yield the known highest probability for the round number indicated.

$$\Omega_P^1 = \texttt{405C0000 04000000}, \quad \Omega_T^2 = \texttt{405C0000 04000000}$$
$$\text{with probability } 1/10,486 \text{ for 5 rounds}$$
$$\Omega_P^2 = \texttt{00000000 19600000}, \quad \Omega_T^4 = \texttt{19600000 00000000}$$
$$\text{with probability } (1/234)^N \approx (2^{-8})^N \text{ for } 2N \text{ rounds}$$

[BS91-1] also pointed out that the following characteristics permit FEAL to be cryptanalyzed within various numbers of rounds.

$$\Sigma_P^1 = \texttt{A2008000 80800000}, \quad \Sigma_T^2 = \texttt{A2008000 80800000}$$
$$\text{with probability } 1/16 \text{ for 5 rounds}$$
$$\Sigma_P^2 = \texttt{80608000 80608000}, \quad \Sigma_T^3 = \texttt{80608000 80608000}$$
$$\text{with probability } (1/4)^{4N} \text{ for } 4N \text{ rounds}$$

Note that since differential attack requires many ciphertext pairs corresponding to plaintext pairs with particular differences, Ω_P and Σ_P, the differential attack is a type of chosen plaintext attack, although the value of plaintexts are not used in the attack procedure directly.

There are four possible types of attack, 3R-attack, 2R-attack, 1R-attack and 0R-attack, depending on the number of additional rounds in the cryptosystem that are not covered by the characteristic itself. A 3R-attack is advisable over a 2R-attack, and both are advisable over a 1R-attack, since characteristic that has higher probability requires fewer ciphertext pairs for the attack. For example, the differential attack on DES reduced to eight rounds, DES(8 round), in [BS90] uses a five-round characteristic, Ω_P^1, with 3R-attack.

To find the right key with a counting scheme in **Step 4**, we need a high probability characteristic and enough ciphertext pairs to guarantee the existence of several right pairs. Usually we relate the number of pairs needed by a counting scheme to the number of right pairs needed. The number of right pairs needed is mainly a function of the ratio between the number of right pairs and the average count in the counting scheme, denoted by S/N. In the DES case, $S/N = \frac{2^k \cdot p}{\alpha \cdot \beta}$ holds, where k-bits of subkey are counted at **Step 4**, α is the average count per counted pair, β is the ratio of the counted to all pairs, and p is the characteristic's probability. When S/N is high enough, only a few right pairs are needed to uniquely identify the right value of the subkey bits.

Biham and Shamir found the suitable characteristic for each round number, for example, (Ω_P^1, Ω_T^1) for DES(8-round), (Ω_P^2, Ω_T^2) for DES(arbitrary round), (Σ_P^1, Σ_T^1) for FEAL8, and (Σ_P^2, Σ_T^2) for FEAL(arbitrary round), and peeled off the subkey from *the last round* using 3R-attack.

2.2 Attacks against Authentication Schemes

There are two frameworks for authentication, digital signature schemes using public key cryptosystems and message authentication code (MAC) schemes based on secret key cryptosystems. In the former case, a mixed type digital signature scheme [DP80] is practical. The signer calculates signature $s = f(h(m))$, where m is a message, h is a public hash function and f is a secret signature function known to the signer, and sends both s and m to the receiver. The receiver checks whether $f^{-1}(s) = h(m)$ holds using h and the public validation function f^{-1}. Here, hash functions are used to compress long messages into short digests to attain high efficiency, and they are not required to be secret. In the latter case, the MAC is produced by the sender using a *secret* authentication key, which is known only to the sender and receiver, and appended to the original message. Upon reception, the receiver uses the authentication key to check whether the received message-MAC pair is valid. Note that while a receiver who doesn't know the secret key can not generate a signature value in a digital signature scheme, he can calculate (forge) the MAC of any message using the authentication key in the MAC scheme. Thus the MAC scheme is applicable only to the case where the sender and receiver trust each other.

There are two kinds of threats in authentication schemes:

(1) forgery of a digital signature: a valid signature-message pair is found using previously used pairs,

(2) determination of secret information: the secret key of the public key cryptosystem or the secret authentication key of the MAC scheme are revealed.

The collision free property of hash functions is discussed in [D87, ZMI90] as a form of Type (1) threat. If an attacker finds a pair of messages, m and m', satisfying $h(m) = h(m')$, where h is a public hash function, then signature value, s, of $h(m)$ is as valid as that of $h(m')$. Thus, he can replace the true message, m, with the invalid message m'; the value, s, remains valid. It is possible to find a pair of such messages using the birthday paradox strategy in computational time, $O(2^{\ell/2})$, when the bit length of hashed value is ℓ. Other attacks of Type (1) have been discussed that addressed the weak key and semi weak key properties of cryptosystems [MOI90], and differential properties [BS91-1].

Type (2) attack means that if the attack succeeds, the authentication system is totally broken, while a Type (1) attack is a kind of ad-hoc forgery. Deriving the secret key of a public key cryptosystem is breaking the cryptosystem itself. Until now, there has not been sufficient discussion of Type (2) attacks on MAC schemes. We will point out the first attack procedure of this type from the viewpoint of differential attack.

3 How to Attack MAC Schemes

3.1 What are the Problems

In the attacks employed in the ciphertext case, it is important for the attacker to use the full length of the ciphertext. On the other hand, only the left half 32-bits of ciphertext in CBC-mode is available as the MAC value. Thus, what happens in the MAC case, where the attacker can not use the full ciphertext, but only partial information of the ciphertext?

The following questions are interesting in the differential attack against MAC schemes to derive the secret authentication key: 1) how many plaintext pairs are needed together with the left-half (partial information) of their ciphertexts, 2) what is the influence of the location of ciphertext information available to an attack, and 3) what is the influence of the bit length of ciphertext information available as the MAC.

3.2 Outline of MAC Attack

Hereafter, we will distinguish the real plaintext (message), M, to be authenticated from P treated as the plaintext in the references in [BS90, BS91-1], and the real ciphertext, C, from the ciphertext, T, where $P = IP(M)$ and $C = IP^{-1}(T)$ in DES, where IP is the initial permutation, and $(P_H, P_L) = (M_H, M_H \oplus M_L)$ and $(C_H, C_L) = (T_H, T_H \oplus T_L)$ in FEAL, where P_H is the left half of P and P_L is the right half of P.

Since the proposed procedure is a chosen plaintext attack, we assume the case, where a plaintext (message) is a single block and the initial value of CBC-mode is public.

Since the right half of the ciphertext can not be used for an attack, we modify the differential attack against MAC schemes, which are based on 0R-attack or 1R-attack introduced by [BS90], to derive the secret subkeys, generated from the authentication key, in *the first round*, as follows:

Step 1: Choose the appropriate pair of real plaintext XOR, ω_P, and real ciphertext XOR, ω_T, that has high characteristic probability and many number of subkey bits whose occurrences can be checked at **Step 4**.

Step 2: Create an appropriate number of real plaintext pairs, (M, M^*), where $M \oplus M^* = \omega_P$, calculate their MAC values, $\gamma = MAC(K, M)$ and $\gamma^* = MAC(K, M^*)$, where K is the authentication key, and keep only the real plaintext pairs (M, M^*) which satisfy

$$\gamma \oplus \gamma^* = \text{the left half of 32-bits of } \omega_T.$$

Step 3: Derive the expected output XOR, A', of the first round function using the differential rules.

Step 4: For each possible subkey value of $K1$, where $K1$ is used by the first round function, count the number of pairs that result with the expected output XOR, A', using this subkey value of $K1$ in the first round, and choose the value that is counted most often as the subkey candidate of $K1$.

Remark

In the DES(r-round)-MAC case, where $r \geq 8$, ω_P is transformed to Ω_P^2 by the initial permutation (IP), that is, $IP(\omega_P) = \Omega_P^2$, and ω_T is transformed to Ω_T^2 by IP, that is, $\omega_T = IP^{-1}(\Omega_T^2)$, in the 0R-attack. ω_T is transformed to a value related to Ω_T^2 by IP, that is, $\omega_T = IP^{-1}(\varphi \oplus (\Omega_T^2)_L, (\Omega_T^2)_H)$, where φ is the output XOR of the last round iteration function with an input, $(\Omega_T^2)_H$, in the 1R-attack. $A' = 00000000$.

In the FEAL N-MAC case, where $N \geq 8$, ω_P equals $(\sigma, 00000000)$, where $\sigma = 80608000$, ω_T equals $(\sigma, 00000000)$ or a value related to $(\sigma, 00000000)$, and $A' = 00800000$.

Note

If there are several candidates for $K1$, we can adopt two strategies to attain high efficiency in order to reduce the number of candidates.

(1) Choose another $\widetilde{\omega}_P$ at **Step 1**, apply the same procedure, and choose the common value between ω_P and $\widetilde{\omega}_P$ cases as the subkey candidate of $K1$.

(2) Derive the expected output XORs, B', of the *second round iteration* function, and count the number of pairs that result with the expected output XOR, B', using the subkey candidate value of $K1$ and newly selected subkey value of $K2$ in the second round. If the counted number is zero, the candidate for $K1$ is discarded.

We can apply the first strategy without the increase in the number of plaintext pairs using quartets which combine the two characteristics [BS90]. The second strategy applied to the case of FEAL-MAC will be described in detail in **Section 5**.

4 Discussion

4.1 DES-MAC

The procedure described is based on 0R-attack or 1R-attack, and can be considered a differential attack against a cryptosystem changing the roles of plaintext and ciphertext. So the discussion of [BS90] is applicable with slight modification. Note that MAC values, which are the partial information of ciphertext, are sufficient in the proposed attack, since the actual plaintext values are not necessary in the original differential attack.

The possibility of subkey value can be checked on some bits of the subkey in the first round entering the S boxes with nonzero input XORs. If we use ω_P corresponding to Ω_P^2, three S boxes, $S1, S2$, and $S3$, yield 18 bits of subkey $K1$.

Since the input XOR is constant, we can not distinguish between several subkey values. However, the number of such values is small and each can be checked later in parallel by the next part of the algorithm (see p.40 line 9 of [BS90]).

Let's consider the case of DES, which is reduced to an even round, where 0R-attack is applicable. Note that the β component of S/N should be 2^{-32}, since the output XOR of the last round iteration is $\Omega_T = (\psi, 0)$, where $\psi = 19600000$, ω_T has specific 64 bits, and the left half 32-bits ω_T are determined definitely. Thus, DES(8-round)-MAC can be broken with 2^{34} pairs, since $S/N = \frac{2^{18}\times(2^{-4})^8}{4^3\times 2^{-32}} = 2^{12}$. DES(10-round)-MAC can be broken with 2^{42} pairs, since $S/N = \frac{2^{18}\times(2^{-4})^{10}}{4^3\times 2^{-32}} = 2^4$. DES(12-round)-MAC can not be broken, since $S/N = \frac{2^{18}\times(2^{-4})^{12}}{4^3\times 2^{-32}} = 2^{-4}$.

Concerning the influence of the location of ciphertext information available as MAC, since the value of the β component of S/N is 2^{-32} constantly (in 0R-attack), the security of DES(even-round)-MAC does not depend on the bit location in the ciphertext used as MAC.

We utilize the following fact in the above discussion: Biham and Shamir observed experimentally that when S/N is about $1-2$, about $20-40$ occurrences of right pairs are sufficient. When the S/N is much higher, even three or four right pairs are usually enough. When the S/N is much smaller than 1, the identification of the right value of subkey bits requires an unreasonably large number of pairs. (see p.23 in [BS90])

Let's consider the case of DES, which is reduced to an odd round, where 1R-attack is applicable. Note that the β component of S/N should be $2^{-20} \sim 2^{-32}$, since the output XOR of the last round iteration is $\Omega_T = (\varphi, \psi)$, where φ contains 20 bits of zeros released by $S4, \ldots, S8$ at the last iteration, ω_T has 12 free bits among 64 bits, and the left half of 32-bits of ω_T might contain them. With careful check on the bit location of MAC, since 7 bits among the 12 free bits are contained in the left half of 32-bits of ω_T, $\beta = 2^{-25}$ holds. Thus, DES(7-round)-MAC can be broken with 2^{26} pairs, since $S/N = \frac{2^{18}\times(2^{-4})^6}{4^3\times\beta} = 2^{13}$. DES(9-round)-MAC can be broken with 2^{34} pairs, since $S/N = \frac{2^{18}\times(2^{-4})^8}{4^3\times\beta} = 2^5$. DES(11-round)-MAC can not be broken, since $S/N = \frac{2^{18}\times(2^{-4})^{10}}{4^3\times\beta} = \frac{1}{2^3}$.

It is also clear that our attack is effective to any DES(N-round)-MAC scheme, where $N \leq 10$; the 32-bits can be randomly selected from among the 64-bits of ciphertext. It is not effective against DES(N'-round)-MAC scheme, where $N' \geq 12$. Only the security of DES(11-round)-MAC depends on the bit location in the ciphertext used as MAC, since $S/N = \frac{2^{18} \times (2^{-4})^{10}}{4^3 \times \beta} = 2^{-8} \sim 2^4$.

Concerning the influence of the bit length of ciphertext information available as the MAC, the value of the β component of S/N is $2^{-\ell}$ in DES(even-round)-MAC if the bit length of MAC is ℓ. For example, DES(8-round)-MAC(24 bit) is breakable, since $S/N = \frac{2^{18} \times (2^{-4})^8}{4^3 \times 2^{-24}} = 2^4$, while DES(8-round)-MAC(18 bit) can not be broken, since $S/N = \frac{2^{18} \times (2^{-4})^8}{4^3 \times 2^{-18}} = 2^{-2}$. The similar discussion holds in DES(odd-round)-MAC. Note that when we use the small bits of MAC, it is easy to find a pair of collision messages, while it is difficult to derive the secret authentication key.

If we use $\widetilde{\omega}_P$ corresponding to $\Omega_P^3 =$ 00196000 00000000, three S boxes, $S3, S4$, and $S5$, give 12 more of the bits in subkey $K1$ (this is in addition to the 18 bits derived with Ω_P^2). This reduces the computational time of the attack procedure.

While 2^{34} plaintext pairs are necessary for differential attack against DES(10-round) in the ciphertext case, 2^{42} pairs are necessary in the MAC case, since 2R-attack and 3R-attack are not applicable to the MAC case. It is an open problem whether 2R-attack and 3R-attack are applicable to MAC schemes.

4.2 FEAL-MAC

The attack procedure is based on 0R-attack or 1R-attack, and can be considered as a differential attack against a cryptosystem changing the roles of plaintext and ciphertext. So the discussion of [BS91-1] is applicable with slight modification.

The possibility of subkey value can be checked using some subkey bits from the first round, where the input XOR, a', equals $\sigma =$ 80608000, and the output XOR, A', equals 00800000.

Biham and Shamir make the following statement in [BS91-1]: "the successfully filtered pairs are used in the process of counting the number of times each possible value of the last actual subkey is suggested, and finding the most popular value. Complicating factors are the small number of bits set in h' [3] (which is a constant defined by the characteristic), and the fact that many values of H' suggest many common values of the last actual subkey. Our (Biham and Shamir's) calculations show that the right value of the last subkey is counted with detectably higher probability than a random value up to $N \leq 31$ round. " (see p.10 line 26 of [BS91-1])

We can apply the above explanation to FEAL-MAC schemes by replacing the last actual subkey with the first actual subkey: h' with a', and H' with A'.

In estimating the sufficient number of plaintext pairs, though Biham and Shamir imply that four right pairs are sufficient if $N \leq 24$ to derive the subset

[3] They employ the notation of an eight round cryptosystem.

key using a *counting method* in the ciphertext attack (see p.11 of [BS91-1]), it is not clear how many pairs are sufficient to derive the subset key in the proposed MAC attack, where the value of A' described in the above is fixed. Our experimental results, described in the next section, confirm that at most 2^6 right pairs are sufficient to derive the subkey $K1$ with a *checking method*, by simply checking whether all pairs that pass the check of **Step 2** also pass **Step4** using A'.

Since FEAL8 has the characteristic probability, 2^{-16}, we can find 2^6 right pairs from 2^{22} pairs of plaintext with high probability in **Step 2**. On the other hand, the probability that a wrong pair is found is 2^{-10}, since the 32 bits are used in **Step 2** to filter right pairs. Thus FEAL8-MAC can be broken with 2^{22} pairs of plaintext with overwhelming probability ($= 1 - \frac{1}{2^{10}}$).

Since FEAL12 has the characteristic probability, 2^{-24}, we can find 2^6 right pairs from 2^{30} pairs of plaintext with high probability in **Step 2**. On the other hand, the probability that a wrong pair is found is 2^{-2} in **Step 2**. Since we confirmed with an experiment that even if one wrong pair is contained among 64 right pairs, we can find a correct bits among the subkey with high probability (99 %), FEAL12-MAC can also be broken with 2^{30} pairs of plaintext with high probability.

On the other hand, since FEAL16 has the characteristic probability, 2^{-32}, it is difficult to find pairs containing 2^6 right pairs and a few wrong pairs from 2^{38} pairs of plaintext at **Step 2**. Since the selected pairs at **Step 2** contain many wrong pairs with high probability, it is necessary to use the counting method instead of the checking method to break FEAL16-MAC. It is an open problem to estimate the sufficient number of pairs for the counting method.

Note that these attacks are applicable to FEAL-MAC only in the chosen plaintext attack, though some known plaintext attacks are pointed out to ciphertext case [CG91, MY92, K91]. It is an open problem to break MAC schemes in the known plaintext attack.

5 Experimental Results

The purpose of this experiment is to estimate the sufficient number of right pairs to derive the subkey $K1$. We will describe experimental results on FEAL8-MAC using the attack technique ***cut off the spread of carry bit***, developed in [MY92]. Here, we adopt the second strategy described in **Section 3**: check whether all pairs that pass the check of **Step 2** also pass **Step 4** using B' of the second round function in addition to the expected output XORs, A'.

For convenience, we use the modified F-function defined by [MY92] in our experiment.

5.1 Notation

We use the following notations in this section.

M : A plaintext input to the FEAL-MAC algorithm

A_M : Output of the first round function (32 bits) corresponding to M

B_M : Output of the second round function (32 bits) corresponding to M

b_M : Input of the second round function (32 bits) corresponding to M

$B[i]$: The i-th bit of B, where $0 \leq i \leq 31$

$K1$: Subkey value used by the first round function (32 bits)

$K2$: Subkey value used by the second round function (32 bits)

$K1[i \sim j]$: The $j - i + 1$ bits data consisting of the i-th, ..., j-th bits of $K1$

$K1[i, j]$: The XORed value of the i-th and j-th bits of $K1$

5.2 Attack Procedure against FEAL-MAC Attack

We select $\omega_P = (\sigma, 00000000)$, where $\sigma = 80608000$, $\omega_T = (\sigma, 00000000)$, and $A' = B' = 00800000$. Hereafter, we will explain how to implement **Step 4** of the MAC attack procedure to reduce the number of candidates for $K1$.

The following procedure selects the 12 bits for $K1$ that influence the 8 bits, 80, of A'.

Step 1: Select 12 bit data of $K1$ using the bits $K1[8 \sim 13], K1[16 \sim 20]$, and $K1[21, 22]$, the remaining bits are determined arbitrarily.

Step 2: Calculate all right pairs, (M, M^*), using the first round function with the selected $K1$, and check whether $A_M \oplus A_{M^*} = 00800000$ holds. If all checks are correct, then let the 12 bit data be a candidate of $K1$.

The following procedure can select more 12 bits of $K1$ and 1 bit of $K2$ in addition to the 12 bits selected above.

Step 1: Select 17 bits of $K1$ using $K1[0 \sim 3], K1[14 \sim 15], K1[23 \sim 28], K2[12 \sim 13], K2[20 \sim 22]$, the remaining bits are fixed arbitrarily.

Step 2: Calculate all right pairs using the first round function and the selected $K1$, and select pairs, (M, M^*), satisfying

$$b[4] \oplus b[12] \oplus b[20] \oplus b[28] = K2[12] \oplus K2[20]. \tag{1}$$

Step 3: Calculate the pairs selected in the above step using the second round function and the selected $K2$, and check whether

$$B_M[16] \oplus B_{M^*}[16] = 0, B_M[17] \oplus B_{M^*}[17] = 0, B_M[23] \oplus B_{M^*}[23] = 1. \tag{2}$$

If all checks are correct for the pairs, then let the value of the 24 bit data be a candidate of $K1$.

Since we can ignore the influence of $K2[8 \sim 11]$ and $K2[16 \sim 19]$ considering equation (1) at **Step 2**, we don't have to guess these bits in the above procedure. This technique, *cut off the spread of carry bit*, was developed by [MY92] to reduce the computational time needed to get the subkey information. Equation (2) corresponds to the fact that $B_M \oplus B_M^*$ should be $B' = 00800000$. Note that since we don't know $B_M[18 \sim 22]$ here, only three bits are checked in equation (2).

After the calculation of these 24 bits of subkey $K1$, we can determine 30 bits by repeating the previous method without the above restrictions of equation (1). Though $K1[7]$ and $K1[31]$ remain undetermined with this procedure, they can be determined by an exhaustive search.

The above procedure was implemented and tested. We have confirmed that 2^6 right pairs are sufficient to derive the subkey $K1$ with the above procedure. We could decrease the number of pairs required by checking the XOR outputs of higher round functions, C', D' and so on.

6 Conclusion and Remarks

We have proposed a modified differential attack which is effective against MAC schemes, where only the left half 32-bits of the ciphertext is available to the attacker. The attack derives the secret authentication key in the chosen plaintext scenario. The procedure is considered as a form of differential attack, 0R-attack or 1R-attack, against a cryptosystem where the roles of plaintext and ciphertext are reversed. We have also pointed out that our procedure is also effective against any MAC scheme even if the 32-bits are randomly selected from among the 64-bits of ciphertext generated by a cryptosystem vulnerable to differential attack in the chosen plaintext scenario.

More exactly, based on the discussion of [BS90], and with slight modification of the S/N ratio, it appears that DES(12-round)-MAC is secure, while DES(8-round)-MAC can be broken with 2^{34} pairs of plaintext in the chosen plaintext scenario. It is clear that our attack is effective against any DES(N-round)-MAC scheme, where $N \leq 10$; the 32-bits can be randomly selected from among the 64-bits of ciphertext.

Concerning the influence of the bit length of ciphertext information available as the MAC, it becomes clear that, for example, DES(8-round)-MAC(24 bit) is breakable, while DES(8-round)-MAC(18 bit) can not be broken.

Based on our experiment and the discussion of [BS91-1], it is clear that FEAL8-MAC can be broken with 2^{22} pairs of plaintext with overwhelming probability and FEAL12-MAC can be broken with 2^{30} pairs with high probability in the chosen plaintext scenario.

There are several open problems:

(1) whether MAC is broken in the known plaintext attack,
(2) whether 2R-attack and 3R-attack are effective against MAC schemes, and
(3) how many pairs of plaintexts are sufficient to break FEAL16-MAC using the counting method.

Acknowledgment

A part of this research was conducted while the first author was visiting the MIT Laboratory for Computer Science. He would like to acknowledge the generous support provided by MIT/LCS. The second author would like to thank Atsuhiro Yamagishi for his kind suggestion and encouragement.

References

[BS90] E. Biham and A. Shamir, "Differential Cryptanalysis of DES-like Cryptosystems," Journal of CRYPTOLOGY, Vol. 4, Number 1, 1991 (The extended abstract appeared at CRYPTO'90)

[BS91-1] E. Biham and A. Shamir, "Differential Cryptanalysis of Feal and N-Hash," EUROCRYPT'91

[BS91-2] E. Biham and A. Shamir, "Differential Cryptanalysis of Snefru, Khafre, REDOC-II, LOKI and Lucifer," CRYPTO'91

[BS92] E. Biham and A. Shamir, "Differential Cryptanalysis of the full 16-round DES," CRYPTO'92

[CG91] A. Tardy-Corfdir, and H. Gilbert, "A known plaintext attack of FEAL-4 and FEAL-6," CRYPTO'91

[D87] I. Damgård, "Collision free hash functions and public key signature schemes," EUROCRYPT'87

[DP80] D. W. Davies and W. L. Price, "The application of digital signatures based on public key cryptosystems," Proceedings of ICC, 1980, pp.525-530

[FIPS77] "Data Encryption Standard." Federal Information Processing Standards Publication 46, National Bureau of Standards, U.S. Department of Commerce, 1977

[ISO8731/1] "Banking-Approved algorithm for message authentication – Part 1: DEA-1."

[ISO8731/2] "Banking-Approved algorithm for message authentication – Part 2: Message authentication algorithm."

[K91] T. Kaneko, "A known plaintext cryptanalytic attack on FEAL-4," Technical Report of the Institute of Electronics, Information and Communication Engineers, ISEC91-25 (1991)

[M93] M. Matsui, "Linear Cryptanalysis Method for DES Cipher," EUROCRYPT'93

[MKOM90] S. Miyaguchi, S. Kurihara, K. Ohta, and H. Morita, "Expansion of FEAL Cipher," NTT Review, Vol. 2, No. 6, 1990

[MOI90] S. Miyaguchi, K. Ohta and M. Iwata, "Confirmation that Some Hash Functions are not Collision Free," EUROCRYPT'90

[MSS88] S. Miyaguchi, A. Shiraishi, and A. Shimizu, "Fast data encryption algorithm FEAL-8," Review of Electrical Communication Laboratories, Vol. 36, No. 4, 1988

[MY92] M. Matsui and A. Yamagishi, "A New Method for Known Plaintext Attack of FEAL Cipher," EUROCRYPT'92

[ZMI90] Y. Zheng, T. Matsumoto and H. Imai, "Structural Properties of One-Way Hash Functions," CRYPTO'90

Cryptanalysis of the CFB mode of the DES with a reduced number of rounds

Bart Preneel*, Marnix Nuttin, Vincent Rijmen**, and Johan Buelens

Katholieke Universiteit Leuven, Laboratorium ESAT-COSIC,
Kardinaal Mercierlaan 94, B–3001 Heverlee, Belgium
bart.preneel@esat.kuleuven.ac.be

Abstract. Three attacks on the DES with a reduced number of rounds in the Cipher Feedback Mode (CFB) are studied, namely a meet in the middle attack, a differential attack, and a linear attack. These attacks are based on the same principles as the corresponding attacks on the ECB mode. They are compared to the three basic attacks on the CFB mode. In 8-bit CFB and with 8 rounds in stead of 16, a differential attack with $2^{39.4}$ chosen ciphertexts can find 3 key bits, and a linear attack with 2^{31} known plaintexts can find 7 key bits. This suggests that it is not safe to reduce the number of rounds in order to improve the performance. Moreover, it is shown that the final permutation has some cryptographic significance in the CFB mode.

1 Introduction

The Data Encryption Standard (DES) was developed in the seventies at IBM (together with NSA) and was published by the National Bureau of Standards in 1977 [8]. Its intended application was sensitive but unclassified data. In spite of the initial controversy, it became the most widespread cryptographic algorithm. Four modes of use of the DES have been specified in national and international standards [9, 11]: Electronic Code Book (ECB), Cipher Block Chaining (CBC), Cipher Feedback (CFB) and Output Feedback (OFB). The DES has been the subject of several studies. One of the first properties that was discovered was the complementation property [10]; it can be exploited to halve the number of operations for an exhaustive key search. Attacks have been described in [6, 7], but the most successful techniques are differential cryptanalysis introduced by E. Biham and A. Shamir [3] and linear cryptanalysis invented by M. Matsui [13]. The first attack which is faster than exhaustive key search was the differential attack of [5]. Most attacks on the DES are applicable to the ECB mode, and some can be extended to the CBC mode [4]. Only one attack was published on the OFB mode [12]: it was shown by R. Jueneman that the size of the feedback

---

* N.F.W.O. postdoctoral researcher, sponsored by the National Fund for Scientific Research (Belgium).

** N.F.W.O. research assistant, sponsored by the National Fund for Scientific Research (Belgium).

variable should be 64 bits. For the time being no evaluation of the DES in the CFB mode has been published [14].

In this volume, K. Ohta and M. Matsui describe a differential attack which is applicable to the m-bit CFB mode for 'large' values of m ($m \geq 24$) [15]. This paper presents attacks that are also applicable for smaller values of m.

In the first part of this paper the CFB mode is described. In Sect. 3 three basic attacks are discussed that depend only on the size of the parameters of the algorithm; they can serve as a point of reference. Section 4 shows how a meet in the middle attack can be applied in the CFB mode. In Sect. 5 the main result is discussed, namely the extension of differential cryptanalysis to the CFB mode. Section 6 discussed the applicability of linear attacks. Finally the conclusions are presented.

2 The CFB mode

This section discusses the CFB mode for a block cipher with a block length of t bits ($t = 64$ in case of the DES). The CFB mode is a stream mode, i.e., the size m of the plaintext blocks can be arbitrarily chosen between 1 and t bits. The scheme that is described here is a simplified version of the more general scheme contained in the standards.

The CFB mode makes use of an internal t-bit register. The state of this register before the encryption or decryption of the ith block is denoted with X_i. First this register is initialized with the starting variable or $X_1 = SV$. The plaintext and ciphertext blocks are denoted with P_i and C_i respectively, and the encryption operation with the secret key K is denoted with $E_K()$.

The encryption of plaintext block i consists of the following two steps:

$$C_i = P_i \oplus \mathrm{rchop}_{t-m}(E_K(X_i))$$
$$X_{i+1} = \mathrm{lchop}_m(X_i) \| C_i \,.$$

Here $\|$ denotes concatenation, rchop_a denotes the function that drops the a rightmost bits of its argument, and lchop_a denotes the function that drops the a leftmost bits of its argument. The decryption operates in a similar way.

The most important property of the CFB mode is that if m is chosen equal to the character size, this mode is *self synchronizing*. This means that if one or more m-bit characters between sender and receiver are lost, automatic re-synchronization occurs after t bits. This is especially important in a communication environment, where m is typically equal to 1 or 8 bits. The price paid for this property is that the performance decreases with a factor t/m. In contrast with the OFB mode, a single bit error is propagated with a factor t.

For the m-bit CFB mode, a known plaintext and a chosen plaintext attack are equivalent: in both cases the cryptanalyst has no control over the input of the block cipher. If the cryptanalyst wants to control this input, like in a chosen plaintext attack on the ECB mode, a chosen ciphertext attack is required. In all cases the cryptanalyst can only observe m output bits. In OFB mode, the number of observable bits is also limited to m, but the most powerful attack is a known plaintext attack.

3 Three basic attacks

The simplest attack is clearly an *exhaustive search* for the key; it is a known plaintext attack. Exploitation of the complementation property requires a chosen ciphertext attack, or more precisely, a sufficient number of pairs of the form $(C_i, P_i), (\overline{C_i}, P_i')$. Even in 1976 there was a debate over the feasibility of an exhaustive search for a 56-bit DES key. It is clear that this attack is becoming more and more realistic. For more details the reader is referred to [16]. The exhaustive attack is only discussed as a reference for other attacks.

Two results will be presented. The first result is an expression for the number of plaintext/ciphertext pairs to determine the key uniquely.

Proposition 1 *Assume one has a block cipher with a k-bit key in m-bit CFB, where every ciphertext bit depends on every key bit and plaintext bit. If one knows M plaintext/ciphertext pairs, the expected number of keys that remains after an exhaustive search is equal to*

$$K_{\exp} = 1 + \frac{2^k - 1}{2^{Mm}} . \tag{1}$$

From this proposition it follows that in order to determine the key uniquely M has to be slightly larger than k/m.

A second result is applicable to the DES with a reduced number of rounds. The DES has 16 rounds; the number of rounds for the reduced version of the DES will be denoted with N. Table 1 indicates how many key bits influence the ciphertext in the case of m-bit CFB with N rounds. It is clear that this depends on the selection of the bits. The standards specify that the leftmost m bits are selected. For the DES, this selection is influenced by IP^{-1}. In 1-bit CFB, the output bit is independent from the operations (and the subkey) in the last round, and for larger values of m the output bits are selected from different S-boxes. It will be shown in Sect. 5 and 6 that differential and linear attacks are very sensitive to these positions. It is remarkable that IP^{-1} has a cryptographic meaning in this context. IP and IP^{-1} were probably introduced to facilitate hardware implementations, and it is easily seen that in ECB and CBC mode they have no security implication (except for the case where the plaintext has a special structure [3]).

A second attack that is relevant is a *comparison attack* [14]: the cryptanalyst searches for t-bit matches between the ciphertext bits. If a match occurs, he knows that the output of the block cipher will be equal in both cases, and hence he knows the exor of two plaintext bits. Note that the position of these plaintext bits cannot be selected. Because of the birthday paradox, such a match will occur after about $2^{t/2+1}$ ciphertext bits. If $t = 64$, and the encryption speed is equal to 2 Mbit/s, the storage requirements are 1 Gigabyte, and it will take about 1.16 hour to find a single match. If one waits for 25 days, one can collect 512 Gigabyte, and one expects about 2^{19} matches. This attack can be thwarted by increasing the frequency of the key change. If more than 2^{42} bits are collected, even triple matches will occur.

A third attack is the *tabulation attack*, which depends only on the size of

Table 1. **The number of key bits that influence the ciphertext in the case of the DES with N rounds in m-bit CFB.**

	number of rounds N				
m	1	2	3	4	5
1	0	6	39	53	56
2	6	45	53	56	56
4	12	50	56	56	56
8	18	52	56	56	56
16	36	55	56	56	56

the input register (and not on the size of the key or the number of rounds). The cryptanalyst will use a huge amount of known plaintexts to build a table of the secret mapping f. After about $2^t \cdot \ln(2^t)$ encryptions of *arbitrary* plaintexts with the unknown key, the secret mapping is completely known. An important difference between this type of attack and the simple exhaustive key search attack is that in this case it is not possible to perform the computations in parallel.

4 A meet in the middle attack

One of the first attacks on the DES with a reduced number of rounds in ECB mode was the meet in the middle attack proposed by D. Chaum and J.-H. Evertse [6]. The attack is faster than exhaustive search for $N \leq 6$. The basic idea is to look for r data bits in a middle round that depend on a limited number s of key bits. First an exhaustive search is performed for these bits, and subsequently the remaining key bits are determined.

In the case of the CFB mode, the probability that a key can be eliminated is significantly smaller, as only a small part of the ciphertext is known. If $N = 3$ and $m = 1$, one can show that the optimal choice is $r = 4$ bits in the middle (namely bits 18, 19, 20, and 21 of the right half of the register in the second round). In this case the subkey has $s = 27$ bits. The probability that a bad subkey survives in one trial is not equal to $1/2^r$ as for the ECB mode, but $h/2^r$, where h is a constant that has to be determined with a computer program (yielding $h = 0.5$). If it is assumed that the probability of survival for different plaintext/ciphertext pairs is independent, one can determine the expected number of remaining keys if M pairs are known:

$$\tilde{K} = 1 + (2^s - 1)\left(1 - \frac{h}{2^r}\right)^M . \quad (2)$$

The expected number of encryptions is equal to $M + (2^r \cdot 2^s)/h \approx 2^{32}$ if $M \ll 2^{32}$. The search for the remaining key bits requires $M + (2^{k-s}\tilde{K})/(1 - 2^{-m})$ encryptions, where k denotes the total number of key bits that influences the output (in this case one finds in Table 1 that $k = 39$). If $M = 256$, the number of operations in the second step is small compared to the first step. The total number of encryptions is a factor 2^7 smaller than for exhaustive search, but about 5 times more plaintext/ciphertext pairs are necessary.

For larger values of m, more ciphertext bits are known, but more key bits come into play as well. If $m = 8$, one can extend the previous approach in a straightforward way. One can however also try to reduce the number of required plaintext/ciphertext pairs by looking to bits 3 and 5 of the output (hence $r = 2$). These bits only depend on $s = 37$ key bits. The number of operations for the first and second step are equal to [6]:

$$M + \frac{2^s}{1 - 2^{-r}} \quad \text{and} \quad M + \frac{2^{n-s} + 2^{n-Mr}}{1 - 2^{r-m}} .$$

The first term corresponds to 2^{36} 3-round DES encryptions, and the second term is equal to 2^{32} if $M = 12$. This means that this is a factor 2^{20} faster than exhaustive search. A comparable improvement was obtained in [6] for $N = 4$ in ECB mode.

These results can be extended partially to 4 or 5 rounds, but the attack becomes more complicated: one cannot simply go backwards, because one has to guess some key bits and part of the ciphertext bits. Note that in ECB mode the improvement for 6 rounds is limited to a factor 4. As these attacks are known plaintext attacks, they are also applicable to m-bit OFB.

5 A differential attack

First it will be explained why a differential attack cannot be applied directly to the CFB mode. Subsequently the required modifications will be discussed, and an attack on 4, 5, and more rounds will be presented. Finally some extensions will be discussed, and several modifications to enhance the security of the DES in the CFB mode will be proposed.

5.1 Why does the conventional differential approach not work?

A differential attack in ECB mode is based on the following principle. The actual values of the input bits of the last round are known (because they are the right half of the ciphertext before IP^{-1}). The output exor of the last round is known with a certain probability (if the input pair is a right pair, the exor can be predicted). Subsequently the exor table and some additional information on the S-boxes allow to determine part of the subkey of the last round.

In the CFB mode only part of the output is known, as indicated in Table 2. The information on the output bits is restricted to exor information. It is clear that a differential attack requires that information on both input and output bits of a single S-box is available. This means that in 1-bit CFB this approach is restricted to the trivial case of 2 rounds. If 3 rounds or more are used, it follows from Table 2 that m has to be at least 3. In the following the differential attack will be described for 8-bit CFB. In this case most information is available on S-box 3, namely 1 input bit and 2 output bits. A reduced exor table can be produced for the bits that are known by adding columns and rows of the original exor table. One could hope to determine information on key bit K_{44} that is exored with input bit a of $S3$. However, it is easy to show that in this situation the output bits will not suggest a particular value for this single key bit (i.e., all values in the reduced exor table are equal).

Table 2. Input bits of S-boxes that are known and output bits that are accessible in the case of m-bit CFB ($m \leq 8$); the 6 inputs bits of an S-box are denoted with a through f, the 4 outputs are denoted with α through δ, and the CFB bits are denoted with the digits 1 through 8.

S-box	known inputs	accessible outputs	S-box	known inputs	accessible outputs
1	$a = 7$		5	$a = 3$	$\alpha = 2$
2	$e = 1$		6	$e = 5$	
3	$a = 1$	$\alpha = 4,\ \beta = 6$	7	$a = 5$	$\alpha = 8$
4	$e = 3$		8	$e = 7$	

5.2 An extended differential attack

The differential attack can be extended to the CFB mode if one also uses the characteristic to predict the input exor of the last round (at least partially). This implies that the reduced exor table is obtained by adding only the columns of the original exor table.

A second property which can be exploited is that if an input bit exor of an S-box equals 0, the output exor reveals some information on the corresponding key bit. Denote the pair of intermediate ciphertext bits corresponding to bit a of $S3$ with (c, c'); the cryptanalyst knows both bits. The corresponding key bit is K_{44}. The unknown input bits to the S-box are $(c \oplus K_i, c' \oplus K_i)$. If $c = c'$, or $c \oplus c' = 0$, these input bits will be equal to $(0,0)$ if $K_i = c = c'$ and will be equal to $(1,1)$ otherwise. One can now divide the reduced exor table into two parts, and distinguish between these two cases. This will reveal some information on key bit K_{44}. If $c \neq c'$, no information can be obtained on K_{44}. Indeed, the input bits to the S-box will be different, and the key bit K_{44} only determines whether they are equal to $(0,1)$ or $(1,0)$. Table 3 gives part of the new exor table for $S3$. The input and output exors of S-box i are denoted with Si'_E and Si'_O respectively, x denotes an unknown bit, and the subscript $_x$ indicates hexadecimal notation.

Information on the key bits can be obtained as follows. The probability that the corresponding key bit K_{44} is equal to 1 can be determined from the observed

Table 3. Part of the reduced exor table for $S3$ where the entries are split based on the actual value of input bits a. Only the most useful input exors are listed.

input exor $S3'_E$	output exor $S3'_O$				quality H
	00xx	01xx	10xx	11xx	
01_x	2	18	18	26	0.719
with $0 \oplus 0$	2	14	10	6	
with $1 \oplus 1$	0	4	8	20	
02_x	2	10	26	26	0.688
with $0 \oplus 0$	0	8	16	8	
with $1 \oplus 1$	2	2	10	18	
04_x	4	8	24	28	0.688
with $0 \oplus 0$	4	8	8	12	
with $1 \oplus 1$	0	0	16	16	
10_x	4	24	12	24	0.625
with $0 \oplus 0$	0	8	8	16	
with $1 \oplus 1$	4	16	4	8	

output exor by applying Bayes' rule[1]. For a right pair, one obtains that :

$$q = \Pr(K_{44} = 1 \mid S3'_O = \alpha\beta xx) = \frac{\Pr(K_{44} = 1) \cdot \Pr(S3'_O = \alpha\beta xx \mid K_{44} = 1)}{\Pr(S3'_O = \alpha\beta xx)} . \tag{3}$$

It is clear that $\Pr(K_{44} = 1) = 1/2$. Both input and output exor are only known with a certain probability. The probability that the input exor is correct is slightly larger, and it is clear that both probabilities are not independent.

The next problem is how to combine the outcome of M pairs in an efficient way. Assume that it follows from pair j that the probability that $K_{44} = 1$ is equal to q^j. Then one defines

$$Q^j = \frac{q^j \cdot Q^{j-1}}{q^j \cdot Q^{j-1} + (1 - q^j) \cdot (1 - Q^{j-1})} \quad \text{for } j = 1, 2, \ldots, M \text{ and } Q^0 = 0.5 . \tag{4}$$

It can be shown that this corresponds to a repeated application of Bayes' rule. If $Q^M > 0.5$, one decides that $K_{44} = 1$. In practice one expects that, after a sufficient number of experiments, Q^M will form a reliable estimate for K_{44} and will be close to 1 (or 0) with high probability.

An important issue is the choice of the characteristic in order to maximize $|q - 0.5|$. This depends on the probability p of the characteristic, the possibility of filtering, and on the difference between the $0 - 0$ and $1 - 1$ entries in the reduced exor table. Let q_i denote the value of q corresponding to a given input and output exor (the same numbering is used as for the values e_i in Table 4).

One can now prove this proposition (the proof will be given in the full paper):

[1] One obtains in fact the exor of the key bit with the corresponding input bit of $S3_E$.

Table 4. A reduced exor table.

input exor S'_E	output exor S'_O			
	00xx	01xx	10xx	11xx
	e_1+e_5	e_2+e_6	e_3+e_7	e_4+e_8
with $0 \oplus 0$	e_1	e_2	e_3	e_4
with $1 \oplus 1$	e_5	e_6	e_7	e_8

Proposition 2 *The number M' of right pairs required to predict a key bit with a probability of error equal to $1-z$ satisfies the following inequality:*

$$M' \leq 64 \cdot \frac{\ln\left(\frac{1}{z}-1\right)}{\ln(\rho)} \quad \textit{with } \rho = \prod_{i=1}^{4}\left(\frac{1}{q_i}-1\right)^{e_i-e_{i+4}}.$$

Note that Proposition 2 can be extended to the case where certain output exors are filtered: one can simply modify the corresponding table entries such that $e_i = e_{i+4}$, yielding $q = 0.5$.

The optimization of the attack, or equivalently the minimization of M is not easy, since M also depends on the properties of the characteristic. A good heuristic measure for the differences in the exor table for a given input exor S'_E is the expression

$$H = \sum_{i=1}^{4} \Pr(S'_O \mid S'_E) \frac{\max(e_i, e_{i+4})}{e_i + e_{i+4}}.$$

Here i indexes the 4 columns corresponding to the 4 possible output exors S'_O. This measure is indicated in Table 3.

5.3 An attack on 4 rounds

It follows from the previous section that the value of $S3'_E = \mathbf{01_x}$ in the last round is optimal. The input exor to the first round is equal to $(\mathbf{40\,08\,00\,00_x}, \mathbf{04\,00\,00\,00_x})$. Then the characteristic has a probability of 1/4 in the first round. In the third round, it is sufficient that the input exor to $S3$ is correct. If the pairs with output exor $00xx$ for $S3$ are filtered, only a fraction of $\frac{2}{64}$ of the right pairs is lost. From this it follows that the fraction $\tilde{p}$ of right pairs after filtering is equal to

$$\frac{\frac{3}{16}\cdot\frac{62}{64}}{\frac{3}{16}\cdot\frac{62}{64}+\frac{12}{16}\cdot\frac{3}{8}} = 0.392.$$

One obtains then with (3) the following equation for q:

$$q = \frac{1}{2} \cdot \frac{\tilde{p}\cdot\frac{14}{30}+(1-\tilde{p})\cdot\frac{1}{3}}{\tilde{p}\cdot\frac{18}{62}+(1-\tilde{p})\cdot\frac{1}{3}} \tag{5}$$

This assumes that the wrong pairs yield a uniform distribution of output exors, which has been confirmed by computer experiments. For $S3'_E = \mathtt{01_x}$ and $\tilde{p} = 0.39$ one obtains for $q = 0.609$, 0.527, and 0.383 for $S3'_O = 01xx$, $10xx$, and $11xx$ respectively. For an error probability of 5% (or $z = 0.95$), Proposition 2 predicts that $M' = 16.6$ and $M = M'/p = 89$, which has been confirmed by computer simulations.

If $m = 8$, a differential attack allows to determine 3 key bits (namely one bit corresponding to $S3$, $S5$, and $S7$). The details will only be discussed for a larger number of rounds.

5.4 An attack on 5 rounds

In order to develop an attack that is extendible to more rounds, an iterative characteristic will be used; this characteristic is probably not optimal for the 5 round case. For derivation of the key bit corresponding to $S3$, an input exor of $\mathtt{01_x}$ is again the best choice. This implies that in the one but last round $S2$ has to receive a non-zero input exor. For the iterative characteristic ϕ [3] (input exor of left halve equals $\mathtt{1B\,60\,00\,00_x}$), the probability in this round is equal to $\frac{55}{128}$, while for ψ (input exor of left halve equals $\mathtt{19\,60\,00\,00_x}$), this probability is equal to $\frac{33}{128}$. This implies that ϕ is preferable (ψ might be used in a quartet structure).

The pairs for which the output exors of bit 1 or 7 are not equal to zero will be filtered; the same holds for the pairs for which the output exor of $S3$ is equal to $00xx$. It is assumed that pairs that do not follow the characteristic in the second round give a uniformly distributed output. For pairs that do not follow the characteristic in round 4, one can filter all those that do not follow the characteristic in $S2$, and $\frac{18}{64}$ of the pairs that do not follow the characteristic in $S1$. This yields the following fraction of right pairs among the filtered ones:

$$\tilde{p} = \frac{\frac{1}{234} \cdot \frac{55}{128} \cdot \frac{62}{64}}{\frac{1}{234} \cdot \frac{55}{128} \cdot \frac{62}{64} + \left(1 - \frac{1}{234}\right) \cdot \frac{3}{16} + \frac{1}{234} \cdot \frac{25}{128} \cdot \frac{46}{64}} = 9.40 \cdot 10^{-3}\,.$$

For an error probability of 5% (or $z = 0.95$), Proposition 2 predicts that 370 000 pairs are sufficient to obtain a key bit (only 1 characteristic has been used). Computer simulations show that the actual number of pairs is even smaller.

For S-boxes 6 and 7, a similar strategy can be followed. The best iterative characteristic for both S-boxes has input exor $\mathtt{00\,00\,1D\,40_x}$ for the left halve. In the one but last round this characteristic has probability $\frac{7}{16}$ for $S5$ and $\frac{63}{256}$ for $S7$ to yield an input exor of $\mathtt{04_x}$ respectively $\mathtt{01_x}$. The fraction of right pairs after filtering is equal to $5.92 \cdot 10^{-3}$ and $3.33 \cdot 10^{-3}$, from which one can estimate that the number of required pairs is equal to 12 and 8.4 million respectively. In both cases one can eliminate those pairs for which the exor of bits 3 and 5 is not equal to 0.

5.5 Six rounds and more

The same characteristics can be used as in the previous section. In order to optimize that attack one can use the ideas of [5] to get around the first round.

The problem here is that the filtering of wrong pairs will be less effective. The estimated number of pairs to find 1 and 3 key bits for 8-bit CFB are indicated in Table 5.

The attack for $N = 7$ (without the optimization to gain an additional round) was implemented as a distributed application on a heterogeneous, non-dedicated farm of 30 DEC workstations, using the PVM (Parallel Virtual Machine) software [1] for interprocess communication. The program was generated and run from the HeNCE (Heterogeneous Network Computer Environment) software [2]. The correct key bits were retrieved from $2^{35.2}$ pairs using a quartet structure; the attack took about 40 hours.

Table 5. Probability of the characteristic and number of pairs to find 1 and 3 key bits in 8-bit CFB.

# rounds N	probability p		# pairs	
	1 bit	3 bits	1 bit	3 bits
6	$9.40 \cdot 10^{-3}$	$3.33 \cdot 10^{-3}$	$2^{18.5}$	$2^{23.0}$
8	$4.05 \cdot 10^{-5}$	$1.15 \cdot 10^{-5}$	$2^{34.2}$	$2^{39.4}$
10	$1.73 \cdot 10^{-7}$	$3.93 \cdot 10^{-8}$	$2^{50.0}$	$2^{55.8}$
16	$1.35 \cdot 10^{-14}$	$1.57 \cdot 10^{-15}$	$2^{97.2}$	$2^{104.7}$

5.6 Extensions

If the number m of feedback bits increases, more key bits can be found (5 if $m \geq 15$). If $m \geq 18$ three output bits of a single S-box are known, which implies that a smaller reduction has to be applied to the exor tables, resulting in a reduction of the required number of chosen ciphertext pairs. If $m \geq 15$, two bits of $S8$ in the last round are known, and the input to the one but last round can be estimated. Only if $m \geq 28$ one obtains in this way information on both input and output of a single S-box, which allows to determine key bits in this round.

This differential attack would be impossible without IP^{-1}. In the absence of IP^{-1} only information on the output of S-boxes of the last round would be available. The security of the DES in 1-bit CFB could be improved if the bit is selected from the left half of the ciphertext. Selecting all the CFB bits from the left half of the ciphertext thwarts the proposed differential attacks for small values of m. Another way to strengthen the DES in the CFB mode against differential attacks could be a redesign of the S-boxes in the last round in order to decrease the difference between the $0-0$ and $1-1$ entries in the reduced exor table. Finally a completely different structure for the computation of the CFB bits from the inputs to the last round could be used.

6 Linear cryptanalysis

This section will summarize the most important results of a linear attack on the DES reduced to 8 rounds in the CFB mode. Additional results will be given in the full paper. The following notation will be used:

$A[i]$ = the i-th bit of A, where the most significant bit has number 1,
$A[i,j,k] = A[i] \oplus A[j] \oplus A[k]$,
$F_i(X, K)$ = the i-th round substitution.

In linear cryptanalysis of the DES [13], one tries to approximate the S-boxes by equations of the form $P[i_1, \ldots, i_p] \oplus C[j_1, \ldots, j_c] = K[l_1, \ldots, l_k]$, where $i_1, \ldots, i_p$, $j_1, \ldots, j_c$, $l_1, \ldots, l_k$ are fixed. This equation holds with probability $p \neq 0.5$. The equations of the different rounds can be combined into a relation that holds for the entire algorithm.

Unlike the differential attack, the published linear attack can be applied directly to the CFB mode. The only limitation is that there are less bits visible from the ciphertext. This reduces the number of useful linear relations. We found the following relations:

$$C[48,56] \oplus P[16,24,42] \oplus F_1(P,K)[16,24] \oplus K[2,3,11,18,35] = 0\,,$$
$$C[16,24] \oplus P[11,48,56] \oplus F_1(P,K)[11] \oplus K[17,18,19,51,52,59,60] = 0\,.$$

They hold with $p = 0.5 + 1.5 \times 2^{-15}$ and $p = 0.5 + 2^{-19}$ respectively. The output bits involved are known if $m \geq 6$. Each relation can be used to determine 7 key bits. For an accuracy of 96 %, 1.78×2^{31} texts are necessary for the first equation and 1.78×2^{39} for the second equation.

7 Conclusions and open problems

Several attacks on the DES in the ECB mode can be extended to the m-bit CFB mode. They are only faster than exhaustive key search if the number of rounds is reduced. A meet in the middle attack on the DES with 3 rounds yields an improvement with a factor 2^{20} over exhaustive search in case of 8-bit CFB or OFB mode. A modified differential attack has been presented that works in m-bit CFB with $m \geq 3$. The most important modifications are that the exor of the input to the S-boxes of the last round are determined based on the characteristic and that the exor table is reduced. Moreover additional information on actual input values is taken into account. The attack is 8 times faster than exhaustive search for 9 rounds or less and 2 times faster for 10 rounds. A linear attack for $m \geq 6$ has been discussed. When the number of rounds of the DES is reduced to 8, 2^{31} known plaintexts are required to determine 7 key bits.

These attacks are completely theoretical in the sense that they pose no threat for the DES with 16 rounds in the m-bit CFB mode (for 'small' m). However, they are of some interest because for small values of m the m-bit CFB mode is very slow: this is an argument to reduce the number of rounds in order to obtain an acceptable performance. An interesting result is that the DES with 8 rounds in 8-bit CFB mode is less secure against these attacks than the DES with 16 rounds in ECB mode, while the first scheme is 4 times slower. It has been shown that in the CFB mode (and the OFB mode) IP^{-1} has a cryptographic meaning.

It would be interesting to extend all these attacks to other iterated block ciphers like IDEA and LOKI91. One of the important differences will be that the known bits are concentrated in a few S-boxes.

References

1. A. Beguelin, J. J. Dongarra, G. A. Geist, R. Mancheck, and V. Sunderam, *"A users' guide to PVM parallel virtual machine"*, Technical report ORNL/TM-11826, Oak Ridge National Laboratory, July 1991.
2. A. Beguelin, J. J. Dongarra, R. Manchek, K. Moore, R. Wade, J. Plank, and V. Sunderam, *"HeNCE: a user's guide"*, Version 1.2, December 1992.
3. E. Biham and A. Shamir, "Differential cryptanalysis of DES-like cryptosystems," *Journal of Cryptology*, Vol. 4, No. 1, 1991, pp. 3–72.
4. E. Biham and A. Shamir, "Differential cryptanalysis of Feal and N-hash," *Advances in Cryptology, Proc. Eurocrypt'91, LNCS 547*, D.W. Davies, Ed., Springer-Verlag, 1991, pp. 1–16.
5. E. Biham and A. Shamir, "Differential cryptanalysis of the full 16-round DES," *Technion Technical Report # 708*, December 1991.
6. D. Chaum and J.-H. Evertse, "Cryptanalysis of DES with a reduced number of rounds," *Advances in Cryptology, Proc. Crypto'85, LNCS 218*, H.C. Williams, Ed., Springer-Verlag, 1985, pp. 192–211.
7. D. Davies, "Investigation of a potential weakness in the DES algorithm," July 1987 (revised January 1990), preprint.
8. FIPS 46, *"Data Encryption Standard,"* Federal Information Processing Standard, National Bureau of Standards, U.S. Department of Commerce, Washington D.C., January 1977.
9. FIPS 81, *"DES Modes of Operation,"* Federal Information Processing Standard, National Bureau of Standards, U.S. Department of Commerce, Washington D.C., December 1980.
10. M. Hellman, R. Merkle, R. Schroeppel, L. Washington, W. Diffie, S. Pohlig and P. Schweitzer, *"Results of an initial attempt to cryptanalyze the NBS Data Encryption Standard,"* Information Systems Lab., Dept. of Electrical Eng., Stanford Univ., 1976.
11. ISO/IEC 10116, *"Information technology - Security techniques -Modes of operation of an n-bit block cipher algorithm,"* 1991.
12. R.R. Jueneman, "Analysis of certain aspects of Output Feedback Mode," *Advances in Cryptology, Proc. Crypto'82*, D. Chaum, R.L. Rivest, and A.T. Sherman, Eds., Plenum Press, New York, 1983, pp. 99–127.
13. M. Matsui, "Linear cryptanalysis method for DES cipher," *Advances in Cryptology, Proc. Eurocrypt'93, LNCS*, Springer-Verlag, to appear.
14. U.M. Maurer, "New approaches to the design of self-synchronizing stream ciphers," *Advances in Cryptology, Proc. Eurocrypt'91, LNCS 547*, D.W. Davies, Ed., Springer-Verlag, 1991, pp. 458–471.
15. K. Ohta and M. Matsui, "Differential attack on message authentication codes," *This Volume.*
16. M. Wiener, "Efficient DES key search," *This Volume.*

Weak Keys for IDEA

Joan Daemen, René Govaerts and Joos Vandewalle

Katholieke Universiteit Leuven, Laboratorium ESAT
Kardinaal Mercierlaan 94, B-3001 Heverlee, Belgium
email: joan.daemen@esat.kuleuven.ac.be

Abstract. Large classes of weak keys have been found for the block cipher algorithm IDEA, previously known as IPES [2]. IDEA has a 128-bit key and encrypts blocks of 64 bits. For a class of 2^{23} keys IDEA exhibits a linear factor. For a certain class of 2^{35} keys the cipher has a global characteristic with probability 1. For another class of 2^{51} keys only two encryptions and solving a set of 16 nonlinear boolean equations with 12 variables is sufficient to test if the used key belongs to this class. If it does, its particular value can be calculated efficiently. It is shown that the problem of weak keys can be eliminated by slightly modifying the key schedule of IDEA.

1 Introduction

At Eurocrypt '90 the block cipher proposal PES (Proposed Encryption Standard) was presented [1]. At Eurocrypt '91 the same authors presented a modification of PES, called IPES (Improved PES) [2]. The reason for this modification were new insights based on differential cryptanalysis [3]. IPES has become commercialized under the name IDEA (International Data Encryption Algorithm).

IDEA is an iterated cipher consisting of 8 similar rounds and a single output transformation. The building blocks of the round function are multiplication modulo $2^{16}+1$, addition modulo 2^{16} and bitwise XOR. IDEA has a 128-bit key and encrypts/decrypts data in blocks of 64 bits.

With exception of the key schedule, the IDEA decryption process is the same as its encryption process. The computational graph of the IDEA algorithm is shown in Fig.1. The encryption round keys are 16-bit substrings of the global key as specified in Table 1. The decryption round keys can be derived from the encryption round keys.

2 Linearities in the Modular Arithmetic Operations

Let x_i denote the i-th bit in the binary representation of the number X, i.e. $X = \sum 2^i x_i$. The bits of $Y = X + Z \bmod 2^n$ are given by

$$y_i = x_i \oplus z_i \oplus c_i \tag{1}$$

with c_i a carry bit that only depends on bits with indices *smaller than* i. The LSB of Y (y_0) is simply equal to $x_0 \oplus z_0$. Propagation of the MSBs of X and Z into Y is restricted to linear propagation (over $\mathbb{Z}_2^{16}$) into the MSB of Y.

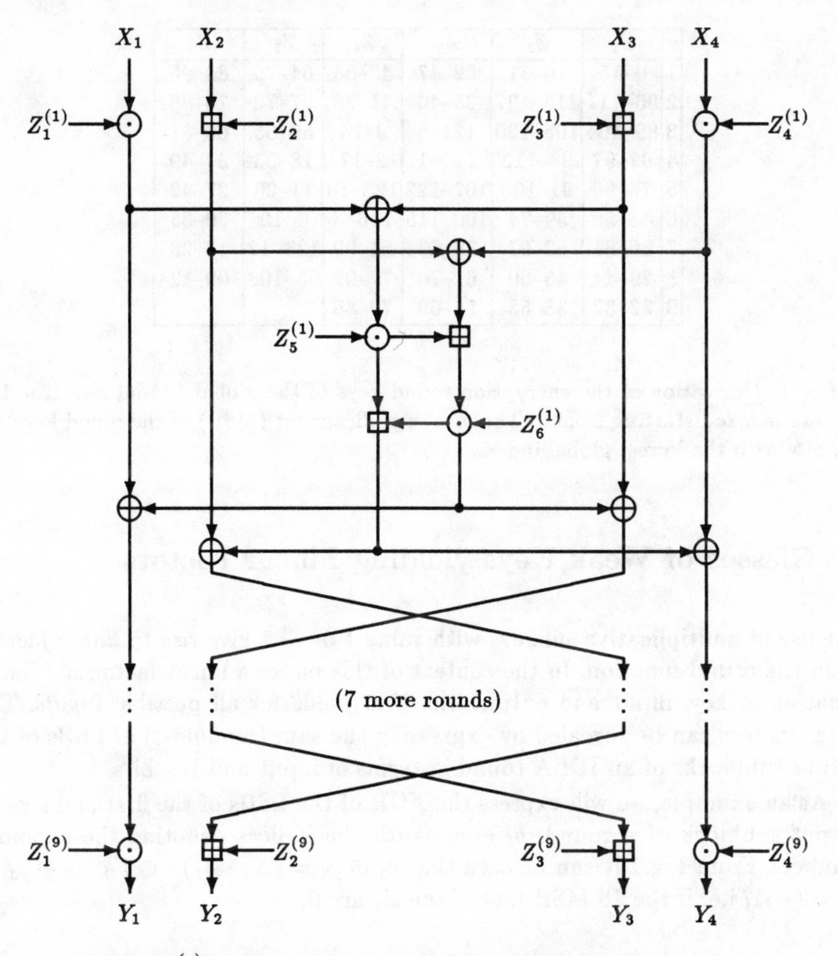

Fig. 1. the encryption process of IDEA.

For the multiplication by -1 (0000 HEX) modulo $2^{16}+1$ as defined in the IDEA block cipher it can easily be checked that

$$-1 \odot A = \bar{A} + 2 \bmod 2^{16} \tag{2}$$

with $\bar{A}$ the bitwise complement of A. Therefore multiplication by -1 inherits the linearity properties of the addition modulo 2^n.

r	Z_1	Z_2	Z_3	Z_4	Z_5	Z_6
1	0–15	16–31	32–47	48–63	64–79	80–95
2	96–111	112–127	25–40	41–56	57–72	73–88
3	89–104	105–120	121–8	9–24	50–65	66–81
4	82–97	98–113	114–1	2–17	18–33	34–49
5	75–90	91–106	107–122	123–10	11–26	27–42
6	43–58	59–74	100–115	116–3	4–19	20–35
7	36–51	52–67	68–83	84–99	125–12	13–28
8	29–44	45–60	61–76	77–92	93–108	109–124
9	22–37	38–53	54–69	70–85	—	—

Table 1. Derivation of the encryption round keys of the global 128-bit key. The key bits are indexed starting from 0. The most significant bit (MSB) of the round keys are the bits with the lowest global index.

3 Classes of Weak Keys yielding Linear Factors

The use of multiplicative subkeys with value 1 or -1 give rise to *linear factors* [4] in the round function. In the context of this paper a linear factor is a linear equation in key, input and output bits that holds for all possible *inputs*. The linear factors can be revealed by expressing the sum (modulo 2) of LSBs of the output subblocks of an IDEA round in terms of input and key bits.

As an example, we will express the XOR of the LSBs of the first and second output subblock of a round: $y_1 \oplus y_2$ (with the indices denoting the subblock number). From Fig.1 it can be seen that $y_1 \oplus y_2 = (X_1 \cdot Z_1)|_0 \oplus 1 \oplus x_3 \oplus z_3$. If $Z_1 = (-)1$, i.e. if the 15 MSB bits of the Z_1 are 0,

$$y_1 \oplus y_2 = x_1 \oplus x_3 \oplus z_1 \oplus z_3 \oplus 1 . \tag{3}$$

If the key bits are considered as (albeit unknown) constants, this linear factor can be interpreted as the propagation of knowledge from $x_1 \oplus x_3$ to $y_1 \oplus y_2$, denoted by $(1,0,1,0) \rightarrow (1,1,0,0)$. Similar factors and their corresponding conditions on subkey blocks can be found for all 15 combinations of LSB output bits and are listed in Table 2.

Multiple-round linear factors can be found by combining linear factors where the involved intermediate terms cancel out. For every round this gives conditions on subkeys that can be converted to conditions on global key bits using Table 1. An example is given in Table 3 for the global linear factor $(1,0,1,0) \rightarrow (0,1,1,0)$. The global key bits whose indices are given in this table must be 0. Since key bits with indices in 26-28, 72-74 or 111-127 don't appear, there are 2^{23} global keys that have this linear factor. This is called a *class of weak keys* since membership can easily be checked by observing some corresponding plaintext-ciphertext combinations.

linear factor	Z_1	Z_4	Z_5	Z_6
$(0,0,0,1) \rightarrow (0,0,1,0)$	-	(−)1	-	(−)1
$(0,0,1,0) \rightarrow (1,0,1,1)$	-	-	(−)1	(−)1
$(0,0,1,1) \rightarrow (1,0,0,1)$	-	(−)1	(−)1	-
$(0,1,0,0) \rightarrow (0,0,0,1)$	-	-	-	(−)1
$(0,1,0,1) \rightarrow (0,0,1,1)$	-	(−)1	-	-
$(0,1,1,0) \rightarrow (1,0,1,0)$	-	-	(−)1	-
$(0,1,1,1) \rightarrow (1,0,0,0)$	-	(−)1	(−)1	(−)1
$(1,0,0,0) \rightarrow (0,1,1,1)$	(−)1	-	(−)1	(−)1
$(1,0,0,1) \rightarrow (0,1,0,1)$	(−)1	(−)1	(−)1	-
$(1,0,1,0) \rightarrow (1,1,0,0)$	(−)1	-	-	-
$(1,0,1,1) \rightarrow (1,1,1,0)$	(−)1	(−)1	-	(−)1
$(1,1,0,0) \rightarrow (0,1,1,0)$	(−)1	-	(−)1	-
$(1,1,0,1) \rightarrow (0,1,0,0)$	(−)1	(−)1	(−)1	(−)1
$(1,1,1,0) \rightarrow (1,1,0,1)$	(−)1	-	-	(−)1
$(1,1,1,1) \rightarrow (1,1,1,1)$	(−)1	(−)1	-	-

Table 2. Linear factors in the round function with conditions on the subkeys.

round	input term	Z_1	Z_5
1	$(1,0,1,0)$	0–14	-
2	$(1,1,0,0)$	96–110	57–71
3	$(0,1,1,0)$	-	50–64
4	$(1,0,1,0)$	82–96	-
5	$(1,1,0,0)$	75–89	11–25
6	$(0,1,1,0)$	-	4–18
7	$(1,0,1,0)$	36–50	-
8	$(1,1,0,0)$	29–44	93–107
9	$(0,1,1,0)$	-	-

Table 3. Conditions on key bits for linear factor $(1,0,1,0) \rightarrow (0,1,1,0)$

4 Classes of Weak Keys yielding Characteristics with Probability 1

In this section differential cryptanalysis [3] is applied where 'difference' is defined by bitwise XOR. The use of multiplicative subkeys with value 1 or -1 gives rise to characteristics with probability 1 in the round function.

A round is executed for a pair of inputs X and X^* with a given XOR $X' = X \oplus X^*$. Let ν be the 16-bit block 8000 (HEX), i.e. the MSB is 1 and all other bits are 0.

Suppose X and X^* only differ in the MSB bit of the 4-th subblock, hence $X'_1 = X'_2 = X'_3 = 0$ and $X'_4 = \nu$. If $Z_4 = (-)1$ this will still be the case after the application of Z_1 to Z_4. The left input to the MA structure is the same for X and X^*. The right input differs by ν. This XOR propagates unchanged through the top right (TR) addition to the bottom right (BR) multiplication by Z_6. If this is equal to $(-)1$, the output XOR is again ν. This difference propagates unchanged through the BL addition and the XORs to the 4 subblocks. The output difference Y' of the round is equal to $(\nu, \nu, \nu, 0)$. Hence if the 15 MSB of both Z_4 and Z_6 are 0, the input XOR $(0, 0, 0, \nu)$ gives rise to the output XOR $(\nu, \nu, \nu, 0)$ with probability 1, denoted by $(0, 0, 0, \nu) \Rightarrow (\nu, \nu, \nu, 0)$. A similar analysis can be made for any other of the 15 possible nonzero input XORs where only the MSB bits of the subblocks are allowed to be 1. The results are listed in Table 4.

characteristic	Z_1	Z_4	Z_5	Z_6
$(0,0,0,\nu) \Rightarrow (\nu,\nu,\nu,0)$	-	$(-)1$	-	$(-)1$
$(0,0,\nu,0) \Rightarrow (\nu,0,0,0)$	-	-	$(-)1$	$(-)1$
$(0,0,\nu,\nu) \Rightarrow (0,\nu,\nu,0)$	-	$(-)1$	$(-)1$	-
$(0,\nu,0,0) \Rightarrow (\nu,\nu,0,\nu)$	-	-	-	$(-)1$
$(0,\nu,0,\nu) \Rightarrow (0,0,\nu,\nu)$	-	$(-)1$	-	-
$(0,\nu,\nu,0) \Rightarrow (0,\nu,0,\nu)$	-	-	$(-)1$	-
$(0,\nu,\nu,\nu) \Rightarrow (\nu,0,\nu,\nu)$	-	$(-)1$	$(-)1$	$(-)1$
$(\nu,0,0,0) \Rightarrow (0,\nu,0,0)$	$(-)1$	-	$(-)1$	$(-)1$
$(\nu,0,0,\nu) \Rightarrow (\nu,0,\nu,0)$	$(-)1$	$(-)1$	$(-)1$	-
$(\nu,0,\nu,0) \Rightarrow (\nu,\nu,0,0)$	$(-)1$	-	-	-
$(\nu,0,\nu,\nu) \Rightarrow (0,0,\nu,0)$	$(-)1$	$(-)1$	-	$(-)1$
$(\nu,\nu,0,0) \Rightarrow (\nu,0,0,\nu)$	$(-)1$	-	$(-)1$	-
$(\nu,\nu,0,\nu) \Rightarrow (0,\nu,\nu,\nu)$	$(-)1$	$(-)1$	$(-)1$	$(-)1$
$(\nu,\nu,\nu,0) \Rightarrow (0,0,0,\nu)$	$(-)1$	-	-	$(-)1$
$(\nu,\nu,\nu,\nu) \Rightarrow (\nu,\nu,\nu,\nu)$	$(-)1$	$(-)1$	-	-

Table 4. XOR propagation in the round function with conditions on the subkeys.

The propagation of a given XOR for multiple rounds can be easily studied by letting the output XOR be the input XOR to the following round. The conditions on the subkeys can be read in Table 4.

An example for the plaintext XOR $(0, \nu, 0, \nu)$ is given in Table 5. It can be seen that for keys with only nonzero bits on positions 26–40, 72–76 and 108–122 the output XOR must be equal to $(0, \nu, \nu, 0)$. This is the largest class we found, comprising a total of 2^{35} keys. Membership can be checked by performing 2 encryptions where the plaintexts have a chosen difference and observing the difference in the ciphertexts. A similar table can be constructed for any input XOR consisting of ν and 0.

round	input xor	Z_4	Z_5
1	$(0,\nu,0,\nu)$	48–62	-
2	$(0,0,\nu,\nu)$	41–55	57–71
3	$(0,\nu,\nu,0)$	-	50–64
4	$(0,\nu,0,\nu)$	2–16	-
5	$(0,0,\nu,\nu)$	123–9	11–25
6	$(0,\nu,\nu,0)$	-	4–18
7	$(0,\nu,0,\nu)$	84–98	-
8	$(0,0,\nu,\nu)$	77–91	93–107
9	$(0,\nu,\nu,0)$	-	-

Table 5. Propagation of plaintext XOR $(0,\nu,0,\nu)$ in IDEA.

5 Expanding Classes of Weak Keys

Classes of weak keys can sometimes be significantly expanded at the cost of some more effort in the checking for membership. Omitting in Table 5 the conditions for the subkeys of round 8 gives rise to the class of 2^{51} keys with nonzero bits on positions 26–40, 72–83 and 99–122. We will show that both checking for membership and calculation of the specific key can be performed efficiently.

5.1 The Membership Test

The input XOR of round 8 is equal to the output XOR of round 7 and is guaranteed to be equal to $(0,0,\nu,\nu)$ by the conditions on the subkeys of the first 7 rounds. Using the fact that $Z_3^{(9)}$, consisting of global key bits 54–69 is 0000 for these keys it can easily be derived that

$$Y_3' \oplus \nu = (Z_1^{(9)^{-1}} \cdot Y_1^*) \oplus (Z_1^{(9)^{-1}} \cdot Y_1) \; . \tag{4}$$

This can be verified by inspecting Fig.2. In (4) only $Z_1^{(9)}$ is unknown. This subkey consists of global key bits 22–37. For the given class only the 12 LSB may differ from 0. If the global key does not belong to the class of weak keys, the probability that (4) has a solution is 1/16. Additional encryptions can be performed to eliminate these solutions. Every pair of encryptions yields an equation for $Z_1^{(9)}$ similar to (4).

5.2 The Determination of the Key

The value of the 12 unknown bits of $Z_1^{(9)}$ are already determined by the membership test. The following step is the determination of the 3 unknown bits of $Z_2^{(9)}$, the 12 unknown bits of $Z_4^{(9)}$ and the 7 unknown bits of $Z_4^{(8)}$. A consistency check can be executed on these bits in the following way. Suppose $Z_2^{(9)}$ and $Z_4^{(9)}$

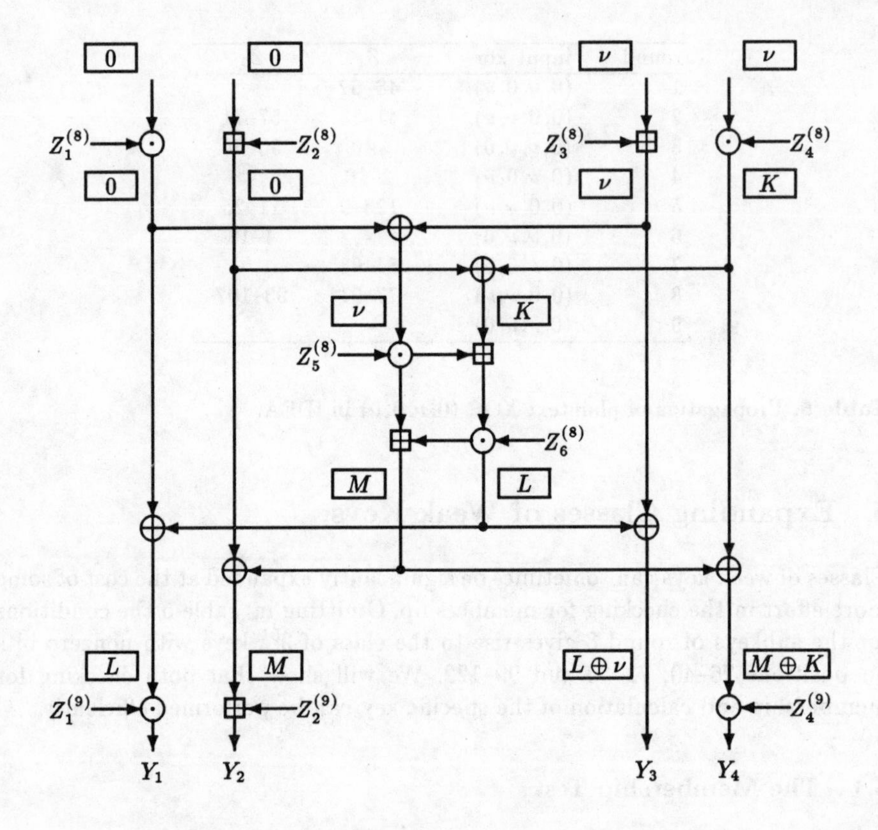

Fig. 2. XOR propagation of $X' = (0, \nu, 0, \nu)$ through the last round of IDEA for keys with only nonzero bits on positions 26–40, 72–83 and 99–122. The XORs are indicated in boxes.

are known. In this case it is possible to calculate the difference that is denoted by K in Fig.2. For this value K there must be a vector A (with MSB 0) such that

$$K = (Z_4^{(8)} \cdot A) \oplus (Z_4^{(8)} \cdot (A \oplus \nu)) = (Z_4^{(8)} \cdot A) \oplus ((Z_4^{(8)} \cdot A) + (Z_4^{(8)} \cdot 2^{15})) \quad (5)$$

For a given vector K it is easy to find the possible values of $Z_4^{(8)}$. Only values of $Z_4^{(8)}$ with the 9 LSB equal to 0 are valid. This information can be calculated in advance for every value of K and stored in an array of 2^{16} lists. The average number of possible $Z_4^{(8)}$ per K value turns out to be smaller than 1. Through this table, the observed value of K specifies a set of possible $Z_4^{(8)}$ values. If the set is empty, the chosen values for $Z_2^{(9)}$ and $Z_4^{(9)}$ must have been wrong. If the set is not empty, the K value resulting from another pair of encryptions (with input XOR at round 8 equal to $(0, 0, \nu, \nu)$) can be observed. The correct value for $Z_4^{(8)}$ must be in the list for both observed values of K. This can be repeated until there is no value for $Z_4^{(8)}$ left. The correct values for $Z_2^{(9)}$ and $Z_4^{(9)}$ are found if there is a value for $Z_4^{(8)}$ that is consistent for all the (say a maximum of 8) encryption pairs. Now 34 bits are fixed. The remaining 17 bits can easily be found by exhaustively trying all remaining 2^{17} possibilities and comparing it with any plaintext-ciphertext pair obtained during the attack.

The complete workload of the key determination is 16 chosen plaintext-difference encryptions, about 2^{15} modular additions, multiplications and table-lookups and 2^{17} key search encryptions.

6 A Modified IDEA Without Weak Keys

In the present specification of IDEA the conditions for weak multiplicative round keys are converted to the condition that global key bits must be 0. In Table 3 and 5 it can be seen that many global key bits appear more than once in the conditions.

Now let $\hat{Z}_i^{(r)} = \alpha \oplus Z_i^{(r)}$ with α a fixed nonzero binary vector. If in IDEA the subkeys $Z_i^{(r)}$ are replaced by $\hat{Z}_i^{(r)}$, the conditions for weak multiplicative keys are converted to the condition that some global key bits must be 0 and some must be 1. The vector α must be chosen such that for all potential multiple-round linear factors and characteristics, the conditions on the subkeys give conflicting conditions on global key bits. Because of the large overlap between subkeys, the exact value of α is not critical. For instance, for $\alpha =$ 0DAE (HEX) no weak keys were found.

7 Conclusions

Large classes of weak keys have been found for the block cipher IDEA. These keys are weak in the sense that it takes only a very small amount of effort to detect their use. It is possible to eliminate the weak key problem by slightly modifying the key schedule of IDEA.

References

[1] X. Lai and J.L. Massey, A Proposal for a New Block Encryption Standard, *Advances in Cryptology–Eurocrypt' 90*, Springer-Verlag, Berlin 1991, pp. 389–404.

[2] X. Lai, J.L. Massey and S. Murphy, Markov Ciphers and Differential Cryptanalysis, *Advances in Cryptology–Eurocrypt' 91*, Springer-Verlag, Berlin 1991, pp. 17–38.

[3] E. Biham and A. Shamir, Differential Cryptanalysis of DES-like Cryptosystems, *Journal of Cryptology*, Springer-Verlag, Vol. 4, No. 1, pp. 3–72, 1991.

[4] D. Chaum, J.-H. Evertse, Cryptanalysis of DES with a Reduced Number of Rounds, Sequences of Linear Factors in Block Ciphers, *Advances in Cryptology, Proceedings of Crypto 85*, pp. 192–211, 1985.

Entity Authentication and Key Distribution

Mihir Bellare[1] and Phillip Rogaway[2]

[1] High Performance Computing and Communications, IBM T.J. Watson Research Center, PO Box 704, Yorktown Heights, NY 10598, USA. e-mail: mihir@watson.ibm.com.

[2] PS LAN System Design, IBM Personal Software Products, 11400 Burnet Road, Austin, TX 78758, USA. e-mail: rogaway@austin.ibm.com

Abstract. We provide the first formal treatment of entity authentication and authenticated key distribution appropriate to the distributed environment. Addressed in detail are the problems of mutual authentication and authenticated key exchange for the symmetric, two-party setting. For each we present a definition, protocol, and proof that the protocol meets its goal, assuming only the existence of a pseudorandom function.

1 Introduction

Entity authentication is the process by which an agent gains confidence in the identity of a communication partner. Though central to computing practice, entity authentication for the distributed environment rests on no satisfactory formal foundations.This is more than an academic complaint; entity authentication is an area in which an informal approach has often lead to work which is at worst wrong, and at best only partially analyzable. In particular, an alarming fraction of proposed protocols have subsequently been found to be flawed (see, e.g., [5, 3]) and the bugs have, in some cases, taken years to discover. It is therefore desirable that confidence in an authentication protocol should stem from more than a few people's inability to break it. In fact, each significant entity authentication goal should be formally defined and any candidate protocol should be proven to meet its goal under a standard cryptographic assumption.

More often than not the entity authentication process is coupled with the distribution of a "session key" which the communicating partners may later use for message confidentiality, integrity, or whatever else. This "authenticated key distribution" goal may be considered even more important in practice than the pure entity authentication goal. As a problem, it is beset with the same foundational difficulties as the entity authentication problem of which it is an extension.

Authentication and authenticated key distribution problems come in many different flavors: there may be two parties involved, or more; the authentication may be unilateral or mutual; parties might (the symmetric case) or might not (the asymmetric case) share a secret key. Here we focus on two version of the the two-party, mutual, symmetric case. In the *mutual authentication* problem the parties, representing processes in a distributed system, engage in a conversation in which each gains confidence that it is the other with whom he speaks. In the *authenticated key exchange* problem the parties also want to distribute a "fresh" and "secret" *session key.*[3]

1.1 Contributions of this Paper

A COMMUNICATION MODEL FOR DISTRIBUTED SECURITY. It has been pointed out in many places that one difficulty in laying foundations for entity authentication and authenticated key distribution protocols has been the lack of a formal communications model for authentication in the distributed environment. Here we specify such a model. To be fully general, we assume that all communication among interacting parties is under the adversary's control. She can read the messages produced by the parties, provide messages of her own to them, modify messages before they reach their destination, and delay messages or replay them. Most importantly, the adversary can start up entirely new "instances" of any of the parties, modeling the ability of communicating agents to simultaneously engage in many *sessions* at once. This gives us the ability to model the kinds of attacks that were suggested by [3]. Formally, each party will be modeled by an infinite collection of oracles which the adversary may run. These oracles only interact with the adversary, they never directly interact with one another. See Section 3.

DEFINITIONS. In the presence of an adversary as powerful as the one we define, it is unclear what it could possibly mean to be convinced that one has engaged in a conversation with a specified partner; after all, every bit communicated has really been communicated to the the adversary, instead. We deal with this problem as follows.

As has often been observed, an adversary in our setting can always make the parties accept by faithfully relaying messages among the communication partners. But this behavior does not constitute a damaging attack; indeed, the adversary has functioned just like a wire, and may as well not have been there. The idea of our definition of a mutual authentication is simple but strong: we formalize that a protocol is secure if the *only* way that an adversary can get a party to accept is by faithfully relaying messages in this manner. In other words, any adversary effectively behaves as a trusted wire, if not a broken one.

To define authenticated key exchange it is necessary to capture a protocol's robustness against the loss of a session key; even if the adversary gets hold of one, this isn't supposed to compromise security beyond the particular session which that key protects. We model this requirement by allowing the adversary to obtain session keys just by asking for them. When this inquiry is made, the

[3] At first glance it might seem unnecessary for two parties who already share a key a to come up with another key α. One reason a new key is useful is the necessity of avoiding cross-session "replay attacks" —messages copied from one session being deemed authentic in another— coupled with an insistence on *not* attempting to carry "state" information (e.g., a message counter) across distinct sessions.

key is no longer *fresh*, and the partner's key is declared unfresh, too. Fresh keys must remain unknown to the adversary, which we define along the lines of formalizations of security for probabilistic encryption [12, 8, 9].

PROTOCOLS. Four protocols are specified. Protocol MAP1, an extension of the 2PP of [3], is a mutual authentication protocol for an arbitrary set I of players. Protocol MAP2 is an extension of MAP1, allowing arbitrary text strings to be authenticated along with its flows. Protocol AKEP1 is a simple authenticated key exchange which uses MAP2 to do the key distribution. Protocol AKEP2 is a particularly efficient authenticated key exchange which introduces the idea of "implicitly" distributing a key; its flows are identical to MAP1, but it accomplishes a key distribution all the same. The primitive required for all of these protocols is a pseudorandom function.

PROOFS OF SECURITY. Assuming that pseudorandom functions exist, each protocol that we give is proven to meet the definition for the task which this protocol is claimed to carry out. The proofs for MAP1 and AKEP1 are given in this paper; the proofs for MAP2 and AKEP2 are omitted because they are essentially identical. The asymptotics implicit in all of our proofs are not so bad as to render the reductions meaningless for cryptographic practice. In other words, if one had a practical method to defeat the entity authentication this would translate into a practical method to defeat the underlying pseudorandom function.

DESIGN FOR PRACTICE. Every protocol presented in this paper is practical. Each is efficient in terms of rounds, communication, and computation. This efficiency was designed into our protocols in part through the choice of the underlying primitive—a pseudorandom function.

From a theoretical perspective, the existence of pseudorandom functions and the existence of many other important cryptographic primitives (e.g., one-way functions, pseudorandom generators, digital signatures) are all equivalent [16, 14, 10, 23].[4] From a practical perspective, pseudorandom functions (with the right domain and range) are a highly desirable starting point for efficient protocols in the symmetric setting. The reason is that beginning with primitives like DES and MD5 one can construct efficient pseudorandom functions with arbitrary domain and range lengths, and these constructions are themselves provably secure given plausible assumptions about DES and MD5. See Section 6 for discussion of these issues.

IMPLEMENTATIONS. A derivative of our AKEP2 is implemented in an IBM prototype of a secure high speed transport protocol. Another derivative of AKEP2 is implemented in an IBM product for Remote LAN Access.

Combining ideas from our proofs and a lemma from [1], we can show that a special case of the 2PP of [3] meets our definition of a secure mutual authentication. (See the end of Section 4 in our full paper for further details.) The 2PP protocol is implemented in an IBM prototype called *KryptoKnight* [20].

---

[4] We remark that the existence of a secure mutual authentication protocol implies the existence of a one-way function, as can be shown using techniques of [15]; thus mutual authentication also exists if and only if one-way functions do.

1.2 History and Related Work

PROVABLE SECURITY. Provable security means providing: (1) a formal definition of the goal; (2) a protocol; (3) a statement of a (standard) assumption; and (4) a proof that the protocol meets its goal given the assumption. The notion emerged in the work of Blum-Micali [4] and Yao [26] (who introduced provably secure pseudorandom generators) and Goldwasser-Micali [12] (who introduced provably secure encryption). A definition for digital signatures (Goldwasser, Micali and Rivest [13]) took slightly longer. We follow in spirit this early foundational work and enable entity authentication to join the ranks of those key primitives having a well-defined goal proven to be achievable under a standard complexity-theoretic assumption.

PROTOCOLS. The number of protocols suggested for entity authentication is too large to survey here; see [5, 17] for some examples.

TOWARDS A MODEL AND DEFINITIONS. Bird, Gopal, Herzberg, Janson, Kutten, Molva and Yung [3] described a new class of attacks, called "interleaving attacks," which they used to break existing protocols. They then suggested a protocol (2PP) defeated by none of these attacks. The recognition of interleaving attacks helped lead us to the formal model of Section 3, and our MAP1 protocol is an extension of 2PP. However, while an analysis such as theirs is useful as a way to spot errors in a protocol, resistance to interleaving attacks does not make a satisfactory notion of security; in particular, it is easy to construct protocols which are insecure but defeated by no attack from the enumeration. When our work was announced, the authors of [3] told us that they understood this limitation and had themselves been planning to work on general definitions; they also told us that the CBC assumption of their paper [3, Definition 2.1] was intended for proving security under a general definition.

Mentioned in the introduction of [3] is an idea of "matching histories." Diffie, Van Oorschot and Wiener [6] expand on this to introduce a notion of "matching protocol runs." They refine this idea to a level of precision adequate to help them separate out what are and what are not "meaningful" attacks on the protocols they consider. Although [6] stops short of providing any formal definition or proof, the basic notion these authors describe is the same as ours and is the basis of a definition of entity authentication. Thus there is a clear refinement of definitional ideas first from [3] to [6], and then from [6] to our work.

RELATION TO OTHER FOUNDATIONAL WORK. Beginning with the paper of Burrows, Abadi and Needham [5], the "logic-based approach" attempts to *reason* that an authentication protocol is correct as it evolves the set of *beliefs* of its participants. This idea is useful and appealing, but it has not been used to define when an arbitrary set of flows constitutes a secure entity authentication. Nor does a correctness proof in this setting guarantee that a protocol is "right," but only that it lacks the flaws in reasoning captured by the underlying logic.

More closely related to our approach is the idea of a *non-transferable proof,* a notion for (asymmetric, unilateral) authentication due to Feige, Fiat and Shamir [7]. Here an (honest) claimant P interacts with a (cheating) verifier $\tilde{V}$, and then a ($\tilde{V}$-conspiring cheating) prover $\tilde{P}$ tries to convince an (honest) verifier V that she ($\tilde{P}$) is really P. This definition accurately models a world of smart-card claimants and untrusted verifiers, but not a distributed system of always-running processes.

PUBLICATION NOTES. A preliminary version of this paper (which included, in addition to the material here, definitions three party authentication) appeared in the proceedings of an IBM internal conference in October 1992. The version of this paper you are now reading has been edited due to page limits. Ask either author for the complete version.

2 Preliminaries

The set of infinite strings is $\{0,1\}^\infty$ and $\{0,1\}^{\leq L}$ is the set of strings of length at most L. The empty string is λ. When a, b, c, ... are strings used in some context, by $a \,.\, b \,.\, c \cdots$ we denote an encoding of these strings such that each constituent string is efficiently recoverable given the encoding and the context of the string's receipt. In our protocols, concatenation will usually be adequate for this purpose. A function is *efficiently computable* if it can be computed in time polynomial in its first argument. A real-valued function $\epsilon(k)$ is *negligible* if for every $c > 0$ there exists a $k_c > 0$ such that $\epsilon(k) < k^{-c}$ for all $k > k_c$. The protocols we consider are two party ones, formally specified by an efficiently computable function Π on the following inputs:

1^k — the "security parameter" — $k \in \mathsf{N}$.
i — the "identity of the sender" — $i \in I \subseteq \{0,1\}^k$.
j — the "identity of the (intended) partner" — $j \in I \subseteq \{0,1\}^k$.
a — the "secret information of the sender" — $a \in \{0,1\}^*$.
κ — the "conversation so far" — $\kappa \in \{0,1\}^*$.
r — the "random coin flips of the sender" — $r \in \{0,1\}^\infty$.

The value of $\Pi(1^k, i, j, a, \kappa, r) = (m, \delta, \alpha)$ specifies:

m — the "next message to send out" — $m \in \{0,1\}^* \cup \{*\}$.
δ — the "decision" — $\delta \in \{\mathsf{A}, \mathsf{R}, *\}$.
α — the "private output" — $\alpha \in \{0,1\}^* \cup \{*\}$.

Here I is a set of *identities* which defines the *players* who can participate in the protocol. Although our protocols involve only two parties, the set of players I could be larger, to handle the possibility (for example) of an arbitrary pool of players who share a secret key. Elements of I will sometimes be denoted A or B (Alice and Bob), rather than i, j; we will switch back and forth irrationally between these notations. We stress that A, B (and i, j) are variables ranging over I (not fixed members of I), so $A = B$ (or $i = j$) is quite possible. Note that the adversary is *not* a player in our formalization. The value a that a player sees is the private information provided to him. This string is sometimes called the *long-lived key* (or LL-key) of a player. In the case of (pure) symmetric authentication, all players $i \in I$ will get the same LL-key, and the adversary will be denied this key. In general, a LL-key generator $\mathcal{G}$ associated to a protocol will determine who gets what initial LL-key (see below). The value "$*$" is supposed to suggest, for m, that "the player sends no message." For δ, it means that "the player has not yet reached a decision." For α, it means "the player does not currently have any private output." The values A and R, for δ, are supposed to suggest "accept" and "reject," respectively. We denote the t-th component of Π (for $t \in \{1, 2, 3\}$) by Π_t. Acceptance usually does not occur until the end of the protocol, although rejection may occur at any time. Some protocol problems, such as mutual authentication, do not make use of the private output; these protocol are concerned only with acceptance or rejection. For others, including key exchange protocols, the private output of a party will be what this party thinks is the key which has been exchanged. It is convenient to assume that once a player has accepted or rejected, this output cannot change. To each protocol is associated its number of moves, R. In general this is a polynomially bounded, polynomial time computable function of the security parameter; in all our protocols, however, it is a constant.

Associated to a protocol is a *long-lived key generator* (LL-key generator) $\mathcal{G}(1^k, \iota, r_G)$. This is a polynomial time algorithm which takes as input a security parameter 1^k, the identity of a party $\iota \in I \cup \{\mathsf{E}\}$, and an infinite string $r_G \in \{0,1\}^\infty$ (coin flips of the generator). For all of the protocols of this paper, the associated LL-key generator will be a *symmetric* one, where for each $i, j \in I$ we have that $\mathcal{G}(1^k, i, r_G) = \mathcal{G}(1^k, j, r_G)$; while, on the other hand, $\mathcal{G}(1^k, \mathsf{E}, r_G) = \lambda$. The value of $\mathcal{G}(1^k, i, r_G)$ will just be a prefix of r_G (that is, a random string). The length of this prefix will vary according to the protocol we consider.

3 A Communication Model for Distributed Security

Formally the adversary E is a probabilistic machine[5] $E(1^k, a_E, r_E)$ equipped with an infinite collection of oracles $\Pi^s_{i,j}$, for $i, j \in I$ and $s \in \mathsf{N}$. Oracle $\Pi^s_{i,j}$ models player i attempting to authenticate player j in "session" s. Adversary E communicates with the oracles via queries of the form (i, j, s, x) written on a special tape. The query is intended to mean that E is sending message x to i, claiming it is from j in session s. Running a protocol Π (with LL-key generator $\mathcal{G}$) in the presence of an adversary E, using security parameter k, means performing the following experiment:

(1) Choose a random string $r_G \in \{0,1\}^\infty$ and set $a_i = \mathcal{G}(1^k, i, r_G)$, for $i \in I$, and set $a_E = (1^k, \mathsf{E}, r_G)$.

(2) Choose a random string $r_E \in \{0,1\}^\infty$ and, for each $i, j \in I$, $s \in \mathsf{N}$, a random string $r^s_{i,j} \in \{0,1\}^\infty$.

(3) Let $\kappa^s_{i,j} = \lambda$ for all $i, j \in I$ and $u \in \mathsf{N}$. (The variable $\kappa^s_{i,j}$ will keep track of the conversation that $\Pi^s_{i,j}$ engages in.)

---

[5] Adversaries can be uniform or non-uniform, and the results of this paper hold in both cases, with uniform adversaries requiring a uniform complexity assumptions and non-uniform adversaries requiring non-uniform ones.

(4) Run adversary E on input $(1^k, a_E, r_E)$, answering oracle calls as follows. When E asks a query (i, j, s, x), oracle $\Pi^s_{i,j}$ computes $(m, \delta, \alpha) = \Pi(1^k, i, j, a_i, \kappa^s_{i,j} \,.\, x, r^s_{i,j})$ and answers with (m, δ). Then $\kappa^s_{i,j}$ gets replaced by $\kappa^s_{i,j} \,.\, x$.

We point out that in response to an oracle call E learns not only the outgoing message but also whether or not the oracle has accepted or rejected. (For convenience of discourse, we often omit mention of the latter.) According to the above, E doesn't learn the oracle's private output. For some problems (such as authenticated key exchange) we will need to give the adversary the power to sometimes learn these private outputs. Such an extension is handled by specifying a new kind of oracle query and then indicating how the experiment is extended with responses to the new class of queries. An adversary is called *benign* if it is deterministic and restricts its action to choosing a pair of oracles $\Pi^s_{i,j}$ and $\Pi^t_{j,i}$ and then faithfully conveying each flow from one oracle to the other, with $\Pi^s_{i,j}$ beginning first. While the choice of i, j, s, t is up to the adversary, this choice is the same in all executions with security parameter k.

In a particular execution of a protocol, the adversary's i-th query to an oracle is said to occur at time $\tau = \tau_i \in \mathbf{R}$. We intentionally do not specify $\{\tau_i\}$, except to demand that $\tau_i < \tau_j$ when $i < j$. Conforming notions of time include "abstract time," where $\tau_i = i$, and "Turing machine time," where $\tau_i =$ the i-th step in E's computation, when parties are realized by interacting Turing machines.

4 Entity Authentication

A central idea in the definition is that of matching conversations. Consider running the adversary E with security parameter k. When E terminates, each oracle $\Pi^s_{i,j}$ has had a certain conversation $\kappa^s_{i,j}$ with E, and it has reached a certain decision $\delta \in \{\mathsf{A}, \mathsf{R}, *\}$. Fix an execution of an adversary E (that is, fix the coins of the LL-key generator, the oracles, and the adversary). For any oracle $\Pi^s_{i,j}$ we can capture its *conversation* (for this execution) by a sequence

$$K = (\tau_1, \alpha_1, \beta_1),\ (\tau_2, \alpha_2, \beta_2),\ \ldots,\ (\tau_m, \alpha_m, \beta_m).$$

This sequence encodes that at time τ_1 oracle $\Pi^s_{i,j}$ was asked α_1 and responded with β_1; and then, at some later time $\tau_2 > \tau_1$, the oracle was asked α_2 and answered β_2; and so forth, until, finally, at time τ_m it was asked α_m and answered β_m. Adversary E terminates without asking oracle $\Pi^s_{i,j}$ any more questions. Suppose oracle $\Pi^s_{i,j}$ has conversation prefixed by $(\tau_1, \alpha_1, \beta_1)$. Then if $\alpha_1 = \lambda$ we call $\Pi^s_{i,j}$ an *initiator* oracle; if α_1 is any other string we call $\Pi^s_{i,j}$ a *responder* oracle. We now define matching conversations. For simplicity we focus on the case where R is odd; the case of even R is analogous and is left to the reader. Explanations follow the formal definition.

Definition 1. (Matching conversations)*Fix a number of moves $R = 2\rho - 1$ and an R-move protocol Π. Run Π in the presence of an adversary E and consider two oracles, $\Pi^s_{A,B}$ and $\Pi^t_{B,A}$, that engage in conversations K and K', respectively.*

(1) *We say that K' is a* matching conversation *to K if there exist $\tau_0 < \tau_1 < \ldots < \tau_R$ and $\alpha_1, \beta_1, \ldots, \alpha_\rho, \beta_\rho$ such that K is prefixed by*

$$(\tau_0, \lambda, \alpha_1), (\tau_2, \beta_1, \alpha_2), (\tau_4, \beta_2, \alpha_3), \ldots, (\tau_{2\rho-4}, \beta_{\rho-2}, \alpha_{\rho-1}), (\tau_{2\rho-2}, \beta_{\rho-1}, \alpha_\rho)$$

and K' is prefixed by

$$(\tau_1, \alpha_1, \beta_1), (\tau_3, \alpha_2, \beta_2), (\tau_5, \alpha_3, \beta_3), \ldots, (\tau_{2\rho-3}, \alpha_{\rho-1}, \beta_{\rho-1}) .$$

(2) *We say that K is a* matching conversation *to K' if there exist $\tau_0 < \tau_1 < \ldots < \tau_R$ and $\alpha_1, \beta_1, \ldots, \alpha_\rho, \beta_\rho$ such that K' is prefixed by*

$$(\tau_1, \alpha_1, \beta_1), (\tau_3, \alpha_2, \beta_2), (\tau_5, \alpha_3, \beta_3), \ldots, (\tau_{2\rho-3}, \alpha_{\rho-1}, \beta_{\rho-1}), (\tau_{2\rho-1}, \alpha_\rho, *) .$$

and K is prefixed by

$$(\tau_0, \lambda, \alpha_1), (\tau_2, \beta_1, \alpha_2), (\tau_4, \beta_2, \alpha_3), \ldots, (\tau_{2\rho-4}, \beta_{\rho-2}, \alpha_{\rho-2}), (\tau_{2\rho-2}, \beta_{\rho-1}, \alpha_\rho)$$

Case (1) defines when the conversation of a responder oracle matches the conversation of an initiator oracle. Case (2) defines when the conversation of an initiator oracle matches the conversation of a responder oracle. Let us paraphrase our definition. Consider an execution in which $\Pi^s_{A,B}$ is an initiator oracle and $\Pi^t_{B,A}$ is a responder oracle. If every message that $\Pi^s_{A,B}$ sends out, except possibly the last, is subsequently delivered to $\Pi^t_{B,A}$, with the response to this message being returned to $\Pi^s_{A,B}$ as its own next message, then we say that the conversation of $\Pi^t_{B,A}$ matches that of $\Pi^s_{A,B}$. Similarly, if every message that $\Pi^t_{B,A}$ receives was previously generated by $\Pi^s_{A,B}$, and each message that $\Pi^t_{B,A}$ sends out is subsequently delivered to $\Pi^s_{A,B}$, with the response that this message generates being returned to $\Pi^t_{B,A}$ as its own next message, then we say that the conversation of $\Pi^s_{A,B}$ matches the one of $\Pi^t_{B,A}$. Note that this second condition is easily seen to imply the first one.

We comment that the party who sends the last flow ($\Pi^s_{A,B}$, above) can't "know" whether or not its last message was received by its partner, so when this oracle accepts accepts, it cannot "know" (assuming this last message to be relevant) whether or not its partner will accept. This asymmetry is an inherent aspect of authentication protocols with a fixed number of moves.

We will say that oracle $\Pi^t_{j,i}$ has a matching conversation with oracle $\Pi^s_{i,j}$ if the first has conversation K', the second has conversation K, and K' matches K. Either party here may be the initiator.

We require that any mutual authentication protocol have $R \geq 3$ rounds. We implicitly make this assumption throughout the remainder of this paper. Let No-Matching$^E(k)$ be the event that there exist i, j, s such that $\Pi^s_{i,j}$ accepted and there is no oracle $\Pi^t_{j,i}$ which engaged in a matching conversation.

Definition 2. (Secure mutual authentication) *We say that Π is a* secure mutual authentication *protocol if for any polynomial time adversary E,*

(1) (Matching conversations $\Rightarrow$ acceptance.) *If oracles $\Pi^s_{A,B}$ and $\Pi^t_{B,A}$ have matching conversations, then both oracles accept.*

(2) (Acceptance $\Rightarrow$ matching conversations.) *The probability of* No-Matching$^E(k)$ *is negligible.*

An oracle's matching partner is unique. Formally, let Multiple-Match$^E(k)$ be the event that some $\Pi^s_{i,j}$ accepts, and there are at least two distinct oracles $\Pi^t_{j,i}$ and $\Pi^{t'}_{j,i}$ which have had matching conversations with $\Pi^s_{i,j}$. The proof of the following is in Appendix C (omitted here due to lack of space; see our full paper).

Proposition 3. *Suppose Π is a secure MA protocol. Let E be any polynomial time adversary. Then the probability of* Multiple-Match$^E(k)$ *is negligible.*

We now proceed to protocols. Let f be a pseudorandom function (PRF) family [10]. Denote by f_a: $\{0,1\}^{\leq L(k)} \rightarrow \{0,1\}^{l(k)}$ the function specified by key a. In general, the length of the key, the length L of the input to f_a, and the length l of the output, are all functions of the security parameter. Here we assume the key length is just k, and, for our first protocol (MAP1) it suffices to assume $L(k) = 4k$ and $l(k) = k$. For any string $x \in \{0,1\}^{\leq L(k)}$ define $[x]_a = (x, f_a(x))$; this will serve as an authentication of message x [10, 11]. For any $i \in I$, $[i\,.\,x]_a$ will serve as i's authentication of message x.

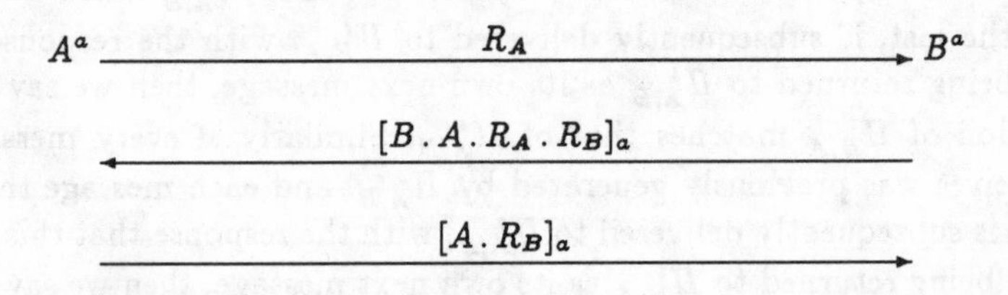

Fig. 1. *Protocol MAP1: a mutual authentication of any two principals, A and B, among a set of principals I who share a key a.*

Our first protocol (called "MAP1," for "mutual authentication protocol one") is represented by Figure 1. Alice (A) begins by sending Bob (B) a random challenge R_A of length k. Bob responds by making up a random challenge R_B of length k and returning $[B\,.\,A\,.\,R_A\,.\,R_B]_a$. Alice checks that this message is of the right form and is correctly tagged as coming from B. If it is, Alice sends Bob the message $[A\,.\,R_B]_a$ and accepts. Bob checks that this message is of the right form and is correctly tagged as coming from A, and, if it is, he accepts. We stress that checking the message is of the right form, for A in the second flow, includes checking that the nonce present in the message is indeed the same nonce she sent in the first flow; similarly for B with respect to checking the third flow. We comment that $A = B$ is permitted; these are any two identities in the set I. The proof of the following Theorem 4 appears in Appendix A.

Theorem 4. *(MAP1 is a secure MA) Suppose f is a pseudorandom function family. Then protocol MAP1 described above and based on f is a secure mutual authentication.*

5 Authenticated Key Exchange

Fix $S = \{S_k\}_{k \in \mathsf{N}}$ with each S_k a distribution over $\{0,1\}^{\sigma(k)}$, for some polynomial $\sigma(k)$. The intent of an AKE will be both to authenticate entities *and* to distribute a "session key" sampled from S_k. When a player accepts, his private output will be interpreted as the session key which he has computed. Formally, the session key α will be defined by Π_3. For simplicity, we assume that an accepting player always has a string-valued private output of the right length (that is, if $\Pi_2 = \mathsf{A}$ then $\Pi_3 \in \{0,1\}^{\sigma(k)}$), while a non-accepting player has a session key of $*$ (that is, if $\Pi_2 \in \{\mathsf{R}, *\}$ then $\Pi_3 = *$).

Compromise of a session key should have minimal consequences. For example, its revelation should not allow one to subvert subsequent authentication, nor should it leak information about other (as yet uncompromised) session keys. To capture this requirement we extend the interaction of the adversary with its oracles by adding a new type of query, as follows: we say that the adversary can learn a session key $\alpha^s_{i,j}$ of an oracle $\Pi^s_{i,j}$ by issuing to the oracle a distinguished reveal query, which takes the form $(i, j, s, \mathsf{reveal})$. The oracles answers $\alpha^s_{i,j}$. To quantify the power of an adversary who can perform this new type of query, we make the following definitions. Initially, each oracle $\Pi^s_{i,j}$ is declared *unopened*, and so it remains until the adversary generates a reveal query $(i, j, s, \mathsf{reveal})$. At this point, the oracle is declared *opened.* We say that an oracle $\Pi^s_{i,j}$ is *fresh* if the following three conditions hold: First, $\Pi^s_{i,j}$ has accepted. Second, $\Pi^s_{i,j}$ is unopened. Third, there is no opened oracle $\Pi^t_{j,i}$ which engaged in a matching conversation with $\Pi^s_{i,j}$. When oracle $\Pi^s_{i,j}$ is fresh, we will also say that "the oracle holds a fresh session key."

We want that the adversary should be unable to understand anything interesting about a fresh session key. This can be formalized along the lines of security of probabilistic encryption; the particular formalization we will adapt is that of (polynomial) indistinguishability of encryptions [12, 8, 9]. We demand that at the end of a secure AKE the adversary should be unable to distinguish a fresh session key α from a random element of S_k. After the adversary has asked all the (i, j, s, x) and $(i, j, s, \mathsf{reveal})$ queries that she wishes to ask, the adversary asks of a fresh oracle $\Pi^s_{i,j}$ a single query (i, j, s, test). The query is answered by flipping a fair coin $b \leftarrow \{0,1\}$ and returning $\alpha^s_{i,j}$ if $b = 0$, or else a random sample from S_k if $b = 1$. The adversary's job is to guess b. To this end, she outputs a bit Guess, and then terminates. Let $\text{Good-Guess}^E(k)$ be the event that Guess $= b$, when the protocol is executed with security parameter k; in other words, this is the probability the adversary has correctly identified whether she was given the real session key or just a sample from S_k. Let

$$advantage^E(k) = \max\left\{\, 0,\ \Pr\left[\text{Good-Guess}^E(k)\right] - \tfrac{1}{2} \right\}.$$

Definition 5. (Authenticated Key Exchange (AKE)) *Protocol* Π *is a* secure AKE *over* $S = \{S_k\}_{k \in \mathsf{N}}$ *if* Π *is a secure mutual authentication protocol, and, in addition, the following are true:*

(1) (Benign adversary $\Rightarrow$ keys according to S_k) *Let B be any benign adversary and let $\Pi^s_{i,j}$ and $\Pi^t_{j,i}$ be its chosen oracles in the experiment with security parameter k. Then both oracles always accept, $\alpha^s_{i,j} = \alpha^t_{j,i}$, and moreover this random variable is distributed according to S_k.*

(2) (Session key is protected) *Let E be any polynomial time adversary. Then $\text{advantage}^E(k)$ is negligible.*

The first condition says that if flows are honestly conveyed then a session key is agreed upon, and this key is properly distributed. The second condition says that the adversary can't tell this session key from a random string of the same distribution.

Since the protocol is assumed to be a secure mutual authentication, we know that if oracles $\Pi^s_{i,j}$ and $\Pi^t_{j,i}$ have matching conversations then they both accept. From the first condition it follows that they will also have the same session key.

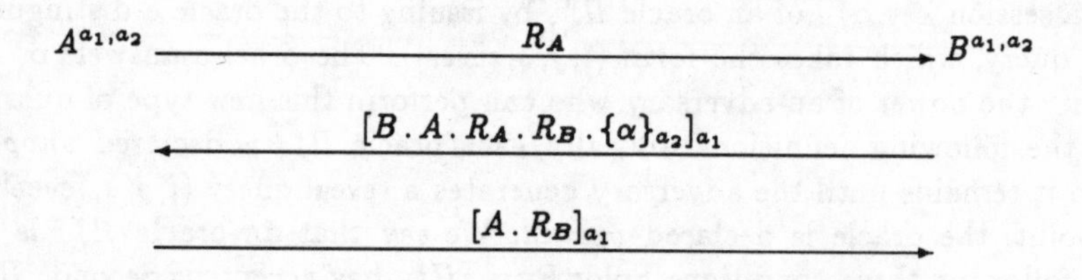

Fig. 2. *Protocol AKEP1: The value α is the session key distributed.*

We now present a protocol for AKE. Let $S = \{S_k\}$ be a family of samplable distributions on $\{0,1\}^{\sigma(k)}$. The parties share a $2k$ bit LL-key which we denote a_1, a_2. The first part, a_1, is taken as the key to the pseudorandom function family f, yielding a PRF $f_{a_1}\colon \{0,1\}^{\leq L(k)} \to \{0,1\}^k$ to be used for message authentication; this time, $L(k) = 5k + \sigma(k)$ will suffice. The second part, a_2, is used as a key to another pseudorandom family f' with the property that $f'_{a_2}\colon \{0,1\}^k \to \{0,1\}^{\sigma(k)}$. A probabilistic encryption of string $\alpha \in \{0,1\}^{\sigma(k)}$ is defined by $\{\alpha\}_{a_2} \stackrel{\text{def}}{=} (r, f'_{a_2}(r) \oplus \alpha)$, with r selected at random. Party B chooses the session key α from S_k and sets Text_2 to be $\{\alpha\}_{a_2}$. The strings Text_1 and Text_3 of MAP2 are set to λ. This protocol, which we call AKEP1, is shown in Figure 2. It is important that a_2 (the key used for encryption) be distinct from a_1 (the shared key used for the message authentication). Formally, the LL-key generator $\mathcal{G}$ provides the parties $i \in I$ with a $2k$-bit shared key. The two keys need not be independent, however; the generator could set $a_i = f_a(i)$ $(i = 1, 2)$ where a is a random k-bit key and f_a is a pseudorandom function. The proof of the following theorem is given in Appendix B.

Theorem 6. *Let $S = \{S_k\}$ be samplable, and suppose f, f' are pseudorandom function families with the parameters specified above. Then the protocol AKEP1 based on f, f' is a secure AKE over S.*

A more efficient (in terms of communication complexity) AKE protocol may be devised by using an "implicit" key distribution. In this case, the flows between A and B are the same as in MAP1 and one (or more) of the parameters already present in its flows (say R_B) is used to define the session key. Specifically, let $S = \{S_k\}$ be a family of distributions given by $S_k = g(U_k)$, for some deterministic, polynomial-time computable function g, where U_k is the uniform distribution on k-bit strings; for example $S_k = U_k$ and g the identity, the most useful choice in practice. Again the parties share a $2k$ bit LL-key a_1, a_2, with a_1 being used as the key in MAP1 (so $L(k) = 4k$). Let f' be a pseudorandom *permutation* family [18]; f'_{a_2} specifies a permutation on $\{0,1\}^k$. Define AKEP2 by having its flows be identical to MAP1 with a_1 being used for message authentication. Each accepting party outputs session key $\alpha = g(f'_{a_2}(R_B))$. This protocol, which we call AKEP2, is shown in Figure 3. Modifying the proof of Theorem 6 we can show the following:

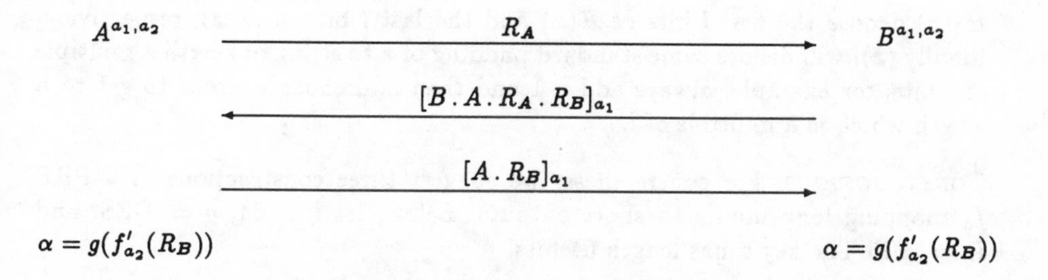

Fig. 3. *Protocol AKEP2: The Implicit Key Exchange Protocol. The value α is the session key "implicitly" distributed.*

Theorem 7. *Let $S = \{S_k\}$ be given by $S_k = g(U_k)$, for some polynomial time g. Suppose f, f' are a pseudorandom function family and a pseudorandom permutation family, with the parameters specified above. Then the protocol AKEP2 based on f, f' is a secure AKE over S.*

6 From Theory to Practice

Cryptographic practice provides good PRFs on particular input lengths l (for example, DES for $l = 64$ [18, 19]). In contrast, our protocols need PRFs for arbitrary input lengths. In devising such PRFs we prefer not to rely purely on heuristics but instead to give provably-correct constructions of arbitrary length PRFs based on fixed length PRFs and collision-free hash functions. The lemmas underlying our constructions are from [1] and are summarized with additional material in Appendix D (omitted here due to lack of space; see our full paper). The exception is the third construction given below; we'll discuss it when we get there.

PRIMITIVES. The algorithm of the DES specifies for each 64 bit key a a permutation DES_a from $\{0,1\}^{64}$ to $\{0,1\}^{64}$. The viewpoint adopted here —suggested by Luby and Rackoff [18, 19]— is to regard DES as a pseudorandom permutation, with respect to practical computation.

The MD5 function [22] maps an arbitrary string x into a 128-bit string $\mathrm{MD5}(x)$. It is intended that this function be a collision-free hash function, with respect to practical computation.

NOTATION. Let g_a denote a PRF of l bits to l bits. Suppose y has length a multiple of l bits, and write it as a sequence of l bit blocks, $y = y_1 \dots y_n$. The cipher block chaining (CBC) operator defines

$$\mathrm{CBC}_a^g(y_1 \dots y_n) = \begin{cases} g_a(y_1) & \text{if } n = 1 \\ g_a(\mathrm{CBC}_a^g(y_1 \dots y_{n-1}) \oplus y_n) & \text{otherwise} \end{cases}$$

Let H denote a collision free hash function of $\{0,1\}^*$ to $\{0,1\}^{2l}$. Let $H_1(x)$ and $H_2(x)$ denote the first l bits of $H(x)$ and the last l bits of $H(x)$, respectively. Finally $\langle x \rangle_l$ will denote some standard padding of x to string of length a multiple of l bits; for example, always add a 1 and then add enough zeroes to get to a length which is a multiple of l.

CONSTRUCTIONS. For concreteness, we suggest three constructions of a PRF f_a mapping long inputs to short outputs. Below, let $l = 64$, $g = \mathrm{DES}$, and $H = \mathrm{MD5}$ The key a has length 64 bits.

(1) *The CBC PRF.* Let $f_a(x)$ be the first $l/2$ bits of $\mathrm{CBC}_a^g(\langle x \rangle_l \,.\, |\langle x \rangle_l|)$, where $|y|$ is the length of y encoded as an l-bit string. This construction is justified by Lemma 12 of Appendix D (omitted).[6]

(2) *The CBC/Hash PRF.* Let $f_a(x)$ be the first $l/2$ bits of $g_a(\, g_a(H_1(x)) \oplus H_2(x)\,) = \mathrm{CBC}_a^g(H(x))$. This construction is justified by Corollary 14 (omitted). In software this is significantly more efficient than the CBC construction, requiring one hash and two DES operations.

(3) *The Pure Hash PRF.* Let $f_a(x)$ be the first $l/2$ bits of $H(x\,.\,a)$. This construction was suggested in [24] as a message authentication code; we suggest the stronger assumption that it is a PRF. However no standard assumption about H of which we are aware can be used to justify the the security of this construction, and it should be viewed more as a heuristic than the two constructions suggested above.[7]

Similar constructions can be given using other primitives; for example the SHA instead of MD5, etc.

---

[6] Lemma 12 does not require us to drop the last $l/2$ bits of the output. We drop them for two reasons. The first is efficiency. The second is specific to DES and will not be discussed here.

[7] See [2] for another viewpoint.

We stress the importance in security considerations of the CBC and Hash Lemmas of Appendix D; the lack of such lemmas has lead in the past to more complex assumptions about the security of CBC and other constructions (e.g., [3, Definition 2.1]).

Acknowledgments

We thank Bob Blakley, Oded Goldreich, Amir Herzberg, Phil Janson, and the member of the CRYPTO 93 committee for all of their comments and suggestions. Especially we acknowledge Oded as suggesting that we formulate the security of fresh keys along the lines of polynomial indistinguishability (instead of the equivalent semantic security formulation we had before); and Amir for suggesting that we give Proposition 3, specify MAP1 in a way that does not assume authenticating entities are distinct agents from a set of cardinality two, and that to avoid the efficiency loss from padding we explicitly specify our pseudorandom functions as acting on $\{0,1\}^{\leq L(k)}$ instead of $\{0,1\}^{L(k)}$.

References

1. M. Bellare, U. Feige, J. Kilian, M. Naor and P. Rogaway, "The security of cipher block chaining," manuscript (1993).
2. M. Bellare and P. Rogaway, "Random oracles are practical: a paradigm for designing efficient protocols," *Proceedings of 1st ACM Conference on Computer and Communications Security*, November 1993.
3. R. Bird, I. Gopal, A. Herzberg, P. Janson, S. Kutten, R. Molva and M. Yung, "Systematic design of two-party authentication protocols," *Advances in Cryptology — Proceedings of CRYPTO 91*, Springer-Verlag, 1991.
4. M. Blum and S. Micali, "How to generate cryptographically strong sequences of pseudo-random bits," *SIAM Journal on Computing* **13**(4), 850-864 (November 1984).
5. M. Burrows, M. Abadi and R. Needham, "A logic for authentication," DEC Systems Research Center Technical Report 39, February 1990. Earlier versions in *Proceedings of the Second Conference on Theoretical Aspects of Reasoning about Knowledge*, 1988, and *Proceedings of the Twelfth ACM Symposium on Operating Systems Principles*, 1989.
6. W. Diffie, P. Van Oorschot and M. Wiener, "Authentication and authenticated key exchanges," *Designs, Codes and Cryptography*, 2, 107–125 (1992).
7. U. Feige, A. Fiat and A. Shamir, "Zero knowledge proofs of identity," *Journal of Cryptology*, Vol. 1, pp. 77–94 (1987).
8. O. Goldreich, "Foundations of cryptography," class notes, Technion University, Computer Science Department, Spring 1989.
9. O. Goldreich, "A uniform complexity treatment of encryption and zero-knowledge," *Journal of Cryptology*, Vol. 6, pp. 21-53 (1993).
10. O. Goldreich, S. Goldwasser and S. Micali, "How to construct random functions," *Journal of the ACM*, Vol. 33, No. 4, 210–217, (1986).

11. O. Goldreich, S. Goldwasser and S. Micali, "On the cryptographic applications of random functions," *Advances in Cryptology — Proceedings of CRYPTO 84*, Springer-Verlag, 1984.
12. S. Goldwasser and S. Micali, "Probabilistic encryption," *Journal of Computer and System Sciences* Vol. 28, 270-299 (April 1984).
13. S. Goldwasser, S. Micali and R. Rivest, "A digital signature scheme secure against adaptive chosen-message attacks," *SIAM Journal of Computing*, Vol. 17, No. 2, 281–308, April 1988.
14. J. Håstad, "Pseudo-random generators under uniform assumptions," *Proceedings of the 22nd Annual ACM Symposium on the Theory of Computing*, ACM (1990).
15. R. Impagliazzo and M. Luby, "One-way functions are essential for complexity based cryptography," *Proceedings of the 30th Annual IEEE Symposium on the Foundations of Computer Science*, IEEE (1989).
16. R. Impagliazzo, L. Levin and M. Luby, "Pseudo-random generation from one-way functions," *Proceedings of the 21st Annual ACM Symposium on the Theory of Computing*, ACM (1989).
17. ISO/IEC 9798-2, "Information technology - Security techniques - Entity authentication - Part 2: Entity authentication using symmetric techniques." Draft 12, September 1992.
18. M. Luby and C. Rackoff, "How to construct pseudorandom permutations from pseudorandom functions," *SIAM J. Computing*, Vol. 17, No. 2, April 1988.
19. M. Luby and C. Rackoff, "A study of password security," manuscript.
20. R. Molva, G. Tsudik, E. Van Herreweghen and S. Zatti, "*KryptoKnight* authentication and key distribution system," ESORICS 92, Toulouse, France, November 1992.
21. R. Needham and M. Schroeder, "Using encryption for authentication in large networks of computers," *Communications of the ACM*, Vol. 21, No. 12, 993–999, December 1978.
22. R. Rivest, "The MD5 message-digest algorithm," IETF Network Working Group, RFC 1321, April 1992.
23. J. Rompel, "One-way functions are necessary and sufficient for secure signatures," *Proceedings of the 22nd Annual ACM Symposium on the Theory of Computing*, ACM (1990).
24. G. Tsudik, "Message authentication with one-way hash functions," *Proceedings of Infocom 92*.
25. P. Van Oorschot, "Extending cryptographic logics of belief to key agreement protocols," *Proceedings of 1st ACM Conference on Computer and Communications Security*, November 1993.
26. Yao, A. C., "Theory and applications of trapdoor functions," *Proceedings of the 23rd Annual IEEE Symposium on the Foundations of Computer Science*, IEEE (1982).

A Proof of Theorem 4

We prove that MAP1 is a secure mutual authentication protocol under the assumption that f is a PRF. The first condition of Definition 2 is easily verified; it merely says that when the messages between A and B are faithfully relayed to one another, each party accepts. We now prove that the second condition holds.

Fix an adversary E. Recall that the domain of our PRF is $\{0,1\}^{\leq L(k)}$ and its range is $\{0,1\}^k$. In the following, Π will denote MAP1. In what follows we will be considering a variety of experiments involving the running of E with its oracles. In order to avoid confusion, we will refer to the experiment of running E with MAP1 (the experiment about which we wish to prove our theorem) as the "real" experiment.

MAP1 WITH A g ORACLE. Let g be a function of $\{0,1\}^{\leq L(k)}$ to $\{0,1\}^k$. Let $[x]_g = (x, g(x))$. MAP1^g denotes the protocol in which, instead of a shared secret a, the parties share an oracle for g, and they compute $[x]_g$ wherever MAP1 asks them to compute $[x]_a$. We define the experiment of running E for MAP1^g to be the same as the experiment of running E for MAP1 except for the following difference. There is no shared secret a; instead, the oracles $\Pi^s_{i,j}$ all have access to a common g oracle and compute their flows according to MAP1^g. Note E is not given access to the g oracle. When $g = f_a$ for randomly chosen a, this experiment coincides with the real experiment. Of interest in our proof is the case of g being a truly random function; we call this the random MAP1 experiment.

THE RANDOM MAP1 EXPERIMENT. In the random MAP1 experiment we select g as a random function of $\{0,1\}^{\leq L(k)}$ to $\{0,1\}^k$, and then run the experiment of running E with MAP1^g. Recall that No-Matching$^E(k)$ denotes the event that there exists an oracle $\Pi^s_{i,j}$ who accepts although no oracle $\Pi^t_{j,i}$ engaged in a matching conversation; we will refer to it also as the event that the adversary is successful. Recall that an initiator oracle is one who sends a first flow (that is, it plays the role of A in Figure 1) while a responder oracle is one who plays the opposite role (namely that of B in the same Figure). Let $T_E(k)$ denote a polynomial bound on the number of oracle calls made by E, and assume wlog that this is at least two.

Lemma 8. *The probability that the adversary E is successful in the random* MAP1 *experiment is at most $T_E(k)^2 \cdot 2^{-k}$.*

Proof: We split the examination of acceptance into two cases.

Claim 1: Fix A, B, s. The probability that $\Pi^s_{A,B}$ accepts without a matching conversation, given that it is an initiator oracle, is at most $T_E(k) \cdot 2^{-k}$.

Proof. Suppose at time τ_0 oracle $\Pi^s_{A,B}$ sent the flow R_A. Let $\mathcal{R}(\tau_0)$ denote the set of all $R'_A \in \{0,1\}^k$ for which there exist τ, t such that $\Pi^t_{B,A}$ was given R'_A as first flow at a time $\tau < \tau_0$. If $\Pi^s_{A,B}$ is to accept, then at some time $\tau_2 > \tau_0$ it must receive $[B\,.\,A\,.\,R_A\,.\,R_B]_g$ for some R_B. If no oracle previously output this flow, the probability that the adversary can compute it correctly is at most 2^{-k}. So consider the case where some oracle did output this flow. The form of the flow implies that the oracle which output it must be a $\Pi^t_{B,A}$ oracle which received R_A as its own first flow. The probability of this event happening before time τ_0 is bounded by the probability that $R_A \in \mathcal{R}(\tau_0)$, and this probability is at most $[T_E(k) - 1] \cdot 2^{-k}$. If it happened after time τ_0 then we would have a matching conversation. We conclude that the probability that $\Pi^s_{A,B}$ accepts but there is no matching conversation is at most $T_E(k) \cdot 2^{-k}$. □

Claim 2: Fix B, A, t. The probability that $\Pi^t_{B,A}$ accepts without a matching conversation, given that it is a responder oracle, is at most $T_E(k) \cdot 2^{-k}$.

Proof. Suppose at time τ_1 oracle $\Pi^t_{B,A}$ received the flow R_A and responded with $[B \,.\, A \,.\, R_A \,.\, R_B]_g$. If $\Pi^t_{B,A}$ is to accept, then at some time $\tau_3 > \tau_1$ it must receive $[A \,.\, R_B]_g$. If no oracle previously output this flow, the probability that the adversary can compute it correctly is at most 2^{-k}. We must now consider the case where some oracle did output this flow. The form of the flow implies that the oracle which output it must be a $\Pi^s_{A,C}$ oracle.

The interaction of a $\Pi^s_{A,C}$ oracle with E has in general the form

$$(\tau_0, \lambda, R'_A), (\tau_2, [C \,.\, A \,.\, R'_A \,.\, R'_B]_g, [A \,.\, R'_B]_g)$$

for some $\tau_0 < \tau_2$. For any such interaction, except with probability 2^{-k}, there is a $\Pi^u_{C,A}$ oracle which output $[C \,.\, A \,.\, R'_A \,.\, R'_B]_g$ at some time. If $(u, C) \neq (t, B)$ then the probability that $R'_B = R_B$ is at most $[T_E(k) - 2] \cdot 2^{-k}$, and thus the probability that the flow $[A \,.\, R'_B]_g$ leads $\Pi^t_{B,A}$ to accept is at most $[T_E(k) - 2] \cdot 2^{-k}$. On the other hand suppose $(u, C) = (t, B)$. It follows that $\tau_0 < \tau_1 < \tau_2 < \tau_3$, $R'_A = R_A$ and $R'_B = R_B$; that is, the conversations match. We conclude that the probability that $\Pi^t_{B,A}$ accepts but there is no matching conversation is at most $T_E(k) \cdot 2^{-k}$. □

The probability that there exists an oracle which accepts without a matching conversation is at most $T_E(k)$ times the bound obtained in the claims, which is $T_E(k)^2 \cdot 2^{-k}$ as desired. □

See our full paper for the argument that these lemmas yield the theorem.

B Proof of Theorem 6

We prove that AKEP1 is a secure authenticated key exchange protocol under the assumption that f, f' are PRFs.

The proof that AKEP1 is a secure mutual authentication protocol is analogous to the proof of Theorem 4 given in Appendix A and is omitted. Condition (1) of Definition 5 is easily verified: the session key α is chosen in AKEP1 according to S_k and so in the presence of a benign adversary the oracles certainly accept, and with this same key. We concentrate on the proof that condition (2) of Definition 5 is satisfied.

Fix an adversary E. Recall that we are using two PRFs: $f_{a_1}\colon \{0,1\}^{\leq L(k)} \rightarrow \{0,1\}^k$ and $f'_{a_2}\colon \{0,1\}^k \rightarrow \{0,1\}^{\sigma(k)}$. The first is for the authentication and the second is to encrypt the session key. In what follows Π will denote AKEP1, and the "real" experiment will denote the experiment of running E for AKEP1.

AKEP1 WITH A g' ORACLE. Let g' be a function mapping: $\{0,1\}^k$ to $\{0,1\}^{\sigma(k)}$. Let $\mathcal{E}_{g'}(\alpha, r) = (r, g'(r) \oplus \alpha)$. Let $\{\alpha\}_{g'}$ be the random variable resulting from picking $r \in \{0,1\}^k$ at random and outputting $\mathcal{E}_{g'}(\alpha, r)$. AKEP1$^{g'}$ denotes the protocol in which the parties share a secret a_1 and an oracle for g'. Whenever

AKEP1 asks them to compute $\{\alpha\}_{a_2}$ they compute $\{\alpha\}_{g'}$. The experiment of running E for AKEP1$^{g'}$ is the same as the experiment of running E for AKEP1 except that the second part of the shared key, namely a_2, is absent, and instead the oracles $\Pi^s_{i,j}$ all have access to a common g' oracle and compute their flows according to AKEP1$^{g'}$. E does not have access to g'. When $g' = f'_{a_2}$ for randomly chosen a_2, this experiment coincides with the real experiment.

THE RANDOM AKEP1 EXPERIMENT. In the random AKEP1 experiment we select g' as a random function of $\{0,1\}^k$ to $\{0,1\}^{\sigma(k)}$, and then run the experiment of running E with AKEP1$^{g'}$. As before, let $T_E(k)$ denote a polynomial bound on the number of oracle calls made by E.

Lemma 9. *In the random AKEP1 experiment, advantage$^E(k)$ is negligible.*

Proof: Let $c > 0$ be a constant. We will show that *advantage*$^E(k) \le k^{-c}$ for all sufficiently large k.

A *view* of E consists of all the oracle queries made by E, the responses to them, and E's own coin tosses; that is precisely what E sees. We denote by *view*(k) the random variable whose value is the view of the interaction of E with its oracles. A particular view will usually be denoted ξ. We will be interested in two properties ξ may possess. If for any accepting oracle there exists an oracle with a matching conversation then we say ξ is *authentic*. If $(r_1, y_1), \ldots, (r_n, y_n)$ denote the encryptions output by oracles in the transcript and $r_1, \ldots, r_n$ are distinct then we say ξ is *non-colliding*. Recall that b denotes the bit flipped in our answer to a test query in the definition of measuring *advantage*$^E(k)$.

Now fix a particular authentic and non-colliding view ξ. Suppose E is pointing to (fresh) oracle $\Pi^s_{A,B}$. Since $\Pi^s_{A,B}$ has accepted and ξ is authentic, there is an oracle $\Pi^t_{B,A}$ which engaged in a matching conversation. This means the encryption for this conversation was selected by one of the oracles (specifically, the one who played the role of the responder). The oracle's being fresh means that any matching partner is unopened. Since ξ is non-colliding it follows that conditioned on *view*$(k) = \xi$, the key $\alpha^s_{A,B}$ is uniformly distributed over S_k, and E's advantage in predicting the bit b is 0.

Let N_k denote the set of non-authentic views and C_k the set of colliding views. We claim that AKEP1$^{g'}$, with g' chosen at random, still remains a secure mutual authentication; the proof of this is analogous to the proof of Theorem 4 and hence is omitted. Based on this claim, we know that the probability of N_k is at most $k^{-c}/2$ for large enough k. On the other hand the probability of C_k is at most $T_E(k)^2 \cdot 2^{-k}$ which is at most $k^{-c}/2$ for large enough k. Combined with the above we conclude that E's advantage is at most k^{-c}. □

See our full paper for the argument that this lemma yields the theorem.

On the Existence of Statistically Hiding Bit Commitment Schemes and Fail-Stop Signatures

Ivan B. Damgård *, Torben P. Pedersen ** and Birgit Pfitzmann ***

Abstract. We show that the existence of a statistically hiding bit commitment scheme with non-interactive opening and public verification implies the existence of fail-stop signatures. Therefore such signatures can now be based on any one-way permutation – the weakest assumption known to be sufficient for fail-stop signatures. We also show that genuinely practical fail-stop signatures follow from the existence of any collision-intractable hash function. A similar idea is used to improve a commitment scheme of Naor and Yung, so that one can commit to several bits with amortized O(1) bits of communication per bit committed to.

Conversely, we show that any fail-stop signature scheme with a property we call the *almost unique secret key property* can be transformed into a statistically hiding bit commitment scheme. All previously known fail-stop signature schemes have this property. We even obtain an equivalence since we can modify the construction of fail-stop signatures from bit commitments such that it has this property.

1 Introduction

In this section, we introduce the two main actors on the scene, fail-stop signatures (FSS) and statistically hiding bit commitments.

Fail-stop signatures were introduced in [16]. Further constructions appear in [13, 14, 5, 6]. A formal definition of the concept and a survey of the recent most efficient schemes will appear in [12].

Before going into the properties of FSS schemes, let us discuss some aspects of ordinary digital signatures: In an application of such signatures, what should happen if someone shows up with a message and a valid looking signature from user A, but A claims that she never signed the message? Suppose the signature scheme is based on a computational problem, P, which everybody accepts cannot be solved in polynomial time. Based on this one could claim that it is not reasonable to assume that the system was broken by an enemy. So either A is

* Aahus University, Matematisk Institut, Ny Munkegade, DK-8000 Aarhus C; Denmark. e-mail: ivan@daimi.aau.dk.

** Aarhus University, Matematisk Institut, Ny Munkegade, DK-8000 Aarhus C, Denmark. e-mail: tppedersen@daimi.aau.dk. Supported in part by the Carlsberg Foundation.

*** Universität Hildesheim, Institut für Informatik, Marienburger Platz 22, D-31141 Hildesheim, Germany. e-mail: pfitzb@informatik.uni-hildesheim.de.

lying, or she must have stored her secret key insecurely, and should therefore be held responsible in either case.

However, this argument sweeps under the rug a very important point: we always have to choose particular instances of the problem for each user, and the discussion should actually refer to how hard this particular instance is to break. If we are using RSA, for example, we have to decide on a size of moduli to use. Even if we believe that factoring is not in polynomial time, this does not answer questions like: "are 512-bit moduli secure enough?". This is a question about the state of the art of practical factoring, and does not have much to do with its complexity theoretic status.

In a practical situation, it is often the case that individual users have only very limited computing power available. This of course limits the size of problem instance they can use, but not the amount of computing power that might be used to break those instances. In such a situation, depending on the practical circumstances, the possibility that A is not lying and someone broke her key, is perhaps not so unreasonable after all.

This raises a natural question: is it possible at all to distinguish between on one hand the case where A is lying or has leaked the secret key, and on the other hand the case where someone with a large (unexpected) amount of computing power has broken the system?

This is precisely what FSS schemes enable us to do. The crucial property that distinguishes FSS from ordinary digital signatures is that there are several possible secret keys corresponding to a given public key. Even an infinitely powerful enemy cannot guess from publicly available information which of the possible secret keys is known to the signer.

Since usage of different secret keys in general leads to different signatures, it is impossible for the enemy to predict which signature the signer would produce on a given message, if it has not already been signed.

Furthermore, from two different signatures on the same message, A can produce what is known as a *proof of forgery*. But if she has only the signature available that she would produce herself, it is not feasible for her to produce such a proof.

Thus if a powerful enemy tries to frame A and submits a message seemingly signed by A, with overwhelming probability the signature will not be the one A would produce herself, and A can therefore respond with a proof of forgery. On the other hand, A cannot falsely repudiate her own signature, if it has in fact not been forged, unless she herself breaks the computational assumption. (Thus even in this case, the proof of forgery correctly indicates that someone with unexpectedly large computing power has broken the scheme.)

In this paper, we show that there is an intimate connection between statistically hiding bit commitment schemes and FSS schemes. A bit commitment scheme is a protocol that party A can conduct with B to commit herself to a bit b without revealing to B (or anyone else) the value of b. At a later time, A can open the commitment and convince B about the value that was chosen originally, i.e., it is not feasible for A to open a commitment to reveal both $b = 0$

and $b = 1$. A commitment scheme is said to be *statistically hiding* if B gets only negligible Shannon-information about b prior to the opening of the commitment. Such bit commitment schemes are extremely important because their existence implies perfect or statistical zero-knowledge arguments for any problem in NP.

Concretely, we show how to construct FSS schemes from any statistically hiding bit commitment scheme with non-interactive opening and public verification (see below for details). This result is also contained more or less implicitly in [13]. Also the result was discussed informally, prior to the work on this paper, by Moti Yung and Birgit Pfitzmann. Our contribution in this respect is to somewhat simplify the construction and to identify the properties needed from the bit commitment scheme.

By the work of Naor et al. ([11]), this means that FSS schemes can be based on any one-way permutation. Before, FSS schemes were only known to follow from the existence of claw-free permutations.

We also show that any collision-intractable hash function can be used to build a secure FSS scheme. If the hash function is efficient, like MD4 [7] or SHA [15], say, then the resulting FSS scheme is practical.

Conversely, we show that any FSS scheme with a property we call the *almost unique secret key property*, can be transformed into a statistically hiding bit commitment scheme. This property means that it is infeasible for a signer to compute more than one significantly different secret key corresponding to her public key; see below for details. All previously known FSS schemes have this property. Finally, we show that the existence of FSS schemes with this property is in fact *equivalent* to the existence of statistically hiding bit commitments with non-interactive opening and public verification.

2 Definitions and Notation

2.1 Fail-Stop Signatures

For the results in this paper, it is sufficient to consider fail-stop signature schemes that allow just one message to be signed. Based on [12] (see also [13]), the definition of such schemes is now sketched.

A fail-stop signature scheme consists of five parts: a protocol for generating the keys, a method for signing, a predicate for verifying signatures (a signature satisfying this predicate is called *acceptable*), a method for constructing proofs of forgery and a predicate for verifying proofs of forgery (a proof satisfying this predicate is called *valid*). The methods for producing signatures and proofs of forgery take the secret key as input, and the public key is input to the computation of the two predicates.

Unlike usual digital signatures, the key generation is a two-party protocol, which is executed by the signer A and a center B trusted by the recipients. This is necessary to ensure that the signer does not generate a pair of keys for which she can prove her own signatures to be forgeries. A or B may reject in key generation, but if both parties are honest, this should only happen with negligible probability. We remark that one can always do without a key center (at the expense of efficiency), by letting every recipient play the role of B.

Obviously, these parts must satisfy that if the keys are generated correctly, then correct signatures and correct proofs of forgery are accepted by the corresponding verification methods. The more interesting parts of the definition are

- *Security for the recipient*: It is infeasible for a polynomially bounded signer to produce an acceptable signature and a valid proof that it is forged.
- *Security for the signer*: It is impossible for a forger with unlimited computing power to produce a signature that the signer cannot prove to be a forgery.

A fail-stop signature scheme usually has two security parameters: k for the security of the recipient and σ for the security of the signer.

To define the security of the recipient in more detail, we consider the following scenario involving the key center, B, and a possibly cheating signer, $\tilde{A}$: First $\tilde{A}$ and B generate a pair of keys, and then $\tilde{A}$ outputs a triple (m, s, pr).

Definition 2.1 *A fail-stop signature scheme is secure for the recipient with respect to the security parameter k if for all $c > 0$ and for all polynomially bounded signers, $\tilde{A}$, the following holds for sufficiently large k: The probability that s is an acceptable signature on m and pr is a valid proof of forgery is at most k^{-c}. This probability is over the random coins of B and $\tilde{A}$.*

In order to define the security of the signer, we consider a cheating center, $\tilde{B}$, possibly with unlimited computing power. As the signer must be secure even if the center cooperates with future recipients, it is sufficient that the center itself cannot construct forgeries that A cannot disavow. Consider the scenario where first A and $\tilde{B}$ execute the key generation protocol. This results in a secret key, sk, and a public key, pk. $\tilde{B}$'s view of this protocol is denoted by $view_{\tilde{B}}$ (random bits and all messages). Then $\tilde{B}$ outputs a pair (m_0, s_0), where s_0 should be an acceptable signature on m_0.

Let SK be the set of possible secret keys given $view_{\tilde{B}}$. SK is equipped with a probability distribution induced by the random coins used by A during the key generation. Then *Good* is defined as the set of pairs $(pk, view_{\tilde{B}})$ such that for all m_0, s_0 and with probability at least $1 - 2^{-\sigma}$ over the choices of possible secret keys sk in SK, A can prove that s_0 is a forgery using sk as the secret key.

Definition 2.2 *A fail-stop signature scheme is secure for the signer with respect to the security parameter σ, if the probability that A accepts the keys and $(pk, view_{\tilde{B}}) \notin Good$ is at most $2^{-\sigma}$. (The probability is over A's coins).*

Intuitively, this means that only with very small probability ($\leq 2^{-\sigma}$) can $\tilde{B}$ make A accept a pair of keys for which she has probability less than $1 - 2^{-\sigma}$ of proving forgeries.

A complete definition also has to take into account chosen message attacks. However, our construction of commitments from FSS does not need security against such attacks, and for the FSS schemes we construct, it is easy to see that such attacks make no difference. Hence, we stick to this somewhat simpler definition.

Intuitively, these definitions imply that a cheating signer cannot compute just any secret key that is possible given the public key. If she could, she could prove her own signatures to be forgeries by signing using one secret key and using a different key in the proof. All fail-stop signature schemes in previous literature have an idealized version of this property: No matter how the (polynomially bounded) signer executes the key generation, she cannot compute two different secret keys that are both possible given the public key. We call this the *unique secret key property.* In the following, we use a relaxed version of it, the *almost unique secret key property*: Although the signer might be able to find more than one secret key fitting a public key, she cannot not find *significantly different ones.* Keys are "not significantly different" if they lead to equal signatures. This can be formalized by introducing a polynomial-time computable mapping κ on the secret keys with the intuitive meaning that $\kappa(sk)$ is the part of sk that makes a difference in the signatures.

Definition 2.3 *A fail-stop signature scheme has the almost unique secret key property if there are a polynomial-time computable predicate Fits and a polynomial time computable mapping κ with the following properties:*

- *If the signer follows the key generation protocol, the resulting secret and public key, sk and pk, always fulfil $Fits(sk, pk) = 1$.*
- *No probabilistic poly-time bounded signer can execute the key generation protocol with the honest key center and compute sk_1, sk_2 such that $\kappa(sk_1) \neq \kappa(sk_2)$ and $Fits(sk_1, pk) = Fits(sk_2, pk) = 1$ with more than superpolynomially small probability.*
- *If sk_1 and sk_2 satisfy $Fits(sk_1, pk) = Fits(sk_2, pk) = 1$ and $\kappa(sk_1) = \kappa(sk_2)$, then for any message, the signature produced with sk_1 equals the one produced with sk_2.*

For a concrete FSS scheme, there will typically exist a function that computes the public key from a secret key. Then $Fits$ can be constructed from this function. Furthermore, note that if κ is the identity, the third property is no restriction, and one just obtains the unique secret key property.

All known schemes (see [13, 14, 5, 6]) have the almost unique secret key property, although one can easily construct artificial schemes without it.

2.2 Bit Commitments

We define a bit commitment scheme as a pair of two-party protocols, namely the *commit* and the *reveal* protocol. They take place between parties A and B, where A is the party committing herself. The participants are modeled in the standard way as interactive probabilistic Turing machines. As before, the *view* of a participant is the bit string consisting of his own coinflips concatenated by all messages sent in the protocol. $\tilde{X}$ will denote any machine playing the role of X in the protocol.

For the commit protocol, A gets as input a bit b. We assume that B knows some a priori information about b, such that $b = 0$ with probability δ, where δ may be different from $1/2$. In addition both parties have access to a security parameter k. The concatenation of all messages sent in the commit protocol is called *the commitment.* In some concrete schemes, it makes sense to define the commitment as a subset or a function of the messages. We have chosen our definition of a commitment for simplicity. A or B may reject in the commit protocol, but if both parties are honest, this should only happen with negligible probability.

For the reveal protocol, A gets as input her view of the commit protocol, while B gets the commitment as input. At the end of the reveal protocol, B outputs *reject*, *accept* 0 or *accept* 1. The intuitive meaning is that either B has detected cheating by A, or he accepts that A has opened the commitment to reveal either 0 or 1.

We will only consider commitment schemes with non-interactive opening, i.e., where the reveal protocol consists of A sending one message to B.

A statistically hiding bit commitment scheme must satisfy two properties:

- *Security Property:* For any $\tilde{B}$, let *bias* denote $\tilde{B}$'s advantage in guessing b given $\tilde{B}$'s view v of the commit protocol, i.e., $bias = |\delta - Prob[b = 0 \mid v]|$. Then the expected value of *bias* is at most 2^{-k}. The probabilities are taken over the coinflips of A.
- *Binding Property:* Let $\tilde{A}$ be any polynomially bounded machine that executes the commit protocol with B, and then outputs two strings s_0, s_1.
  Let $p(\tilde{A}, k)$ be the probability that B outputs *accept* b on input s_b in the reveal protocol, for both $b = 0$ and $b = 1$. The probability is taken over the coinflips of B and $\tilde{A}$. Then $p(\tilde{A}, k)$ is superpolynomially small as a function of k.

We require an exponential decrease of the bias in the security property. However, this is not a significant restriction, since: standard "XOR-ing" techniques can be used to improve weaker schemes such that they satisfy the definition. Moreover, most practical examples known in fact have a bias of 0.

Note that we have built into the model two properties that the bit commitment scheme must satisfy in order for the result in the next section to work: first non-interactive opening must be possible, as mentioned; secondly B must be able to verify the opening based on the commitment only. This means that anyone who trusts that a given commitment is the result of a conversation with B can verify the opening without knowing B's coinflips. Hence the term *public verification.* This property is necessary in the construction of a FSS scheme to ensure that everybody can verify signatures.

3 Fail-Stop Signatures from Bit Commitments and Hash Functions

In this section, we assume that we are given a statistically hiding bit commitment scheme as defined above, and will use this to build an FSS scheme. The basic idea is very similar to Lamport and Diffie's one-time signatures.

We will only show how to sign a 1-bit message – this easily generalizes to any number of bits. Recall that an FSS scheme uses two security parameters, k and σ.

KEY GENERATION
In this phase, the key center B and user A execute 4σ instances of the commit protocol, which is executed with security parameter k' such that $k' \geq k$ and $k' \geq 2\sigma + 4$. For each instance, A chooses randomly and uniformly the bit to commit to. The resulting commitments are organized in 2σ pairs called $(C_{0,i}, C_{1,i})$, $i = 1, \ldots, 2\sigma$. The public key is the set of commitments, while the secret key is the set of (4σ) strings known by A that will open the commitments. A stops and rejects the keys, if she detects cheating during any of the commit protocols.

SIGNING
The signature on a bit b consists of the 2σ strings that A would send to open the commitments $C_{b,i}$, $i = 1, \ldots, 2\sigma$.

VERIFICATION
To verify the signature on a bit b, one verifies that the 2σ strings in the signature open correctly the commitments $C_{b,i}$, $i = 1, \ldots, 2\sigma$.

PROOF OF FORGERY
Given an acceptable signature S on a bit b, A generates her own signature on b. For $i = 1, \ldots, 2\sigma$, she tries to find an i for which the i'th bit opened in S is different from the i'th bit opened in her own signature. If such an i is found, she outputs i and the two strings used to open this commitment. If not, she fails to generate a proof of forgery.

VALIDATING PROOF OF FORGERY
A triple, (i, s_1, s_2) proves that S is a forged signature on b, if s_1 is the i'th string in S $(1 \leq i \leq 2\sigma)$, $s_1 \neq s_2$, and s_1 and s_2 can be used to open $C_{b,i}$ to reveal different bits.

Theorem 3.1 *The signature scheme outlined above based on a statistically hiding bit commitment scheme with public verification and non-interactive opening is a secure fail-stop signature scheme.*

Proof sketch: We first prove security for the signer. Let *Acc* denote the event A accepts the keys. As the event $G = Good$, we take the event that B cannot guess any bit committed to with probability better than 5/8. For a single commitment, by the security property and Markov's rule, the probability that B's guess is better than 5/8, is at most $8 \cdot 2^{-k'} \leq 2^{-2\sigma-1}$. Therefore we get that

$$Prob[Acc, \neg G] \leq 1 - (1 - 2^{-2\sigma-1})^{4\sigma} < 4\sigma 2^{-2\sigma-1} \leq 2^{-\sigma}.$$

This shows the first part of security for the signer. Next, assume G occurs. Then, even an infinitely powerful enemy cannot predict in which way A will open any commitment with probability better than 5/8. To predict A's signature, one must guess the contents of 2σ commitments, which can be done with probability at most $(5/8)^{2\sigma} \leq 2^{-\sigma}$. Therefore the probability that the algorithm for generating a proof of forgery fails when given a false signature (i.e., a signature generated by anyone else than A) is less than $2^{-\sigma}$. This implies security for the signer.

Note that it does not help B if he first gets a signature from A (e.g., if A signs the bit 1 and B wants to forge the signature on 0), because the commitments are chosen independently.

Furthermore, it is clear that any algorithm that would allow A to generate a proof of forgery by herself would also allow her to break the binding property of the commitment scheme. This implies security for the recipient. ■

Corollary 3.2 *If one-way permutations exist, then there exists a secure FSS scheme.*

Proof sketch: In [11], a statistically (in fact perfectly) hiding bit commitment scheme is constructed from any one-way permutation. It is easy to check that this commitment scheme has the properties of public verification and non-interactive opening. ■

It is clear that the signature scheme we just constructed is a one-time signature scheme, and therefore not very efficient. If the commitment scheme we use needs interaction for every new commitment, however, there does not seem to be a way around this.

However, with a (perhaps) stronger assumption we can do much better, namely the assumption that collision-intractable (collision-free) hash functions exist.

Assume we have a family, H, of collision-intractable hash functions, such that functions in the family map $(k + 2\sigma + 1)$-bit inputs to k-bit outputs. Functions in the family can be computed easily and can be efficiently selected at random, but the probability that a poly-time bounded enemy can find collisions for a member of H is superpolynomially small in k.

Note that we can build such a family with the right input length from any collision-intractable family by fixing some input bits if the input length is too large, and using the iterative construction of [4] if inputs are too short. Now, we have the following observation:

Lemma 3.3 *Let h be any function from $k + 2\sigma + 1$ bits to k bits. Then, when x is uniformly chosen, the probability that the preimage of $h(x)$ has size at least 2^σ is at least $1 - 2^{-\sigma-1}$.*

Proof: Let the degree of a point in the image of h be the size of its preimage under h. Since h maps into the set of k-bit strings, at most $2^k \cdot 2^\sigma$ elements can be preimages of elements of degree $\leq 2^\sigma$. Hence a uniformly chosen x is such a preimage with probability at most $2^{k+\sigma}/2^{k+2\sigma+1}$. ■

We can use this result to build a simple FSS scheme along the lines of the one described in detail above: To generate keys, B chooses a hash function h from the family, sends it to A, and A chooses two preimages (her secret key) x_0, x_1 and sends the public key $h(x_0), h(x_1)$ to B. The signature on bit b will be x_b. A proof of forgery for a signature on bit b will be two different preimages of $h(x_b)$.

This scheme is secure for the recipient by the collision-intractability of H, and secure for the signer by Lemma 3.3: the event *Good* is that both parts of the public key have preimages of size at least 2^σ. (If more than one bit is signed, the probabilities that one of the preimages is too small accumulate, but this can be countered by letting the size of the inputs to the hash function grow logarithmically with the number of such public keys used.)

This is still just a one-time signature scheme. But since we have a collision-intractable hash function, we can use $2k$ pairs from the public key to authenticate the hashed image of any number of new pairs, and thus make an arbitrary number of signatures in a tree-like structure, in the style of Merkle's signature schemes [8, 9]. We can get the public key even shorter by hashing the original pairs down to k bits. Similarly, messages of arbitrary length can be also signed using only k pairs by hashing them first.

False signatures can then also be generated by finding new messages colliding under h with already signed ones. But this is not a problem for generating proofs of forgery: the signer can show the collision as a proof (which by collision-intractability she could not generate herself).

Note that, since the function h can be the same for all signers, it makes sense not to count the description of h as a part of the public key. Thus we have shown:

Theorem 3.4 *If collision-intractable hash functions exist, there exists a secure fail-stop signature scheme where an arbitrary number of messages can be signed, and the length of the public key is just the security parameter k.*

Since extremely efficient hash functions exist in practical applications (MD4, SHA, etc..), this shows that really practical FSS schemes can be constructed based on conventional cryptography only.

4 Efficient Statistically Hiding Commitments

Naor and Yung [10] have shown that a statistically hiding bit commitment scheme can be built from collision-intractable hash functions. This scheme needs interaction only in an initialization phase, after which both committing and opening are non-interactive.

We now sketch how to modify the Naor-Yung scheme to get more efficient commitments, where several bits can be committed to at once. The amortized number of bits of communication per bit committed to is only O(1). Our scheme makes use of families of universal hash functions [3]. These functions are interesting because they emulate some properties of random functions, although they have much shorter descriptions, and can therefore be efficiently used in protocols. The standard example of a family of universal hash functions from n-bit strings to $i \leq n$-bit strings are the functions that map x to $ax + b|_i$, where $|_i$ means that we take only the most significant i bits, and where $a, b \in GF(2^n)$. Thus each member of the family is characterized by a choice of a, b. This family is 2-universal, which means that for any 2 fixed inputs x_1, x_2, the images $f(x_1), f(x_2)$ are uniformly and independently distributed i-bit strings, when f is uniformly chosen from the family.

Let t denote the number of bits to be committed to and k the security parameter of the scheme, and let H be a family of collision-intractable hash functions constructed such that the input length is $2k + t$ whenever the output length is k. Consider the following commitment scheme:

INITIALIZATION PHASE
B chooses at random a function $h \in H$ with output length k bits. He sends h to A.

COMMIT PROTOCOL
A chooses at random a $(2k + t)$-bit string x, and a 2-universal hash function f from $2k+t$ bits to t bits. Let $\bar{b} = b_1, \ldots, b_t$ be the t-bit string A wants to commit to. She then sends f, $h(x)$ and the bitwise XOR $C = \bar{b} \oplus f(x)$ to B.

REVEAL PROTOCOL
1. A sends $\bar{b}$ and x to B.
2. B checks that h maps x to $h(x)$, and compares C to $\bar{b} \oplus f(x)$. If OK, he accepts the opening, otherwise he rejects.

A formal proof of security for this commitment scheme would require a generalization of the definition in Section 2.2 to commitments to many bits. We have omitted this for simplicity, and therefore only sketch the proof below.

Theorem 4.1 *The scheme described above is a statistically hiding commitment scheme, under the assumption that H is a family of collision-intractable hash functions. It allows commitment to t bits by a commitment of size $5k + 3t$ bits.*

Proof sketch: The size of commitments is clear from the description above.

The binding property is trivial from the collision-intractability of H. For the security property, the privacy amplification theorem of [1] (see also [2]) says that, over the choice of x and f, B's expected information about $f(x)$ (and therefore about $\bar{b}$) given by knowledge of f, h, and $h(x)$ is at most $2^{-k}/\ln 2$. ■

5 Bit Commitments from Fail-Stop Signatures

The main idea in our construction of bit commitments from FSS schemes is to use the key generation protocol between user A and key center B as the commit protocol, and to think of the resulting public key as the commitment and the secret key as the string that can open the commitment.

If the FSS scheme has the almost unique secret key property, it is obvious that A is committed to any value that can be computed from $\kappa(sk)$, where sk is the secret key. There are two major difficulties, however: First, the distribution of the secret key held by A given the public key is not necessarily uniform. So we need a way to assign a value to the secret key known by A in such a way that B has essentially no information about it, given the public key. This is done by using universal hash functions [3] and the extended privacy amplification result of [1]. Secondly, the definition of FSS schemes somewhat counterintuitively allows the key generation to lead to a secret key that can be guessed by a dishonest key center. This may happen for keys with the strange property that the signer can prove that (her own) signatures made with this secret key are forgeries. Such keys are not necessarily unlikely if the key center is dishonest. Hence we must provide a way for the signer (now the committer) to exclude these keys.

We now give a more detailed description of the construction:

COMMIT PROTOCOL

1. A and B execute the key generation protocol of the FSS scheme with security parameters (k, σ), where $\sigma = 4k + 4$, and k equals the security parameter for the bit commitment scheme we are building. Here B plays the role of the key center. If A or B reject in the key generation, the commit protocol stops. Otherwise let sk be the resulting secret key and pk the public key.
2. A signs the message, "0" (consisting of one 0-bit) using sk. She runs the algorithm for generating proofs of forgery on the resulting signature. If this results in a proof of forgery, she stops. Otherwise she continues.
3. A chooses and sends to B a random 2-universal hash function h with a 1-bit image.
4. Let b be the bit A wants to commit to. Then A sends $c = h(\kappa(sk)) \oplus b$ to B.

OPENING

1. A sends b and sk to B.
2. B verifies the secret key, by checking that $Fits(sk, pk) = 1$. He then compares b with $c \oplus h(\kappa(sk))$. If they are equal, he outputs *accept b*, if not, he outputs *reject*.

Theorem 5.1 ***If the above construction is based on a secure FSS scheme with the almost unique secret key property, the result is a statistically hiding bit commitment scheme (with non-interactive opening and public verification).***

Proof: First note that the possibility of stopping in Step 2 does not prevent an honest A and B from completing the protocol: security for the recipient implies that the scheme almost never stops in Step 2 if A and B are honest.

The binding property is clear from the almost unique secret key property of the FSS scheme: if the committer could open the commitment in two different ways, she would know two secret keys satisfying the predicate *Fits* and with different κ-images.

For the security property, we need the following notation: Let *Acc* be the event that A accepts the key generation, i.e., does not stop in Step 1, and let U be the event that A does not stop in Step 1 or 2. Finally let G be the event that the public key produced, *pk*, and $view_{\tilde{B}}$ are in the set *Good*, as defined before Definition 2.2.

The extended privacy amplification theorem from [1] deals with collision entropies, instead of Shannon entropies. The collision entropy, or Renyi entropy, of a distribution is defined as minus the logarithm base 2 of the sum of the squared probabilities. For a binary distribution, like that of $h(\kappa(sk))$, with probabilities p and $1-p$, the collision entropy is

$$R(p) = -\log_2(p^2 + (1-p)^2).$$

This is a value between 0 and 1, like the Shannon entropy. It is therefore natural to define the *collision information* to be $1 - R$.

Let $I_{\tilde{B}}$ denote the collision information obtained by $\tilde{B}$ about $h(\kappa(sk))$ during the commit protocol, and let $E_{\tilde{B}}$ be its expected value, taken over the random choices of A. For an event X, $E_{\tilde{B}}(X)$ denotes the expected information given that X occurs. Then we have

$$\begin{aligned} E_{\tilde{B}} &= Prob[\neg U]E_{\tilde{B}}(\neg U) + Prob[U,G]E_{\tilde{B}}(U,G) + Prob[U,\neg G]E_{\tilde{B}}(U,\neg G) \\ &\leq 0 + Prob[U,G]E_{\tilde{B}}(U,G) + Prob[Acc,\neg G] \\ &\leq Prob[U,G]E_{\tilde{B}}(U,G) + 2^{-\sigma} \end{aligned}$$

by the security for the signer, and since A does not reveal anything at all if the commit protocol is aborted in Step 1 or 2.

The rest of the proof proceeds in 3 parts: We first show that in most cases, the best guess at the significant part of the secret key from the point of view of $\tilde{B}$ still has a rather small probability of being correct. Secondly, we derive with the extended privacy amplification theorem that in most cases, an enemy has very little collision information about $h(\kappa(sk))$. Finally, we derive an upper bound on the advantage an enemy has in guessing the content of the commitment.

Part 1 Let SK be the random variable denoting the secret key of A, and let sk_{max} denote a secret key such that $\kappa(sk_{max})$ has maximal probability given $v := (pk, view_{\tilde{B}})$, U and G. We now show that *on average* over the possible v's and given U and G, this maximal probability is upper bounded:

$$Prob[\kappa(SK) = \kappa(sk_{max}) \mid U, G] \leq 2^{-\sigma} Prob[U,G]^{-1}. \qquad (*)$$

For this, it is sufficient to show that

$$Prob[\kappa(SK) = \kappa(sk_{max}), U, G] \leq 2^{-\sigma}.$$

To do this, we consider the following attack by B^* on the FSS scheme.

1. B^* executes the key generation protocol with A in the same way as $\tilde{B}$ did.
2. B^* finds sk_{max} and uses it to make a signature on the message "0".

Let F denote the event that A fails to prove this forgery. Note that the distribution of the keys after Step 1 of this attack and of the commit protocol are equal. Furthermore, $U \subset Acc$, and whenever Acc and $\kappa(SK) = \kappa(sk_{max})$ occur, U implies F by definition of the almost unique secret key property. This gives us

$$\begin{aligned}
Prob[\kappa(SK) = \kappa(sk_{max}), U, G] &= Prob[\kappa(SK) = \kappa(sk_{max}), U, G, Acc] \\
&\leq Prob[F, \kappa(SK) = \kappa(sk_{max}), G, Acc] \\
&= Prob[F, Acc, \kappa(SK) = \kappa(sk_{max}) \mid G] Prob[G] \\
&\leq Prob[F \mid G] \\
&\leq 2^{-\sigma}.
\end{aligned}$$

The final inequality follows from the security for the signer of the FSS scheme. This finishes the proof of (*).

Now let V be a random variable denoting $v = (pk, view_{\tilde{B}})$ and M the set of cases where the probability of the best guess is much larger than on average, and also the event that such a case occurs:

$$M := \{v \mid Prob[\kappa(SK) = \kappa(sk_{max}) \mid G, U, V = v] \geq Prob[U, G]^{-1} 2^{-\sigma/2}\}.$$

By Markov's rule, the average inequality $(*)$ implies that

$$Prob[M \mid U, G] \leq 2^{-\sigma/2}.$$

We split the expected information according to whether M occurs or not.

$$\begin{aligned}
E_{\tilde{B}}(U, G) &\leq Prob[M \mid U, G] + Prob[\neg M \mid U, G] E_{\tilde{B}}(U, G, \neg M) \\
&\leq 2^{-\sigma/2} + \sum_{v \notin M} Prob[V = v \mid U, G] E_{\tilde{B}}(U, G, V = v).
\end{aligned}$$

Part 2 Whenever $v \notin M$, the extended privacy amplification lemma, Lemma 5.2 below, immediately implies that the information $E_{\tilde{B}}(U, G, V = v)$ is small:

$$E_{\tilde{B}}(U, G, V = v) \leq Prob[U, G]^{-1} 2^{-\sigma/2} \frac{2}{\ln 2} < Prob[U, G]^{-1} 2^{2-\sigma/2}.$$

Substituting this into the two equations above where $E_{\tilde{B}}$ was partitioned gives

$$\begin{aligned}
E_{\tilde{B}}(U, G) &< 2^{-\sigma/2} + \sum_{v \notin M} Prob[V = v \mid U, G] Prob[U, G]^{-1} 2^{2-\sigma/2} \\
&\leq 2^{-\sigma/2} + Prob[U, G]^{-1} 2^{2-\sigma/2}
\end{aligned}$$

and thus

$$E_{\tilde{B}} \leq Prob[U, G] 2^{-\sigma/2} + 2^{2-\sigma/2} + 2^{-\sigma} < 2^{3-\sigma/2}.$$

Part 3 Let *Bias* be the random variable denoting $\tilde{B}$'s advantage β in guessing $h(\kappa(sk))$. (This also bounds the advantage in guessing b.) From the definition of the collision entropy for binary distributions one sees

$$R(1/2+\beta) = -\log_2(\frac{1}{2}+2\beta^2) = 1-\log_2(1+4\beta^2) \leq 1-4\beta^2$$

for $|\beta| \leq \frac{1}{2}$, i.e., $4\beta^2 \leq 1$. This implies

$$\beta \leq \frac{1}{2}\sqrt{1-R(1/2+\beta)}.$$

Thus we have shown the following pointwise inequality between the random variables *Bias* and the collision information $I_{\tilde{B}}$:

$$Bias \leq \frac{1}{2}\sqrt{I_{\tilde{B}}}.$$

Applying the general formula $E(X) \leq \sqrt{E(X^2)}$ to $X = \sqrt{I_{\tilde{B}}}$ yields

$$E(Bias) \leq \frac{1}{2}E(\sqrt{I_{\tilde{B}}}) \leq \frac{1}{2}\sqrt{E(I_{\tilde{B}})} = \frac{1}{2}\sqrt{E_{\tilde{B}}} < 2^{1-\frac{\sigma}{4}}.$$

In the last inequality, the result of Part 2 was used.

As the security parameter of the FSS scheme, σ, equals $4k+4$, this shows that the commitment scheme has the security property. ∎

Remark: From the proof of the security property, it is clear that we do not need to hash all the way down to a 1-bit value to wipe out the enemy's information. Therefore we can commit to more than one bit in one commitment.

Lemma 5.2 *Let S be a random variable with a given distribution $\{p_i \mid i = 1,2,\ldots,2^n\}$ on the set of n-bit strings. If there is an $\alpha > 0$ such that $p_i \leq \alpha$ for each i, and a random 2-universal hash function mapping n bits to 1 bit is chosen, then the expected collision information about the image $h(S)$ is at most $\alpha\frac{2}{\ln 2}$.*

Proof: Let $N = 2^n$. Then

$$\sum_{i=1}^{N} p_i^2 \leq \sum_{i=1}^{N} p_i\alpha = \alpha.$$

Hence the collision entropy R of the given distribution is

$$R \geq -\log_2(\alpha).$$

Theorem 5 of [1] shows that if an unbounded enemy knows at most l bits of collision information about an n-bit string (defined as $n-R$), and if the string is hashed down to $n-l-s$ bits, the enemy's expected collision information about the result is at most $2^{-s}/\ln(2)$ bits. In our case, we hash down to 1 bit, and thus $s = R-1 \geq -\log_2(\alpha)-1$. Hence, the enemy's expected collision information E about $h(S)$ is

$$E \leq \alpha \cdot \frac{2}{\ln(2)}.$$

∎

6 Equivalence

The FSS scheme constructed from bit commitments in Section 3 does not necessarily have a unique secret key property: for example, more than one bit string may be acceptable as opening a commitment as a 1. In the following, we modify the scheme so that it has at least the almost unique secret key property. Together with Theorem 5.1, this yields the equivalence between FSS schemes with the almost unique secret key property and statistically hiding bit commitment schemes with non-interactive opening and public verification. The only changes to the protocol are:

- In key generation, A makes an additional commitment to every bit in the strings that will open the commitments $C_{b,i}$. These *secondary commitments* belong to the public key, and the strings that open them belong to the secret key.
- The security parameter σ is increased a little (since the secondary commitments may give a small amount of extra information).

It is clear that this scheme is still secure. For the almost unique secret key property, $Fits(sk, pk)$ is defined to mean that the strings in sk open all the commitments of pk correctly, and the significant part, $\kappa(sk)$, of sk, consists of those strings that open the original commitments. Clearly, finding two secret keys fitting the same public key, but with different κ-images, would mean opening at least one secondary commitment in both ways. Moreover, two secret keys sk_1, sk_2 with $\kappa(sk_1) = \kappa(sk_2)$ obviously lead to the same signatures.

7 Conclusion

We have shown that the existence of FSS schemes with the almost unique secret key property is equivalent to the existence of bit commitment schemes with non-interactive opening. In addition, we have shown how to construct efficient FSS schemes from any collision-intractable hash function.

Acknowledgement

It is a pleasure to thank *Moti Yung* for interesting discussions about all sorts of statistically hiding schemes, and in particular, one that lead to one of the independent discoveries of the construction of fail-stop signature schemes from bit commitments. We also thank *Michael Waidner* for letting us extend a joint result that has never really been published.

References

1. C. H. Bennett, G. Brassard, C. Crépeau, U. Maurer: *Privacy Amplification Against Probabilistic Information.* In preparation.

2. C. H. Bennett, G. Brassard, J.-M. Robert: *Privacy Amplification by Public Discussion.* SIAM Journal on Computing, vol 17, no. 2, 1988, pp. 210–229.
3. J. L. Carter, M. N. Wegman: *Universal Classes of Hash Functions.* Journal of Computer and System Sciences 18, 1979, pp. 143–154.
4. I. B. Damgård: *A Design Principle for Hash Functions.* Proceedings of Crypto'89, LNCS 435, pp. 416–427, 1990.
5. E. van Heyst, T. P. Pedersen: *How to Make Efficient Fail-Stop Signatures.* Presented at Eurocrypt'92, Balatonfüred, Hungary, 1992.
6. E. van Heyst, T. P. Pedersen, B. Pfitzmann: *New Constructions of Fail-Stop Signatures and Lower Bounds.* Presented at Crypto'92, Santa Barbara, 1992.
7. R.Rivest: *The MD4 message-digest algorithm*, Proc. of Crypto 90.
8. R. C. Merkle: *Protocols for Public Key Cryptosystems.* In: Secure Communications and Asymmetric Cryptosystems, AAAS Selected Symposium 69, G. J. Simmons (ed.); Westview Press, Boulder 1982, pp. 73–104.
9. R. C. Merkle: *A digital signature based on a conventional encryption function.* Proceedings of Crypto'87, LNCS 293, Springer-Verlag, Berlin 1988, pp. 369–378.
10. M. Naor, M. Yung: *Universal One-Way Hash Functions and their Cryptographic Applications.* Proceedings of 21st STOC, pp. 33–43, 1989.
11. M. Naor, R. Ostrovsky, R. Venkatesan, M. Yung: *Perfect Zero-Knowledge Arguments for NP Can Be Based on General Complexity Assumptions.* Presented at Crypto'92, Santa Barbara, 1992.
12. T. P. Pedersen, B. Pfitzmann: *Fail-Stop Signatures.* Manuscript, February 1993.
13. B. Pfitzmann, M. Waidner: *Formal Aspects of Fail-Stop Signatures.* Internal report 22/90, Fakultät für Informatik, Universität Karlsruhe.
14. B. Pfitzmann, M. Waidner: *Fail-Stop Signatures and their Application.* Securicom 91, Paris, pp. 145 – 160.
15. *Specifications for a Secure Hash Standard*, Federal Information Processing Standards Publication YY, 1992.
16. M. Waidner, B. Pfitzmann: *The Dining Cryptographers in the Disco: Unconditional Sender and Recipient Untraceability with Computationally Secure Serviceability.* Proceedings of Eurocrypt'89, LNCS 434, page 690, 1990.

Joint Encryption and Message-Efficient Secure Computation

Matthew Franklin*[1] and Stuart Haber[2]

[1] Columbia University, New York, NY 10027

[2] Bellcore, 445 South Street, Morristown, NJ 07960-6438

Abstract. This paper connects two areas of recent cryptographic research: secure distributed computation, and group-oriented cryptography. We construct a probabilistic public-key encryption scheme with the following properties:

- It is easy to encrypt using the public keys of any subset of parties, such that it is hard to decrypt without the cooperation of every party in the subset.
- It is easy for any private key holder to give a "witness" of its contribution to the decryption (e.g., for parallel decryption).
- It is "blindable": From an encrypted bit it is easy for anyone to compute a uniformly random encryption of the same bit.
- It is "xor-homomorphic": From two encrypted bits it is easy for anyone to compute an encryption of their xor.
- It is "compact": The size of an encryption does not depend on the number of participants.

Using this joint encryption scheme as a tool, we show how to reduce the message complexity of secure computation versus a passive adversary (gossiping faults).

1 Introduction

This paper connects two areas of recent cryptographic research: secure distributed computation, and group-oriented cryptography. The problem of securely evaluating an arbitrary boolean circuit under cryptographic assumptions has been much studied, beginning with the work of Yao [14] and Goldreich, Micali, and Wigderson [8]. The notion of group-oriented cryptography, in which the power of a secret key holder is distributed over a number of participants, was introduced by Desmedt [3].

Practical implementations of group-oriented public-key encryption were given by Desmedt and Frankel [4]. (See also the related notion of fair public-key encryption [12].) We extend their implementations to achieve additional useful properties. Our scheme, which we call "additive joint encryption," can then be used to reduce the message complexity of cryptographic multi-party circuit evaluation.

* Paritally supported by an AT&T Bell Laboratories Ph.D. Scholarship. Part of this work done during a summer internship at Bellcore

An additive joint encryption scheme enables a group of parties to have individual public keys, such that a message can be encrypted using the keys of any subset of parties. It is easy for any party to "withdraw" from an encryption, and the cooperation of all (current) participants is needed to decrypt. It is easy for each participant to give a "witness" of its contribution to the decryption, so that full decryption can occur in parallel (i.e., anyone can decrypt after seeing all the witnesses). Encrypted messages can be blinded (i.e., replaced by a random encryption of the same message), and they are xor-homomorphic (i.e., from the encryption of two bits it is easy for anyone to compute an encryption of their exclusive-or).

We present an implementation of additive joint encryption with the critical property that the size of an encrypted bit is independent of the number of parties participating in the encryption. This "compact" implementation is a construction that combines El-Gamal public-key encryption and schemes based on quadratic residues and non-residues.

We demonstrate the use of our encryption scheme as a tool for designing efficient multi-party cryptographic protocols (i.e., communication via broadcast channels only). We show that, using additive joint encryption, a factor of n can be gained in the number of bits broadcast by n parties to securely compute a circuit of size C. Specifically, in the privacy setting (against a passive adversary), only $O(nC)$ encrypted bits of communication are needed, as opposed to $O(n^2C)$ encrypted bits using existing methods.

For specific functions, further gains are possible by exploiting particular connections between properties of additive joint encryption and properties of the functions themselves. We illustrate this for the problem of comparing bit-strings.

Definitions and models are given in Sections 5.2 and 5.3. In Section 5.4, we describe the construction of our new encryption scheme. In Sections 5.5 and 5.6, we demonstrate the application of our scheme to reducing the message complexity of secure multi-party computation. Conclusions and some open problems are given in Section 5.7.

2 Model

We assume that there are n parties, each of which is a probabilistic polynomial time Turing Machine (read-only input tape, write-only output tape, random tape, one or more work tapes). The parties communicate by means of a broadcast channel, which can be modeled as an additional tape (communication tape) for each machine that is write-only for its owner and read-only for everyone else. When a party writes a message to this tape, we may say that the message has been "broadcast" or "posted." A protocol begins with all n parties in their start states, and ends when all have reached their final states. The output of the protocol is the (common) value written on the output tapes of the processors.

We are concerned with the message complexity of a protocol. This is measured as the total number of bits written on the communication tapes during the execution of the protocol. Since our protocols are cryptographic, we will state the message complexity in terms of the number of *encrypted bits* written on the communication tapes. For the protocols we consider, this is all or most of the communication that occurs, and it is also a convenient measure independent of advances in either encryption methods or cryptanalytic techniques.

We will say that a protocol is "private" if its execution reveals no useful information to any subset of (polynomial bounded) gossiping processors. More specifically, anything that is efficiently computable from the views of a subset S of participants is also computable from just the output of the protocol together with the inputs and private keys of S.

3 Additive Joint Encryption

In this section we give definitions for additive joint encryption, and then a naive implementation for which the size of an encrypted bit depends on the number of participating parties.

3.1 Definition

A *joint encryption scheme* for $[1 \cdots n]$ is a collection of encryption functions $\{E_S : S \subseteq [1 \cdots n]\}$ and a collection of partial decryption functions $\{D_i : i \in [1 \cdots n]\}$ such that $D_i(E_S(M)) = E_{S-\{i\}}(M)$ for all messages M and all $i \in S$, and such that it is easy to compute M from $E_\emptyset(M)$. We use the notation $D_S(c)$ to stand for the result of applying to c the decryption functions corresponding to each element of S.

A joint encryption scheme is *public-key* if each E_S is easy to compute, while computing each D_i requires a different trapdoor; in fact, our implementation has the stronger "upward conversion" property: $E_{S \cup S'}(M)$ is easy to compute from $E_S(M)$ for any subset S'. A joint public-key encryption scheme is probabilistic if each E_S is probabilistic; we write $E_S(M)$ to denote a uniformly random choice of possible encryptions of M. A probabilistic public-key joint encryption scheme is *secure* if each E_S is GM-secure [10] (computational indistinguishability of ciphertexts, even given the decryption functions $\{D_i : i \in S'\}$ for any $S' \subset S$).

A probabilistic joint encryption scheme is *blindable* if, given any subset $S \subseteq [1 \cdots n]$, and an encryption $C = E_S(M)$, it is possible to sample efficiently from the uniform distribution on the set $\{C' : D_S(C') = M\}$ of all possible encryptions of M (although it suffices for our purposes to be able to sample efficiently from any computationally indistinguishable distribution). We write *blind*(C) to denote a random choice from this set.

A joint encryption scheme is *xor-homomorphic* if, given any messages M, M' in $\{0,1\}^k$, any subset $S \subseteq [1 \cdots n]$, and any encryptions $E_S(M), E_S(M')$, it is easy to compute $E_S(M \oplus M')$.

A joint encryption scheme is *witnessed* if there are functions $\{W_i : 1 \leq i \leq n\}$ such that each W_i is hard to compute without the trapdoor for D_i, and such that $D_i(E_S(M))$ can be easily computed from $E_S(M)$ and $W_i(E_S(M))$ for any M and any $i \in S$, but no other $D_i(E_S(M'))$ can be easily computed from $E_S(M')$ and

$W_i(E_S(M))$. If every participant in an encryption provides a witness in parallel, the decryption can be computed much faster than by the use of (inherently sequential) partial decryption functions.

Finally, we use the term *additive joint encryption scheme* to denote a secure, blindable, xor-homomorphic, witnessed probabilistic public-key joint encryption scheme.

Desmedt and Frankel [4] consider "threshold" encryption, which is essentially a public-key joint encryption scheme for which any sufficiently large subset of parties can decrypt a message. We do not include this property in our definition, because it is not needed for our main application, secure circuit evaluation. In Section 5.4, we explain how to add threshold decryption capability to our implementation using their techniques.

Notation: We may abuse the xor symbol by extending it to encryptions in the obvious way. When $\hat{x} = E_S(x), \hat{y} = E_S(y)$ are encryptions, we may write $\hat{x} \oplus \hat{y}$ to denote $E_S(x \oplus y)$.

3.2 Naive Implementation Based on Xor Shares

A naive implementation of additive joint encryption can be based on any blindable and xor-homomorphic probabilistic public-key encryption scheme. For example, each encryption function e_i could be an instance of the scheme due to Goldwasser and Micali [10], based on quadratic residues and nonresidues, using a distinct modulus N_i. Let d_i be the decryption function corresponding to e_i.

An additive joint encryption scheme can be constructed as follows. Encryption is given by $E_S(b) = (b', [e_j(b_j) : j \in S])$, where $b_j \in_R \{0,1\}$ for all $j \in S$, and where $b' = b + \sum_{j \in S} b_j \bmod 2$. Decryption is given by $D_i(\alpha, [\beta_j : j \in S]) = (\alpha \oplus d_i(\beta_i), [\beta_j : j \in S - \{i\}])$ whenever $i \in S$ (or just $d_i(\beta_i)$ can be given as a decryption witness).

Note that the size of an encryption in this scheme grows as the number of participants increases. We seek a compact scheme for which the size of an encryption is independent of the number of parties.

4 Compact Additive Joint Encryption

In this section, we show how additive joint encryption can be implemented compactly, i.e., such that the size of an encrypted bit does not depend on the number of parties participating in the encryption.

4.1 Intuition of El-Gamal Based Scheme

As shown by Desmedt and Frankel [4], it is possible to construct a joint encryption scheme from El-Gamal's method of public-key encryption [6]. Although blindable, this scheme is not xor-homomorphic, and thus not an additive joint encryption scheme. However, we can convert this into an additive joint encryption scheme for which the size of an encryption is independent of the number of parties.

The basic idea is to use El-Gamal with a composite modulus (whose factorization is unknown), encrypting zeros and ones as El-Gamal encryptions of random quadratic residues and non-residues, respectively (all with Jacobi symbol +1). This almost works as is, since blinding and xor can be achieved by component-wise multiplication of the two parts of an El-Gamal encryption. Unfortunately, although these products preserve the correct quadratic character (residue or non-residue) of the encrypted values, the parties—who don't know the factorization of the modulus—will be unable to make use of them. The parties cannot compute the quadratic character of an El-Gamal decryption, unless the entire history of blindings is stored and revealed. It wouldn't help to give the factorization of the modulus to the parties, since that would allow any party to decrypt on its own.

This problem can be solved by accompanying each encrypted bit with an encryption of the witness that allows its decryption. When s^2 is used to encrypt a zero, or $-s^2$ is used to encrypt a one, then the encryption of s^2 or $-s^2$ is accompanied by an El-Gamal encryption of s. This accompanying information makes decrypted values identifiable as residues or non-residues without knowing the factors of N. We give details in the next section.

4.2 Details of El-Gamal based Scheme

The public key is $[N, g^{x_1} \bmod N, \cdots, g^{x_m} \bmod N]$, where $g \in Z_N^*$, $N = pq$, $p \equiv q \equiv 3 \bmod 4$, although the prime factors p, q are unknown. The trapdoor information for D_i is x_i. Encryption of a zero is given by

$$E_S(0) = [g^r \bmod N, g^{r'} \bmod N, s^2(\prod_{j\in S} g^{x_j})^r \bmod N, s(\prod_{j\in S} g^{x_j})^{r'} \bmod N]$$

for $r, r', s \in_R Z_N^*$. Encryption of a one is given by

$$E_S(1) = [g^r \bmod N, g^{r'} \bmod N, -s^2(\prod_{j\in S} g^{x_j})^r \bmod N, s(\prod_{j\in S} g^{x_j})^{r'} \bmod N]$$

for $r, r', s \in_R Z_N^*$. Decryption is given by $D_i([\alpha, \beta, \gamma, \delta]) = [\alpha, \beta, \gamma\alpha^{-x_i} \bmod N, \delta\beta^{-x_i} \bmod N]$. Note that the fourth component of $E_\emptyset(b)$ enables the quadratic character of the third component (and hence the value of b) to be computed easily.

If $E_S(b) = [\alpha, \beta, \gamma, \delta]$ and $E_S(b') = [\alpha', \beta', \gamma', \delta']$, then

$$E_S(b \oplus b') = [\alpha\alpha' \bmod N, \beta\beta' \bmod N, \gamma\gamma' \bmod N, \delta\delta' \bmod N];$$

thus this scheme is xor-homomorphic. If $[\alpha, \beta, \gamma, \delta]$ is a joint encryption using E_S, and $r, r' \in_R Z_N^*$, then $[\alpha g^r \bmod N, \beta g^{r'} \bmod N, \gamma(\prod_{j\in S} g^{x_j})^r \bmod N,$ $\delta(\prod_{j\in S} g^{x_j})^{r'} \bmod N]$ is a (nearly) uniformly random joint encryption of the same value (from a computationally indistinguishable distribution when g has large order); thus this scheme is blindable. If $E_S(b) = [\alpha, \beta, \gamma, \delta]$, then $D_i(E_S(b))$ can be easily computed from $[\alpha^{-x_i} \bmod N, \beta^{-x_i} \bmod N]$ for any $i \in S$; thus this scheme is witnessed. The size of each encrypted bit is four elements of Z_N^*, independent of the number of participants; thus this scheme is compact.

We note that our implementation can be modified to include the property of threshold decryption, i.e., encryption such that any sufficiently large subset of parties can decrypt. This can be done, for example, by incorporating the modified shadow generation scheme based on Lagrange interpolation developed by Desmedt and Frankel [4] (which is possible when g and N are chosen as described in the next section).

4.3 Security of El-Gamal Based Scheme

Theorem 1. *If El-Gamal encryption with a composite modulus is GM-secure, then our compact encryption scheme is GM-secure.*

Proof. Suppose, for purposes of establishing a contradiction, that El-Gamal encryption with a composite modulus is GM-secure while our compact additive joint encryption scheme is not GM-ssecure. Then it would be easy to distinguish between composite El-Gamal encryptions of $+1$ and -1, since these can be easily converted into random additive joint encryptions of one and zero, as follows.

Let $(g^r \bmod N, (-1)^b g^{rx} \bmod N)$ be a composite El-Gamal encryption of $(-1)^b$ (using El-Gamal public key $g^x \bmod N$). Then $[g^r \bmod N, g^{r'} \bmod N, s^2(-1)^b g^{rx} \bmod N, sg^{r'x} \bmod N]$ is an additive joint encryption $E_S(b)$ for random $r', s \in_R Z_N^*$ (e.g., using joint public key

$$[N, r_1, \cdots, r_{|S|-1}, (g^x \prod_{i<|S|} r_i^{-1} \bmod N)]$$

for random $r_1, \cdots, r_{|S|-1}$).

However, our encryption scheme (and composite El-Gamal) is not GM-secure if composite quadratic character (residue vs. non-residue) is easy to compute. An attacker sees $g^x \bmod N, \alpha = g^r \bmod N, \beta = g^{r'} \bmod N, \gamma = (-1)^b s^2 g^{rx} \bmod N, \delta = sg^{r'x} \bmod N$, where b is the value of the encrypted bit (and where $x = \sum_{i \in S} x_i$). Let $QR_N(v) = 0$ if v is a quadratic residue modulo N and 1 otherwise. If $QR_N(\cdot)$ is easy to compute, then the attacker can determine $b = (QR_N(\alpha) * QR_N(g^x \bmod N)) \oplus QR_N(\gamma)$. We do not know whether the additional information available to the attacker makes the GM-security of our scheme (and composite El-Gamal) strictly weaker than the difficulty of computing quadratic character modulo N.

The security of the original El-Gamal public-key encryption scheme reduces to the difficulty of breaking an instance of the Diffie-Hellman key exchange scheme [5] (i.e., a problem that is no more difficult than but not known to be equivalent to the discrete log problem). McCurley [11] showed how El-Gamal encryption with a composite modulus (and a careful choice of g and N) can be secure against an adversary who could break the Diffie-Hellman key exchange, or could factor the modulus, but not both. However this was a proof of security in the sense that no polynomial time algorithm can invert a non-negligible fraction of ciphertexts, and not GM-security (computational indistinguishability of ciphertexts).

We note that if g and N are chosen as suggested by McCurley, then the technical condition for incorporating threshold decryption into our scheme can be met. Specifically, McCurley's proof is based on the choices $g = 16$, $N = pq$, $p = 8r + 3$, $q = 8s - 1$ (where r, s have special structure), and this meets the condition of Desmedt and Frankel [4] for their modified shadow generation scheme based on Lagrange interpolation (i.e., that g have odd order in Z_N^*).

5 Message-Efficient General Secure Computation

Several solutions have been found to the problem of securely evaluating an arbitrary boolean circuit under cryptographic assumptions, beginning with the work of Yao [14] and Goldreich, Micali, and Wigderson [8]. We focus on the message complexity (i.e., number of encrypted bits of communication) of such protocols in the privacy setting.

5.1 Previous Approaches to Private Computation

Previously, the lowest message complexity known for n parties to privately evaluate a circuit of size C under reasonable cryptographic assumptions was $O(n^2C)$ encrypted bits of communication. This same complexity was achievable using either of the main techniques for secure circuit evaluation in the cryptographic setting: the "gate-by-gate" approach or the "circuit-scrambling" approach.

In the gate-by-gate approach, each gate of the circuit is computed by having each pair of the n parties perform a private two-party protocol. In the protocol of Galil, Haber, and Yung [7], with efficiency improvements by Goldreich and Vainish [9], each two-party protocol is a single instance of "One out of Two Oblivious Transfer" (1-2-OT). It is possible to implement two-party 1-2-OT privately using a constant number of encrypted bits under a cryptographic assumption (e.g., three encrypted bits suffice under the assumption that composite quadratic character is hard). This gives a total message complexity of $O(n^2C)$ encrypted bits.

In the circuit-scrambling approach, each party takes a turn modifying the truth tables of the gates of the circuit. In the protocol of Chaum, Damgård, and van de Graaf [2], each party can randomly permute the rows, and can randomly complement certain of the rows and columns of each truth table. Records of each party's modifications are preserved in the form of bit commitments, which accompany the scrambled circuit as it passes from party to party (to enable circuit evaluation after the nth party has finished scrambling). Each party contributes a constant number of bit commitments for each gate (e.g., one bit commitment for each truth table row), and so the scrambled circuit as it passes from party i

to party $i+1$ includes $O(iC)$ bit commitments. When each bit commitment is a single encryption, this gives a total message complexity of $O(n^2C)$ encrypted bits.

5.2 Reduced "Gate-by-Gate" Message Complexity

A gain of $O(n)$ in the message complexity of secure computation can be achieved via compact additive join encryption using either a gate-by-gate approach or a circuit-scrambling approach. We describe the gate-by-gate approach in detail in this section.

Theorem 2. *Under the assumption that compact additive joint encryption is possible, any boolean circuit with C gates can be privately evaluated by n parties using $O(nC)$ encrypted bits of communication.*

Proof. No communication is required for each NOT gate. Each AND gate requires two rounds of communication, and message complexity $4n$ encrypted bits (actually, three encryptions and two decryption witnesses per party, where each witness is half the length of an encryption for the El-Gamal based scheme).

The protocol begins with encryptions of the input bits on the shared tape. We show how the encrypted output of any gate can be computed in a constant number of rounds from its encrypted inputs. For a NOT gate, the output can be found without any communication by XORing the encrypted input with a default encryption of a one.

For an AND gate, suppose the encrypted gate inputs are $E_{[1\cdots n]}(x) = \hat{x}$ and $E_{[1\cdots n]}(y) = \hat{y}$. Each party i chooses $b_i, c_i \in_R \{0,1\}$, and broadcasts $\hat{b}_i = E_{[1\cdots n]}(b_i)$ and $\hat{c}_i = E_{[1\cdots n]}(c_i)$. With no communication, the parties can then find $\hat{x}' = E_{[1\cdots n]}(x \oplus b_1 \oplus \cdots \oplus b_n)$ and $\hat{y}' = E_{[1\cdots n]}(y \oplus c_1 \oplus \cdots \oplus c_n)$. The parties broadcast decryption witnesses for $\hat{x}', \hat{y}'$ to find $x' = x + \sum_{1 \leq j \leq n} b_j \bmod 2$ and $y' = y + \sum_{1 \leq j \leq n} c_j \bmod 2$. Let $z' = x' \wedge y'$.

For every $1 \leq i, j \leq n$, party i can find $E_{[1\cdots n]}(b_i \wedge c_j)$ by either encrypting a zero (if $b_i = 0$) or by blinding $\hat{c}_j$ (if $b_i = 1$). Similarly, each party i can find $E_{[1\cdots n]}(b_i \wedge y)$ and $E_{[1\cdots n]}(x \wedge c_i)$. Each party broadcasts a blinded encryption of the XOR of all of these encrypted values (in parallel with the previous broadcast). Now an encryption of the XOR of all received encrypted XOR's, together with an encryption of z', is equal to the encryption of $z = x \wedge y$ (i.e., by the distributivity of AND over XOR: $u \wedge (v \oplus w) = (u \wedge v) \oplus (u \wedge w)$).

When the last gate in the circuit has been computed, all parties know a joint encryption of the circuit output. At this point, the parties broadcast decryption witnesses to enable all of them to compute the actual circuit output.

For privacy, we need to argue that no subset S of parties learns anything about the other parties' inputs beyond what is implied by a knowledge of the inputs of S and the circuit output. It suffices to show that the distribution of transcripts of protocol executions, as viewed by S, can be simulated by a polynomial time machine that has access to the inputs and decryption functions of S, such that the simulated distribution and the actual distribution are computationally indistinguishable. Excluding the decryption witnesses for the circuit output, the transcript of an execution of the protocol gives a number of joint encryptions for related values, and a number of decryption witnesses for uniformly random values. The simulator computes joint encryptions of uniformly random values to substitute for all of the joint encryptions in the transcript except the last one, and substitutes a uniformly random joint encryption of the output for the joint encryption of the output of the last gate, and substitutes decryption witnesses for uniformly random values for all of the decryption witnesses. By standard cryptographic arguments, the computational indistinguishability of these distributions follows from the computational indistinguishability of individual joint encryptions.

5.3 Reduced "Circuit-Scrambling" Message Complexity

To get the same gain in message complexity with a circuit-scrambling approach, the bit commitments are additive joint encryptions, but only a single commitment accompanies each truth table row as it passes from party to party. The single commitment at a row represents the xor of modifications performed by all parties. When a compact scheme is used, the size of the scrambled circuit doesn't increase from scramble to scramble.

Although the order of magnitude of the message complexity is the same, note that the multiplicative constant is better for our methods using the gate-by-gate approach. In the gate-by-gate approach, four encrypted bits are needed per AND gate (counting one decryption witness as half an encryption), and no communication is needed per NOT gate. In the circuit-scrambling approach, the same four encrypted bits are needed per AND gate, while two encrypted bits are needed per NOT gate (i.e., one bit commitment for each truth table row).

5.4 Measures of Message Complexity

In this section, we have shown that the message complexity of secure computation can be decreased by a linear factor in the number of participants. This gain has been computed under a "broadcast" measure of message complexity. Specifically, if one party posts a bit to the publicly readable bulletin board, then the protocol is charged one bit. The same charge applies no matter how many of the other parties ever read that posted bit.

It is reasonable to consider an alternative "readership" measure of message complexity, in which the number of readers of a message is relevant. By this measure, the protocol is charged k bits if a single posted bit is read by k of the other parties.

The linear gain in message complexity is maintained with respect to the readership measure for the "circuit-scrambling" protocol described in Section 5.3. This is because most broadcasts are only used to pass the scrambled circuit from one party to the next, i.e., to be read by only one other party. However, the "gate-by-gate" protocol loses the linear gain with respect to the readership measure. Posting messages that are read by all other parties seems to be an essential feature of this approach.

6 Customized Secure Protocol for Bit-String Comparison

In this section, we show further application of our encryption method by describing a novel protocol for n parties to privately compare two encrypted bit-strings. The message complexity of this protocol has the same order of magnitude as would be achieved by computing a comparison circuit using our general techniques, although a small constant factor (roughly three) is saved. We believe that this protocol is of interest because it demonstrates that for practical applications (where constant factors count) useful gains in communication complexity can come from customizing cryptographic tools to the specific secure computational task at hand. Note that a constant factor is also gained by this protocol with respect to the readership measure of message complexity.

Before giving our comparison protocol, we need to develop a tool to randomly permute pairs of encrypted inputs with low message complexity.

6.1 Shuffle Gate Computation

A "shuffle gate" has two main inputs x, y, a control input c, and two outputs α, β. When $c = 0$, the inputs pass through the gate unchanged: $\alpha = x$ and $\beta = y$. When $c = 1$, the inputs are flipped as they pass through the gate: $\alpha = y$ and $\beta = x$. A shuffle gate can be represented as a circuit with six AND and OR gates. Using our gate-by-gate approach, $24n$ encrypted bits of communication are needed to privately evaluate a shuffle gate.

By contrast, a uniformly random shuffle gate can be privately computed directly at a cost of only $2n$ encrypted bits of communication using the El Gamal based scheme. Let $\hat{x}, \hat{y}$ be the encryptions of the main inputs. Each party i chooses a uniformly random $c_i \in_R \{0,1\}$. We want x, y to be flipped only if the xor of all of the c_i values is one. Each party i posts two encrypted values as follows: two encrypted zeros if $c_i = 0$, and two encryptions of $x \oplus y$ if $c_i = 1$. By xoring all of these posted pairs to the input pair $\hat{x}, \hat{y}$, each party gets an encryption of the appropriate output pair.

6.2 Details of the Comparison Protocol

Intuitively, the comparison protocol works on l-bit strings in $l - 1$ rounds, where each round reduces the length of the encrypted bit strings by one. The reduction is done in such a way that the result of comparing the two decrypted bit-strings is preserved after each round. Specifically, the parties remove the leading bits if these bits are equal, and remove the next-to-leading bits if the leading bits are unequal. Of course, it would violate privacy if any proper subset of parties could determine which of these two actions occurred in any round.

Our protocol guarantees that the correct action occurs obliviously (i.e., so that no proper subset of the parties can detect which action occurred) by repeatedly using the shuffle gate construction described in the preceding section. In each round, two shuffle gates are controlled by the same control bit γ. The decision about which pair of bits to discard each round depends critically on the value of the control bit for that round. The details are given in Figure 1.

Initially, $\tilde{x} = E_{[1\cdots n]}(x)$, $\tilde{y} = E_{[1\cdots n]}(y)$.

1. **In Round i, $1 \le i \le l-1$, the following messages are sent:**
   (a) **Each party j $(1 \le j \le n)$ posts $[\alpha_{1j}, \beta_{1j}, \alpha_{2j}, \beta_{2j}; \gamma_j] =$**
      i. $[blind(\tilde{x}.i \oplus \tilde{x}.(i+1)), blind(\tilde{y}.i \oplus \tilde{y}.(i+1)), blind(\tilde{x}.i \oplus \tilde{x}.(i+1)),$ $blind(\tilde{y}.i \oplus \tilde{y}.(i+1)); E_{[1\cdots n]}(1)]$ **with prob $\frac{1}{2}$.**
      ii. $[E_{[1\cdots n]}(0), E_{[1\cdots n]}(0), E_{[1\cdots n]}(0), E_{[1\cdots n]}(0); E_{[1\cdots n]}(0)]$ **with prob $\frac{1}{2}$.**
   (b) **Each party j $(1 \le j \le n)$ posts a witness for $D_j(\tilde{x}.i \oplus \tilde{y}.i \oplus \gamma)$, where $\gamma = \gamma_1 \oplus \cdots \oplus \gamma_n$.**
   (c) **Before the start of Round $i+1$, each party (without communication) locally replaces $\tilde{x}.(i+1), \tilde{y}.(i+1)$ with**
      i. $blind(\alpha_{21} \oplus \cdots \oplus \alpha_{2n} \oplus \tilde{x}.(i+1)), blind(\beta_{21} \oplus \cdots \oplus \beta_{2n} \oplus \tilde{y}.(i+1))$ **if $D_{[1\cdots n]}(\tilde{x} \oplus \tilde{y} \oplus \gamma) = 0$.**
      ii. $blind(\alpha_{11} \oplus \cdots \oplus \alpha_{1n} \oplus \tilde{x}.i), blind(\beta_{11} \oplus \cdots \oplus \beta_{1n} \oplus \tilde{y}.i)$ **if $D_{[1\cdots n]}(\tilde{x} \oplus \tilde{y} \oplus \gamma) = 1$.**
2. **In Round l, the following messages are sent:**
   (a) **Each party j $(1 \le j \le n)$ posts a witness for $D_j(\tilde{x}.l)$.**
   (b) **Without communication, each party computes the final answer to be $v = D_{[1\cdots n]}(\tilde{x}.l)$. (If $v = 1$ then $x \ge y$ else $x \le y$.)**

Fig. 1. Message-Efficient Multi-party Comparison Scheme

All messages for this protocol, except for the last round, are witnesses for joint encryptions of uniformly random values, or joint encryptions of related values. Privacy follows from an argument similar to that of Theorem 5.2.

Theorem 3. *Under the assumption that compact additive joint encryption is possible, private multi-party comparison of two l-bit strings is possible with communication* $4(l-1)n$ *encrypted bits and* ln *decryption witnesses.*

7 Summary and Open Problems

The message complexity of secure distributed computation can be reduced by extending techniques from group-oriented cryptography. We show how gains in multi-party evaluation of general circuits can be achieved by augmenting a joint encryption scheme to support blinding, witnessing, and adding ciphertexts *without* increasing the length of the ciphertexts.

We would like to find compact implementations of additive joint encryption based on other, possibly weaker, intractability assumptions, and to find other applications for such encryption schemes. In addition, we would like to explore other ways to improve the secure evaluation of specific useful functions by exploiting special properties of customized encryption methods. It would also be interesting to reduce the computational resources required for secure computation in other settings, possibly tolerating stronger adversaries.

References

1. D. Beaver, "Secure multiparty protocols and zero-knowledge proof systems tolerating a faulty minority," J. Cryptology (1991) 4: 75-122.
2. D. Chaum, I. Damgård, and J. van de Graaf, "Multiparty computations ensuring privacy of each party's input and correctness of the result," Crypto 1987, 87-119.
3. Y. Desmedt, "Society and group oriented cryptography: A new concept," Crypto 1987, 120-127.
4. Y. Desmedt and Y. Frankel, "Threshold cryptosystems," Crypto 1989, 307-315.
5. W. Diffie and M. Hellman, "New directions in cryptography," IEEE Transactions on Information Theory, 22(6):644-654, 1976.
6. T. El-Gamal, "A public key cryptosystem and a signature scheme based on discrete logarithms," IEEE Transactions on Information Theory, 31:469-472, 1985.
7. Z. Galil, S. Haber, and M. Yung, "Cryptographic computation: secure fault-tolerant protocols and the public-key model," Crypto 1987, 135-155.
8. O. Goldreich, S. Micali, and A. Wigderson, "How to play any mental game," STOC 1987, 218-229.
9. O. Goldreich and R. Vainish, "How to solve any protocol problem – an efficiency improvement," Crypto 1987, 73-86.
10. S. Goldwasser and S. Micali, "Probabilistic encryption," JCSS, 28(2):270:299, 1984.
11. K. McCurley, "A key distribution system equivalent to factoring," J. Crypt., 1(2):95-105, 1988.
12. S. Micali, "Fair public-key cryptosystems," Crypto 1992, 3.11-3.24 (preproceedings abstracts).
13. S. Micali and P. Rogaway, "Secure Computation," Crypto 1991, 392-404.
14. A. Yao, "How to generate and exchange secrets," FOCS 1986, 162-167.

Cryptographic Primitives Based on Hard Learning Problems

Avrim Blum*[1] Merrick Furst**[1] Michael Kearns[2] Richard J. Lipton[3]

[1] Carnegie Mellon University, Pittsburgh PA 15213
[2] AT&T Bell Labs, Murray Hill NJ 07974
[3] Princeton University, Princeton, NJ 08544

1 Introduction and Motivation

Modern cryptography has had considerable impact on the development of computational learning theory. Virtually every intractability result in Valiant's model [13] (which is *representation-independent* in the sense that it does not rely on an artificial syntactic restriction on the learning algorithm's hypotheses) has at its heart a cryptographic construction [4, 9, 1, 10]. In this paper, we give results in the reverse direction by showing how to construct several cryptographic primitives based on certain assumptions on the difficulty of learning. In doing so, we develop further a line of thought introduced by Impagliazzo and Levin [6].

As we describe, standard definitions in learning theory and cryptography do not appear to correspond perfectly in their original forms. However, we show that natural modifications to standard learning definitions can yield the desired connections. The particular cryptographic primitives we consider are pseudorandom bit generators, one-way functions, and private-key cryptosystems. We give transformations of hard learning problems into these cryptographic primitives with the desirable property that the complexity of the resulting primitive is not much greater than that of the hard-to-learn functions and distributions. In particular, our constructions are especially adept at preserving the degree of parallelism inherent in the hard functions and distributions.

Note that while it is well-known that some of the primitives above imply the existence of others (for instance, the equivalence of bit generators and one-way functions) [14, 7], we are interested in the separate results because the equivalences between primitives often do not preserve complexity measures such as circuit depth (parallelism). For instance, it is not known how to construct a bit generator in $\mathcal{NC}$ given a one-way function in $\mathcal{NC}$. One of the main potential benefits of this line of research is that as "simple" function classes (for instance, DNF formulae) continue to elude efficient learning, our belief in the intractability of learning such classes increases, and we can exploit this intractability to obtain simpler cryptographic primitives.

In addition to generic transformations, we describe a very simple pseudorandom bit generator based on the assumption that the class of parity functions

* Supported in part by an NSF Postdoctoral Fellowship.
** Supported in part by NSF grant CCR-9119319.

is hard to learn in the presense of random noise (an assumption similar to the intractability of decoding random linear codes). This generator is quite similar to a proposed one-way function due to Goldreich, Krawczyk and Luby [3], who then obtain a generator by running the one-way function through a generic transformation. We show that the output of this one-way function is already pseudorandom. This stronger assertion is apparently already known to some researchers in the cryptography community as a "folk theorem".

1.1 The Apparent Necessity of Average-Case Assumptions

In most learning theory models, a learning algorithm is required for *every* value of n to learn *all* functions over $\{0,1\}^n$ that meet some (usually strong) constraints known to the algorithm. For instance, we might ask that the learning algorithm be able to learn any DNF formula over $\{0,1\}^n$ with at most n^2 terms. Under such a definition, a failed DNF learning algorithm might still always be able to learn any n^2-term DNF provided n was even, and might learn all but a small scattered set of n^2-term DNF formulae when n was odd. A "hard" learning problem with such an algorithm seems not especially useful for cryptography.

We thus introduce an average-case model of learning that reduces these discrepancies between hardness for learning and the conditions required by cryptography, but otherwise preserves both the spirit and technical aspects of many existing models in learning theory. We then proceed to demonstrate that this worst-case/average-case discrepancy was essentially the only barrier to a natural correspondence between hard learning problems and many common cryptographic primitives.

2 Preliminaries

2.1 Learning Models

The learning models we consider are models of learning boolean functions from *labeled examples*.

Throughout the paper, $\mathcal{F}_n$ will denote a class of boolean functions over $\{0,1\}^n$, so each $f \in \mathcal{F}_n$ is a mapping $f : \{0,1\}^n \to \{0,1\}$. We assume that each function in $\mathcal{F}_n$ is represented using some fixed and reasonable *representation scheme*. A representation scheme for $\mathcal{F}_n$ is a pair $(\mathcal{R}_n, \mathcal{E}_n)$, where $\mathcal{R}_n \subseteq \{0,1\}^{r(n)}$ for some fixed polynomial $r(n)$, and $\mathcal{E}_n : \mathcal{R}_n \to \mathcal{F}_n$ is an onto mapping.

We interpret a string $\sigma \in \mathcal{R}_n$ as a *representation* of the function $\mathcal{E}_n(\sigma) \in \mathcal{F}_n$. Note that a function $f \in \mathcal{F}_n$ may have many representations in $\mathcal{R}_n$. Also, since $\mathcal{R}_n$ contains only $r(n)$-bit strings, we are insisting that each function in $\mathcal{F}_n$ have a "short" representation. We shall see that the computational details of *evaluating* the representations are relevant to our study. We define $\mathcal{F} = \{\mathcal{F}_n\}$ and $(\mathcal{R}, \mathcal{E}) = \{(\mathcal{R}_n, \mathcal{E}_n)\}$.

In our average-case learning models, the unknown target function will be generated according to some fixed distribution $\mathcal{P}_n$ over the function class $\mathcal{F}_n$ from the distribution *ensemble* $\mathcal{P} = \{\mathcal{P}_n\}$ over $\mathcal{F}$. When we have fixed a representation scheme $(\mathcal{R}, \mathcal{E})$ for $\mathcal{F}$, sometimes we instead prefer to think of $\mathcal{P}_n$ as a distribution over representations $\mathcal{R}_n$, which implicitly defines a distribution over $\mathcal{F}_n$.

Once a target function $f \in \mathcal{F}_n$ is generated according to $\mathcal{P}_n$ (which we shall denote $f \in \mathcal{P}_n$), a learning algorithm will receive access to labeled examples of f selected according to some fixed distribution $\mathcal{D}_n$ over the input space $\{0,1\}^n$ from the distribution ensemble $\mathcal{D} = \{\mathcal{D}_n\}$. Each example is a pair $\langle x, f(x)\rangle$ where x is drawn randomly according to $\mathcal{D}_n$ (denoted $x \in \mathcal{D}_n$). If $S = x_1, \ldots, x_m$ is a sequence of inputs from $\{0,1\}^n$, we use $\langle S, f\rangle$ to denote the sequence $\langle x_1, f(x_1)\rangle, \ldots, \langle x_m, f(x_m)\rangle$ of labeled examples of f.

Definition 1. Let $\mathcal{F}$ be a class of boolean functions, $\mathcal{P}$ a distribution ensemble over $\mathcal{F}$, and $\mathcal{D}$ a distribution ensemble over $\{0,1\}^*$. For any $0 < \epsilon < 1/2$, we say that $\mathcal{F}$ is *ϵ-predictable on average with respect to $\mathcal{P}$ and $\mathcal{D}$* if there exists a polynomial time algorithm M (taking a labeled sample $\langle S, f\rangle$ and a test input $\tilde{x}$) and a polynomial $m(n)$ such that for infinitely many n,

$$\Pr_{f\in\mathcal{P}_n, S\in\mathcal{D}_n^{m(n)}, \tilde{x}\in\mathcal{D}_n}[M(\langle S, f\rangle, \tilde{x}) = f(\tilde{x})] \geq 1 - \epsilon.$$

We call M an *ϵ-prediction algorithm* (for $\mathcal{F}$ with respect to $\mathcal{P}$ and $\mathcal{D}$), and the function $m(n)$ is the *sample size* of M.

Note that in this definition, a smaller value of ϵ places a greater demand on the learning algorithm. We will want in particular to consider two extreme cases of predictability as follows:

Definition 2. Let $\mathcal{F}$ be a class of boolean functions, $\mathcal{P}$ a distribution ensemble over $\mathcal{F}$, and $\mathcal{D}$ a distribution ensemble over $\{0,1\}^*$. We say that $\mathcal{F}$ is *weakly predictable on average with respect to $\mathcal{P}$ and $\mathcal{D}$* if there exists some fixed polynomial $q(n)$ such that $\mathcal{F}$ is $(1/2 - 1/q(n))$-predictable on average with respect to $\mathcal{P}$ and $\mathcal{D}$. We say that $\mathcal{F}$ is *strongly predictable on average with respect to $\mathcal{P}$ and $\mathcal{D}$* if for any polynomial $q(n)$, $\mathcal{F}$ is $1/q(n)$-predictable on average with respect to $\mathcal{P}$ and $\mathcal{D}$.

We will also consider these same learning models when the learning algorithm is provided with *membership queries*. Here the definitions of prediction on average remain unchanged, but in addition to random labeled examples, the learning algorithm may receive the value $f(x)$ on inputs x of its own choosing; the test input $\tilde{x}$ is drawn after all queries are made to prevent the algorithm from cheating by querying $f(\tilde{x})$.

2.2 Measuring the Complexity of Representation Schemes and Distributions

In order to quantify the complexity of our proposed cryptographic primitives, we need to define complexity measures for the representation scheme $(\mathcal{R}, \mathcal{E})$ and the distribution ensembles $\mathcal{P}$ and $\mathcal{D}$. As we have mentioned, we use uniform circuit families to allow the most precise statements and to emphasize the preservation of parallelism in our constructions. We begin with the straightforward case of the input distribution ensemble $\mathcal{D}$.

Definition 3. Let $\mathcal{D}$ be a distribution ensemble over $\{0,1\}^*$, and let $D = \{D_n\}$ be a uniform circuit sequence, where D_n takes $d(n)$ input bits for some polynomial $d(n)$ and outputs n bits. We say that $\mathcal{D}$ is *generated by* D if for every n, the output distribution of D_n is exactly $\mathcal{D}_n$: that is, if we choose w uniformly at random from $\{0,1\}^{d(n)}$ then $D_n(w) \in \{0,1\}^n$ is distributed according to $\mathcal{D}_n$.

We wish to formulate a similar notion for the generation of the distribution $\mathcal{P}$. For our cryptographic constructions, given a fixed representation scheme $(\mathcal{R}, \mathcal{E})$, it will be easier to think of the function distribution $\mathcal{P}_n$ as being over the set $\mathcal{R}_n$ (rather than over function class $\mathcal{F}_n$ itself), which then implicitly defines a distribution over $\mathcal{F}_n$ under $\mathcal{E}_n$. In this case we can define what it means for $\mathcal{P}$ to be generated by a circuit sequence.

Definition 4. Let $\mathcal{P}$ be a distribution ensemble over $\mathcal{R}_n \subseteq \{0,1\}^{r(n)}$, and let $P = \{P_n\}$ be a uniform circuit sequence, where P_n takes $p(n)$ input bits and outputs $r(n)$ bits for polynomials $p(n)$ and $r(n)$. We say that $\mathcal{P}$ is *generated by* P if for every n, the output distribution of P_n is exactly $\mathcal{P}_n$: that is, if we choose w uniformly at random from $\{0,1\}^{p(n)}$ then $P_n(w) \in \mathcal{R}_n$ is distributed according to $\mathcal{P}_n$.

A distribution $\mathcal{P}$ over $\mathcal{R}$ that allows very rapid generation of function representations will not be especially useful to us if the representation of f makes it expensive to compute $f(x)$. Thus, we make the following definition:

Definition 5. Let $(\mathcal{R}, \mathcal{E})$ be an encoding scheme for $\mathcal{F}$ such that $\mathcal{R}_n \subseteq \{0,1\}^{r(n)}$, and let $E = \{E_n\}$ be a uniform circuit sequence, where E_n takes $r(n) + n$ input bits and outputs a single bit. We say that $(\mathcal{R}, \mathcal{E})$ can be *evaluated by* E if for every n, on inputs $\sigma_f \in \mathcal{R}_n$ such that $f = \mathcal{E}_n(\sigma_f) \in \mathcal{F}_n$ and $x \in \{0,1\}^n$, we have $E_n(\sigma_f, x) = f(x)$.

Finally, we will need to define circuit sequences that are formed from other circuit sequences in certain ways. If $C = \{C_n\}$ is a circuit sequence, we define $C^m = \{C_n^m\}$ to be the sequence of m-fold replication of the circuits in C. More precisely, the circuit C_n^m takes $m \cdot n$ inputs, and consists of m disjoint "copies" of C_n: on inputs $x_1, \ldots, x_m \in \{0,1\}^n$, the output $C_n^m(x_1, \ldots, x_m)$ is the concatenation of $C_n(x_1), \ldots, C_n(x_m)$.

If C and D are sequences of circuits, we define the sequence $C \circ D = \{C_n \circ D_n\}$ as follows: the circuit $C_n \circ D_n$ has $q(n)$ inputs for some polynomial $q(n)$. Some or all of these inputs feed the fixed circuit D_n, whose outputs (along with some or all of the inputs) in turn feed the fixed circuit C_n. (Note that $C_n \circ D_n$ is technically a *set* of circuits since we have not specified exactly how the inputs are wired.) Similarly, if C,D and E are sequences of circuits, the sequence $C \circ (D, E)$ is that in which the circuit $C_n \circ (D_n, E_n)$ has some or all of its inputs fed to the fixed circuits D_n and E_n in parallel, whose outputs (along with some or all of the inputs) in turn feed the fixed circuit C_n. So, replication increases the circuit width, and the composition operation $\circ$ increases circuit depth.

2.3 Discussion

Several widely studied function classes, such as DNF formulas, are believed to be hard to learn even when the distribution $\mathcal{D}$ is uniform. There has been less *formal* work studying what hard function distributions look like because distribution $\mathcal{P}$ is not used in the standard models. The following, however, is a distribution on DNF formulas that seems to defy all known methods of attack, and we believe that any method that could even weakly predict such functions over a uniform $\mathcal{D}$ would require profoundly new ideas.

Select at random two disjoint sets $A, B \subset \{1, \ldots, n\}$ each of size $\log n$. On input x, compute the parity of the bits indexed by A and the majority function of the bits indexed by B, and output the exclusive-or of the two results. This function can be represented by a polynomial size DNF formula (or decision tree) since the truth table has only $2^{2 \log n} = n^2$ entries. Because this $\mathcal{P}$ distribution seems hard even to weakly predict over uniform $\mathcal{D}$, it could be used for our bit generator in Section 3.1 (though learning becomes easy if membership queries are allowed, so the generator of Section 3.2 does not apply).

3 General Results

3.1 A Bit Generator Based on Hardness for Weak Prediction

We begin by showing that a function class hard to weakly predict on average can be used to create a CSPRBG whose circuit depth is comparable to that of the function class plus that of generating the hard distributions.

Definition 6. A *cryptographically strong pseudorandom bit generator (CSPRBG)* is a uniform circuit sequence $\mathcal{G} = \{\mathcal{G}_n\}$, where $\mathcal{G}_n$ takes n bits of input and produces $g(n) > n$ bits of output, with the following property: for any polynomial time algorithm T that produces a boolean output and any polynomial $q(n)$, there exists an n_0 such that for all $n \geq n_0$,

$$\left| \Pr_{y \in \{0,1\}^{g(n)}} [T(y) = 1] - \Pr_{x \in \{0,1\}^n} [T(\mathcal{G}_n(x)) = 1] \right| \leq 1/q(n).$$

We call the function $e(n) = g(n) - n$ the *expansion* of $\mathcal{G}$.

We first consider the special case in which the input distribution ensemble is uniform.

Theorem 7. *Let $\mathcal{F}$ be a class of boolean functions, $(\mathcal{R}, \mathcal{E})$ a representation scheme for $\mathcal{F}$, $\mathcal{P}$ a distribution ensemble over $\mathcal{R}$, and $\mathcal{U}$ the uniform distribution ensemble over $\{0,1\}^*$. Let $(\mathcal{R}, \mathcal{E})$ be evaluated by the circuit sequence E,* *and let $\mathcal{P}$ be generated by the uniform circuit sequence P. Then if $\mathcal{F}$ is not weakly predictable on average with respect to $\mathcal{P}$ and $\mathcal{U}$, there is a CSPRBG computed by the uniform circuit sequence $E^{m(n)} \circ P$ for some fixed polynomial $m(n)$.*

Proof. (Sketch) Informally, for any n, the bit generator $\mathcal{G}_n$ behaves as follows: it takes as input a random bit string, and uses some of these bits to generate a function $f \in \mathcal{F}_n$ according to the distribution $\mathcal{P}_n$. The rest of the input bits are used directly as m inputs, $x_1, \ldots, x_m \in \{0,1\}^n$ for f. The output of the generator consists of $x_1, \ldots, x_m$ followed by the m bits $f(x_1), \ldots, f(x_m)$.

More formally, $\mathcal{G}_n$ takes as input a random string of $p(n) + m \cdot n$ bits. Here $p(n)$ is the number of random input bits required by the circuit P_n for generating a representation $\sigma_f \in \mathcal{R}_n$ of a function $f = \mathcal{E}_n(\sigma_f) \in \mathcal{F}_n$, where σ_f is distributed according to $\mathcal{P}$; m will be determined by the analysis. The generator feeds the first $p(n)$ input bits into P_n to obtain σ_f. The remaining $m \cdot n$ random input bits are regarded as m random vectors $x_1, \cdots, x_m \in \{0,1\}^n$. For each i the generator then feeds σ_f and x_i to a parallel copy of E_n to obtain $E_n(\sigma_f, x_i) = f(x_i)$. The output of the generator is then $x_1, \ldots, x_m, f(x_1), \ldots, f(x_m)$. It is easy to verify that $\mathcal{G}$ is computed by the circuit sequence $E^m \circ P$. Since our generator produces $g(n) = m \cdot n + m$ output bits, we obtain expansion provided that $m > p(n)$.

We now argue that $\mathcal{G}$ is in fact a CSPRBG. For contradiction, suppose that $\mathcal{G}$ is not, and let T be a polynomial time algorithm such that

$$\left| \Pr_{y \in \{0,1\}^{g(n)}} [T(y) = 1] - \Pr_{x \in \{0,1\}^n} [T(\mathcal{G}(x)) = 1] \right| \geq 1/q(n)$$

for some polynomial $q(n)$. For each i, let t_i denote the probability that T outputs 1 when its first i input bits are the first i bits of $\mathcal{G}(x)$ on random x, and the remaining input bits of T are truly random. Then we have $|t_0 - t_{g(n)}| \geq 1/q(n)$, and by a standard "probability walk" argument there must be an $1 \leq i < g(n)$ such that $|t_i - t_{i+1}| \geq 1/(n \cdot q(n))$ (note that in fact i must be larger than $m \cdot n$ since the first $m \cdot n$ bits of $\mathcal{G}_n$'s output are in fact truly random, having simply been copied from the input). Furthermore, we can find such an i (with high probability) by performing repeated experiments with T using random draws from $\mathcal{P}_n$ and $\mathcal{D}_n$, and once such an i is found we can "center the bias" to produce an efficient algorithm T' such that $t'_i \geq 1/2 + 1/q'(n)$ and $t'_{i+1} \leq 1/2 - 1/q'(n)$ for some polynomial $q(n)$.

Algorithm T' thus has the following property: suppose we draw a function f randomly according to $\mathcal{P}_n$, and draw m random n-bit vectors $x_1, \ldots, x_m$ and we give T' the inputs $x_1, \ldots, x_m$ along with $f(x_1), \ldots, f(x_i)$, followed by an input bit b_{i+1} that is *either* $f(x_{i+1})$ or a truly random bit, followed by $g(n) - i - 1$ random bits. Then T' can determine with probability significantly better than random guessing whether its $i + 1$st input bit b_{i+1} is $f(x_{i+1})$ or a truly random bit; we interpret an output of 1 as a guess that b_{i+1} is random, and an output of 0 as a guess that $b_{i+1} = f(x_{i+1})$.

Now suppose we have access to random examples according to $\mathcal{U}_n$ of a target function f drawn according to $\mathcal{P}_n$, and we also have a random test input $\tilde{x}$.

Suppose we give to T' the random strings $x_1, \ldots, x_i, \tilde{x}, x_{i+2}, \ldots, x_m$ followed by $f(x_1), \ldots, f(x_i)$, followed by an undetermined input bit b_{i+1}, followed by $g(n) - i - 1$ random bits. Then it is easy to show by a simple averaging argument that the following strategy yields an algorithm for weakly predicting $\mathcal{F}$ with respect to $\mathcal{P}$ and $\mathcal{U}$: with the other inputs as specified, we run T' both with $b_{i+1} = 0$ and $b_{i+1} = 1$. If both inputs cause an output of 0, or both inputs cause an output of 1, we flip a coin to predict $f(\tilde{x})$. Otherwise, we predict that $f(\tilde{x})$ is the value of b_{i+1} that caused T' to output 0. □

The following simple lemma relates the hardness of learning a function class with respect to *some* input distribution ensemble to the hardness of learning a related function class with respect to the uniform distribution ensemble.

Lemma 8. *Let $\mathcal{F}$ be a class of boolean functions, $\mathcal{P}$ a distribution ensemble over $\mathcal{F}$, and $\mathcal{D}$ a distribution ensemble over $\{0,1\}^*$. Let the ensemble $\mathcal{D}$ be generated by the circuit sequence D, and let $\mathcal{F} \circ D$ denote the class of functions obtainable by composing a function in $\mathcal{F}_n$ with the circuit D_n. Then for any ϵ, if $\mathcal{F}$ is not ϵ-predictable on average with respect to $\mathcal{P}$ and $\mathcal{D}$, then $\mathcal{F} \circ D$ is not ϵ-predictable on average with respect to $\mathcal{P}$ and the uniform ensemble $\mathcal{U}$ over $\{0,1\}^*$.*

Proof. Immediate; we are simply letting the computation of the hard distribution ensemble $\mathcal{D}$ be part of the target function. □

From Theorem 7 and Lemma 8, we can now easily obtain a bit generator from a learning problem that is hard with respect to some input distribution ensemble.

Corollary 9. *Let $\mathcal{F}$ be a class of boolean functions, $(\mathcal{R}, \mathcal{E})$ a representation scheme for $\mathcal{F}$, $\mathcal{P}$ a distribution ensemble over $\mathcal{R}$, and $\mathcal{D}$ a distribution ensemble over $\{0,1\}^*$. Let $(\mathcal{R}, \mathcal{E})$ be evaluated by the circuit sequence E, let $\mathcal{P}$ be uniformly generated by the circuit sequence P, and let $\mathcal{D}$ be uniformly generated by the circuit sequence D. Then if $\mathcal{F}$ is not weakly predictable on average with respect to $\mathcal{P}$ and $\mathcal{D}$, there is a CSPRBG computed by the uniform circuit sequence $(E \circ D)^{m(n)} \circ P$ for some fixed polynomial $m(n)$.*

3.2 Improved Expansion via Nisan-Wigderson

The pseudorandom generator just described takes $p(n) + m \cdot n$ truly random input bits to $m \cdot n + m$ output bits, giving expansion $e(n) = m - p(n)$. While we can let $e(n)$ attain any desired value by choosing $m = m(n)$ as large as necessary, the *expansion ratio* (the number of output bits divided by the number of input bits) will not exceed $1 + 1/n$. There are standard methods which can be used to amplify the expansion of any generator. However, these methods iterate the generator and therefore significantly increase the resulting circuit depth.

By applying a result due to Nisan and Wigderson [12] we can improve our generator to obtain a much greater expansion ratio without a correspondingly large increase in circuit depth (or size). However, in order to prove security, we will need to assume that the class of functions is hard to weakly predict on average even when the learning algorithm is provided with membership queries. We now describe how this stronger intractability assumption allows us to modify our generator to obtain an expansion ratio on the order of n^2 rather than just $O(1)$ as before.

The new CSPRBG $\mathcal{G}_n$ takes as input a random string of $p(n) + n^2$ bits. As before, $p(n)$ is the number of random input bits required by the circuit P_n. (For simplicity we assume that the hard distribution $\mathcal{D}_n$ over the inputs $\{0,1\}^n$ is the uniform distribution; similar improvements can be given for the general case as was done above.) Call the additional n^2 input bits $v = v_1, \ldots, v_{n^2}$. It is described by Nisan and Wigderson [12] how to uniformly construct a family of n^4 sets $S_1, \ldots, S_{n^4}$ such that: (1) $S_i \subset \{v_1, \ldots, v_{n^2}\}$; (2) $|S_i| = n$ for all i; and (3) $|S_i \bigcap S_j| \leq \log n$ for all $i \neq j$.

Our new generator will work as follows: as before, the first $p(n)$ bits are used to generate a function $f \in \mathcal{F}_n$ according to the distribution $\mathcal{P}_n$, and the remaining n^2 bits v are copied to the first n^2 output bits. Now, however, if we let f_i denote the function f applied to the subset of $v_1, \ldots, v_{n^2}$ indicated by the set S_i, the (n^2+i)th output bit is $f_i(v)$. Thus, $\mathcal{G}_n$ takes $p(n)+n^2$ input bits and gives $g(n) = n^2 + n^4$ output bits, for an expansion ratio of $\Omega(n^2)$. Note that the sets S_i are fixed as part of the generator description and thus are known to any potential adversary.

We now sketch the argument that if $\mathcal{F}$ is not weakly predictable on average even with membership queries, then $\mathcal{G}_n$ is a CSPRBG. To see this, suppose the contrary that $\mathcal{G}_n$ is not a CSPRBG. By standard arguments, there is an $1 \leq i \leq n^4$ and a polynomial time algorithm T such that

$$\Pr_{f \in \mathcal{P}_n, v \in \{0,1\}^{n^2}} [T(v, f_1(v), \ldots, f_i(v)) = f_{i+1}(v)]$$

exceeds $1/2 + 1/q(n)$ for some polynomial $q(n)$.

The function f_{i+1} only depends on the bits in S_{i+1} which without loss of generality we call $v_1, \ldots, v_n$. By an averaging argument we can find a *fixed* setting $z = z_{n+1}, \ldots, z_{n^2} \in \{0,1\}^{n^2-n}$ of the remaining bits $v_{n+1}, \ldots, v_{n^2}$ such that

$$\Pr_{f \in \mathcal{P}_n, v \in \{0,1\}^{n}} [T(vz, f_1(v,z), \ldots, f_i(v,z)) = f_{i+1}(v,z)]$$

exceeds $1/2 + 1/q(n)$ for some polynomial $q(n)$.

Now we describe how a learner who can make membership queries can weakly predict $\mathcal{F}$ on average. By the Nisan-Wigderson construction, each f_j other than f_{i+1} is actually a function of only $\log n$ bits in $v_1, \ldots, v_n$ (all other bits have been fixed) and we know which $\log n$ bits since the sets S_i are fixed as part of the generator description. Thus the entire truth table of each f_j is of size polynomial in n, and can be determined by making only n membership queries to f (the queries simply let the variables in $S_j \bigcap \{v_1, \ldots, v_n\}$ assume all n possible settings while the remaining variables have their values fixed according to z). To predict $f(v)$ on a challenge input v the learner looks up the values of $f_j(vz)$ for $1 \leq j \leq i$ and then outputs $T(vz, f_1(vz), \ldots, f_i(vz))$. This weakly predicts $f_{i+1}(vz) = f(v)$, and by contradiction proves our assertion that $\mathcal{G}_n$ is a CSPRBG.

3.3 A One-Way Function Based on Hardness for Strong Prediction

We now show that under the weaker assumption that a class of functions is hard to strongly predict on average, we can construct a one-way function whose circuit depth is comparable to that of the function class plus that of the hard distributions. A related result is given by Impagliazzo and Levin [6].

Definition 10. Let $F = \{F_n\}$ be a uniform sequence of circuits $F_n : \{0,1\}^n \to \{0,1\}^{s(n)}$ for some polynomial $s(n)$. We say that F is a *one-way function* if there exists a polynomial $q(n)$ such that for any polynomial time algorithm T, there exists an n_0 such that for all $n \geq n_0$,

$$\Pr_{x\in\{0,1\}^n}[F_n(T(F_n(x))) \neq F_n(x)] \geq 1/q(n).$$

Theorem 11. *Let $\mathcal{F}$ be a class of boolean functions, $(\mathcal{R},\mathcal{E})$ a representation scheme for $\mathcal{F}$ ($\mathcal{R}_n \subseteq \{0,1\}^{r(n)}$), $\mathcal{P}$ a distribution ensemble over $\mathcal{R}$, and $\mathcal{D}$ a distribution ensemble over $\{0,1\}^*$. Let $(\mathcal{R},\mathcal{E})$ be evaluated by the circuit sequence E, $\mathcal{P}$ be generated by the circuit sequence P, and $\mathcal{D}$ be generated by the circuit sequence D. Then if $\mathcal{F}$ is not strongly predictable on average with respect to $\mathcal{P}$ and $\mathcal{D}$, there is a one-way function F computed by the uniform circuit sequence $E^{m(n)} \circ (P, D^{m(n)})$ for some fixed polynomial $m(n)$.*

Proof. (Sketch) The construction is similar to that of the bit generator; the main differences are the polynomial $m(n)$ and the analysis. The input to the one-way function F_n will consist of $p(n)+m\cdot d(n)$ bits, where $p(n)$ is the number of inputs to P_n and $d(n)$ is the number of inputs to D_n. The first $p(n)$ bits are fed to P_n to produce a representation $\sigma_f \in \mathcal{R}_n$ of a function $f \in \mathcal{F}_n$. The remaining $m \cdot d(n)$ input bits are regarded as m blocks $w_1, \ldots, w_m \in \{0,1\}^{d(n)}$. Each w_i is fed to a parallel copy of D_n in order to produce an $x_i \in \{0,1\}^n$; if the w_i are selected randomly, then the x_i are distributed according to $\mathcal{D}_n$. Finally, each x_i is given along with σ_f to a parallel copy of E_n in order to obtain $E_n(\sigma_f, x_i) = f(x_i)$. The output of F_n is then $x_1, \ldots, x_m$ followed by $f(x_1), \ldots, f(x_m)$.

We begin the analysis by noting that if $v, v' \in \{0,1\}^{p(n)}$ and $w_1, \ldots, w_m$, $w'_1, \ldots, w'_m \in \{0,1\}^{d(n)}$ are such that

$$F_n(v, w_1, \ldots, w_m) = F_n(v', w'_1, \ldots, w'_m)$$

we must have $D_n(w_i) = D_n(w'_i)$ for all i. Let $x_i = D_n(w_i)$. Now although it is not necessarily true that $v = v'$, if we let $\sigma_f = P_n(v)$ and $\sigma_{f'} = P_n(v')$ be the representations in $\mathcal{R}_n$ of the functions $f, f' \in \mathcal{F}_n$, by construction of F_n it must be the case that $f(x_i) = f'(x_i)$ for all i.

Let $q(n)$ be such that $\mathcal{F}$ is not $3/q(n)$-predictable (with respect to $\mathcal{P}$ and $\mathcal{D}$). For any fixed $f \in \mathcal{F}_n$, the probability over m random examples from $\mathcal{D}_n$ that there exists a function $f' \in \mathcal{F}_n$ agreeing with f on those examples, but that has error greater than $1/q(n)$ with respect to f and $\mathcal{D}$, is at most $|\mathcal{F}_n|(1-1/q(n))^m$. This probability is smaller than $1/q(n)$ for $m = \Omega(q(n)[\log|\mathcal{F}_n| + \log q(n)])$, (which is polynomial in n since $\log|\mathcal{F}_n| \leq r(n)$). We set m such that this is the case.

Thus, suppose that F is not a one-way function. Then there exists an algorithm T with probability at least $1-1/q(n)$ of finding an inverse for $F_n(v, w_1, \ldots, w_m)$ on random inputs. So given a test input $\tilde{x} \in \mathcal{D}_n$ and m examples $\langle x_i, f(x_i)\rangle$ of f drawn according to $\mathcal{D}_n$, we can simply give the string $x_1, \ldots, x_m, f(x_1), \ldots, f(x_m)$ to T. There is then probability at least $1-1/q(n)$ that T returns $v', w'_1, \ldots, w'_m$ such that $F_n(v', w'_1, \ldots, w'_m) = x_1, \ldots, x_m, f(x_1), \ldots, f(x_m)$. If this occurs, then by the argument above there is probability at least $1 - 1/q(n)$ that $P_n(v')$ represents a function f' with error at most $1/q(n)$, and we can simply compute $f'(\tilde{x})$. The probability that $f'(\tilde{x}) = f(\tilde{x})$ is at least $1 - 3/q(n)$, violating the assumption that $\mathcal{F}$ is not $3/q(n)$-predictable. □

3.4 A Private-Key Cryptosystem Based on Hardness for Weak Prediction

In section we describe a simple and natural mapping from hard learning problems to private-key cryptosystems. This mapping has the property that if a simple function representation (for instance, DNF formulas) is hard to learn over a simple distribution (for instance, uniform) then encrypting and decrypting are both easy. Define the function $E_{n,\sigma}(x) = E_n(\sigma, x)$. For example, if $(\mathcal{R}, \mathcal{E})$ is a representation for DNF formulas, then E_n is a circuit that takes a string representing a DNF formula f and some x, and produces $f(x)$ as output. $E_{n,\sigma}$, however, is just a depth-2 circuit.

A *private-key cryptosystem* is a tuple $(\mathbf{G}, \mathbf{E}, \mathbf{D})$ of three probabilistic polynomial time algorithms. [4] The key generator $\mathbf{G}$ takes as input 1^n, and outputs a key k of length n. The encryption algorithm $\mathbf{E}$ takes as input a message m and a key k, and produces ciphertext as output. The decryption algorithm $\mathbf{D}$ takes as input ciphertext and a key and produces a message. We require that $\mathbf{D}(\mathbf{E}(m, k), k) = m$.

The private-key cryptosystems we will discuss are probabilistic schemes that encrypt one bit at a time, encrypting each bit independently from the previous ones, as in [5]. We use standard notions of chosen plaintext and/or ciphertext attack. Since we encrypt bits independently, we may say a system is secure if for any polynomial time algorithm T, polynomial $q(n)$, and sufficiently large n, the probability that after performing its allowed type of attack T correctly guesses the decryption of an encrypted random bit is less than $1/2 + 1/q(n)$.

Suppose $\mathcal{F}$ is a class of boolean functions, $(\mathcal{R}, \mathcal{E})$ a representation scheme for $\mathcal{F}$, $\mathcal{P}$ a distribution ensemble over $\mathcal{R}$, and $\mathcal{D}$ a distribution ensemble over $\{0,1\}^*$. In addition, suppose $(\mathcal{R}, \mathcal{E})$ is evaluated by the uniform circuit sequence E, and $\mathcal{P}$ and $\mathcal{D}$ are generated by the uniform circuit sequences P and D respectively. Assume that the probability a random example from $\mathcal{D}_n$ is positive for a random

[4] Here we depart from our policy of describing primitives as uniform circuit families since we intend to describe the private-key system informally.

function f from $\mathcal{P}_n$ is in the range $[\frac{1}{2} - \frac{1}{n}, \frac{1}{2} + \frac{1}{n}]$ (if this is not the case then $\mathcal{F}$ is weakly predictable).

The cryptosystem based on $\mathcal{F}, \mathcal{P}$, and $\mathcal{D}$ is as follows. The key generator **G** uses input 1^n to generate P_n. **G** then feeds $p(n)$ random bits into P_n to produce a string $\sigma \in \mathcal{R}_n$. String σ is the key given to the encryption and decryption algorithms (so, technically, the security parameter is $r(n)$). Let f be the function represented by σ.

The encryption algorithm **E** begins by generating circuits D_n and E_n, and evaluating E_n on the private key to create $E_{n,\sigma}$. It encrypts a 1 by sending a random (according to $\mathcal{D}$) positive example of f and encrypts a 0 by sending a random (according to $\mathcal{D}$) negative example of f. This requires running D_n on $d(n)$ random bits to create an example x, and then computing $E_{n,\sigma}(x)$ to see if the example is of the appropriate type, repeating the procedure if it is not. Notice that the expected number of calls to D_n and $E_{n,\sigma}$ is just $2 + o(1)$ by our assumption on $\mathcal{D}$. (This could be improved by encrypting many bits at a time).

Decryption is even simpler than encryption. The decryption algorithm **D** begins by generating $E_{n,\sigma}$, and then decrypts strings x by computing $E_{n,\sigma}(x)$.

Theorem 12. *If $\mathcal{F}$ is not weakly predictable on average with respect to $\mathcal{P}$ and $\mathcal{D}$, then the cryptosystem $(\mathbf{G}, \mathbf{E}, \mathbf{D})$ described above is secure against chosen plaintext attack. If furthermore $\mathcal{F}$ is not weakly predictable with membership queries with respect to $\mathcal{P}$ and $\mathcal{D}$, then $(\mathbf{G}, \mathbf{E}, \mathbf{D})$ is secure against chosen plaintext and ciphertext attack.*

Proof. (Sketch) Suppose algorithm M can break $(\mathbf{G}, \mathbf{E}, \mathbf{D})$ with chosen plaintext attack for infinitely many n, and asks for the encryption of $m(n)$ plaintext bits. The learning algorithm simply requests $3m(n)$ labeled examples (which with high probability will result in at least $m(n)$ positive examples and at least $m(n)$ negative examples) and uses them to simulate **E** for M's plaintext queries. It then feeds the test input $\tilde{x}$ to M and uses M's response as the prediction. Since $r(n)$ is polynomial in n, and M responds correctly with probability at least $1/2 + 1/poly(r(n))$, this will be a weak prediction algorithm (for $\mathcal{F}$ with respect to $\mathcal{P}$ and $\mathcal{D}$).

If M makes chosen ciphertext queries, the learning algorithm, if it is allowed membership queries, can answer these in the obvious way as membership and ciphertext queries are equivalent here. Thus, if $(\mathbf{G}, \mathbf{E}, \mathbf{D})$ is vulnerable to chosen plaintext and ciphertext attack, then $\mathcal{F}$ is weakly predictable with membership queries with respect to $\mathcal{P}$ and $\mathcal{D}$. □

4 A Bit Generator Based on Parity Functions with Noise

Let $\mathcal{S}_n$ denote the set of all parity functions over $\{0,1\}^n$; specifically, for each of the 2^n subsets $S = \{x_{i_1}, \ldots, x_{i_k}\} \subseteq \{x_1, \ldots, x_n\}$ there is a function $f_S \in \mathcal{S}_n$ defined by

$$f_S(x_1, \ldots, x_n) = x_{i_1} \oplus \cdots \oplus x_{i_k}.$$

In this section we describe a simple bit generator whose security is based on the assumption that parity functions are hard to learn *in the presence of classification noise*. This means that we add a parameter $0 < \eta < 1/2$ to our learning model called the *noise rate*, and now a learning algorithm, rather than always receiving a labeled example $\langle x, f(x) \rangle$ of the target function f, will instead receive a *noisy* labeled example $\langle x, \ell \rangle$. Here $\ell = f(x)$ with probability $1 - \eta$ and $\ell = \neg f(x)$ with probability η where this choice is made independently for each requested example. Our specific hardness assumption is as follows.

The Parity Assumption: For some fixed constant $0 < \eta < 1/2$, there is *no* algorithm taking δ and n as input that runs in time polynomial in $1/\delta$ and n, and for infinitely many n does the following: Given access to noisy examples from the uniform distribution $\mathcal{U}_n$ on $\{0,1\}^n$ classified by an arbitrary $f_S \in \mathcal{S}_n$ with noise rate η, it produces the set S with probability at least $1 - \delta$.

Note that in comparison with our definitions in earlier sections, we have decreased the strength of our assumption in a number of ways. First, on those values of n for which a learning algorithm "succeeds", it must succeed for *every* function in $\mathcal{S}_n$. We have thus eliminated the distribution $\mathcal{P}_n$. Second, the learning algorithm must now actually find the target concept with arbitrarily high confidence. Third, the learning algorithm must learn with a possibly large rate of noise in the labels.

The parity assumption is closely related to the problem of efficiently decoding random linear codes, which is a long-standing open problem. It is known that the problem of finding the parity function that minimizes the number of disagreements with an input set of labeled examples is $\mathcal{NP}$-hard [2] (and easy to show it is MAX-SNP hard), and this optimization problem was used by McEliece [11] as the core of a proposed public-key cryptosystem with informal security arguments in which the matrix of examples must be carefully chosen. Recent results [8] provide some evidence in favor of the parity assumption by proving that parity functions cannot be learned using a certain class of statistical algorithms that include all known noise-tolerant learning algorithms in the Valiant model.

Based on the (unproven) parity assumption, we propose a simple pseudorandom bit generator. This generator is quite similar to a proposed one-way function due to Goldreich, Krawczyk and Luby [3], who then obtain a generator by running the one-way function through a generic transformation. Essentially, our contribution is to prove that the output of this one-way function is already pseudorandom. This stronger assertion is apparently already known to some researchers in the cryptography community as a "folk theorem".

The input seed to the generator $\mathcal{G}$ consists of $s(n) = n + m \cdot n + \mathcal{H}(\eta) \cdot m$ random bits, regarded as a block s_f of n bits, followed by m blocks $x_1, \ldots, x_m$ of n bits each, followed by a block s_r of $\mathcal{H}(\eta) \cdot m$ bits. Here m will be determined by the analysis, and $\mathcal{H}(\eta)$ is the binary entropy of the noise rate η in the parity assumption.

The block s_f encodes a parity function $f_S \in \mathcal{S}_n$; each 1 in s_f indicates a variable included in the subset S. As in our previous generator, $x_1, \ldots, x_m$ are regarded as inputs to f_S. The block s_r encodes a *longer* m-bit noise vector s'_r, where s'_r has exactly $\lfloor \eta \cdot m \rfloor$ 1's. Such an encoding can be shown to require s_r to be of length only $\mathcal{H}(\eta) \cdot m$, and this encoding can be done in a number of standard ways. Thus, if we randomly choose s_r among all $\mathcal{H}(\eta) \cdot m$-bit vectors, then we randomly choose s'_r from among all m-bit vectors with exactly $\lfloor \eta \cdot m \rfloor$ 1's.

$\mathcal{G}$ works as follows: it first uses s_f to obtain the represented parity function f_S. It then expands s_r to obtain s'_r. It next computes $f(x_1), \ldots, f(x_m)$, and for each i lets $\ell_i = f(x_i)$ if the ith bit of s'_r is 0, and lets $\ell_i = \neg f(x_i)$ if the ith bit of s'_r is 1. The output of $\mathcal{G}$ is $x_1, \ldots, x_m$, followed by $\ell_1, \ldots, \ell_m$. Note that the output of $\mathcal{G}$ is a noisy sample of f_S, but where the number of noisy labels is *exactly* $\lfloor \eta \cdot m \rfloor$, rather than determined by m flips of a coin of bias η as in the parity assumption; this discrepancy will be dealt with in the proof of the coming theorem.

$\mathcal{G}$ takes $s(n) = n + m \cdot n + \mathcal{H}(\eta) \cdot m$ random bits as input, and outputs $t(n) = m \cdot n + m$ bits, for expansion $t(n) - s(n) = m(1 - \mathcal{H}(\eta)) - n$. Thus we get expansion provided we choose $m > (1/(1 - \mathcal{H}(\eta)) \cdot n$, which can be done so long as $\eta < 1/2$.

Theorem 13. *Under the parity assumption, $\mathcal{G}$ is a pseudorandom bit generator.*

Proof. (Sketch) The overall strategy is to show that the parity assumption in fact implies the stronger assumption (the ***strong parity assumption***) that parity functions are not even weakly predictable on average (in the presence of noise rate η) with respect to the uniform distribution $\mathcal{P}_n$ over $\mathcal{S}_n$ and the uniform distribution $\mathcal{U}_n$ over $\{0,1\}^n$. The security of $\mathcal{G}$ can then be shown from arguments similar to that for the general bit generator outlined earlier. We must additionally show that the difficulty of learning with noise rate η implies the difficulty of learning when the m examples requested by the learning algorithm contain *exactly* $\lfloor \eta \cdot m \rfloor$ errors.

To see that the parity assumption implies the strong parity assumption, consider an algorithm A that contradicts the latter. To break the parity assumption, we first randomize S by XORing the labels given to examples with the classification given by some *random* parity function S'. Given such examples, A will perform noticeably better than guessing on a test example $\tilde{x}$. However, instead of giving to A these examples, we give to A examples in which we have replaced the ith bit x_i by a random bit in each. If x_i was irrelevant to the classification (not in $S \Delta S'$) the resulting distribution is indistinguishable from the original one, and A will still do better than random guessing. However, if $x_i \in S \Delta S'$, then there is now no correlation between the examples and the labels (the noise does not affect this) and so A cannot do better than random guessing. Thus, by repeating this process we can determine which variables are relevant to the classification and recover S, breaking the parity assumption.

Finally, we must address the fact that our generator is always injecting a fixed fraction of label errors rather than using a coin of bias η. Suppose that the learning problem became easy provided that the noise model always injected *exactly* $\lfloor \eta \cdot m \rfloor$ errors into any sample of size m requested by the learning algorithm.

We could then run this "fixed-fraction" algorithm many times on its requested sample size m, where the m examples come from a source with probability η of noise independently on each example. The probability that exactly $\eta \cdot m$ labels are noisy is certainly $\Omega(1/m)$ regardless of the value of η. So one of the runs of the fixed-fraction algorithm must succeed with high probability, and we can determine which run by hypothesis testing. □

Acknowledgements

We are grateful to Russell Impagliazzo and Steven Rudich for many insightful comments and suggestions on this research.

References

1. Dana Angluin and Michael Kharitonov. When won't membership queries help? In *Proceedings of the Twenty-Third Annual ACM Symposium on Theory of Computing*, pages 444–454, May 1991.
2. E. Berlekamp, R. McEliece, and H. van Tilborg. On the inherent intractability of certain coding problems. *IEEE Transactions on Information Theory*, 24, 1978.
3. O. Goldreich, H. Krawczyk, and M. Luby. On the existence of pseudorandom generators. In *29th Annual Symposium on Foundations of Computer Science*, pages 12–21, October 1988.
4. Oded Goldreich, Shafi Goldwasser, and Silvio Micali. How to construct random functions. *Journal of the ACM*, 33(4):792–807, October 1986.
5. Shafi Goldwasser and Silvio Micali. Probabilistic encryption. *JCSS*, 28(2):270–299, April 1984.
6. R. Impagliazzo and L. Levin. No better ways to generate hard NP instances than picking uniformly at random. In *31st Annual Symposium on Foundations of Computer Science*, October 1990.
7. R. Impagliazzo, L. Levin, and M. Luby. Pseudorandom generation from one-way functions. In *Proceedings of the Twenty First Annual ACM Symposium on Theory of Computing*, May 1989.
8. Michael Kearns. Efficient noise-tolerant learning from statistical queries. In *Proceedings of the Twenty-Fifth Annual ACM Symposium on the Theory of Computing*, May 1993.
9. Michael Kearns and Leslie G. Valiant. Cryptographic limitations on learning Boolean formulae and finite automata. In *Proceedings of the Twenty First Annual ACM Symposium on Theory of Computing*, pages 433–444, May 1989. To appear, *Journal of the Association for Computing Machinery*.
10. M. Kharitonov. Cryptographic hardness of distribution-specific learning. In *Proceedings of the Twenty-Fifth Annual ACM Symposium on the Theory of Computing*, May 1993.
11. R. J. McEliece. *A Public-Key System Based on Algebraic Coding Theory*, pages 114–116. Jet Propulsion Lab, 1978. DSN Progress Report 44.
12. N. Nisan and A. Wigderson. Hardness vs. randomness. In *29th Annual Symposium on Foundations of Computer Science*, pages 2–12, October 1988.
13. L. G. Valiant. A theory of the learnable. *Communications of the ACM*, 27(11):1134–1142, November 1984.
14. Andrew C. Yao. Theory and applications of trapdoor functions. In *23rd Annual Symposium on Foundations of Computer Science*, pages 80–91, 1982.

Extensions of Single-term Coins

Niels Ferguson

CWI, PO Box 94079 Amsterdam, The Netherlands. E-mail: niels@cwi.nl

Abstract. We show how the electronic cash scheme in [Fer93a] can be extended to provide n-spendable coins. Furthermore, we show how observers can be incorporated in the protocols to provide prior restraint against double spending by the user, instead of just detection after the fact.

1 Introduction

In [Fer93a, Fer93b] a coin system is presented that is an order of magnitude more efficient and simpler than earlier systems for electronic cash. In this paper we show two new extensions of this scheme. The first one is the construction of n-spendable coins (or n-show credentials) as opposed to the usual 1-spendable coins. These coins can be spent up to n times without the user being identified, but spending the coin a $(n+1)$'th time reveals the user's identity. The second extension is the incorporation of observers [CP93a] which allows prior restraint against double-spending while still maintaining all other properties of the coin system. Similar results for a different electronic cash scheme based on discrete logarithms are described in [Bra94, Bra93].

2 Multi-spendable coins

Under some circumstances it might be useful to allow the user to spend a specific coin several times. A 5-spendable coin might, for example, be used to represent a 5-trip subway ticket. The same effect can of course be achieved using five 1-spendable coins, but there are a few differences. First of all, a single 5-spendable coin requires less storage than five 1-spendable coins. On the other hand, the uses of the multi-spendable coin can easily be linked together by the Bank, so the unlinkability is lost. This makes n-spendable coins less useful for electronic cash applications, but for the subway fare the linkability does not provide any problems. It even has the slight advantage of allowing the subway company to gather statistical data on the use of the 5-trip fare.

2.1 Original payment protocol

The coins in [Fer93a] are based on two RSA signatures. The factorization of the RSA modulus is known only to the bank. The user Alice has three numbers C, A, and B of a special form and two signatures $(C^k A)^{1/v}$ and $(C^U B)^{1/v}$. Here, k is a random number known to the user, U is the user's identity and v is a prime

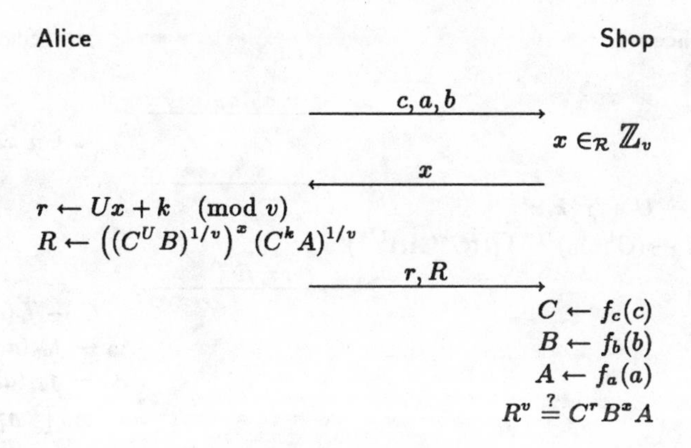

Fig. 1. Payment protocol for 1-spendable coin

large enough to make a birthday attack modulo v impossible (128 bits). The numbers C, A, and B are all images of oneway functions on the base numbers c, a and b. Thus $C = f_c(c)$, $A = f_a(a)$ and $B = f_b(b)$. The payment protocol for this coin system is shown in figure 1. When Alice wants to pay a coin to a shop, she first sends c, a and b. The shop then generates a random challenge x which it sends to the user. Finally, Alice sends the number $Ux + k \pmod v$ and the signature $(C^{Ux+k} B^x A)$, which she can easily construct from her two given signatures. The main aim of this protocol is to catch Alice if she spends the same coin twice. If Alice spends a coin only once, then only C, A, B, and $Ux+k$ are revealed. As nobody else has any knowledge about C, A, B or k (this is ensured by the withdrawal protocol), these four numbers do not identify her. If she spends the same coin twice, she will receive two different challenges with high probability. If she answers them both, then Alice's identity U can easily be determined from the two answers. Note that the computations in the exponents are done modulo v. In the above protocol Alice has to apply a correction factor to R (which is not shown) to get $(C^{(Ux+k) \bmod v} B^x A$ instead of $C^{Ux+k} B^x A$. This is accomplished by dividing R by a proper power of C. In the rest of this paper we will assume implicitly that all computations of exponents are done modulo v, and that the necessary corrections are applied to the resulting signatures.

2.2 Achieving n-spendability

We now convert these coins to n-spendable coins. The 1-spendable case uses a secret sharing line. We generalize this to use a higher degree polynomial to hide the identity U [Sha79]. For an n-spendable coin Alice stores $n+2$ numbers of a special form: C, A_0, ...,A_n. She also receives $n+1$ signatures during the withdrawal protocol: $(C^U A_0)^{1/v}$, $(C^{k_1} A_1)^{1/v}$, ...,$(C^{k_n} A_n)^{1/v}$. The modified

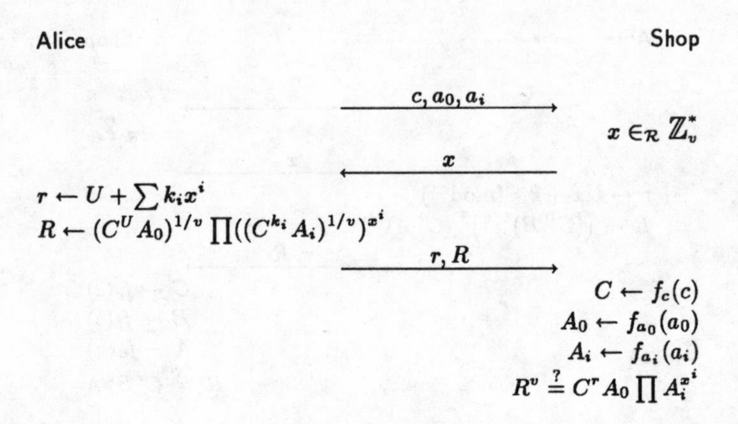

Fig. 2. Payment protocol for n-spendable coin.

payment protocol is shown in figure 2. In this figure all occurrences of i are assumed to be over the range $1, \ldots, n$. Alice starts by sending c, $a_0, \ldots, a_n$ to the shop. The shop replies as before with a random challenge x. Finally Alice sends a point on the polynomial back to the shop, together with an RSA signature that proves it is the correct point. i.e. she sends $r := U + \sum_{i=1}^{n} k_i x^i \pmod{v}$ and the signature $(C^r A_0 \prod_{i=1}^{n} A_i^{x^i})^{1/v}$.

If Alice spends this coin l times, then she must reveal l points on the polynomial $\sum_{i=1}^{n} k_i x^i + U$. As long as $l \leq n$ this does not reveal any information about U (assuming that the challenge $x = 0$ is excluded). As soon as Alice reveals $n+1$ points on the polynomial, the entire polynomial can easily be constructed, thereby revealing her identity U.

As always, the payment protocol is the simple part of a coin system. The withdrawal protocol is much more complicated. We show how the withdrawal protocol from [Fer93a] can be modified in a fairly straightforward manner for the n-spendable case (figure 3). It becomes the original withdrawal protocol if $n = 1$ is substituted. (In that case, A_0 takes the function of B and A_1 takes the function of A when compared with figure 1.) Again, all occurrences of i should be read as running over the range 1 to n. For a description of the workings of this protocol the reader is referred to [Fer93a] or [Fer93b]. The numbers C, A_0 and A_i are of the form $f_c(c) := c g_c^{f(h_c^c)}$, $f_{a_0}(a_0) := a_0 g_{a_0}^{f(h_a^{a_0})}$ and $f_{a_i}(a_i) := a_i g_{a_i}^{f(a_i)}$ respectively, where $f()$ is a suitable oneway function and the g's are publicly known elements of large order in the multiplicative RSA group. The numbers h_c and h_a are elements of order ω in $\mathbb{Z}_p^*$ where ω is the RSA-modulus and p is a prime with $p \bmod \omega = 1$.

The withdrawal protocol in figure 3 provides unconditional unlinkability between coins. For every transcript that the Bank gets from a withdrawal protocol

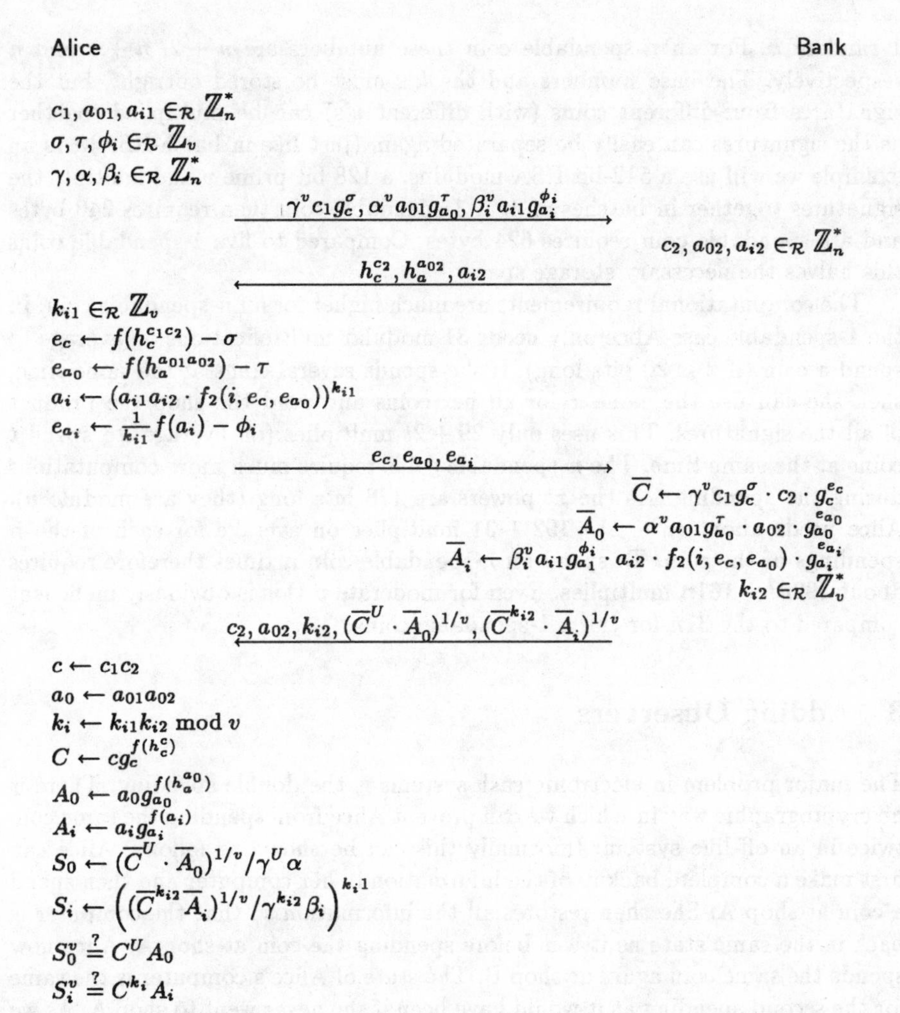

Fig. 3. n-spendable coin withdrawal protocol

and for every legal coin, there is exactly one possible set of choices out of the random choices that Alice can make that would result in her getting that specific coin from the given transcript.

2.3 Efficiency

The n-spendable coins are in some respects more efficient than n 1-spendable coins. For a 1-spendable coin Alice must store 3 base numbers, 2 signatures and

1 random k. For an n-spendable coin these numbers are $n+2$, $n+1$ and n respectively. The base numbers and the k's must be stored outright, but the signatures from different coins (with different v's) can be multiplied together as the signatures can easily be separated again (just like in batch-RSA). As an example we will use a 512-bit RSA modulus, a 128 bit prime v and multiply the signatures together in batches of 4. A 1-spendable coin then requires 240 bytes and a 5-spendable coin requires 624 bytes. Compared to five 1-spendable coins this halves the necessary storage space.

The computational requirements are much higher for an n-spendable coin. In the 1-spendable case Alice only needs 31 modular multiplications on average to spend a coin (if x is 20 bits long). If she spends several coins at the same time, then she can use the same x for all her coins and send the shop the product of all the signatures. This uses only $29 + 2t$ multiplies (on average) to spend t coins at the same time. The n-spendable coins require much more computations during the spending. As the x^i powers are 128 bits long (they are modulo v), Alice needs about $(n-1) * 192 + 31$ multiplies on average for each of the n spendings of the coin. To spend an n-spendable coin n times therefore requires about $192n^2 - 161n$ multiplies. Even for moderate n this is obviously inefficient compared to the $31n$ for the n 1-spendable coins.

3 Adding Observers

The major problem in electronic cash systems is the double spending. There is no cryptographic way in which we can prevent Alice from spending the same coin twice in an off-line system. Informally this can be shown as follows: Alice can first make a complete backup of the information in her computer and then spend a coin at shop A. She then restores all the information so that the computer is back in the same state as it was before spending the coin at shop A. Alice now spends the same coin again at shop B. The state of Alice's computer is the same for the second spending as it would have been if she never went to shop A. As we are talking about an off-line system, the state of shop B's computer after Alice spent her coin at shop A is the same as the state of shop B's computer before Alice spent her coin at shop A. As both the participants in the second spending are in the same state as they would have been if Alice had not been at shop A at all, the second payment completes successfully. Alice has succeeded in spending the same coin twice.

The only cryptographic protection against this attack is to detect the double-spending and to identify the user who did it. This is the way which is taken by all electronic cash systems [CFN90, CdBvH+90, vA90, OO92, Fer93a, Bra93, FY93].

An observer [CP93a, Cha92, Cra92, BCC+93, CP93b] is a tamper-resistant module that is incorporated in the user's computer. This is done in such a way that all communications to and from the observer is done via the user's computer. The observer is produced by a central authority and has its own native digital signature scheme. If the observer is incorporated in the electronic cash protocols in such a way that it prevents double-spending, then this provides prior restraint against the double-spending fraud. If Alice succeeds in breaking the tamper-resistance of the observer, she can double-spend a coin but will still be caught by the underlying coin scheme. The combination of an electronic cash system with observers provides the best of both worlds: the security is only dependent on cryptographic assumptions and Alice is prevented from double-spending her coins by a tamper-resistant device.

We show how observers can be incorporated in the coin scheme of [Fer93a]. As the underlying randomized blind signature protocol for these coins is the same as the 'validator' protocols from [Cra92] and [BCC+93] we use the same type of construction.

Instead of giving all the information of a coin to Alice, we will keep a vital part of it in the observer. Alice will store the values c, a, b, the signature $(C^k A)^{1/v}$ and the blinding factor β. The observer will store the signature $S := (\beta^v C^U B)^{1/v}$. The modified payment protocol is shown in figure 4. The protocol is basically the same as in figure 1 except that Alice no longer sends the signature $(C^r B^x A)^{1/v}$ but executes a Guillou-Quisquater identification protocol [GQ88] to prove that she knows a root of $X := (C^r B^x A)$. This is enough to convince the shop of the accuracy of r. Alice doesn't know the root of X by herself, but (as the protocol demonstrates) Alice and her observer together can convince the shop. (From the shops point of view, a standard GQ protocol is executed.) The way in which Alice and the observer cooperate in producing the proof is related to the diverted 'meta' protocols in [OO90]. Alice does a lot of additional blinding on the messages to and from the observer; these serve to prevent shared information [Cra92]. It is assumed that the observer might one day be returned to the central authority. In that case we still want to maintain the unlinkability of the payments. The blinding being done by Alice ensures that the observers transcript of a payment protocol cannot be linked to the shops transcript of the payment protocol. Some minor additional modifications have been left out for clarity. For a full discussion of these issues, and proofs of the properties of the second part of the protocol, we refer the reader to [Cra92] and [BCC+93].

It remains to be shown how we ensure the distribution of the information over Alice and her observer during the withdrawal protocol. We start again with the withdrawal protocol from [Fer93a] (see figure 3 with $n = 1$, and substitute $A_0 = B$, $A_1 = A$). We want to achieve the following aims:

- Alice can only conduct a withdrawal protocol with the help of an observer.
- At the end of the protocol, Alice is left with c, a, b and $(C^k A)^{1/v}$ and β while the observer gets $\beta(C^U B)^{1/v}$. The observer should not get any information about the values that Alice gets, and Alice should get no information about the signature that the observer gets.

Alice and the observer create a mutually random number η such that Alice knows η^v and the observer knows η. This is done using the elementary protocol shown in figure 5 which is a slight modification to the coin tossing protocol by Blum [Blu82].

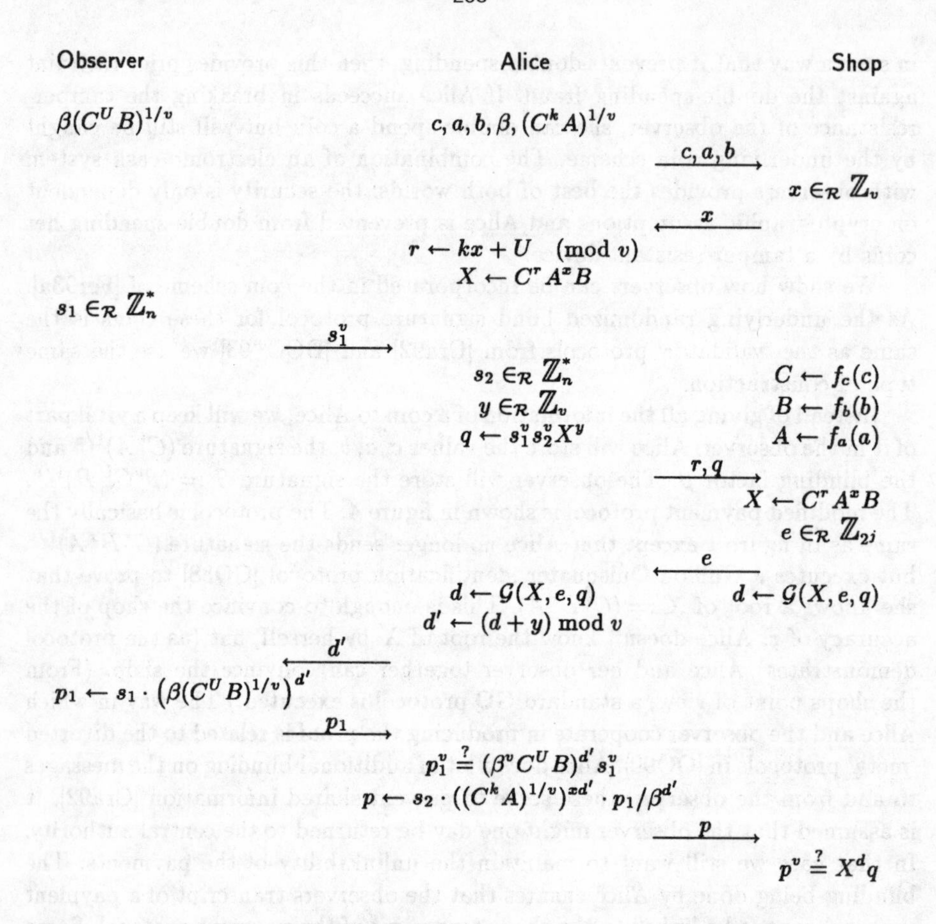

Fig. 4. Payment protocol with observer

The observer signs a message consisting of η^v and the data in the third transmission using its native digital signature scheme. This signature, together with η^v is also sent to the bank in the third transmission. The bank verifies the digital signature to ensure that η was indeed created in cooperation with an observer (and that therefore Alice only knows η^v). In the final message the bank sends the signature $(\overline{C}^U \overline{B}\eta^v)^{1/v}$ instead of $(\overline{C}^U \overline{B})^{1/v}$. Alice divides this by γ^U giving $\eta(C^U \overline{B})^{1/v}$ which is equal to $\beta\eta(C^U B)^{1/v}$. She sends this number to the observer who divides out the η factor resulting in $\beta(C^U B)^{1/v}$. As Alice has no knowledge about η except its v'th power, she cannot compute the signature $(C^U B)^{1/v}$ that she needs to spend the coin by herself. During the one spending that the observer will allow, the observer only executes a Guillou-Quisquater protocol with Alice proving the knowledge of the root $\beta(C^U B)^{1/v}$ which does

Observer		Alice
$\lambda \in_{\mathcal{R}} \mathbb{Z}_n^*$		
$c \leftarrow \text{commit}(\lambda^v)$		
	$\xrightarrow{c}$	
		$\xi \in_{\mathcal{R}} \mathbb{Z}_n^*$
	$\xleftarrow{\xi}$	
$\eta \leftarrow \xi\lambda$		
	$\xrightarrow{\lambda^v}$	
		$c \stackrel{?}{=} \text{commit}(\lambda^v)$
		$\eta^v \leftarrow \lambda^v \xi^v$

Fig. 5. Creation of η

not help her in an attempt to spend the coin a second time.

There are a few other subtle changes necessary to the withdrawal protocol to get all the desired properties. These are not described here but will be included in the full paper. (For some of the details, see [Cra92, BCC+93]).

4 Discussion

We have shown two extensions to the single term off-line coins. This coin scheme was the first of what promises to be a new class of far more efficient electronic cash schemes (see for example [Bra93, Bra94]). The efficient implementation of coins eliminate the necessity of using checks [CFN90, vA90] with all the related organizational and security problems of refunds (e.g. [Hir93]). Although checks are often more efficient from the cryptographers point of view, they are more complicated. There is also a difficulty in finding a simple and consistent user-interface for them. Checks need to be of different denomination (if all checks were of large denominations then the user would lose too much money if she ever lost her computer), but it is hard for the average user to predict which sequence of payments will be possible with a given set of checks[1]. Efficient coins solve this problem as it is easy to collect a set of coins for which the computer can say "You have $ 12.30 and I guarantee that you can make any 7 payments (as long as the total amount is below $ 12.30)". Users are already used to the concept of a limited number of payments as exhibited by checkbooks. For practical applications the storage requirements of coin systems are unfortunately still on the large side.

Although n-spendable coins seem attractive and are more storage-efficient,

[1] To protect the user's privacy, there can be no change given by the shop

it still remains to be seen if the gained storage is worth the extra computational complexity and linkability problems.

The incorporation of observers into electronic cash protocols improves the overall functionality of the system. Banks do not like to allow their customers to cheat them and then attempt to recover the loss from the perpetrators afterwards. With observers providing the prior-restraint, the security of electronic cash is now better in all respects than any other means of payments (unless factoring is easy :–).

References

[BCC⁺93] Stefan Brands, David Chaum, Ronald Cramer, Niels Ferguson, and Torben Pedersen. Transaction systems with observers. Technical report, CWI (Centre for Mathematics and Computer Science), Amsterdam, 1993. To appear.

[Blu82] M. Blum. Coin flipping by telephone. In *Proc. 24th IEEE Compcon*, pages 133–137, 1982.

[Bra93] Stefan Brands. An efficient off-line electronic cash system based on the representation problem. Technical Report CS-R9323, CWI (Centre for Mathematics and Computer Science), Amsterdam, 1993. Anonymous ftp: ftp.cwi.nl:/pub/CWIreports/AA/CS-R9323.ps.Z.

[Bra94] Stefan Brands. Electronic cash systems based on the representation problem in groups of prime order. In *Proceedings of CRYPTO '93*, 1994. To appear.

[CdBvH⁺90] David Chaum, Bert den Boer, Eugène van Heyst, Stig Mjølsnes, and Adri Steenbeek. Efficient off-line electronic checks. In J.-J. Quisquater and J. Vandewalle, editors, *Advances in Cryptology—EUROCRYPT '89*, Lecture Notes in Computer Science, pages 294–301. Springer-Verlag, 1990.

[CFN90] David Chaum, Amos Fiat, and Moni Naor. Untraceable electronic cash. In S. Goldwasser, editor, *Advances in Cryptology—CRYPTO '88*, Lecture Notes in Computer Science, pages 319–327. Springer-Verlag, 1990.

[Cha92] David Chaum. Achieving electronic privacy. *Scientific American*, pages 96–101, August 1992.

[CP93a] David Chaum and Torben Pedersen. Wallet databases with observers. In *Advances in Cryptology—CRYPTO '92*, 1993. To appear.

[CP93b] R.J.F. Cramer and T.P. Pedersen. Improved privacy in wallets with observers. In *Proceedings of EUROCRYPT '93*, 1993. To appear.

[Cra92] Ronald Cramer. Shared information in the moderated setting. Master's thesis, Rijksuniversiteit Leiden, Netherlands, August 1992.

[Fer93a] Niels Ferguson. Single term off-line coins. In *Proceedings of EUROCRYPT '93*, 1993. To appear.

[Fer93b] Niels Ferguson. Single term off-line coins. Technical Report CS-R9318, CWI (Centre for Mathematics and Computer Science), Amsterdam, 1993. Anonymous ftp: ftp.cwi.nl:/pub/CWIreports/AA/CS-R9318.ps.Z.

[FY93] Matthew Franklin and Moty Yung. Secure and efficient off-line digital money. In A. Lingas, R. Karlsson, and S. Carlsson, editors, *Automata, Languages and Programming, 20th International Colloquium, ICALP 93, Lund, Sweden*, Lecture Notes in Computer Science 700, pages 265–276. Springer-Verlag, 1993.

[GQ88] Louis C. Guillou and Jean-Jacques Quisquater. A practical zero-knowledge protocol fitted to security microprocessor minimizing both transmission and memory. In Christoph G. Günther, editor, *Advances in Cryptology—EUROCRYPT '88*, Lecture Notes in Computer Science, pages 123–128. Springer-Verlag, 1988.

[Hir93] Rafael Hirschfeld. Making electronic refunds safer. In *Advances in Cryptology—CRYPTO '92*, 1993. To appear.

[OO90] Tatsuaki Okamoto and Kazuo Ohta. Divertible zero knowledge interactive proofs and commutative random self-reducibility. In J.-J. Quisquater and J. Vandewalle, editors, *Advances in Cryptology—EUROCRYPT '89*, Lecture Notes in Computer Science, pages 134–149. Springer-Verlag, 1990.

[OO92] Tatsuaki Okamoto and Kazuo Ohta. Universal electronic cash. In J. Feigenbaum, editor, *Advances in Cryptology—CRYPTO '91*, Lecture Notes in Computer Science, pages 324–337. Springer-Verlag, 1992.

[Sha79] Adi Shamir. How to share a secret. *Communications of the ACM*, 22(11):612–613, 1979.

[vA90] Hans van Antwerpen. Off-line electronic cash. Master's thesis, Eindhoven University of Technology, department of Mathematics and Computer Science, 1990.

Untraceable Off-line Cash in Wallet with Observers

(Extended abstract)

Stefan Brands

CWI, PO Box 4079 Amsterdam, The Netherlands. E-mail: brands@cwi.nl

Abstract. Incorporating the property of untraceability of payments into off-line electronic cash systems has turned out to be no easy matter. Two key concepts have been proposed in order to attain the same level of security against double-spending as can be trivially attained in systems with full traceability of payments.

The first of these, one-show blind signatures, ensures traceability of double-spenders after the fact. The realizations of this concept that have been proposed unfortunately require either a great sacrifice in efficiency or seem to have questionable security, if not both.

The second concept, wallets with observers, guarantees prior restraint of double-spending, while still offering traceability of double-spenders after the fact in case tamper-resistance is compromised. No realization of this concept has yet been proposed in literature, which is a serious problem. It seems that the known cash systems cannot be extended to this important setting without significantly worsening the problems related to efficiency and security.

We introduce a new primitive that we call restrictive blind signatures. In conjunction with the so-called representation problem in groups of prime order this gives rise to highly efficient off-line cash systems that can be extended at virtually no extra cost to wallets with observers under the most stringent of privacy requirements. The workload for the observer is so small that it can be performed by a tamper-resistant smart card capable of performing the Schnorr identification scheme.

We also introduce new extensions in functionality (unconditional protection against framing, anonymous accounts, multi-spendable coins) and improve some known constructions (computional protection against framing, electronic checks).

The security of our cash system and all its extensions can be derived directly from the security of the well-known Schnorr identification and signature schemes, and the security of our new primitive.

1 Introduction

It is clear that the level of efficiency and security attainable in an off-line electronic cash system with fully traceable payments always outperforms that attainable in a system with the additional property of privacy of payments. This is caused by the urgent need to protect against account-holders who double-spend their electronic cash, since hardly anything is easier to copy than digital information.

In literature, various realizations have been proposed for untraceable off-line electronic cash. Untraceability is an important asset, but one should not ignore the fact that it is hard to realize at little cost. For this reason, we present in Section 2 an analysis of the cost it takes in terms of efficiency and security to incorporate untraceability of payments. We argue that known realizations of untraceable off-line cash systems offering only traceability of double-spenders after the fact already require a large sacrifice in either efficiency or (provability of) security, if not both. More seriously, no realizations of untraceable off-line cash systems have been proposed yet that can offer prior restraint of double-spending, whereas this property can be trivially attained in fully traceable off-line systems. Thirdly, various other useful extensions in functionality seem hard to achieve in the known systems.

To overcome these drawbacks, we propose in Section 4 the primitive of restrictive blind signatures, and use it in combination with the representation problem in groups of prime order (described in Section 3) to create untraceable off-line electronic cash systems that can offer not only traceability of double-spenders after the fact (Section 5), but more importantly also prior restraint of double-spending under the most stringent of privacy requirements (Section 6). These systems are almost as efficient as fully traceable off-line systems.

In addition, three new extensions can be realized: unconditional protection against framing, anonymous accounts, and multispendable coins. We refer the interested reader to [2, 3] for this. The new approach also supports a better construction for two known extensions, computational protection against framing and electronic checks. The extension to computational protection against framing is incorporated in Sections 5 and 6, and we refer to [2] for a description of the extension to checks.

All the statements we make in this extended abstract have been fully proven. We refer to [2, 3] for these proofs.

2 The cost of incorporating untraceability

2.1 Privacy-compromising systems

An off-line electronic cash system with full traceability of payments can be simply realized using only the basic cryptographic concept of a digital signature. Each coin is represented by a unique piece of digital information with a corresponding digital signature of the bank. If an account-holder ever double-spends then he will be identified by the bank *after* the corresponding payment transcripts have been deposited, if only the bank conscientiously keeps track for each coin to which account-holder it issued that coin. Since off-line cash systems are a medium for low-value payments only (high-value payments are made on-line), this traceability after the fact by itself will discourage many account-holders from double-spending.

If the bank in addition has the payment devices of its account-holders manufactured such that they are tamper-resistant, then a level of prior restraint of double-spending is attained that can only be withstood by an organization with the capabilities of a national laboratory. By maintaining the database concerning issued coins, the bank can still trace a double-spender after the fact in case he unexpectedly breaks the tamper-resistance of the payment device.

Such a system can be realized very efficiently. In each of the three protocols for withdrawal, payment, and deposit of a coin, only one signed number has to be transmitted (in practice other information will be sent along as well, such as signatures that serve as receipts), and a computational effort for each type of participant to verify the validity of the signature of the bank is required. The bank has to compute a digital signature for each coin it issues, and maintain a database with information about the coins issued to the account-holders. This database has to be searched on a regular basis to find out if double-spending has occurred.

The level of security of the system is also very satisfactory. In principle, the bank can use a digital signature proposed by [1], which is provably secure against adaptively chosen message attacks (assuming the existence of one-way permutations). However, since these signatures grow in size, and require quite some computational effort, they are inefficient for practical use in systems such as cash systems, where enormous amounts of signatures are routinely produced and verified. In practice one hence must inevitably sacrifice some provability of security and use e.g. signatures of the Fiat/Shamir type ([13]), such as Schorr signatures ([15]).

Although the system sketched thus far is highly satisfactory from both the efficiency and security points of view, it does not protect the interests of the account-holders. As we discussed, by the very nature of the system the bank has to maintain databases to keep track of the information issued in executions of the withdrawal protocol and the deposited payment transcripts. Since a payment transcript encompasses the withdrawn coin, per definition the entire payment history of all account-holders is stored in computer files by the bank. Hence, not only is such a system not privacy-protecting, it in fact is the extreme opposite. This can have considerable social and political impact (see e.g. [5, 6, 16]). Henceforth, we will refer to such a system as a privacy-compromising system.

2.2 Privacy-protecting cash systems

Two ingenious key concepts have been developed to enable the incorporation of full untraceability of payments while maintaining the level of security against double-spending of the privacy-compromising system.

Concept I. The first key concept is ***one-show blind signatures***, introduced in [8]. One-show blind signatures enable traceability of a payment if and only if the account-holder double-spent the coin involved in that payment. That is, traceability after the fact can be accomplished only for double-spenders.

Realizing this concept has turned out to be no easy matter. For traceability, the identity of the account-holder must be encoded into the withdrawn information, whereas this information is not known to the bank by virtue of the blind signature property needed to achieve untraceability of payments. If we forget about the even less efficient theoretical constructions proposed for this (although they seem to guarantee the same level of security as can be achieved in privacy-compromising systems), then only cut-and-choose withdrawal protocols seem to remain. These still cause an enormous overhead in computational and communication complexity that we believe is unacceptable.

Our new primitive, *restrictive* blind signature schemes, in combination with the representation problem in groups of prime order, allows us to construct a three-move withdrawal protocol (i.e. no cut-and-choose) in which the computational effort required by the bank is almost equivalent to that required to compute Schnorr signatures. In the payment protocol, only two (!) modular multiplications are required of the account-holder in order to pay. The database that must be maintained by the bank is almost of the same size as that in the privacy-compromising system.

We refer to [3] for an overview of the cryptographic literature on untraceable off-line electronic cash systems. We confine ourselves here to the remark that in concurrent work ([11]), a system offering traceability after the fact is proposed that also does not use a cut-and-choose withdrawal protocol. Unfortunately, its security seems highly questionable. This is caused by the use of many unspecified one-way hash functions, nested within one another up to four levels deep, and a strange construction to create an element with an order equal to the order of the multiplicative group modulo a composite.

As we show, there is no need at all to resort to such "ad hoc" constructions. In fact, our approach allows for greater efficiency, security, and extendibility in functionality.

Concept II. The privacy-compromising system offered prior restraint of double-spending. Using the second key concept (see [7]), *wallets with observers*, this can also be achieved in privacy-protecting systems. In this setting, a tamper-resistant device that takes care of prior restraint of double-spending, called an *observer*, is embedded into the payment device of the account-holder in such a way that a payment can only be successfully executed if the observer cooperates. The ensemble of payment device and observer is called a wallet. In order to guarantee the untraceability of payments, the embedding must be such that any message the observer sends to the outside world passes through the payment device. This enables the payment device to recognize attempts of the observer to leak information (*outflow*) related to its identity, and vice versa (*inflow*).

If the observer stores all information it receives during the period it is embedded within the payment device, it might still be that the bank can trace payments to account-holders afterwards by comparing this information with the deposited payment transcripts (and possibly also its view in executions of the withdrawal protocol). Mutually known information which enables traceability is called *shared information*; it comprises both inflow and outflow. This concern, although not specifically in the context of off-line cash systems, was raised in [10]. Although it might seem unrealistic to worry about the development of shared information, it is not hard to construct withdrawal and payment protocols with no inflow and outflow, whereas all payments can be traced virtually effortlessly by the bank once the observer is handed in. A trivial example of development of shared information without inflow or outflow is a payment protocol in which the observer and the payment device generate mutually at random a number, known to both, which the payment device sends to the shop.

As in the privacy-compromising system, the bank should not rely solely on tamper-resistance. If an account-holder unexpectedly breaks the tamper-resistance of the observer and double-spends, then he should still be traceable after the fact. This implies that the first concept acts as a safety net, and hence a realization of the second concept must be an extension of a realization of the first concept. For this reason, we refer to a system realizing the first concept as a basic cash system.

Contrary to the first concept, no realizations of an untraceable off-line cash system satisfying these conditions have been proposed yet. It seems that the known cash systems that provide realizations of the first concept cannot be extended to this important setting without worsening the problems related to efficiency and security. Our system can be extended to meet all the requirements of the second concept at virtually no extra cost in efficiency and security. Only a minor modification of the basic system is required. The workload for the observer is so small that it can be performed by a smart card capable of performing the Schnorr identification scheme.

Recently ([12]), Ferguson sketched how to extend his basic system ([11]) to wallets with observers; however, this seems to significantly worsen the problems related to security present in his basic system, as well as efficiency. As with the first concept, we show that there is no need for ad hoc constructions.

3 The representation problem in groups of prime order

All arithmetic in this article is performed in a group G_q of prime order q for which polynomial-time algorithms are known to multiply, invert, determine equality of elements, test membership, and randomly select elements. There is a vast variety of groups known to satisfy these requirements.

Definition 1. Let $k \geq 2$. A *generator-tuple* of length k is a k-tuple $(g_1, \ldots, g_k)$ with $g_i \in G_q \setminus \{1\}$ and $g_i \neq g_j$ if $i \neq j$. For any $h \in G_q$, a *representation* of h *with respect to* a generator-tuple $(g_1, \ldots, g_k)$ is a tuple $(a_1, \ldots, a_k)$, with $a_i \in \mathbb{Z}_q$ for all $1 \leq i \leq k$, such that $\prod_{i=1}^{k} g_i^{a_i} = h$.

Usually, it will be clear with respect to what generator-tuple a representation is taken, and we will not mention it. If $h = 1$, one representation immediately springs to mind, namely $(0, \ldots, 0)$. We call this the *trivial* representation.

Proposition 2. *For all $h \in G_q$ and all generator-tuples of length k there are exactly q^{k-1} representations of h.*

This simple result implies that the density of representations of h is negligible with respect to the set of size q^k containing all tuples $(a_1, \ldots, a_k)$. Therefore, any polynomial-time algorithm that applies an exhaustive search strategy in this set to find one has negligible probability $1/q$ of success. The following result shows that there is no essentially better strategy, assuming the Discrete Log assumption.

Proposition 3. *Assuming that it is infeasible to compute discrete logarithms in G_q, there cannot exist a number $h \in G_q$ and a polynomial-time algorithm that, on input a randomly chosen generator-tuple $(g_1, \ldots, g_k)$, outputs a (nontrivial if $h = 1$) representation of h with nonnegligible probability of success.*

Since the difference between two distinct representations of any number $h \in G_q$ is a nontrivial representation of 1, we get the following important result.

Corollary 4. *Assuming that it is infeasible to compute discrete logarithms in G_q, there cannot exist a polynomial-time algorithm that, on input a generator-tuple $(g_1, \ldots, g_k)$ chosen at random, outputs a number $h \in G_q$ and two different representations of h with nonnegligible probability of success.*

We next define the *representation problem in groups of prime order* (using a standard specification format).

Name: Representation problem in groups of prime order.
Instance: A group G_q, a generator-tuple $(g_1, \ldots, g_k)$, $h \in G_q$.
Problem: Find a representation of h with respect to $(g_1, \ldots, g_k)$.

Although our electronic cash system can be implemented with any group G_q that satisfies the listed conditions, and for which no feasible algorithms are known to compute discrete logarithms, we will for explicitness assume henceforth that G_q is the unique subgroup of order q of some multiplicative group $\mathbb{Z}_p^*$, for a prime p such that $q|(p-1)$.

4 Restrictive blinding in groups of prime order

In order to explain the notion of restrictive blinding, we give a high-level overview of the basic cash system, in which the primitive is put to use. In the following, (g_1, g_2) is a randomly chosen generator-tuple.

In setting up an account, the bank generates a unique number $u_1 \in_{\mathcal{R}} \mathbb{Z}_q$ which is registered together with the identity of the account-holder with the newly created account. When the account-holder wishes to withdraw a coin from his account, the bank multiplies $I = g_1^{u_1}$ by g_2. Hence, the account-holder knows the representation $(u_1, 1)$ of the number $m = Ig_2$ with respect to (g_1, g_2). During the three-move withdrawal protocol, the account-holder will blind m to a number A, such that he ends up with a signature of the bank corresponding to A. A and the signature will be unconditionally untraceable to any specific execution of the withdrawal protocol. By construction of the payment protocol, the account-holder at this stage *must* know a representation (x_1, x_2) of A with respect to (g_1, g_2) in order to be able to pay.

Here, the role of the *restrictive* blind signature protocol becomes clear.

Definition 5. Let $m \in G_q$ (in general, it can be a vector of elements) be such that the receiver at the start of a blind signature protocol knows a representation $(a_1, \ldots, a_k)$ of m with respect to a generator-tuple $(g_1, \ldots, g_k)$. Let $(b_1, \ldots, b_k)$ be the representation the receiver knows of the blinded number A of m after the protocol has finished. If there exist two functions I_1 and I_2 such that

$$I_1(a_1, \ldots, a_k) = I_2(b_1, \ldots, b_k),$$

regardless of m and the blinding transformations applied by the receiver, then the protocol is called a *restrictive* blind signature protocol. The functions I_1 and I_2 are called *blinding-invariant functions* of the protocol *with respect to* $(g_1, \ldots, g_k)$.

Intuitively, one can think of it as being a protocol in which the receiver can blind the "outside" of the message m (and signature), but not its internal structure.

For the application to untraceable off-line cash systems, in which the bank must be able to identify a payer if and only if he double-spends, we construct the payment protocol such that the account-holder not only has to reveal A and the signature, but also some additional information about the representation he knows of A. This additional information must be such that one such piece of information does not reveal any Shannon information about u_1 (the internal structure), whereas knowledge of two such pieces enables the bank to extract this number in polynomial time.

Clearly, if the account-holder in the payment protocol is able to also blind the internal structure of m, then he will not be identified after the fact when double-spending. Hence, it is absolutely essential that the receiver is restricted in the blinding manipulations he can perform, which explains the terminology *restrictive* blind signature scheme.

5 The basic cash system

In this section, we describe the most basic form of the cash system, involving only signed information (coins) of one value. We denote the bank by $\mathcal{B}$, a generic account-holder by $\mathcal{U}$, and a generic shop by $\mathcal{S}$. Although $\mathcal{U}$ will be a payment device (such as a smart card, palmtop or personal computer) in a practical implementation, we will often identify $\mathcal{U}$ with the account-holder.

The setup of the system. The setup of the system consists of $\mathcal{B}$ generating at random a generator-tuple (g, g_1, g_2), and a number $x \in_{\mathcal{R}} \mathbb{Z}_q^*$.

$\mathcal{B}$ also chooses two suitable collision-intractable (or even better, correlation-free one-way, as defined in [14]) hash functions $\mathcal{H}$, $\mathcal{H}_0$, with

$$\mathcal{H} : G_q \times G_q \times G_q \times G_q \times G_q \rightarrow \mathbb{Z}_q^*$$

and, for example,

$$\mathcal{H}_0 : G_q \times G_q \times \text{SHOP-ID} \times \text{DATE/TIME} \rightarrow \mathbb{Z}_q.$$

The function $\mathcal{H}$ is used for the construction and verification of signatures of $\mathcal{B}$, and the function $\mathcal{H}_0$ specifies in what way the challenges must be computed in the payment protocol. $\mathcal{B}$ publishes the description of G_q (which is p, q in the specific case of $G_q \subset \mathbb{Z}_p^*$), the generator-tuple (g, g_1, g_2), and the description of $\mathcal{H}$, $\mathcal{H}_0$ as its public key. The secret key of $\mathcal{B}$ is x.

The format of $\mathcal{H}_0$ assumes that each shop $\mathcal{S}$ has a unique identifying number $I_{\mathcal{S}}$ (this can be its account number at $\mathcal{B}$) known to at least $\mathcal{B}$ and $\mathcal{S}$; we denote above the set of all such numbers by SHOP-ID. The input from SHOP-ID ensures that two different shops with overwhelming probability will generate different challenges. The input from the set DATE/TIME is a number representing the date and time of transaction, which guarantees that the same shop will generate different challenges per payment. We stress that the format of $\mathcal{H}_0$ is just exemplary; other formats might do as well.

$\mathcal{B}$ also sets up two databases. One is called the account database and is used by the bank to store information about account-holders (such as their name and address), the other is called the deposit database and is used to store relevant information from deposited payment transcripts.

A signature $\text{sign}(A, B)$ of $\mathcal{B}$ on a pair $(A, B) \in G_q \times G_q$ consists of a tuple $(z, a, b, r) \in G_q \times G_q \times G_q \times \mathbb{Z}_q$ such that

$$g^r = h^{\mathcal{H}(A,B,z,a,b)} a \quad \text{and} \quad A^r = z^{\mathcal{H}(A,B,z,a,b)} b.$$

A *coin* is a triple $A, B, \text{sign}(A, B)$. If an account-holder knows a representation of both A and B with respect to (g_1, g_2), then we will simply say that he knows a representation of the coin.

Opening an account. When $\mathcal{U}$ opens an account at $\mathcal{B}$, $\mathcal{B}$ requests $\mathcal{U}$ to identify himself (by means of, say, a passport). $\mathcal{U}$ generates at random a number $u_1 \in_{\mathcal{R}} \mathbb{Z}_q$, and computes $I = g_1^{u_1}$. If $g_1^{u_1} g_2 \neq 1$, then $\mathcal{U}$ transmits I to $\mathcal{B}$, and keeps u_1 secret. $\mathcal{B}$ stores the identifying information of $\mathcal{U}$ in the account database, together with I. We will refer to I as the account number of $\mathcal{U}$. The uniqueness of the account number is essential, since it enables $\mathcal{B}$ to uniquely identify $\mathcal{U}$ in case he double-spends.

$\mathcal{B}$ computes $z = (Ig_2)^x$, and transmits it to $\mathcal{U}$. Alternatively, $\mathcal{B}$ publishes g_1^x and g_2^x as part of his public key, so that $\mathcal{U}$ can compute z for himself.

The withdrawal protocol. When $\mathcal{U}$ wants to withdraw a coin, he first must prove ownership of his account. To this end, $\mathcal{U}$ can for example digitally sign a request for withdrawal, or identify himself by other means. Then the following withdrawal protocol is performed:

Step 1. $\mathcal{B}$ generates at random a number $w \in_{\mathcal{R}} \mathbb{Z}_q$, and sends $a = g^w$ and $b = (Ig_2)^w$ to $\mathcal{U}$.

Step 2. $\mathcal{U}$ generates at random three numbers $s \in_{\mathcal{R}} \mathbb{Z}_q^*$, $x_1, x_2 \in_{\mathcal{R}} \mathbb{Z}_q$, and uses them to compute $A = (Ig_2)^s$, $B = g_1^{x_1} g_2^{x_2}$, and $z' = z^s$. $\mathcal{U}$ also generates at random two numbers $u, v \in_{\mathcal{R}} \mathbb{Z}_q$, and uses them to compute $a' = a^u g^v$ and $b' = b^{su} A^v$. He then computes the challenge $c' = \mathcal{H}(A, B, z', a', b')$, and sends the blinded challenge $c = c'/u \bmod q$ to $\mathcal{B}$.

Step 3. $\mathcal{B}$ sends the response $r = cx + w \bmod q$ to $\mathcal{U}$, and debits the account of $\mathcal{U}$.

$\mathcal{U}$ accepts if and only if $g^r = h^c a$ and $(Ig_2)^r = z^c b$. If this verification holds, $\mathcal{U}$ computes $r' = ru + v \bmod q$.

$\mathcal{U}$		$\mathcal{B}$
		$w \in_{\mathcal{R}} \mathbb{Z}_q$
		$a \leftarrow g^w$
$s \in_{\mathcal{R}} \mathbb{Z}_q^*$	$\xleftarrow{a,\, b}$	$b \leftarrow (Ig_2)^w$
$A \leftarrow (Ig_2)^s$		
$z' \leftarrow z^s$		
$x_1, x_2, u, v \in_{\mathcal{R}} \mathbb{Z}_q$		
$B \leftarrow g_1^{x_1} g_2^{x_2}$		
$a' \leftarrow a^u g^v$		
$b' \leftarrow b^{su} A^v$		
$c' \leftarrow \mathcal{H}(A, B, z', a', b')$		
$c \leftarrow c'/u \bmod q$	$\xrightarrow{c}$	
$g^r \stackrel{?}{=} h^c a$	$\xleftarrow{r}$	$r \leftarrow cx + w \bmod q$
$(Ig_2)^r \stackrel{?}{=} z^c b$		
$r' \leftarrow ru + v \bmod q$		

Proposition 6. *If $\mathcal{U}$ accepts in the payment protocol, then $A, B, (z', a', b', r')$ is a coin of which he knows a representation.*

Proposition 7. *Assume that it is infeasible to existentially forge Schnorr signatures, even when querying the prover in the Schnorr identification protocol polynomially many times. Then it is infeasible to existentially forge a coin, even when performing the withdrawal protocol polynomially many times and with respect to different account numbers.*

In other words, the number of coins in circulation can never exceed the number of executions of the withdrawal protocol. This obviously is an important fact, since one should not be able to create his own money. In fact, as Lemma 8 shows, the task is even much more difficult, since one in addition has to know a representation of a coin in order to be able to spend it.

Assumption 1. *The withdrawal protocol is a restrictive blind signature protocol (with $m = Ig_2$) with blinding invariant functions I_1 and I_2 with respect to (g_1, g_2) defined by $I_1(a_1, a_2) = I_2(a_1, a_2) = a_1/a_2 \bmod q$.*

Although this assumption is stronger than the Diffie-Hellman assumption, there are convincing arguments based on partial proofs that suggest that breaking it requires breaking either the Schnorr scheme or the Diffie-Hellman assumption. For an extensive discussion, we refer to [3].

The payment protocol. When $\mathcal{U}$ wants to spend his coin at $\mathcal{S}$, the following protocol is performed:

Step 1. $\mathcal{U}$ sends $A, B, \text{sign}(A, B)$ to $\mathcal{S}$.
Step 2. If $A \neq 1$, then $\mathcal{S}$ computes challenge $d = \mathcal{H}_0(A, B, I_{\mathcal{S}}, \text{date/time})$, where date/time is the number representing date and time of the transaction. $\mathcal{S}$ sends d to $\mathcal{U}$.
Step 3. $\mathcal{U}$ computes the responses $r_1 = d(u_1 s) + x_1 \bmod q$ and $r_2 = ds + x_2 \bmod q$, and sends them to $\mathcal{S}$.

$\mathcal{S}$ accepts if and only if $\text{sign}(A, B)$ is a signature on (A, B), and $g_1^{r_1} g_2^{r_2} = A^d B$.

$\mathcal{U}$		$\mathcal{S}$
	$A, B, \text{sign}(A,B)$ $\longrightarrow$	$A \overset{?}{\neq} 1$
		$d \leftarrow \mathcal{H}_0(A, B, I_{\mathcal{S}}, \text{date/time})$
	d $\longleftarrow$	
$r_1 \leftarrow d(u_1 s) + x_1 \bmod q$		
$r_2 \leftarrow ds + x_2 \bmod q$	(r_1, r_2) $\longrightarrow$	
		Verify $\text{sign}(A, B)$
		$g_1^{r_1} g_2^{r_2} \overset{?}{=} A^d B$

If $\mathcal{U}$ has access to a clock and the capability of looking up the identifying information I_S of $\mathcal{S}$ (which seems more plausible when the payment device is, say, a personal computer dialing in via a modem, then in the case where it is a smart card), this protocol can be collapsed to one single move since $\mathcal{U}$ can then compute d himself. In any case, it is of no importance to $\mathcal{U}$ whether d is correctly determined. This is only of concern to $\mathcal{S}$, since the bank will not accept the payment transcript in the deposit protocol if d is not of the correct form.

Lemma 8. *If $\mathcal{U}$ in the payment protocol can give correct responses with respect to two different challenges, then he knows a representation of both A and B with respect to (g_1, g_2).*

Since $\mathcal{H}_0$ is a randomizing hash function, this result implies that the probability that $\mathcal{S}$ accepts in the payment protocol, whereas $\mathcal{U}$ does not know a representation of both A and B with respect to (g_1, g_2), is negligible. Together with the completeness of the payment protocol this implies the following.

Corollary 9. *$\mathcal{U}$ can spend a coin if and only he knows a representation of it.*

The deposit protocol. After some delay in time (since the system is off-line), $\mathcal{S}$ sends the to $\mathcal{B}$ the payment transcript, consisting of $A, B, \text{sign}(A,B), (r_1, r_2)$ and date/time of transaction.

If $A = 1$, then $\mathcal{B}$ does not accept the payment transcript. Otherwise, $\mathcal{B}$ computes d using the identifying number of the shop $I_\mathcal{S}$ sending the payment transcript, and the supplied date/time of transaction. $\mathcal{B}$ then verifies that $g_1^{r_1} g_2^{r_2} = A^d B$ and that $\text{sign}(A,B)$ is a signature on (A,B). If not both verifications hold, then $\mathcal{B}$ does not accept the payment transcript. Otherwise, $\mathcal{B}$ searches its deposit database to find out whether A has been stored before. There are two possibilities:

- A has not been stored before. In that case, $\mathcal{B}$ stores $(A, \text{ date/time}, r_1, r_2)$ in its deposit database as being deposited by $\mathcal{S}$, and credits the account of $\mathcal{S}$. Note that not the entire payment transcript need be deposited.
- A is already in the deposit database. In that case, a fraud must have occurred. If the already stored transcript was deposited by $\mathcal{S}$, and date/time are identical to that of the new payment transcript, then $\mathcal{S}$ is trying to deposit the same transcript twice. Otherwise (the challenges are different), the coin has been double-spent. Since $\mathcal{B}$ now has at its disposal a pair (d, r_1, r_2) from the new transcript and a pair (d', r_1', r_2') from the deposited information (where $\mathcal{B}$ computes d' from the date/time of transaction of the stored information and the identifying number $I_\mathcal{S}$ of the shop who deposited the transcript), it can compute

$$g_1^{(r_1 - r_1')/(r_2 - r_2')} .$$

$\mathcal{B}$ then searches its account database for this account number; the corresponding account-holder is the double-spender. The number $(r_1 - r_1')/(r_2 - r_2') \bmod q$ serves as a proof of double-spending; it is equal to $\log_{g_1} I$, with I the account number of the double-spender.

Since not even $\mathcal{B}$ needs to know a non-trivial representation of 1 with respect to (g, g_1, g_2) in order to perform the withdrawal protocol, there obviously cannot exist an adaptively chosen message attack that enables account-holders to know more than one representation of a coin (assuming that there are polynomially many account-holders and shops). Therefore, we get the following:

Proposition 10. *If Assumption 1 holds, then the computation that $\mathcal{B}$ performs in the deposit protocol in case of double-spending, results in the account number of the double-spender.*

We next prove, informally speaking, that the privacy of payments of account-holders who follow the protocols and do not double-spend is protected unconditionally.

Proposition 11. *For any $\mathcal{U}$, for any possible view of $\mathcal{B}$ in an execution of the withdrawal protocol in which $\mathcal{U}$ accepts, and for any possible view of $\mathcal{S}$ in an execution of the payment protocol in which the payer followed the protocol, there* *is exactly one set of random choices that $\mathcal{U}$ could have made in the execution of the withdrawal protocol such that the views of $\mathcal{B}$ and $\mathcal{S}$ correspond to the withdrawal and spending of the same coin.*

An immediate consequence of this proposition is the following.

Corollary 12. *Assuming the Discrete Log assumption, if $\mathcal{U}$ follows the protocols and does not double-spend, $\mathcal{B}$ cannot compute a proof of double-spending.*

That is, $\mathcal{U}$ is computationally protected against a framing.

In [14], Okamoto described a signature protocol that is structurally equivalent to our payment protocol. Existential forgery of these signatures is a harder task than existential forgery of Schnorr signatures.

Proposition 13. *Existential forgery of payment transcripts is a harder task than existential forgery of Okamoto signatures.*

The following two results imply that no additional encryption of messages that are transmitted is needed anywhere in our system.

Proposition 14. *Wire tapping an execution of the withdrawal protocol does not result in a coin.*

Proposition 15. *Wire tapping an execution of the payment protocol with $\mathcal{S}$ does not result in a payment transcript that can be deposited to another account than that of $\mathcal{S}$.*

6 Prior restraint of double-spending

We describe how to extend our basic cash system to the setting of wallets with observers in such a way that not even shared information can be developed. Even if the tamper-resistance is broken (and the account-holder can simulate the role of the observer), we still have the same level of security as in the original system, in fact the protocols reduce completely to those of the basic system. In particular, if one breaks the tamper-resistance and, as a result, can double-spend, one will still be identified after the fact.

The setup of the system. This is the same as in the basic cash system.

Opening an account. When $\mathcal{U}$ opens an account at $\mathcal{B}$, $\mathcal{B}$ requests $\mathcal{U}$ to identify himself (by means of, say, a passport). $\mathcal{U}$ generates at random a number $u_1 \in_{\mathcal{R}} \mathbb{Z}_q$, and computes $g_1^{u_1}$. $\mathcal{U}$ transmits $g_1^{u_1}$ to $\mathcal{B}$, and keeps u_1 secret. $\mathcal{B}$ stores the identifying information of $\mathcal{U}$ in the account database, together with $g_1^{u_1}$.

$\mathcal{B}$ then provides $\mathcal{U}$ with an observer $\mathcal{O}$, with stored in its (ROM) memory a randomly chosen number $o_1 \in \mathbb{Z}_q^*$ which is unknown to $\mathcal{U}$. We will denote $g_1^{o_1}$ by $A_{\mathcal{O}}$. $\mathcal{B}$ computes $I = A_{\mathcal{O}}(g_1^{u_1})$ and $z = (Ig_2)^x$, and transmits $A_{\mathcal{O}}$ and z to $\mathcal{U}$. $\mathcal{U}$ stores $u_1, A_{\mathcal{O}}, z$.

We will refer to I as the account number. This number will perform the role that I performed in the basic cash system. Note that, contrary to the basic cash system, $\mathcal{U}$ by himself does not know $\log_{g_1} I$.

The withdrawal protocol. When $\mathcal{U}$ wants to withdraw a coin from his account, he first must prove ownership of his account, as in the basic cash system. Then the following withdrawal protocol is performed:

Step 1. $\mathcal{O}$ generates at random a number $o_2 \in \mathbb{Z}_q$, and computes $B_{\mathcal{O}} = g_1^{o_2}$. He then sends $B_{\mathcal{O}}$ to $\mathcal{U}$. Although this step is part of the protocol, $\mathcal{O}$ can send $B_{\mathcal{O}}$ to $\mathcal{U}$ at any time before Step 3.

Step 2. $\mathcal{B}$ generates at random a number $w \in_{\mathcal{R}} \mathbb{Z}_q$, and sends $a = g^w$ and $b = (Ig_2)^w$ to $\mathcal{U}$.

Step 3. $\mathcal{U}$ generates at random four numbers $s \in_{\mathcal{R}} \mathbb{Z}_q^*$, $x_1, x_2, e \in_{\mathcal{R}} \mathbb{Z}_q$, and uses them to compute $A = (Ig_2)^s$, $B = g_1^{x_1} g_2^{x_2} A_{\mathcal{O}}^{es} B_{\mathcal{O}}$, and $z' = z^s$. $\mathcal{U}$ also generates at random two numbers $u, v \in_{\mathcal{R}} \mathbb{Z}_q$, and uses them to compute $a' = a^u g^v$ and $b' = b^{su} A^v$. He then computes the challenge $c' = \mathcal{H}(A, B, z', a', b')$, and sends the blinded challenge $c = c'/u \bmod q$ to $\mathcal{B}$.

Step 4. $\mathcal{B}$ sends the response $r = cx + w \bmod q$ to $\mathcal{U}$, and debits the account of $\mathcal{U}$.

$\mathcal{U}$ accepts if and only if $g^r = h^c a$ and $(Ig_2)^r = z^c b$. If this verification holds, $\mathcal{U}$ computes $r' = ru + v \bmod q$.

$\mathcal{O}$		$\mathcal{U}$		$\mathcal{B}$
				$w \in_{\mathcal{R}} \mathbb{Z}_q$
$o_2 \in_{\mathcal{R}} \mathbb{Z}_q$				$a \leftarrow g^w$
$B_{\mathcal{O}} \leftarrow g_1^{o_2}$	$\xrightarrow{B_{\mathcal{O}}}$	$s \in_{\mathcal{R}} \mathbb{Z}_q^*$	$\xleftarrow{a,b}$	$b \leftarrow (Ig_2)^w$
		$A \leftarrow (Ig_2)^s$		
		$z' \leftarrow z^s$		
		$x_1, x_2, e \in_{\mathcal{R}} \mathbb{Z}_q$		
		$B \leftarrow g_1^{x_1} g_2^{x_2} A_{\mathcal{O}}^{se} B_{\mathcal{O}}$		
		$u, v \in_{\mathcal{R}} \mathbb{Z}_q$		
		$a' \leftarrow a^u g^v$		
		$b' \leftarrow b^{su} A^v$		
		$c' \leftarrow \mathcal{H}(A, B, z', a', b')$		
		$c \leftarrow c'/u \bmod q$	$\xrightarrow{c}$	
		$g^r \stackrel{?}{=} h^c a$	$\xleftarrow{r}$	$r \leftarrow cx + w \bmod q$
		$(Ig_2)^r \stackrel{?}{=} z^c b$		
		$r' \leftarrow ru + v \bmod q$		

If we concentrate on $\mathcal{O}$ and $\mathcal{U}$ as one party, then this is exactly the basic withdrawal protocol. Hence, Propositions 6 (with "$\mathcal{O}$ and $\mathcal{U}$ together know" substituted for "he knows") and 7 hold.

The payment protocol. When $\mathcal{U}$ wants to pay with the withdrawn information at $\mathcal{S}$, the following protocol is performed:

Step 1. $\mathcal{U}$ sends $A, B, \text{sign}(A, B)$ to $\mathcal{S}$.
Step 2. If $A \neq 1$, $\mathcal{S}$ computes challenge $d = \mathcal{H}_0(A, B, I_{\mathcal{S}}, \text{date/time})$, and sends it to $\mathcal{U}$.
Step 3. $\mathcal{U}$ computes $d' = s(d + e) \bmod q$, and sends this to $\mathcal{O}$.
Step 4. If o_2 is still in memory, then $\mathcal{O}$ computes the response $r_1' = d'o_1 + o_2 \bmod q$ and send it to $\mathcal{U}$. (If o_2 has already been erased, then $\mathcal{O}$ e.g. locks up.) Then $\mathcal{O}$ erases o_2 from its memory.
Step 5. $\mathcal{U}$ verifies that $g_1^{r_1'} = A_{\mathcal{O}}^{d'} B_{\mathcal{O}}$. If this verification holds, he computes $r_1 = r_1' + d(u_1 s) + x_1 \bmod q$ and $r_2 = ds + x_2 \bmod q$. He then sends (r_1, r_2) to $\mathcal{S}$.

$\mathcal{S}$ accepts if and only if $\text{sign}(A, B)$ is a signature on (A, B), and $g_1^{r_1} g_2^{r_2} = A^d B$.

$\mathcal{O}$		$\mathcal{U}$		$\mathcal{S}$
			$\xrightarrow{A, B, \text{sign}(A,B)}$	$A \stackrel{?}{\neq} 1$
				$d \leftarrow \mathcal{H}_0(A, B, I_{\mathcal{S}}, \text{date/time})$
			$\xleftarrow{d}$	
	$\xleftarrow{d'}$	$d' \leftarrow s(d+e) \bmod q$		
o_2 still in memory?				
$r_1' \leftarrow d'o_1 + o_2 \bmod q$				
	$\xrightarrow{r_1'}$	$g_1^{r_1'} \stackrel{?}{=} A_{\mathcal{O}}^{d'} B_{\mathcal{O}}$		
		$r_1 \leftarrow r_1' + d(u_1 s) + x_1 \bmod q$		
		$r_2 \leftarrow ds + x_2 \bmod q$		
			$\xrightarrow{(r_1, r_2)}$	
				Verify $\text{sign}(A, B)$
				$g_1^{r_1} g_2^{r_2} \stackrel{?}{=} A^d B$

As in the basic system, if $\mathcal{U}$ has a clock and the capability of looking up the identifying information $I_{\mathcal{S}}$ of $\mathcal{S}$, the protocol can be collapsed to one move from $\mathcal{U}$ to $\mathcal{S}$.

Since $\mathcal{U}$ by himself does not know a representation of I, it is easy to prove that he cannot know a representation of the coin by himself if the basic withdrawal protocol is a restrictive blind signature protocol. From Lemma 8 we hence get:

Proposition 16. *Assuming the tamper-resistance of $\mathcal{O}$ cannot be broken, $\mathcal{U}$ cannot spend a coin without cooperation of $\mathcal{O}$.*

Due to the important fact that $\mathcal{O}$ in the ensemble of withdrawal and payment protocols in effect performs exactly the Schnorr identification protocol, proving knowledge of $\log_{g_1} A_{\mathcal{O}}$, this result should hold even after polynomially many executions of the protocols.

We next investigate the privacy of the account-holders in this system.

Proposition 17. *If $\mathcal{U}$ follows the protocols, and does not double-spend, then no shared information can be developed between $\mathcal{O}$, $\mathcal{B}$, and all shops $\mathcal{S}$ in executions of the withdrawal and payment protocols he takes part in.*

Informally speaking, the privacy of payments of an account-holder who follows the protocols and does not double-spent is unconditionally protected, even if his observer's contents can be examined afterwards by the bank. If we encode denominations in g_2, then the property of no shared information even relates to the value of the coin.

The deposit protocol. This is exactly the same as in the basic system. In case the coin was double-spent, the number $(r_1 - r_1')/(r_2 - r_2') - o_1 \bmod q$ serves as a proof of double-spending.

Proposition 18. *If the tamper-resistance of $\mathcal{O}$ is broken (enabling $\mathcal{U}$ to simulate its role), then still the same level of security as in the basic cash system is guaranteed. In particular, if $\mathcal{U}$ double-spends, he will be identified after the fact.*

This follows immediately from the fact that the protocols in that case reduce to those of the basic cash system (view $\mathcal{U}$ and $\mathcal{O}$ as one entity).

7 Concluding remarks

In practice, the random number generator of $\mathcal{U}$ can be a quite simple pseudo-random bit generator. In that case, it might be preferable to reconstruct A, B, x_1,, x_2, s at payment time from the intermediary state of the generator.

The random number generators of $\mathcal{O}$ and $\mathcal{B}$ on the other hand must be cryptographically strong, since $\mathcal{U}$ can heavily analyze their outputs; preferably, $\mathcal{B}$'s pseudo-random numbers should be combined with numbers obtained from physical randomness (e.g. noise generators).

Certain security aspects can be straightforwardly strengthened by using the idea of [4]. However, this modification does not seem to increase the plausibility of Assumption 1, whereas it requires more computations of the payment device.

The only thing left open in mathematically proving the security of our system and its extensions to as great an extent as the current state of knowledge in cryptography seems to allow, is proving that the particular blind signature protocol we used is a restrictive one, without assuming non-standard assumptions. We do not know how to do this, although there are convincing partial proofs (see [3]) that suggest that breaking it requires breaking the Diffie-Hellman key assumption.

Nevertheless, this is not a serious problem; recently ([3]), we have come up with various other (even more efficient) restrictive blind signature schemes in groups of prime order, of which we can rigorously prove for fixed m that the security is equivalent to that of the Schnorr signature scheme. As should be obvious, any restrictive blind signature scheme can be substituted for the particular one used in this abstract, requiring only some minor modifications to the protocols. In [3], we also describe similar constructions based on the representation problem in RSA-groups and restrictive blind signature schemes related to the Guillou/Quisquater signature scheme.

8 Acknowledgements

This system, in particular the extension to wallets with observers, is being studied by the European ESPRIT project CAFE. Various members of the project provided feedback on earlier versions of my technical report [2]. I am grateful to Ronald Cramer, Torben Pedersen and Berry Schoenmakers for their comments. Torben also provided part of the partial proofs that suggest that breaking Assumption 1 requires breaking the Schnorr signature scheme or the Diffie-Hellman assumption.

I especially want to thank David Chaum. This work was greatly inspired by his innovative work on untraceable electronic cash.

References

1. Bellare, Micali, "How To Sign Given Any Trapdoor Function," Proceedings of Crypto '88, Springer-Verlag, pages 200–215.
2. Brands, S., "An Efficient Off-line Electronic Cash System Based On The Representation Problem," CWI Technical Report CS-R9323, April 11, 1993.
3. Brands, S., "Untraceable Off-Line Cash Based On The Representation Problem," manuscript. To be published as a CWI Technical Report in Januari/Februari 1994.
4. Brickell, E. and McCurley, K., "An Interactive Identification Scheme Based On Discrete Logarithms And Factoring," Journal of Cryptology, Vol. 5 no. 1 (1992), pages 29–39.
5. Chaum, D., "Achieving Electronic Privacy," Scientific American, August 1992, pages 96–101.
6. Chaum, D., "Security Without Identification: Transaction Systems To Make Big Brother Obsolete," Communications of the ACM, Vol. 28 no. 10, October 1985, pages 1020–1044.
7. Chaum, D., "Card-computer moderated systems," (unpublished), 1989.
8. Chaum, D., Fiat, A. and Naor, M., "Untraceable Electronic Cash," Proceedings of Crypto '88, Springer-Verlag, pages 319–327.
9. Chaum, D. and Pedersen, T., "Wallet Databases With Observers," Preproceedings of Crypto '92.
10. Cramer, R. and Pedersen, T., "Improved Privacy In Wallets With Observers', Preproceedings of EuroCrypt '93.
11. Ferguson, N., "Single Term Off-Line Coins", Preproceedings of EuroCrypt '93.
12. Ferguson, N., "Extensions Of Single-Term Off-Line coins," these proceedings.
13. Fiat, A. and Shamir, A., "How To Prove Yourself: Practical Solutions To Identification And Signature Problems," Proceedings of Crypto '86, Springer-Verlag, pages 186–194.
14. Okamoto, T., "Provably Secure And Practical Identification Schemes And Corresponding Signature Schemes," Preproceedings of Crypto '92.

15. Schnorr, C.P., "Efficient Signature Generation By Smart Cards," Journal of Cryptology, Vol. 4 no. 3 (1991), pages 161–174.
16. "No Hiding Place / Big Brother Is Clocking You," The Economist, August 7th–13th 1993.

Discreet Solitary Games

Claude Crépeau[1*] and Joe Kilian[2]

[1] LIENS (CNRS URA 1327),
45 rue d'Ulm, 75230 Paris CEDEX 05, FRANCE.
e-mail: crepeau@dmi.ens.fr.
[2] NEC Research Institute,
4 Independence Way, Room 1C01,
South Brunswick, NJ 08540, USA.
e-mail: joe@research.nj.nec.com.

Abstract. Cryptographic techniques have been used intensively in the past to show how to play multiparty games in an adversarial scenario. We now investigate the cryptographic power of a deck of cards in a solitary scenario. In particular, we show how a person can select a random permutation satisfying a certain criterion *discreetly* (without knowing which one was picked) using a simple deck of cards. We also show how it is possible using cards to play games of partial information such as POKER, BRIDGE and other cards games in solitary.

1 Introduction

It's nearly Christmas time and you have to buy presents for your family and friends. Indeed, among certain families there is a more economical approach to this situation than buying one present per person: each member of a group picks the name of another member and becomes responsible for buying that person a present. That way everybody gets something but each person buys a single present. Traditionally, the one person each member is responsible for is allocated at random using the "names-in-the-hat" technique: each person puts its name in a common hat and then everybody picks a name at random from the hat. If by accident one picks his own name, he puts it back, otherwise he is responsible for the present of the person he picked. To put it abstractly, the goal is for the n persons involved to pick a random permutation π in a way that each of them p_i knows nothing but $\pi(i)$.

Indeed the "names-in-the-hat" technique leaks some information since participant p_i who picks his own name learns that $p_1, ..., p_{i-1}$ did not pick his name. In order for this technique not to leak information whatsoever, it is necessary to start from scratch each time someone picks his own name. One can check fairly easily that a random permutation of n elements will have no fixed point with probability roughly $\frac{1}{e}$. Therefore, a completely secret permutation should be found after roughly e trials.

* This work was performed while visiting NEC Research Institute in the spring of 1992.

Now consider the scenario where the members of this family cannot be gathered in a room to do the "names-in-the-hat" technique, for instance if some of them live abroad in several different countries. How can such a permutation be chosen locally, for instance by a single person without that person knowing the chosen permutation but knowing that it has no fixed point?

1.1 Related Work

Cryptography and card playing have a long history of connections. There has been substantial work on implementing card games using cryptography [SRA81, GM82, BF83, FM85, Yun85, Cré86, Cré87, GMW87, CCD88, BOGW88, RB]. Conversely, a number of researchers have considered mechanisms for implementing cryptographic primitives based on card games. Winkler [Win81a, Win81b, Win83] shows how two bridge players can securely communicate during the bidding process. More recently, Fischer, Paterson and Rackoff [FPR91] and Fischer and Wright [FW92, FW93] give a number of secret-key exchange protocols based on random card deals. Den Boer [den90] gives a protocol by which two parties may securely compute the AND function, based on the ability to make an *oblivious cut* on a deck of cards. We also base our protocol on oblivious cuts, and use a modified form of the secure AND protocol as a subroutine.

The novel contribution of our work is to provide a new scenario of a single person using cryptographic techniques as building blocs for playing sophisticated solitary games.

1.2 The Scenario

We consider this question in an *honest but non-oblivious* scenario. One person is going to be responsible for picking this permutation and will do this following a protocol we describe (it does not make much sense to try to prevent someone to cheat himself). At no point this person will be asked to forget information it has seen or can deduce from what it saw. Nevertheless we assume that an operation such as choosing a secret random cyclic shift of a set of objects (you may think of randomly cutting a deck of cards) is available to that person in order to create (unknown) randomness in his data. We will show how this person can pick a random permutation on the numbers $1, 2, ..., n$ and verify that it has no fixed point, learning no information whatsoever about which permutation was chosen. We qualify this process of "discreet". We also generalize this problem to the extent that we show how any "solitary game" can be played "discreetly" as long as there exist some polynomial size circuit to describe it.

We work with the following alphabet (each value can be thought as a suit in a deck of cards):

$$\left\{\boxed{\clubsuit}, \boxed{\heartsuit}, \boxed{\diamondsuit}, \boxed{\spadesuit}, \boxed{?}\right\}$$

The value $\boxed{?}$ representing any of the first four but face down on the table. We assume that all copies of one of these 5 elements are indistinguishable from one another. We define two notations on the basic elements for the rest of this paper:

Definition 1 $(c_1c_2...c_k)$. For any symbols $c_1, c_2, ...c_k$, we write $(c_1c_2...c_k)$ to represent the elements of $\{c_1c_2...c_k, c_2c_3...c_kc_1, ..., c_kc_1...c_{k-1}\}$, that is any cyclic permutation of $c_1c_2...c_k$.

indeed $(c_1c_2...c_k)$ is the equivalence class of strings equivalent up to a cyclic permutation.

Definition 2 $\langle c_1c_2...c_k\rangle$. For any symbols $c_1, c_2, ...c_k$, we write $\langle c_1c_2...c_k\rangle$ to express the fact that this string is obtained by a random cyclic shift, meaning that it is replaced by a random element of $(c_1c_2...c_k)$.

for instance,

⟨[♡][♣][◇][♠]⟩ → [◇][♠][♡][♣].

2 A Solution to the "no-fixed Point" Problem

For the solution to our first problem we use the following trivial coding:

[♣] = 0, [♡] = 1.

A sequence of n bits is associated to each participant p_i. Initially, to p_i associate the sequence of n [♣] except in position i where it is a [♡].

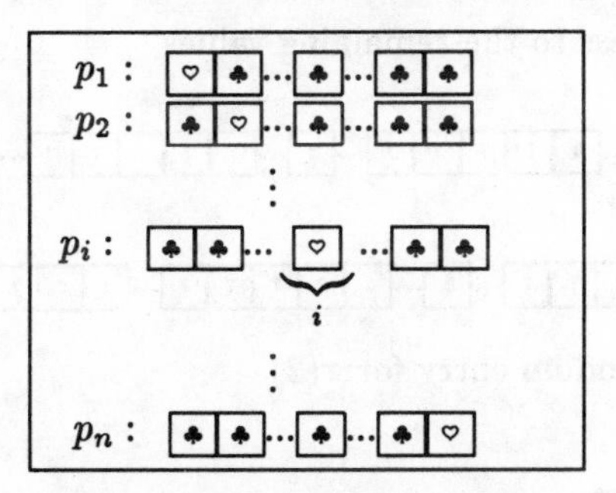

Then construct a long sequence by putting each of these sequences side by side, separated by markers made of $\frac{n}{2}$ [◇]'s followed by $\frac{n}{2}$ [♠]'s

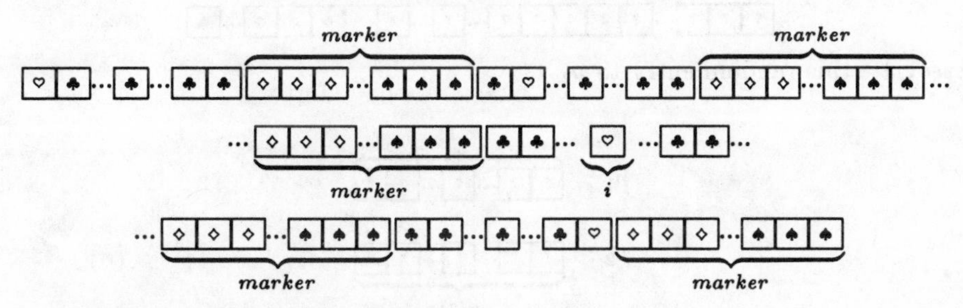

Apply them a random cyclic shift

⟨[?][?]...[?]...[?][?][?][?][?]...[?][?][?][?][?]...[?]...[?][?][?][?][?]...[?][?][?]...

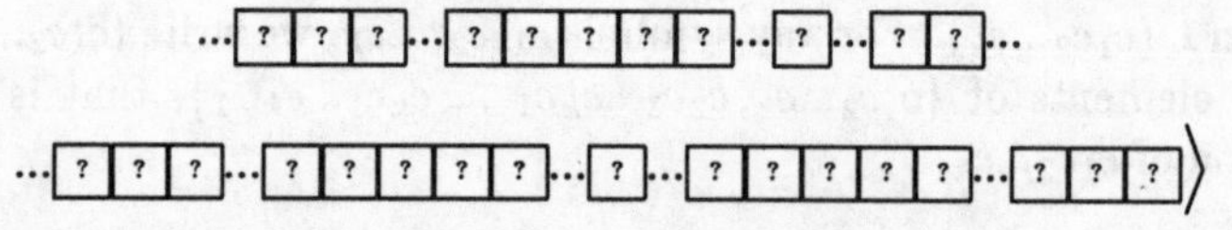

If the first element of the result is a ♡ or a ♠ then hide it and generate another random cyclic shift until you find a ◇ or ♠. When the first element is a ◇ then open values forward until you show $\frac{n}{2}$ ♠'s and then enough values backwards to show $\frac{n}{2}$ ◇'s. When you find a ♠ first, proceed in the reverse order.

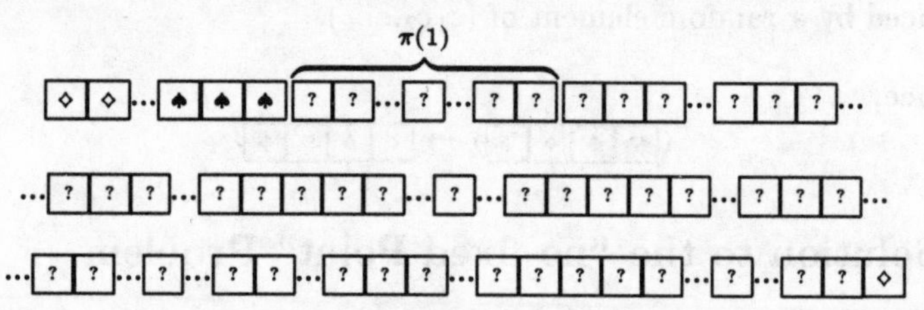

Get rid of the marker and extract the random entry for $\pi(1)$ located in the n values following the opened marker and associate it to p_1.

$\pi(1)$

p_1 : ? ? ... ? ... ? ?

Apply the same process to the remaining values

in order to select a random entry for $\pi(2)$.

$\pi(2)$

Associate this random entry to p_2

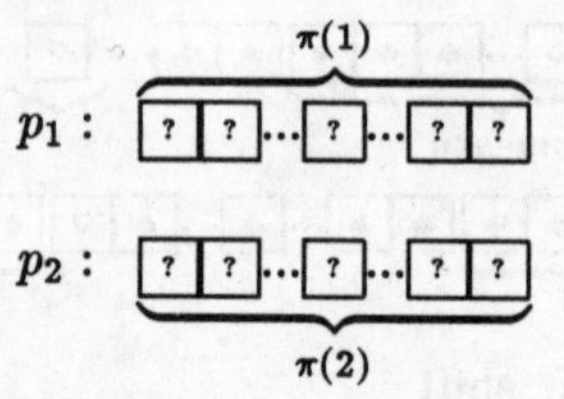

Repeat this process $n-2$ more times an obtain values for $\pi(3), ..., \pi(n)$ and associate each $\pi(i)$ to each p_i.

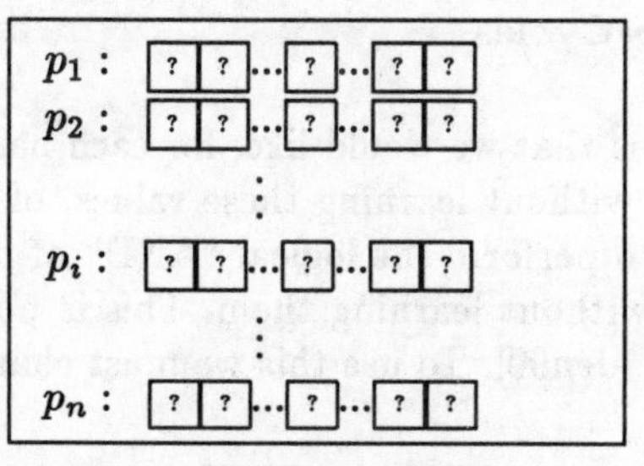

and check that the permutation π generated has no fixed point by opening the diagonal of this table and checking that it contains only ♣'s

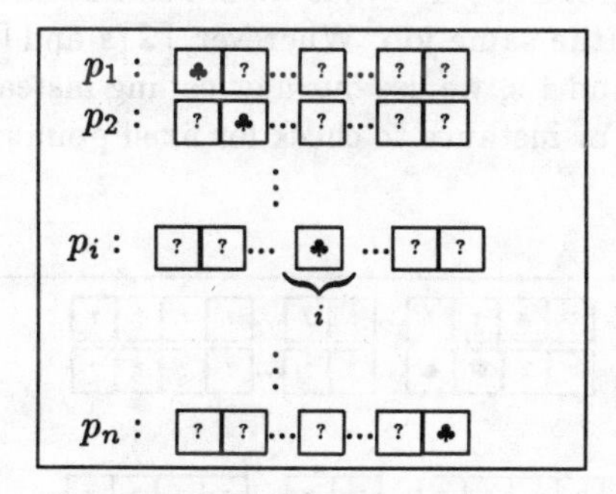

After doing so, you can put each sequence of values in an envelope and mail them out to the participants, telling them the correspondence between the n possible sequences and the people. You know for sure that nobody will get its own name and you know nothing at all about the permutation except for that fact.

3 A More Elaborate Problem: No Short Cycles

We now generalize the "no-fixed point" problem in non-trivial ways that will lead us to developing a general theory about what can be done discreetly by oneself.

Suppose that in order to make the exchange more diversified we disallow short cycles of length at most k in the permutation, for some constant k. This constraint makes the problem much more complicated. We first deal, in an ad hoc fashion, with the case $k = 2$ and build tools useful in the general scenario with $k > 2$.

It is very easy to see that for any $k < n$ the probability that a random permutation will have no cycle of length at most k is at least $\frac{1}{n}$. This is because the number of permutations with a single cycle of length n is $(n-1)!$, while the total number of permutations is $n!$. Therefore we can generate permutations with no cycles up to length k simply by picking a random permutation and checking it. On average, after at most n trials, one will work.

3.1 Detecting Two-Cycles

The basic observation is that we would like, for each pair i, j, to check whether $\pi(i) = j$ and $\pi(j) = i$ without learning these values, of course. Basically, what we need is to be able to perform the logical "AND" of positions (i, j) and (j, i) from the table above without learning them. This is possible using den Boer's "Match Making" trick [den90]. To use this we must change our coding to

[♡][♣] = 0, [♣][♡] = 1.

This does not change much about what we did so far, except that we use twice as many values to do the same job. Wherever, [♣]'s and [♡]'s were used in the past to represent 0's and 1's, we use our new coding instead (we also double the size of the markers). For instance to check for fixed points now involves opening two values per entry

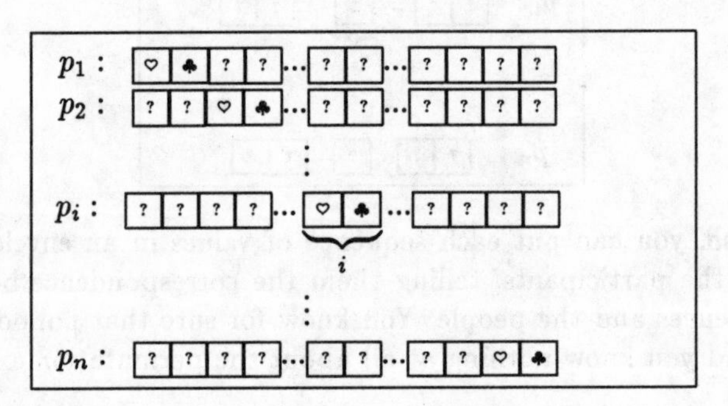

DenBoer's trick is used as follows to compute the logical "AND" of two secret bits coded as above.

Let b_0 and b_1 be two secret bits for which we would like to find out their logical "AND". Put b_0, b_1 and a [♣] side by side

[?][?] [?][?] [♣]
b_0 b_1

After hiding the [♣], swap b_1's values and randomly shift the 5 values cyclically.

⟨[?][?] [?][?] [?]⟩ (⇌ over b_1)

If the resulting sequence has its two [♡]'s side by side, ([♣][♡][♡][♣][♣]) it means that $b_0 = b_1 = 1$ and otherwise ([♣][♡][♣][♡][♣]) it means that at least one of them was a 0. We can use this trick in order to check whether $\pi(i) = j$ and $\pi(j) = i$

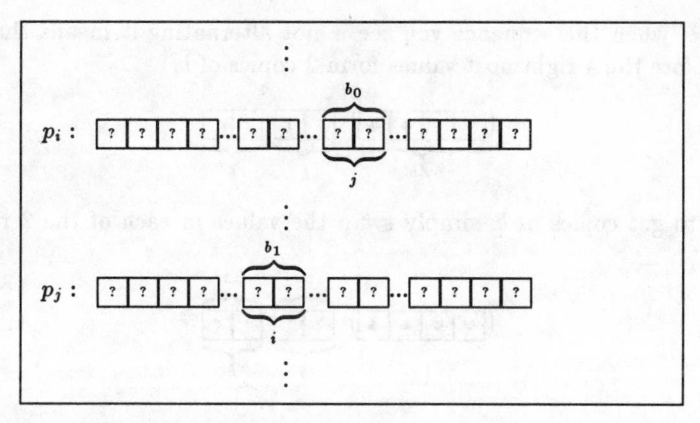

Unfortunately, doing so for a single pair i, j will destroy the data in a non-recoverable way. Therefore we need a mechanism to duplicate a bit in order to compare copies of bits and save some copy for further use.

3.2 Copying a Bit

Here is how one can make secret copies of a bit: Starting from a bit b at the left, put an alternation of 6 ♡'s and ♣'s to its right.

After hiding the 6 rightmost values, apply a random cyclic shift to them.

Because of the alternation that was put there in the beginning you know that each of the 3 pairs on the right represent a same bit b', but no longer know the value of b' because of the random cyclic shift.

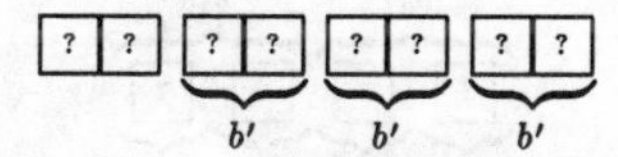

Now randomly shift the 4 leftmost values.

Open the 4 leftmost values; if the sequence you see is alternating then it means that $b = b'$ and therefore the 4 rightmost values form 2 copies of b.

Otherwise, when the sequence you see is not alternating it means that $b \neq b'$ and therefore the 4 rightmost values form 2 copies of $\bar{b}$.

In order to get copies of b, simply swap the values in each of the 2 rightmost pairs.

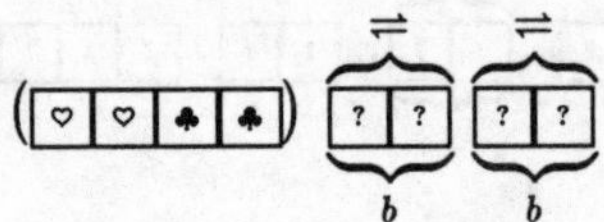

3.3 Detecting k-Cycles

In order to detect cycles longer than 2, it is necessary to perform more complicated computations on the secret bits. Indeed, it is nice to be able to evaluate "AND" gates but, as it is, we cannot even use the result of such a gate in a further secret computation because the answer is not in the same format as the data (a bit represented by a pair of values). Therefore we now show another tool to compute a pair of values (discreetly) that will satisfy some relation with two original pair of values (for instance the later represents the "AND" of the former). Equipped with such a tool, we can easily check for k-cycles of any length simply by designing a circuit that checks the length of all the cycles is at least k. (this is at most an n^3 algorithm)

3.4 Evaluating Logical Gates

Starting with two secret bits b_0, b_1, we show how to create a new secret bit for $b_0 \wedge b_1$. First, by the result of section 3.2 we can easily make copies of b_0, b_1 which we will use later in order to preserve the originals. Call x_0, y_0, x_1, y_1 the values of copies of b_0 and b_1 as follows:

b_0 b_1

? ? ? ?

x_0 y_0 x_1 y_1

Then, generate a set of 4 closed values as an alternation of ◇ and ♠:

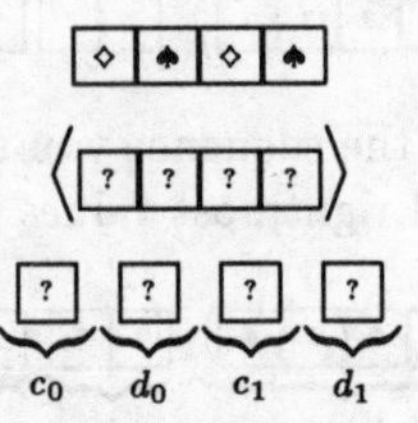

and build two decks as follows:

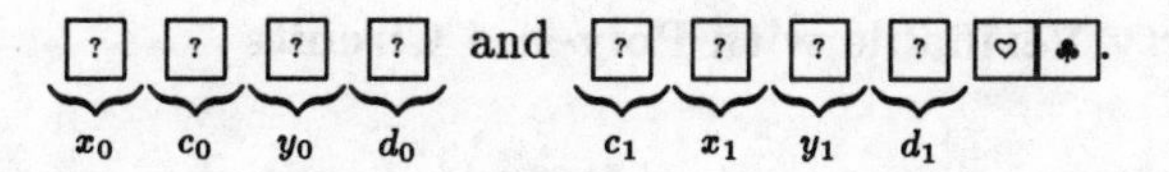

After random cyclic shift of the first four

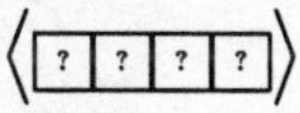

$\langle$? | ? | ? | ? $\rangle$

two possibilities may occur when opened:

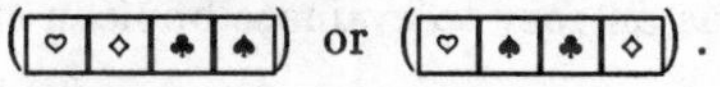

$\langle$ ♡ | ◇ | ♣ | ♠ $\rangle$ or $\langle$ ♡ | ♠ | ♣ | ◇ $\rangle$.

Repeat random cyclic shifts of the second deck

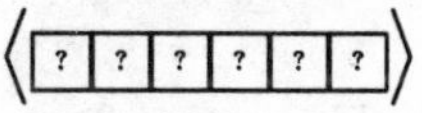

$\langle$? | ? | ? | ? | ? | ? $\rangle$

until the first value on top is the same as the value following the ♡ in the first deck:

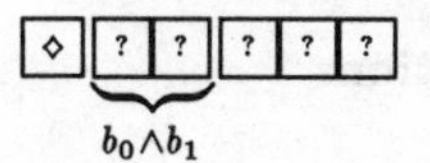

in the first case or

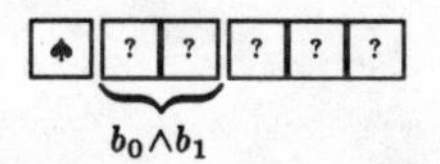

in the second case. The next two values after that will contain $b_0 \wedge b_1$.

Similar techniques will also work to build other gates. To do "OR" gates replace

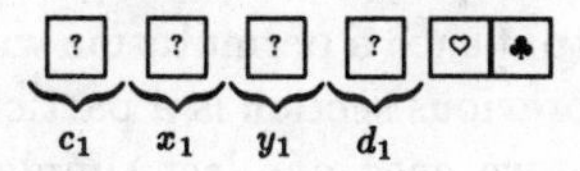

by

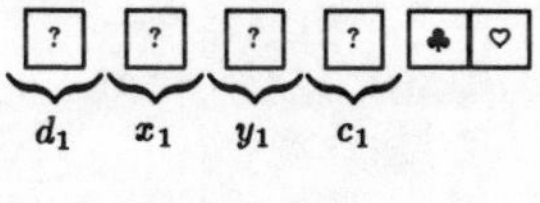

? ? ? ? ♣ ♡ (d_1, x_1, y_1, c_1)

and for "XOR" gates by

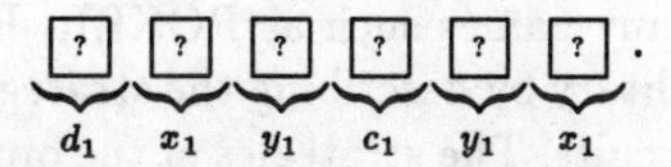

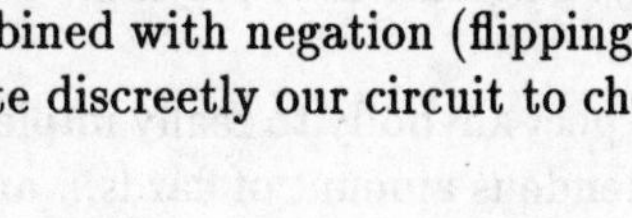

These operations combined with negation (flipping the values representing a bit) will suffice to simulate discreetly our circuit to check k-cycles.

3.5 Property Verifiable with Poly-size Circuits

It is clear that any similar property that can be described as an easy to evaluate circuit can be applied to a random permutation. Our techniques therefore allow us to generate discreetly permutations satisfying a certain condition $\mathcal{C}$ as long as

- there is a non-negligible probability of picking a random permutation satisfying $\mathcal{C}$
- $\mathcal{C}$ can be described as an easy to evaluate boolean circuit.

4 General Results

The tools that we have developed in section 3 are indeed very powerful. We now explain some of the things we can do with them.

4.1 Generating Permutations

Let's first focus on our earlier problem of generating permutations with a specific property.

4.2 Property Constructible with Poly-size Circuits

An extension of the result of the previous section is that it is possible to generate permutations satisfying certain properties as long as we can find a probabilistic boolean circuit that will output such a permutation with the correct distribution. What we suggested in the previous section is a particular case of this technique. In order to accomplish this we need one last simple tool: secret random bits. (Generating a random bit is simple: apply a random cyclic shift to a pair of hidden [♣] and [♡].)

4.3 Playing Games

Finally, we observe that any games such as POKER, BRIDGE and other card games can be played in solitary by describing the strategies of one's opponents as (probabilistic) boolean circuits. The strategies of the opponents can therefore be applied discreetly and played against as if playing with real opponents without learning extra information.

Of course, we do not expect anybody to really implement this idea with cards since it would take a tremendous amount of cards... and time...

5 Remarks and Open Questions

- A two-symbol alphabet suffices to achieve our result.(with a more complex construction)
- We have considered more general primitives such as *shuffling* but conjecture that the same result is impossible in that model.
- Another question is about the minimal number of values that must be randomly shifted through our protocols. We believe 4 suffice (even with a two-symbol alphabet).
- Several of the techniques used in section 3 and 4 are Las Vegas: it may take many iterations before the proper condition is met. This can be avoided at the price of using much longer sequences. (more details available in the final paper) One last open question is to achieve the same result keeping everything efficient.

Acknowledgments

Claude would like to thank Claire Kenyon, whose family-in-law, spread around the world, inspired this research. We would like to thank Charles H. Bennett for his highest enthusiasm in the early days of our result.

References

[BF83] I. Bárány and Z. Füredi. Mental poker with three or more players. *Information and Control*, 59:84–93, 1983.

[BOGW88] M. Ben-Or, S. Goldwasser, and A. Wigderson. Completeness theorems for fault-tolerant distributed computing. In *Proc. 20th ACM Symposium on Theory of Computing*, pages 1–10, Chicago, 1988. ACM.

[CCD88] D. Chaum, C. Crépeau, and I. Damgård. Multi-party unconditionally secure protocols. In *Proc. 20th ACM Symposium on Theory of Computing*, Chicago, 1988. ACM.

[Cré86] C. Crépeau. A secure poker protocol that minimizes the effects of player coalitions. In H. C. Williams, editor, *Advances in Cryptology: Proceedings of Crypto '85*, volume 218 of *Lecture Notes in Computer Science*, pages 73–86. Springer-Verlag, 1986.

[Cré87] C. Crépeau. A zero-knowledge poker protocol that achieves confidentiality of the players' strategy or how to achieve an electronic poker face. In A. M. Odlyzko, editor, *Advances in Cryptology: Proceedings of Crypto '86*, volume 263 of *Lecture Notes in Computer Science*, pages 239–247. Springer-Verlag, 1987.

[den90] B. denBoer. More efficient match-making and satisfiability: The five card trick. In *Advances in Cryptology: Proceedings of Eurocrypt '89*, volume 434 of *Lecture Notes in Computer Science*, pages 208–217. Springer-Verlag, 1990.

[FM85] S. Fortune and M. Merrit. Poker protocols. In G. R. Blakley and D. C. Chaum, editors, *Advances in Cryptology: Proceedings of Crypto '84*, volume

196 of *Lecture Notes in Computer Science*, pages 454–464. Springer-Verlag, 1985.

[FPR91] M. J. Fischer, M. S. Paterson, and C. Rackoff. Secret bit transmission using a random deal of cards. In *Distributed Computing and Cryptography*, pages 173–181. American Mathematical Society, 1991.

[FW92] M. J. Fischer and R. N. Wright. Multiparty secret exchange using a random deal of cards. In *Advances in Cryptology: Proceedings of Crypto '91*, volume 576 of *Lecture Notes in Computer Science*, pages 141–155. Springer-Verlag, 1992.

[FW93] M. J. Fischer and R. N. Wright. An Efficient Protocol for Unconditionally Secure Secret Key Exchange, In *Proc. 4th Annual Symposium on Discrete Algorithms*, January, 1993, 475–483.

[GM82] S. Goldwasser and S. Micali. Probabilistic encryption and how to play mental poker keeping secret all partial information. In *Proc. 14th ACM Symposium on Theory of Computing*, pages 365–377, San Francisco, 1982. ACM.

[GMW87] O. Goldreich, S. Micali, and A. Wigderson. How to play any mental game, or: A completeness theorem for protocols with honest majority. In *Proc. 19th ACM Symposium on Theory of Computing*, pages 218–229, New York City, 1987. ACM.

[RB] T. Rabin and M. Ben-Or. Verifiable Secret Sharing and Multiparty Protocols with Honest Majority, In *Proc., 21st ACM Symposium on Theory of Computing*, 1989.

[SRA81] A. Shamir, R. L. Rivest, and L. M. Adleman. Mental poker. In D. Klarner, editor, *The Mathematical Gardner*, pages 37–43. Wadsworth, Belmont, California, 1981.

[Yun85] M. Yung. Cryptoprotocols: Subscription to a public key, the secret blocking and the multi-player mental poker game. In G. R. Blakley and D. C. Chaum, editors, *Advances in Cryptology: Proceedings of Crypto '84*, volume 196 of *Lecture Notes in Computer Science*, pages 439–453. Springer-Verlag, 1985.

[Win81a] P. Winkler. Cryptologic techniques in bidding and defense: Parts I, II, III and IV. In *Bridge Magazine*, April–July 1981.

[Win81b] P. Winkler. My night at the Cryppie club. In *Bridge Magazine*, 60–63, August 1981.

[Win83] P. Winkler. The advent of cryptology in the game of bridge. In *Cryptologia*, 7(4):327–332, October 1983.

On Families of Hash Functions via Geometric Codes and Concatenation

Jürgen Bierbrauer[1], Thomas Johansson[2], Gregory Kabatianskii[3], Ben Smeets[2]

[1] Mathematisches Institut der Universität, Im Neuenheimer Feld 288, 69 Heidelberg, Germany

[2] Dept. of Information Theory, University of Lund, Box 118, S-221 00, Lund, Sweden †

[3] Inst. for Problems of Information Transmission, Russian Academy of Sciences, Ermolovoy 19, Moscow, GSP-4, Russia

Abstract. In this paper we use coding theory to give simple explanations of some recent results on universal hashing. We first apply our approach to give a precise and elegant analysis of the Wegman-Carter construction for authentication codes. Using Reed-Solomon codes and the well known concept of concatenated codes we can then give some new constructions, which require much less key size than previously known constructions. The relation to coding theory allows the use of codes from algebraic curves for the construction of hash functions. Particularly, we show how codes derived from Artin-Schreier curves, Hermitian curves and Suzuki curves yield good classes of universal hash functions.

1 Introduction

The concept of universal classes of hash functions was introduced by Carter and Wegman in [1]. It has found numerous applications of which we mention only cryptography, complexity theory, search algorithms and associative memory (see the Introduction in [2]). Three essentially different applications of universal hashing to authentication codes, [3], have been described in [4],[5] and [6]. Two of them are concerned with authentication without secrecy, the third (in [5]) is a novel use of universal classes of hash functions for error detection and information reduction in a system which guarantees integrity and secrecy.

In this paper, we present a detailed analysis of constructions of families of almost strongly universal hash functions proposed by Wegman and Carter [4] and recently, by Stinson [7],[6]. Our analysis is based on a recently discovered relationship between families of hash functions (or authentication codes (A-codes)) and error-correcting codes [8].

In Section 2 we give a simple explanation of previous results on the above mentioned constructions using the theory of concatenated codes, [10]. In the next section we present various improvements by using the concatenation of the

† These authors was supported by the TFR grant 222 92-662

well-known Reed-Solomon codes (RS-codes) and by using the powerful algebraic geometry codes (AG-codes) derived from algebraic curves. Finally we present some numerical results.

2 Universal hash functions and codes

In this section we recall some definitions and results from [7], [6]. We start by reformulating some of the results given by Stinson in a coding theoretic language and then proceed with introducing some additional notation.

Definition 1. Let $\epsilon > 0$. A multiset H of n functions from a set A to a q-set B is *$\epsilon-$almost universal$_2$* (short: AU_2) if for every pair $a_1, a_2 \in A$, $a_1 \neq a_2$ the number $d_H(a_1, a_2) = |\{h \in H\,;\, h(a_1) = h(a_2)\}| \leq \epsilon n$.

Consider now a q-ary code V of length n, $(q = |B|,\, n = |H|)$, whose codewords have the form $v = (h_1(a), ..., h_n(a))$, $a \in A$. It is rather clear (see also [8], [9]) that the property of ϵ-almost universal$_2$ is equivalent to the property that the minimal code distance $d(V)$ of the code V is not less than $n(1-\epsilon)$. So we have

Lemma 2. *Let $\varepsilon > 0$, $q = |B|$ and $n = |H|$. Then the following are equivalent:*

(i) H is an $\epsilon - AU_2$ class of hash functions from A to B.
(ii) The words $v = (h_1(a), ..., h_n(a))$, $a \in A$ form a q-ary code of length n with minimum distance d, where $1 - \frac{d}{n} \leq \epsilon$.

Stinson's analysis and improvements of the Wegman and Carter construction are based on two constructions, composition 1 and 2, which can be considered as the construction of concatenated codes (see [10]). The corresponding theorems on the performance of these constructions can be reformulated now in a more familiar manner.

In [7, Theorem 5.5] the following is shown:

Theorem 3 (Composition 1). *Let H_1 be an $\varepsilon_1 - AU_2$ from A_1 to B_1 and let H_2 be an $\varepsilon_2 - AU_2$ from B_1 to B_2. Then $H = H_1 \times H_2$ is an $\varepsilon - AU_2$ from A_1 to B_2 with $\varepsilon \leq \varepsilon_1 + \varepsilon_2 - \varepsilon_1\varepsilon_2$.*

In our language, this is a concatenation of two codes. If D is the distance of the concatenated code, then it is a well known fact that $D \geq d_1 d_2$. This gives $D \geq (1-\varepsilon_1)(1-\varepsilon_2)n_1 n_2$ and $\varepsilon \leq \varepsilon_1 + \varepsilon_2 - \varepsilon_1\varepsilon_2$. Thus this result is only a reformulation of the distance property of concatenated codes!

We recall from [7] also the notion of almost strongly universal hash functions.

Definition 4. Let $\epsilon > 0$. A multiset H of n functions from a set A to a q-set B is *$\epsilon-$almost strongly universal$_2$* (short: ASU_2) if

1. for every $a \in A$ and $y \in B$ the number of elements of H mapping $a \mapsto y$ is n/q,

2. for every pair $a_1, a_2 \in A$, $a_1 \neq a_2$, and every pair $y_1, y_2 \in B$ the number of elements of H affording the operation $a_1 \mapsto y_1, a_2 \mapsto y_2$ is $\leq \epsilon \cdot n/q$.

The notion of ASU_2 is clearly a generalization of orthogonal arrays of strength 2. In fact, an orthogonal array is obtained if in 2. of Definition 4 we always have equality. Condition 1 then follows automatically. Hence ASU_2 may be described as *generalized orthogonal arrays.* This link between ASU_2–classes and orthogonal arrays has been observed in earlier work. The relation between ASU_2–classes (or equivalently authentication codes), and error-correcting codes is already described in [8].

Now, also Theorem 5.6 in [7] can be reformulated as the product of "distances":

Theorem 5 (Composition 2). *Let H_1 be an $\varepsilon_1 - AU_2$ from A_1 to B_1 and let H_2 be an $\varepsilon_2 - ASU_2$ from B_1 to B_2. Then $H = H_1 \times H_2$ is an $\varepsilon - ASU_2$ from A_1 to B_2 with $\varepsilon \leq \varepsilon_1 + \varepsilon_2 - \varepsilon_1\varepsilon_2$.*

In our terms the theorem states that the concatenation of two codes, where the inner code additionally has the A-code properties, gives a code that satisfies the A-code properties with $\varepsilon_1 + \varepsilon_2 - \varepsilon_1\varepsilon_2$. Using our coding theoretic notation, we give a new proof, found in the appendix.

The families of hash functions which are used in Wegman&Carter and in Stinson's constructions can also be described using well-known codes. For example, it is obvious that the codes of Theorem 5.1, [7], are RS-codes with two information symbols. The codes of Theorems 5.2 and 5.3 in the same paper can be obtained in the same manner, see our Lemma 10, which is a generalization of these two.

Stinson's construction consists of two ingredients: the AU_2 classes (or error-correcting codes) and the ASU_2 classes.

For the first ingredient consider any linear code over a finite field. We fix the ground-field $\mathbb{F}_q$ and the relative minimum distance d/n of such a q-ary code. In fact the minimum distance has to be extremely large, as $\epsilon = 1 - \frac{d}{n}$ should be small. For a fixed number Q of codewords we ask for the minimum length of such a code. That is, we want a code with the highest possible rate.

Definition 6. Let natural numbers q, Q, and the real number $\epsilon, 0 < \epsilon < 1$ be given. Define $m(\epsilon, q, Q)$ as the minimum length n of a q-ary code with Q code-words and minimum distance d satisfying $d/n \geq 1 - \epsilon$.

This is a rather unusual question in coding theory. Unusual is also the fact that we are only interested in q-ary codes with relatively large q.

Similarly for the ASU_2 classes:

Definition 7. Define $m_A(\epsilon, q, Q)$ as the minimum number of functions of an $\epsilon - ASU_2$ class of hash functions from a Q-set to a q-set.

Using the above terminology, Stinson's compositions give:

Lemma 8. *With $m(\varepsilon, q, Q)$ and $m_A(\varepsilon, q, Q)$ as defined above we have*

(i) For composition 1: $m(\epsilon_1 + \epsilon_2, q, Q) \leq m(\epsilon_2, q, Q_1) \cdot m(\epsilon_1, Q_1, Q)$.
(ii) For composition 2: $m_A(\epsilon_1 + \epsilon_2, q, Q) \leq m_A(\epsilon_2, q, Q_1) \cdot m(\epsilon_1, Q_1, Q)$.
(iii) Cartesian product, [6]: $m(\epsilon, q^i, Q^i) \leq m(\epsilon, q, Q)$.

Example 1. Starting from the 2-dimensional RS-code and using composition 1 and the Cartesian product recursively one obtains

$$m(\frac{i}{q}, q, q^{2^i}) \leq q^i,$$

for every prime-power q and every $i \geq 1$. This is a construction of Stinson's, [6, Theorem 6.1], expressed in different words. □

The k-dimensional Reed-Solomon code yields $m(\frac{k-1}{q}, q, q^k) \leq q$. The Singleton bound shows that we actually have equality:

Theorem 9.

$$m(\frac{k-1}{q}, q, q^k) = q$$

for every prime power q and $k \geq 2$.

What the ASU_2–classes are concerned we may use the following lemma:

Lemma 10 "projection hashing". *Let π be some $\mathbb{F}_{q_0}$-linear map from $\mathbb{F}_Q$ on $\mathbb{F}_q$, where $Q = q_0^n, q = q_0^m$ and q_0 a prime power. Then the following family of hash functions $H = \{h_{a,b}\,;\, h_{a,b}(x) = \pi(ax) + b\}$, where $a, x \in \mathbb{F}_Q, b \in \mathbb{F}_q$ is $\epsilon - ASU_2$ with $\epsilon = 1/q$.*
Consequently $m_A(\frac{1}{q_0^m}, q_0^m, q_0^n) \leq q_0^{m+n}$ for every prime-power q_0 and $n \geq m$.

Proof. Proof follows from Theorem 11 below. □

Remark 1: We obtain the same ASU_2's if we take the family of orthogonal arrays constructed in [12, page 363].
Remark 2: The case n=m=1 stems from a 2-dimensional Reed-Solomon code.
Remark 3: This lemma can be generalized as is done in Theorem 11. The first author gave a generalization via orthogonal arrays $OA_{q^{(t-1)(n-m)}}(t, q^n, q^m)$, with $t \geq 2$.

3 The evaluation of parameters of Wegman&Carter's construction

Wegman and Carter proposed in [4] the following method for constructing an authentication code. Let A be a set binary words of length a' and B a set of binary words of length b'. Divide a word $a \in A$ in segments of length s, where the parameter s will be chosen later in a proper way, and apply to each segment some (but the same) hash function from family $\widetilde{H}$ of [7, Theorem 5.2], where $q = 2^s$. As a result one has again binary words but only halve as long. Repeat this procedure

ν times with arbitrary hash functions $h_1, ..., h_\nu$ until we get a word of length s. And the last step consists in taking some (e.g. the low-order) b' bits from this word. Wegman&Carter and later Stinson investigated the important parameters of the A-code thus obtained: the probability of successful impersonation[5] P_I, the probability of successful substitution P_S and the size of the key (logarithm of the number of hash functions). Wegman&Carter's construction was interesting because of their basic observation that by increasing P_S beyond P_I ($P_S \leq 2 \cdot P_I$ say), the source space can be dramatically enlarged. In fact, in [8], it was shown that when $P_S > P_I$ the source space grows exponentially in the key size! In what follows we always have $P_I = 2^{-b'}$.

For our purpose it is more convenient to represent the construction by two "stages". The first stage consists of $\nu - 1$ concatenations. As a result of this stage we have a family of hash functions from A to B^*, where B^* is a set of binary words of length $2s$. Or, in other words, we have q^*-ary code ($q^* = 2^{2s}$), which is a result of $\nu - 1$ concatenations. According to Theorem 3 we have got the following inequality for the corresponding value of ϵ for this code $\varepsilon_1 = \epsilon^* \leq 1 - (1 - \widetilde{\epsilon})^{\nu-1}$ where $\widetilde{\epsilon} = 1/q$, see [7, Theorem 5.2].

The second stage consists of application of a hash function from $\widetilde{H}$ to a $2s$-bit word and then taking b'-bits from the resulting word. The performance of such family of hash functions is given by Lemma 10.

Combining these two stages we have got exactly(!) the parameters of the Wegman&Carter construction. Namely $\epsilon = \epsilon_1 + \epsilon_2 - \epsilon_1 \cdot \epsilon_2$, where $\epsilon_1 = 1 - (1 - (1/2^s))^{\nu-1}$, $\epsilon_2 = 1/2^{b'}$, $\nu = \log_2 a'/s$. The number of hash functions, n, (or length of the corresponding code) equals $2^{(\nu-1)3s+2s+b'}$ (this slightly better than in the original paper as the authors used a rough estimate, i.e., Q^2 instead of Qq as we have from Lemma 10). Thus we can confirm the correctness of [4] and refute the remark in [7, page 83].

Example 2. (see [7]). Let $s = 23$, $b' = 20$, $\nu = 7$. Then the W&C construction gives $a' = 23 \cdot 2^7$, $\epsilon_1 \leq (3/4)2^{-20}$, $\varepsilon \leq \varepsilon_1 + \varepsilon_2 \leq 2^{-19}$, and the number of hash functions equals 2^{480}. □

The disadvantage of the original W&C construction is the usage of A-codes within the first stage as it is enough to use only ordinary codes. This observation immediately leads us to replacing the A-codes of [7, Theorem 5.2] by codes of [7, Theorem 5.1]. It decreases the number of function to $2^{(\nu-1)s+2s+b'}$ without decreasing the final ϵ. In particular, one gets 2^{204} as the number of hash functions for the considered example, like in [6]. However we can do better as we will show in the next section.

4 Construction of families of hash functions via RS codes

Before we describe our construction we first prove the following theorem:

---

[5] Stinson uses here P_{d_0} and P_{d_1}. We keep the original notation of Simmons [3].

Theorem 11. *Let Q, q, p, π be the same as in Lemma 10. Then the following family of hash functions*

$$H = \{h_{x,y} : h_{x,y}(a_1, ..., a_k) = \pi(xa_1 + x^2a_2 + ... + x^k a_k) + y$$
$$\text{where } x, a_1, a_2, ..., a_k \in \mathbb{F}_Q \text{ and } y \in \mathbb{F}_q\}$$

is ϵ–ASU_2 with $\epsilon = k/q$ and $|A| = Q^k, |H| = qQ$. Thus we have $m_A(k/q, q, Q^k) \leq qQ$.

Proof. It is clear that for any $a \in \mathbb{F}_Q, z \in \mathbb{F}_q$, the number of hash functions $h : h(a) = z$ is the same and equals n/q, where $n = |H|$. We now calculate the maximal number of hash functions such that $h(a) = z, h(b) = z'$, where $a, b \in \mathbb{F}_q^k$, $z, z' \in \mathbb{F}_q$. Saying in other words, we are interested in the evaluation of the maximal number of solutions of the corresponding system of two algebraic equations. This system is equivalent to the following system $h(a) = z, \pi(c_1x + c_2x^2 + ... + c_kx^k) = w$, where $c = a - b, w = z - z'$. According to Bezout's Theorem the number of solutions of the second equation is not greater than $k|\text{Kern}\pi|$, where $|\text{Kern}\pi| = |\{u \in \mathbb{F}_q ; \pi(u) = 0\}| = Q/q$. □

Remark 1: For $k = 1$ one has Lemma 10. This theorem easily gives some of the results found in [6].
Remark 2: This construction can be explained in coding theoretic language starting with Reed-Solomon codes.

A natural application of RS-codes is their concatenation as inner codes together with ASU_2-codes of Lemma 10.

Proposed construction: We propose to construct $\epsilon - ASU_2$ classes of hash functions for authentication in the following way: Concatenate an $\epsilon_1 - AU_2$ class which is obtained from an RS-code over $\mathbb{F}_Q$ with an $\epsilon_2 - ASU_2$ class from Lemma 10. According to Theorem 5 we get an $(\epsilon_1 + \epsilon_2) - ASU_2$ class.

In detail, it can be described as follows: Let $q = 2^r$ and $Q = 2^{r+s}$. Choose an RS-code over $\mathbb{F}_Q$ with $n = Q$ and $k = 1 + 2^s$. The size of the message space is $|M| = Q^{1+2^s} = 2^{(r+s)(1+2^s)}$. This is the $\epsilon_1 - AU_2$ class and $\epsilon_1 = 1 - d/n = 1 - (2^{r+s} - 2^s)/2^{r+s} = 1/2^r$. From Lemma 10 we have an $\epsilon_2 - ASU_2$ class from $\mathbb{F}_Q$ to $\mathbb{F}_q$, with $\varepsilon_2 = 1/2^r$. The concatenation of these two gives the desired $\epsilon - ASU_2$ class, where $\epsilon \leq 2/2^r$. The size of the key space is then Q^2q. Note there is only one (!) concatenation in this construction. Note also that this works in any characteristic. The result is

$$m_A(\frac{2}{q^r}, q^r, q^{(r+s)(1+q^s)}) \leq q^{3r+2s}.$$

Example 3. Let us show how the construction works by giving a numerical example for the case considered in Example 2. Take a Q-ary RS-code with $Q = 2^{27}, k = 1 + 2^7 = 129$. This is an AU_2-code with $\epsilon = 2^{-20}$. Application of the concatenation construction with codes, the Lemma with $Q = 2^{27}$, $q = 2^{20}$ gives an ASU_2 code with $|A| = 2^{27 \cdot 129}$, $|H| = 2^{74}$, $\epsilon \leq 2^{-19}$. This is case $q = 2, r = 20$, $s = 7$ above. □

5 The use of geometric codes

We want to show how more sophisticated classes of linear codes, in particular of codes defined on algebraic curves, may be used to improve Stinson's bound considerably (see Example 1). It is natural in our context to use the machinery of geometric codes in the following form:

Theorem 12 (Canonical construction). *Let $q \geq 9$ be a quadratic prime power and let K be a function field of transcendence-degree 1 (equivalently: an algebraic curve) over the field $\mathbb{F}_q$ of constants, $P_0, P_1, \ldots P_n$ rational points of K. Consider the divisors $D = P_1 + P_2 \ldots + P_n, G = mP_0$. Let $m_1 = 0, m_2, \ldots, m_k, \ldots$ be the pole-orders of P_0. Consider the code*

$$C_k = C(D, m_k P_0)$$

of functions which are everywhere holomorphic except for a pole of degree $\leq m_k$ at P_0, evaluated at $P_1, \ldots P_n$ (this is the L-construction of [13]). Then C_k has dimension k and minimum distance $\geq n - m_k$. Hence

$$m(\frac{m_k}{n}, q, q^k) \leq n.$$

If moreover $m_k - 1$ is a Weierstraß gap, then

$$m(\frac{m_{k-1}}{n}, q, q^k) \leq n.$$

We need curves with many rational points and at least one rational Weierstraß-point whose gaps are as large as possible. In fact, Reed-Solomon codes result from the canonical construction when applied to the rational curve. In [14] and [15] a class $K_q^{(r)}$ of function fields defined over an arbitrary finite field $\mathbb{F}_q$ of constants is studied, where $r \geq 2$. Here $K_q^{(r)}$ is a tower of Artin-Schreier extensions of the rational function field. The following facts are to be found in [15]: The number N_1 of rational points of $K_q^{(r)}$ is $N_1 = q^r + 1$. There is a rational Weierstraß-point P_0 whose semigroup of pole-orders is

$$\sum_{i=1}^{r} q^{r-i}(q+1)^{i-1}\mathbb{N}_0.$$

This yields improvements upon the Stinson-bound valid for all sufficiently large prime-powers. In fact we can get a precise asymptotic statement. Upon using a well-known inequality between binomials and the binary entropy-function H:

$$2^{mH(l/m)}/(m+1)^2 \leq \binom{m}{l} \leq 2^{mH(l/m)}$$

(see [16]) the following is obtained:

Theorem 13. *Let q_0 be the unique positive solution of the equation*

$$H(q_0) = q_0.$$

For every $\epsilon > 0$ and sufficiently large i we have

$$m(\frac{i}{q}, q, q^{2^i}) \leq q^r,$$

where $r = \lfloor (i-1)(1-q_0)/q_0 - \epsilon \rfloor$ and q is an arbitrary prime-power, $q > (i-1)(r-1)$.

The numerical values are

$$q_0 = .7729\ldots, (1-q_0)/q_0 \approx .2938$$

We note that the same number q_0 appears in the theory of *Sperner capacity*, a recently discovered extension of the concept of *Shannon capacity* of a graph (see [17]).

For small values of i and a quadratic ground-field we obtain improvements by means of *Hermitian codes.* Consider the *Hermitian curve* defined by the equation $X^{q+1} + Y^{q+1} + Z^{q+1}$ over the field $\mathbb{F}_{q^2}$ of constants. This curve has genus $\binom{q}{2}$ and $q^3 + 1$ rational points. These form the well-known Hermitian unital. They are all Weierstraß-points. The semigroup of pole-orders of any of them is $q\mathbb{N}_0 + (q+1)\mathbb{N}_0$. In particular the integers between wq and w(q+1) are pole orders. Let us call w the ***weight*** of such a pole-order. If $w < q$, then a pole order of weight w doesn't have any other weight. The number of pole orders of weight $\leq w$ is then $1 + 2 + \ldots + (w+1) = \binom{w+2}{2}$.

Lemma 14. *Let (m_k) be the pole orders of the Hermitian curve over $\mathbb{F}_{q^2}$. If $w < q$, then*

$$m_{\binom{w+2}{2}} = w(q+1),$$

$$m_{\binom{w+1}{2}+1} = w \cdot q > m_{\binom{w+1}{2}} + 1.$$

We may use the construction of the preceding section and replace the RS-codes by Hermitian codes. If we choose $k = q^s + 1$ and use the canonical construction in its strengthend form, the following is obtained:

Example 4. We want bounds on $m_A(2^{-19}, 2^{20}, 2^{2^{28}})$. This is case $q = 2$, $r = 20$ above. We have

$$m_A(2^{-19}, 2^{20}, 2^{2^{28}}) \leq m(\frac{2^{28}}{q^3}, q^2, 2^{2^{28}}) \cdot m_A(2^{-20}, 2^{20}, 2^{32}),$$

by Lemma 8, where $q = 2^{16}$. The second factor above is bounded by 2^{52} (Lemma 10). Use the canonical construction for the Hermitian curve, where $k = 2^{28}/32 =$

$2^{23}, w = 2^{12} - 1$. As $\binom{w+2}{2} > k$, it follows from Lemma 14 that $m_k < w(q+1) = (2^{12} - 1)(2^{16} + 1) < 2^{28}$. By the canonical construction

$$m(\frac{2^{28}}{q^3}, q^2, 2^{2^{28}}) \leq m(m_k/q^3, q^2, (q^2)^k) \leq q^3.$$

Thus

$$m_A(2^{-19}, 2^{20}, 2^{2^{28}}) \leq q^3 2^{52} = 2^{100}$$

and we may thus choose $s = 12$. □

In characteristic 2 we get further improvements by using a family of curves which admit the Suzuki groups as automorphism groups. This family is studied in [18]. Let $q = 2^{2f+1}, q_0 = 2^f$. The curve is defined over $\mathbb{F}_q$ by the homogeneous equation

$$X^{q_0}(Z^q + ZX^{q-1}) = Y^{q_0}(Y^q + YX^{q-1}),$$

has $q^2 + 1$ rational points and a Weierstraß-point whose semigroup of pole-orders is

$$q\mathbb{N}_0 + (q + q_0)\mathbb{N}_0 + (q + 2q_0)\mathbb{N}_0 + (q + 2q_0 + 1)\mathbb{N}_0.$$

The number of pole-orders of weight $w < q_0$ is $\binom{w+2}{2} + \binom{w+1}{2} = (w+1)^2$. Via the canonical construction we obtain:

Theorem 15. *Let* $q = 2^{2f+1} \geq 128$. *Then*

$$m(\frac{i}{q}, q, q^{2^i}) \leq q^2 \; (i = 3, 4, 5, 6, 7).$$

If we use Suzuki codes in the same spirit as we used RS-codes and Hermitian codes above, we get

$$m_A(\frac{1}{2^{r-1}}, 2^r, 2^{(r+s)\{1+(2^s+1)(2^s+2)(2^{s+1}+3)/6\}} \leq 2^{4r+3s} \text{ if } s < r, s + r \text{ odd}.$$

The last statement of Theorem 12 follows from a recent result of A. Garcia et.al. ([19]). We give an application of this strengthened form of the canonical construction when applied to Hermitian curves:

Theorem 16. *Let q be a quadratic prime-power. Then*

$$m(\frac{1}{q} + \frac{1}{q^{3/2}}, q, q^4) \leq q^{3/2}.$$

Observe that there is not much of a difference between probabilities $\frac{1}{q}$ and $\frac{1}{q} + \frac{1}{q^{3/2}}$. For practical purposes the statement above should therefore be interpreted as

$$m(\approx \frac{1}{q}, q, q^4) \leq q^{3/2}.$$

It is natural to conjecture that all the Deligne-Lusztig curves will yield good codes and good classes of hash functions. In the case of the Ree curves we have not yet been able to verify this as the Weierstraß-points and their pole orders seem to be unknown.

6 Numerical Results

In order to illustrate our results we proceed as in Section 6 of [6]: Let $|A| = 2^{a'}$, $|B| = 2^{b'}$. This means that we have an a'-bit source and we want to use b'-bit authenticators. The cases we tabulate include those given in [6], with improved values for the necessary length of key. Here $P_S = 2^{-19}$. In most cases the AU_2 class is produced by Reed-Solomon codes (Section 4). Only in the case $a' = 2^{28}$, $b' = 20$ we use a Hermitian code, Example 4, where we get 100 instead of 106 bits of key!

length of source	length of authenticator	length of key		
a'	b'	s	new	[6]
2^8	20	24	68	135
2^{12}	20	28	76	236
2^{16}	20	32	84	332
2^{20}	20	35	90	445
2^{24}	20	39	98	-
2^{28}	20	32	100	-
2^8	40	43	126	255
2^{12}	40	47	134	346
2^{16}	40	51	142	612
2^{20}	40	55	150	805

Table 1. Table with source and key size for A-codes with $P_I = 2^{-b}$ and $P_S \leq 2 \cdot P_I$.

We can compare these results with a lower bound based on the q-twisting technique and the Varshamov-Gilbert bound which gives a Varshamov-Gilbert type bound for A-codes. This bound is an existence result and tells us that there exist A-codes with given P_I and $P_S \geq P_I$ with a certain number of source states (as function of the number of keys). For example, using the results of another paper, [20], which discusses this in more detail we have that for the situation in Examples 2 and 3 with $P_I = 2^{-20}$ and $P_S \leq 2P_I$ there exist classes which require only 52 bits for the key size.

7 Conclusion

We have shown that using coding theory we can easily reformulate and prove results in [4] and [7]. In particular, the concepts of geometric codes and concatenated codes gives us powerful tools. We also gave a simple analysis of the Wegman&Carter construction by our approach and suggested some improvements. Finally the idea of concatenation was used to get a new type of construction which requires considerably less key size than previously known. The construction using RS-codes is surpassed by the one using AG-codes for very large source sizes. In our table this happened when we authenticate 33MB(yte) source strings. Further development on algebraic geometry might improve this in favor of the AG-codes.

A Proof of Theorem 5

Proof. We have a Q-ary code ($Q = |B_1|$) V of length n_1 ($n_1 = |H_1|$) and with distance $d = (1-\varepsilon_1)n_1$ and a q-ary code W of size Q ($|H_2| = Q$) of length n_2 with the special property

$$\forall w \neq w' \in W, \forall \alpha, \beta \in \{0,1,\ldots,q-1\} \quad |\{i\,;\, w_i = \alpha, w'_i = \beta\}| \leq \varepsilon_2 \frac{n_2}{q}$$

and

$$\forall w \in W, \forall \alpha \in \{0,1,\ldots,q-1\} \quad |\{i\,;\, w_i = \alpha\}| = \frac{n_2}{q}.$$

We form the concatenated code C from V and W by replacing symbols from the Q-ary alphabet in the codewords of V by the corresponding codewords from code W. Let us now compute

$$\left|\{j\,;\, c_j = \alpha, c'_j = \beta\}\right|, \quad \alpha, \beta \in \{0,1,\ldots,q-1\},$$

where $\underline{c} = \phi(\underline{v}), \underline{c}' = \phi(\underline{v}')$,

$$\phi(\underline{v}) = (\rho(v_1), \rho(v_2), \ldots, \rho(v_{n_1}))$$

and where

$$\rho : \{0,1,\ldots,q-1\} \rightarrow W$$

is a bijective map. The index j of c_j can be considered as a pair (j_1, j_2), where

$$0 \leq j_1 < n_1, 0 \leq j_2 < n_2.$$

Let us first consider the case $\alpha = \beta$. Then the set $\left\{j\,;\, c_j = c'_j = \alpha\right\}$ consists of all $j = (j_1, j_2)$ such that

a) $v_{j_1} = v'_{j_1}$ and $\rho(v_{j_1})_{j_2} = \alpha$

b) $v_{j_1} \neq v'_{j_1}$ and $\rho(v_{j_1})_{j_2} = \rho(v'_{j_1})_{j_2} = \alpha$.

We have that the number of indices satisfying a) is $(n_1 - d(\underline{v}, \underline{v}'))\frac{n_2}{q}$. The number of indices satisfying b) is $\leq d(\underline{v}, \underline{v}')\varepsilon_2 \frac{n_2}{q}$. Since a) and b) count disjoint situations the total number satisfies

$$\begin{aligned}\left|\{j\,;\, c_j = c'_j = \alpha\}\right| &\leq \frac{n_2}{q}(n_1 - d + d\varepsilon_2) \\ &= \frac{n_1 n_2}{q}(1 - (1-\varepsilon_1)(1-\varepsilon_2)) \\ &= \frac{n_1 n_2}{q}(\varepsilon_1 + \varepsilon_2 - \varepsilon_1\varepsilon_2).\end{aligned}$$

For the case $\alpha \neq \beta$ we can not have situation a) but only b) and the contribution is less for this case. Thus we have proved that $\varepsilon \leq \varepsilon_1 + \varepsilon_2 - \varepsilon_1\varepsilon_2$. □

References

1. J.L. Carter, M.N. Wegman, "Universal Classes of Hash Functions", *J.Computer and System Sci.*, Vol. 18, 1979, pp. 143-154.
2. D.R. Stinson, *Combinatorial techniques for universal hashing,* University of Nebraska-Lincoln. Department of Computer Science and Engineering, 1990.
3. G.J. Simmons, "A survey of Information Authentication", in *Contemporary Cryptology, The science of information integrity,* ed. G.J. Simmons, IEEE Press, New York, 1992.
4. M.N. Wegman, J.L. Carter, "New hash functions and their use in authentication and set equality", *J. Computer and System Sciences,* Vol. 22, 1981, pp. 265-279.
5. C.H. Bennett, G. Brassard, J-M. Roberts, " Privacy amplification by public discussion", *SIAM J.Comput.*, Vol. 17:2, 1988, pp. 210-229.
6. D.R. Stinson, "Universal Hashing and Authentication Codes", to appear in IEEE Transactions on Information Theory. This is a final version of [7].
7. D. R. Stinson, "Universal hashing and authentication codes" *Proceedings of Crypto 91,* Santa Barbara, USA, 1991, pp. 74-85.
8. T. Johansson, G. Kabatianskii, B. Smeets, "On the relation between A-codes and codes correcting independent errors" *Proceedings Eurocrypt'93,* to appear.
9. J. Bierbrauer, "Universal hashing and geometric codes", manuscript.
10. G.D. Forney, Jr., *Concatenated Codes,* M.I.T. Press, Cambridge, MA., 1966.
11. J. Bierbrauer, "Construction of orthogonal arrays", to appear in *Journal of Statistical Planning and Inference.*
12. T. Beth, D. Jungnickel, H. Lenz, *Design Theory,* Bibliographisches Institut, Zürich 1985.
13. M.A. Tsfasman, S.G. Vlăduţ , *Algebraic-Geometric codes,* Kluwer Academic Publ., Dordrecht/Boston/London, 1991.
14. R. Pellikaan, B.Z. Shen, and G.J.M. van Wee, " Which linear codes are algebraic-geometric?", *IEEE Trans. Information Theory,* Vol. 37, 1991, pp. 583-602.
15. B.H. Matzat, "Kanonische Codes auf einigen Überdeckungskurven", *Manuscripta Mathematica,* Vol. 77, 1992, pp. 321-335.
16. L. Gargano, J. Körner, U. Vaccaro, "Sperner capacities", to appear in *Graphs and Combinatorics.*
17. L. Gargano, J. Körner, U. Vaccaro, "Capacities: from information theory to extremal set theory", to appear in *Journal of the AMS.*
18. J.P. Hansen, H. Stichtenoth, "Group Codes on Certain algebraic curves with many rational points",*AAECC,* Vol. 1, 1990, pp. 67-77.
19. A. Garcia, S.J. Kim, R.F. Lax, "Consecutive Weierstrass gaps and minimum distance of Goppa codes", *Journal of Pure and Applied Algebra,* Vol. 84, 1993, pp. 199-207.
20. G. Kabatianskii, B. Smeets, T. Johansson, "Bounds on the size of a-codes and families of hash functions via coding theory", manuscript.

On the Construction of Perfect Authentication Codes that Permit Arbitration

Thomas Johansson

Department of Information Theory, Lund University
Box 118, S-221 00 Lund, Sweden

Abstract. [1] Authentication codes that permit arbitration are codes that unconditionally protect against deceptions from the outsiders and additionally also protect against some forms of deceptions from the insiders. Simmons introduced this authentication model and he also showed a way of constructing such codes, called the Cartesian product construction. We present a general way of constructing such codes and we also derive two specific classes of such codes. One that is perfect in the sense that it meets the lower bounds on the size of the transmitter's and the receiver's keys and one that allows the number of source states to be chosen arbitrarily large.

1 Introduction

The purpose of traditional authentication codes is to protect the transmitter and the receiver from active deceptions from a third party, often called the opponent. The attacks are of two different types, impersonation and substitution. A model for this case has been developed and many different ways of constructing such codes have been proposed, [1] - [7]. However, the model is restricted in the sense that the transmitter and the receiver must both trust each other in not cheating, since they are using the same key. But it is not always the case that the two communicating parties want to trust each other. In fact, it may be that the transmitter sends a message and then later denies having sent it. Or the other way around, the receiver may claim to have received a message that was never sent by the transmitter.

Inspired by this problem Simmons has introduced an extended authentication model, referred to as the *authentication model with arbitration*, [8], [9]. Here caution is taken both against deceptions from the outsiders (opponent) and also against some forms of deceptions from the insiders (transmitter and receiver). The model includes a fourth person, called the *arbiter*. The arbiter has access to all key information and is by definition not cheating. Codes which take caution against all these kinds of deceptions are called *authentication codes that permit arbitration*, or simply A^2-codes. One proposed construction of A^2-codes is the Cartesian product construction due to Simmons, [8]. In [10] lower bounds on

[1] This work was supported by the TFR grant 222 92-662

the probability of success for the different kinds of deceptions were given. Also lower bounds on the number of messages and the number of encoding rules for a fixed probability of deception were given. The A^2-codes which meet these lower bounds with equality are referred to as *equitably perfect* A^2-codes.

It is easily checked that codes obtained from the Cartesian product construction are not equitably perfect. In fact, the size of the keys grows exponentially with the number of source states. In this paper we consider the problem of constructing more efficient classes of A^2-codes. In Section 2 we give a detailed description of the model of authentication with arbitration. In Section 3 we introduce a general technique to construct A^2-codes and in Section 4 we use this to give two constructions, one that is equitably perfect and one that allows the number of source states to be chosen arbitrarily large.

2 The model of authentication with arbitration

In this model there are four different participants. These are the *transmitter*, the *receiver*, the *opponent* and the *arbiter*. The transmitter wants to transmit some information, which we call a *source state*, to the receiver in such a way that the receiver can recover the transmitted source state and also verify that the transmitted message came from the legal transmitter. This is done by mapping a source state S from the set $\mathcal{S}$ of possible source states to a message M from the set $\mathcal{M}$ of possible messages. The message is then transmitted over the channel. The mapping from $\mathcal{S}$ to $\mathcal{M}$ is determined by the transmitters secret encoding rule E_T chosen from the set $\mathcal{E_T}$ of possible encoding rules. Thus we assume that the transmitter uses a mapping f such that:

$$f : S \times E_T \rightarrow M, \tag{1}$$

$$f(s, e_t) = f(s', e_t) \Rightarrow s = s'. \tag{2}$$

To be able to uniquely determine the source state from the transmitted message, we have property (2). The opponent has access to the channel in the sense that he can either impersonate a message or substitute a transmitted message for another. When the receiver receives a message that was transmitted, he must check whether this message is valid or not. For this purpose we assume that the receiver uses a mapping g from his own secret encoding rule E_R taken from the set $\mathcal{E_R}$ of possible encoding rules and from the messages $\mathcal{M}$, that determine if a message is valid and if so also the source state.

$$g : M \times E_R \rightarrow S \cup \{\text{FRAUD}\}, \tag{3}$$

$$P(e_t, e_r) \neq 0, f(s, e_t) = m \Rightarrow g(m, e_r) = s. \tag{4}$$

Since all messages generated by the transmitter are valid messages and since the receiver must be able to determine which of the source states that was transmitted, the property (4) must hold for all possible pairs (E_T, E_R). However, in general not all pairs (E_T, E_R) will be possible.

The arbiter is the supervisory person who has access to all information, including E_T and E_R. However, he does not take part in any communication activities on the channel and his only task is to solve disputes between the transmitter and the receiver whenever such occur. As said before, the arbiter is by definition not cheating. This is an assumption which can be removed if we want to consider an even more general model of authentication, where the arbiter may also cheat. See [11] and [12] for details.

There are five different kinds of attacks to cheat which are possible in this model. The attacks are the following:

I, Impersonation by the opponent. The opponent sends a message to the receiver and succeeds if the message is accepted by the receiver as authentic.

S, Substitution by the opponent. The opponent observes a message that is transmitted and substitutes this message for another. The opponent succeeds if the receiver accepts this other message as authentic.

T, Impersonation by the transmitter. The transmitter sends a message to the receiver. The transmitter succeeds if the message is accepted by the receiver as authentic and if the message is not one of those messages that the transmitter can generate due to his own encoding rule.

$\mathbf{R_0}$, Impersonation by the receiver. The receiver claims to have received a message from the transmitter. The receiver succeeds if the message could have been generated by the transmitter due to his encoding rule.

$\mathbf{R_1}$, Substitution by the receiver. The receiver receives a message from the transmitter but claims to have received another message. The receiver succeeds if this other message could have been generated by the transmitter due to his encoding rule.

In all these possible attacks to cheat it is understood that the cheating person is using an optimal strategy when choosing a message, or equivalently, that the cheating person chooses the message that maximizes his chances of success. For each way of cheating, we denote the probability of success with P_I, P_S, P_T, P_{R_0} and P_{R_1}. The *overall probability of deception* is denoted P_D and is defined to be

$$P_D = \max(P_I, P_S, P_T, P_{R_0}, P_{R_1}).$$

The setup of the encoding rules may be done in several ways. One possible way is by letting the receiver choose his own encoding rule E_R and then secretly pass this on to the arbiter. The arbiter then constructs the encoding rule E_T and pass this on to the transmitter. Another way is to do the other way around and a third way is to allow the arbiter to construct both the encoding rules.

A traditional A-code is sometimes denoted $A(\mathcal{S}, \mathcal{M}, \mathcal{E}, f)$, where f is the authentication map $f : \mathcal{S} \times \mathcal{E} \mapsto \mathcal{M}$. In similar manner we denote an A^2-code as $A^2(\mathcal{S}, \mathcal{M}, \mathcal{E}_T, \mathcal{E}_R, f, g)$, where f is the transmitter's map given in (1) and g is the receiver's map given in (3). In [8] Simmons defined an authentication code to be *equitable* if the probabilities of success for all types of deceptions are the same, i.e., if $P_I = P_S = P_T = P_{R_0} = P_{R_1}$. In [10] it was shown that if an A^2-code provides $P_D = \frac{1}{q}$, then the cardinality of the sets of encoding rules must satisfy

$$|E_R| \geq q^3 \text{ and } |E_T| \geq q^4.$$

An A^2-code with $P_D = \frac{1}{q}$ is then defined to be *equitably perfect* if $|E_R| = q^3$ and $|E_T| = q^4$.

3 A general construction of A^2-codes

Let us for a moment consider the problems that occur in authentication with arbitration. Consider first the two deceptions from the receiver, R_0 and R_1, where he claims to have received a message that the transmitter never sent. We can think of a solution to this problem if we assume that the transmitter must add a "signature" to the source state S, that is to be transmitted. If the receiver now claims to have received a message from the transmitter he must also be able to produce the transmitter's signature.

This signature is actually nothing abstract but can be accomplished from a traditional A-code without secrecy. This code is a mapping from the source states and the encoding rules to the messages that has the form

$$Acode : S \times E \to M = (S, \alpha).$$

Let $\alpha = \alpha(S, E)$ be the signature. The transmitter maps the source state S into another "source state", Z, that also includes the signature α. This new source state, $Z = (S, \alpha)$, can now be transmitted in a second A-code without secrecy in order to protect against impersonation and substitution attacks from the opponent. Since this code only has to protect the original source state we can assume that the messages generated by the transmitter are of the form

$$M = (S, \alpha(S, E_T), \beta(S, E_T)) = (S, \alpha, \beta).$$

When the receiver checks a message for authenticity, he only checks whether β is correct. If the receiver claims to have received a message that the transmitter never sent, then he must be able to produce the signature α.

This concatenation of two normal A-codes gives protection against I, S, R_0 and R_1, but it realizes no protection against cheating from the transmitter. The transmitter cheats by sending a message that does not contain his own signature and succeeds if the message is accepted as authentic. In order to make this cheating difficult we introduce a modification for the receiver in the second A-code. Let this second A-code for the receiver protect both S and α. Then the received messages will have the form

$$M = (S, \alpha, \gamma) = (S, \alpha, \gamma(S, \alpha, E_R)).$$

This means that the receiver accepts all values of S and α and then checks that $\gamma = \gamma(S, \alpha, E_R)$. If properly generated messages are to be accepted and α is the transmitter's signature, we must have that

$$\forall S, \ \ \beta(S, E_T) = \gamma(S, \alpha, E_R), \text{ if } P(E_T, E_R) \neq 0. \tag{5}$$

Also E_T and E_R must be chosen in such a way that (5) always holds when the setup of the encoding rules is done. If the transmitter now tries to cheat by changing his signature he must also determine the change in $\gamma(S, \alpha, E_R)$ which might be difficult.

We give a concrete example of these arguments.

Example 1. This example is based on the signature function $\alpha(s) = as + b$, where $a, b, s \in \mathbb{F}_2$. Assume that $S = s$, $E_T = (e_1, e_2, e_3, e_4)$ and $E_R = (f_1, f_2, f_3)$ where $s, e_i, f_i \in \mathbb{F}_2$. Let the transmitter's signature function be $\alpha(S, E_T) = e_1 + se_2$ and let $\beta(S, E_T) = e_3 + se_4$. Thus the transmitter generates messages as $M = (s, e_1 + se_2, e_3 + se_4)$. For the receiver, let $\gamma(S, \alpha, E_R) = f_1 + \alpha f_2 + sf_3$. The receiver then accepts messages of the form $M = (s, \alpha, f_1 + \alpha f_2 + sf_3)$. Also, the encoding rules must have been chosen in such a way that $\beta(S, E_T) = \gamma(S, \alpha, E_R)$, or

$$e_3 + se_4 = f_1 + (e_1 + se_2)f_2 + sf_3. \tag{6}$$

Equivalently, this can be written

$$e_3 = f_1 + e_1 f_2, \tag{7}$$

$$e_4 = f_3 + e_2 f_2. \tag{8}$$

For this A^2-code the authentication matrix for the receiver is

	S	Message $M = (s, \alpha, f_1 + \alpha f_2 + sf_3)$ 000	001	010	011	100	101	110	111
$E_R = (f_1, f_2, f_3)$	000	0	-	0	-	1	-	1	-
	001	0	-	0	-	-	1	-	1
	010	0	-	-	0	1	-	-	1
	011	0	-	-	0	-	1	1	-
	100	-	0	-	0	-	1	-	1
	101	-	0	-	0	1	-	1	-
	110	-	0	0	-	-	1	1	-
	111	-	0	0	-	1	-	-	1

and the parameters for the code are

$$|S| = 2, \ |M| = 8, \ |E_R| = 8, \ |E_T| = 16.$$

By inspection we can check that the probabilities of the different kinds deceptions are

$$P_I = P_S = P_{R_0} = P_{R_1} = P_T = \frac{1}{2}.$$

□

4 Some specific constructions of A^2-codes

We have described an abstract way of modeling the problems in authentication with arbitration. We now deal with the problem of giving specific constructions of A^2-codes. Let us first give some preliminary definitions and results in the construction of traditional authentication codes. We consider first the case when

$|\mathcal{S}|P_D \leq 1$. Let $|\mathcal{S}| = q^n$, $|\mathcal{M}| = q^{n+m}$ and $|\mathcal{E}| = q^{2m}$, where $n \leq m$. Let $S = s$ and let the messages and the encoding rules consist of two parts,

$$M = (m_1, m_2), \tag{9}$$

$$E = (e_1, e_2). \tag{10}$$

We now assume that $s, m_1 \in \mathbb{F}_{q^n}$ and $m_2, e_1, e_2 \in \mathbb{F}_{q^m}$. Define an arbitrary injective mapping ϕ such that it maps a source state s from $\mathbb{F}_{q^n}$ to $\mathbb{F}_{q^m}$,

$$\phi : \mathbb{F}_{q^n} \mapsto \mathbb{F}_{q^m}, \ \phi(s) = \hat{s},$$

where $s \in \mathbb{F}_{q^n}$ and $\hat{s} \in \mathbb{F}_{q^m}$.

From these definitions we can state the following:

Theorem 1. *Let a traditional authentication code generate messages M of the form $M = (m_1, m_2)$, where $m_1 = s$ and $m_2 = e_1 + \hat{s}e_2$. This A-code is Cartesian (no secrecy) and provide $P_I = P_S = \frac{1}{q^m}$ if $n \leq m$. Moreover, it has parameters*

$$|\mathcal{S}| = q^n, \ |\mathcal{M}| = q^{n+m}, \ |\mathcal{E}| = q^{2m}.$$

Proof. The fact that the code is Cartesian is clear and the cardinality parameters are obvious. We have to prove that $P_I = P_S = \frac{1}{q^m}$.

Impersonation: A message M can be written as the sum of two independent parts, $M = (s, \hat{s}e_2) + (0, e_1)$. Thus success in impersonation is equivalent to the problem of guessing the correct value of e_1, which is done with probability q^{-m}.

Substitution: The opponent has observed the message $M = (s, e_1 + \hat{s}e_2)$. Now he replaces this with another message M', which must correspond to another source state s'. Then $s' = s + c$, where $c \neq 0$, and since the mapping ϕ is injective we have that $\hat{s}' = \hat{s} + \hat{c}$, where $\hat{c} \neq 0$. We can write the message M' as

$$M' = (s', e_1 + \hat{s}'e_2) = (s, e_1 + \hat{s}e_2) + (c, \hat{c}e_2) = M + (c, 0) + (0, \hat{c}e_2).$$

Thus success in substitution is equivalent to the problem of guessing the value of $\hat{c}e_2$ for any $\hat{c} \neq 0$. But since $\hat{c}e_2$ runs through $\mathbb{F}_{q^m}$ for $\hat{c} \neq 0$ as e_2 runs through $\mathbb{F}_{q^m}$, the correct value is guessed with probability q^{-m}. □

We now have a simple construction of A-codes, which in fact is the best possible for this case. Our aim is to generalize this construction in such a way that we obtain A^2-codes. Assume that we want to construct an A^2-code with $|\mathcal{S}| = q^n$ and $P_D = \frac{1}{q^m}$. Assume that $n \leq m$. Let the parameters for the A^2-code be the following:

$$|\mathcal{S}| = q^n, \ |\mathcal{M}| = q^{n+2m}, \ |\mathcal{E}_\mathcal{T}| = q^{4m}, \ |\mathcal{E}_\mathcal{R}| = q^{3m},$$

Consider the message and the encoding rules as consisting of several parts. Write

$$M = (m_1, m_2, m_3), \tag{11}$$

$$E_T = (e_1, e_2, e_3, e_4), \tag{12}$$

$$E_R = (f_1, f_2, f_3), \tag{13}$$

where $s, m_1 \in \mathbb{F}_{q^n}$ and $m_2, m_3, e_1, e_2, e_3, e_4, f_1, f_2, f_3 \in \mathbb{F}_{q^m}$.

Construction I: Let an A^2-code with $n \leq m$ be constructed as follows: The transmitter generates messages of the form $M = (m_1, m_2, m_3)$, where $m_1 = s$, $m_2 = e_1 + \hat{s}e_2$ and $m_3 = e_3 + \hat{s}e_4$. The receiver accepts all messages $M = (m_1, m_2, m_3)$ which has $m_3 = f_1 + \hat{m}_1 f_2 + m_2 f_3$. The encoding rules have been chosen in such a way that

$$e_3 + \hat{s}e_4 = f_1 + \hat{s}f_2 + (e_1 + \hat{s}e_2)f_3, \tag{14}$$

or equivalently,

$$e_3 = f_1 + e_1 f_3, \tag{15}$$

$$e_4 = f_2 + e_2 f_3. \tag{16}$$

From the way the encoding rules are chosen we check the following properties,

Lemma 2. *Let $\mathcal{E}_T \circ \mathcal{E}_R$ denote the set of all possible pairs of encoding rules (E_T, E_R). Then*

$$|\mathcal{E}_T \circ \mathcal{E}_R| = q^{5m}.$$

Also the transmitter has no knowledge about f_3 and the receiver has no knowledge about the pair (e_1, e_2). Expressed in terms of entropy we have $H(F_3|E_T) = m \log q$ and $H(E_1, E_2|E_R) = 2m \log q$.

Let us give the parameters of this construction.

Theorem 3. *Construction I gives a Cartesian A^2-code which has the following parameters for $n \leq m$:*

$$|\mathcal{S}| = q^n,\ |\mathcal{M}| = q^{n+2m},\ |\mathcal{E}_R| = q^{3m},\ |\mathcal{E}_T| = q^{4m}.$$

The probabilities of deceptions are

$$P_I = P_S = P_{R_0} = P_{R_1} = P_T = \frac{1}{q^m}.$$

Proof. The cardinality of the different sets is the number of possible values and is thus easily checked. Also, the code is Cartesian. Let us find the probabilities of success for the different kinds of deceptions.

Impersonation by the opponent, I: The opponent sends a message M and hopes for it to be authentic. The messages accepted by the receiver can be written in independent parts as

$$M = (s, \alpha, \hat{s}f_2 + \alpha f_3) + (0, 0, f_1).$$

In order to succeed the opponent must guess the value of f_1 and this is done with probability q^{-m}. Thus $P_I = q^{-m}$.

Substitution by the opponent, S: The opponent has observed a message M and substitutes this for another message M'. The substitution attack must include a change of the source state. Assume that the new source state is s', which can be written as $s' = s + c$, where $c \neq 0$. Since the map ϕ is injective we also have that $\hat{s}' = \hat{s} + \hat{c}$, where $\hat{c} \neq 0$. The observed message can be written as $M = (s, \alpha, f_1 + \hat{s}f_2 + \alpha f_3)$, where $\alpha \in \mathbb{F}_{q^m}$. The message M' is then of the form

$$M' = (s + c, \alpha + c', f_1 + (\hat{s} + \hat{c})f_2 + (\alpha + c')f_3),$$

where $c' \in \mathbb{F}_{q^m}$. This can be rewritten in independent parts as

$$M' = M + (c, c', c'f_3) + (0, 0, \hat{c}f_2).$$

Since $\hat{c} \neq 0$ and the last part is independent of the other two parts we have that success in substitution is equivalent to the problem of guessing the value of $\hat{c}f_2$ for any $\hat{c} \neq 0$. But $\hat{c}f_2$ runs through $\mathbb{F}_{q^m}$ as f_2 runs through $\mathbb{F}_{q^m}$, so the correct value is guessed with probability q^{-m}. Thus $P_S = q^{-m}$.

Impersonation by the receiver, R_0: The receiver claims to have received the message M. He succeeds if the signature α is correct. From Theorem 1 and Lemma 2 it follows that the probability of success is $P_{R_0} = q^{-m}$.

Substitution by the receiver, R_1: The receiver receives the message M but claims to have received another message M' corresponding to another source state s'. As before he succeeds if the signature α in M' is correct. From Theorem 1 and Lemma 2 it again follows that the probability of success is $P_{R_1} = q^{-m}$.

Impersonation by the transmitter, T: The transmitter sends a message M and then denies having sent it. He succeeds if the message contains a different signature from his own and is accepted by the receiver as authentic. The message received can be written as

$$M = (s, \alpha + c', f_1 + \hat{s}f_2 + \alpha f_3) + (0, 0, c'f_3),$$

where $c' \neq 0$. As before, the correct value of $c'f_3$ is guessed with probability q^{-m}, i.e., $P_T = q^{-m}$. □

Corollary 4. *Construction I is an equitably perfect A^2-code.*

Remark: A construction very similar to this was also found in [12], where the construction additionally also protected against attacks from the arbiter.

Let us give an example of how this construction works.

Example 2. Assume that we want to construct an A^2-code with the properties that $|\mathcal{S}| = 2$ and $P_D = \frac{1}{2^2}$. Following Construction I we find that $\mathbb{F}_{q^m} = \mathbb{F}_{2^2}$ and that $\mathbb{F}_{q^n} = \mathbb{F}_2$. Let the mapping ϕ map the elements of $\mathbb{F}_2$ to the subfield $\{0, 1\}$ in $\mathbb{F}_{2^2}$, i.e., 0 maps to 0 and 1 maps to 1. The encoding rules are chosen in such a way that (14) holds. The transmitter generates messages as

$$M = (m_1, m_2, m_3) = (s, e_1 + \hat{s}e_2, e_3 + \hat{s}e_4).$$

The receiver receives messages of the form $M = (m_1, m_2, m_3)$ and checks that

$$m_3 = f_1 + \hat{m}_1 f_2 + m_2 f_3.$$

The number of messages and the number of encoding rules are

$$|\mathcal{M}| = 32, \quad |\mathcal{E}_{\mathcal{R}}| = 64, \quad |\mathcal{E}_{\mathcal{T}}| = 256.$$

□

We have obtained a general construction for the case $n \leq m$. If we consider the same construction for the case $n > m$ we see that it is now not possible for the map ϕ to be injective. If ϕ is not injective there exist two source states s, s' that map to the same $\hat{s}$. These two source states would have the same last part in the message for all encoding rules and thus the probability of substitution becomes 1. However, with some modifications we can get a construction that can be used for the case $n > m$. In order for the construction to provide the same probability of deception we must increase the number of encoding rules. Thus the construction for the case $n > m$ will not be perfect.

As before we first give a construction of a traditional authentication code where $n \geq m$. We use the same notation as in (9)-(10), but now $s, m_1, e_2 \in \mathbb{F}_{q^n}$ and $m_2, e_1 \in \mathbb{F}_{q^m}$. Also we need a mapping ϕ, $\phi : \mathbb{F}_{q^n} \mapsto \mathbb{F}_{q^m}$ with the property that the number of $x \in \mathbb{F}_{q^n}$ such that $\phi(x) = y$ is the same for all $y \in \mathbb{F}_{q^m}$. We also assume that ϕ has the homomorphism property that $\phi(x)+\phi(x') = \phi(x+x')$. Then we state the following:

Theorem 5. *Let a traditional authentication code with $n \geq m$ generate messages M of the form $M = (m_1, m_2)$ where $m_1 = s$ and $m_2 = e_1 + \phi(se_2)$. This A-code is Cartesian (no secrecy) and provide $P_I = P_S = \frac{1}{q^m}$. Moreover, it has parameters*

$$|\mathcal{S}| = q^n, \ |\mathcal{M}| = q^{n+m}, \ |\mathcal{E}| = q^{n+m}.$$

Proof. The cardinality parameters are obvious and the A-code is Cartesian. For the different kinds of deceptions we have:

Impersonation, I: Write the message as $M = (s, \phi(se_2)) + (0, e_1)$. The value of e_1 is guessed with probability q^{-m}. Thus $P_I = q^{-m}$.

Substitution, S: The opponent has observed $M = (s, e_1 + \phi(se_2))$. Now he substitutes this message for another message, which has $s' \neq s$. The message M' is then written as

$$M' = (s', e_1 + \phi(s'e_2)) = (s, e_1 + \phi(se_2)) + (c, \phi(ce_2)) = M + (c, 0) + (0, \phi(ce_2)).$$

Since $c \neq 0$, ce_2 take any value in $\mathbb{F}_{q^n}$ with the same probability and $\phi(ce_2)$ is guessed with probability q^{-m}. Thus $P_S = q^{-m}$. □

We now give a construction of A^2-codes with $n \geq m$. We make one simplification, namely that $m = 1$. The notation is the same as in (11)-(13), but for this case we have $s, m_1, e_2, e_4, f_2 \in \mathbb{F}_{q^n}$ and $m_2, m_3, e_1, e_3, f_1, f_3 \in \mathbb{F}_q$. We also need a mapping ϕ, $\phi : \mathbb{F}_{q^n} \mapsto \mathbb{F}_q$ with the property that the number of $x \in \mathbb{F}_{q^n}$ such that $\phi(x) = y$ is the same for all $y \in \mathbb{F}_q$. We choose a specific ϕ.

Since $\mathbb{F}_{q^n}$ is an extension field of $\mathbb{F}_q$ any element $x \in \mathbb{F}_{q^n}$ can be written as $x = r_0 + r_1\alpha + \ldots + r_{n-1}\alpha^{n-1}$, where $r_i \in \mathbb{F}_q$, $i = 0, 1, \ldots, n-1$, and α is a root of an irreducible polynomial of degree n over $\mathbb{F}_q$. Define ϕ as

$$\phi : r_0 + r_1\alpha + \ldots + r_{n-1}\alpha^{n-1} \mapsto r_0.$$

From these definitions we can verify the homomorphism property. Assume that $x, x' \in \mathbb{F}_{q^n}$ and $y \in \mathbb{F}_q$. Then

$$\phi(x) + \phi(x') = \phi(x + x'), \tag{17}$$

$$\phi(x)y = \phi(xy). \tag{18}$$

We give the promised construction:

Construction II: Let an A^2-code with $n \geq m$ be constructed as follows: The transmitter generates messages of the form $M = (m_1, m_2, m_3)$, where $m_1 = s$, $m_2 = e_1 + \phi(se_2)$ and $m_3 = e_3 + \phi(se_4)$. The receiver accepts all messages $M = (m_1, m_2, m_3)$ which have $m_3 = f_1 + \phi(sf_2) + m_2 f_3$. The encoding rules have been chosen in such a way that

$$e_3 + \phi(se_4) = f_1 + \phi(sf_2) + (e_1 + \phi(se_2)) f_3, \tag{19}$$

or equivalently,

$$e_3 = f_1 + e_1 f_3,$$
$$\phi(se_4) = \phi(sf_2) + \phi(se_2) f_3.$$

But from the properties (17) and (18) this is the same as

$$e_3 = f_1 + e_1 f_3,$$
$$\phi(se_4) = \phi(s(f_2 + e_2 f_3)).$$

If we choose the encoding rules as

$$e_3 = f_1 + e_1 f_3, \tag{20}$$
$$e_4 = f_2 + e_2 f_3, \tag{21}$$

we know that (19) holds.

Lemma 6. *If the encoding rules are chosen as in (20) and (21) then*

$$|\mathcal{E}_T \circ \mathcal{E}_\mathcal{R}| = q^{2n+3}.$$

Also, the transmitter has no knowledge about f_3 and the receiver has no knowledge about the pair (e_1, e_2).

Theorem 7. *Construction II gives a Cartesian A^2-code with the following parameters for $n \geq m$:*

$$|\mathcal{S}| = q^n, \ |\mathcal{M}| = q^{n+2}, \ |\mathcal{E}_\mathcal{R}| = q^{n+2}, \ |\mathcal{E}_T| = q^{2n+2}.$$

The probabilities of deceptions are

$$P_I = P_S = P_{R_0} = P_{R_1} = P_T = \tfrac{1}{q}.$$

Proof. We determine the probabilities of success for the different kinds of deceptions.

Impersonation by the opponent, I: The message M received by the receiver can be written as

$$M = (s, \alpha, f_1 + \phi(sf_2) + \alpha f_3) = (s, \alpha, \phi(sf_2) + \alpha f_3) + (0, f_1).$$

The probability of guessing the correct value of f_1 is q^{-1} and thus $P_I = q^{-1}$.

Substitution by the opponent, S: The opponent has observed the message $M = (s, \alpha, f_1 + \phi(sf_2) + \alpha f_3)$. Now the opponent substitutes this for another message M' which correspond to a source state s', where $s' \neq s$. Write s' as $s' = s + c$, where $c \neq 0$ and $c \in \mathbb{F}_{q^n}$. The message M' is written as

$$M' = (s', \alpha + c', f_1 + \phi(s'f_2) + (\alpha + c')f_3)$$

where $c' \in \mathbb{F}_q$. But this is rewritten in independent parts as

$$M' = (s + c, \alpha + c', f_1 + \phi((s + c)f_2) + (\alpha + c')f_3) = M + (c, c', c'f_3) + (0, 0, \phi(cf_2)).$$

The probability of guessing the correct value of $\phi(cf_2)$ is q^{-1}.

Impersonation by the receiver, R_0: The receiver claims to have received the message $M = (s, \alpha, \beta)$ and succeeds if α is correct. The message generated by the transmitter is written as

$$M = (s, e_1 + \phi(se_2), m_3) = (s, \phi(se_2), m_3) + (0, e_1, 0).$$

By Lemma 6, the probability of guessing the value of e_1 is q^{-1} and $P_{R_0} = q^{-1}$.

Substitution by the receiver, R_1: The receiver receives a message M but claims to have received another message M' with another source state. If $M = (s, e_1 + \phi(se_2), m_3)$ we can write M' as

$$M' = M + (c, 0, m_3' - m_3) + (0, \phi(ce_2), 0)$$

where $c \neq 0$ and $c \in \mathbb{F}_{q^n}$. As before, the value of $\phi(ce_2)$ is guessed with probability q^{-1}, i.e. $P_{R_1} = q^{-1}$.

Impersonation by the transmitter, T: The transmitter is able to generate a message M but sends the message M' with a different signature α'. The signature is written as $\alpha' = \alpha + c'$ where $c' \neq 0$ and $c' \in \mathbb{F}_q$. Then M' can be written in independent parts as

$$M' = (s, \alpha + c', f_1 + \phi(sf_2) + (\alpha + c')f_3) = M + (0, c', c'f_3).$$

The value of $c'f_3$ is by Lemma 6 guessed with probability q^{-1} and $P_T = q^{-1}$. □

We end this section by giving a small example of how the last construction works.

Example 3. Assume that we want to construct an A^2-code with the properties that $|\mathcal{S}| = 2^2$ and $P_D = \frac{1}{2}$. The elements of $\mathbb{F}_{2^2}$ are written as $r_0 + r_1\alpha$, where $r_0, r_1 \in \mathbb{F}_2$ and $\alpha^2 + \alpha + 1 = 0$. Assume that the encoding rules have the following values,

$$f_1 = f_3 = e_1 = 1,\ f_2 = e_2 = 1 + \alpha.$$

From (20) we have that

$$e_3 = f_1 + e_1 f_3 = 1 + 1 * 1 = 0,$$

and from (21) it follows that

$$e_4 = f_2 + e_2 f_3 = (1 + \alpha) + (1 + \alpha) * 1 = 0.$$

Assume that the transmitter wishes to communicate the source state $s = \alpha$. He then generates the message

$M = (s, e_1 + \phi(se_2), e_3 + \phi(se_4)) = (\alpha, 1 + \phi(\alpha(1+\alpha)), 0 + \phi(\alpha * 0)) = (\alpha, 0, 0)$.

When the receiver receives this message he checks that

$$m_3 = f_1 + \phi(m_1 f_2) + m_2 f_3 = 1 + \phi(\alpha(1+\alpha)) + \phi(0 * 1) = 0.$$

Since m_3 was correct the message is accepted as authentic as it should be. □

5 Acknowledgement

Y. Desmedt is greatly acknowledged for making the author pay attention to [12].

References

1. E.N. Gilbert, F.J. MacWilliams and N.J.A. Sloane, "Codes which detect deception", *Bell Syst. Tech. J.*, Vol. 53, 1974, pp. 405–424.
2. G.J. Simmons, "Authentication theory/coding theory", in *Advances in Cryptology, Proceedings of CRYPTO 84*, G.R. Blakley and D. Chaum, Eds. Lecture notes in Computer Science, No. 196. New York, NY: Springer, 1985, pp. 411–431.
3. J.L. Massey,"Contemporary Cryptology, An Introduction", in *Contemporary Cryptology, The Science of Information Integrity*, G.J Simmons , Ed., IEEE Press, 1991, pp. 3-39.
4. G.J. Simmons, "A survey of Information Authentication", in *Contemporary Cryptology, The science of information integrity*, ed. G.J. Simmons, IEEE Press, New York, 1992.
5. D.R. Stinson, "The combinatorics of authentication and secrecy codes", Journal of Cryptology, Vol. 2, no 1, 1990, pp. 23-49.
6. D. R. Stinson, "Universal hashing and authentication codes" *Proceedings of Crypto 91*, Santa Barbara, USA, 1991, pp 74-85.
7. T. Johansson, G. Kabatianskii, B. Smeets, "On the relation between A-codes and codes correcting independent errors" *Proceedings Eurocrypt'93*, to appear.
8. G.J. Simmons,"A Cartesian Product Construction for Unconditionally Secure Authentication Codes that Permit Arbitration", in *Journal of Cryptology*, Vol. 2, no. 2, 1990, pp. 77-104.
9. G.J. Simmons, "Message authentication with arbitration of transmitter/receiver disputes", in *Proceedings of Eurocrypt '87*, D. Chaum and W.L. Price, Eds., Amsterdam, The Netherlands, April 13-15, 1987, pp. 151-165. Berlin: Springer-Verlag, 1988.
10. T. Johansson, "Lower Bounds on the Probability of Deception in Authentication with Arbitration", in *Proceedings of 1993 IEEE International Symposium on Information Theory*, San Antonio, USA, January 17-22, 1993, p. 231.
11. E.F. Brickell D.R. Stinson, "Authentication codes with multiple arbiters", in *Proceedings of Eurocrypt '88*, C.G Günter, Ed., Davos , Switzerland, May 25-27, 1988, pp. 51-55, Berlin: Springer-Verlag, 1988.
12. Y. Desmedt, M. Yung, "Asymmetric and Securely-Arbitrated Unconditional Authentication Systems", submitted to IEEE Transactions on Information Theory. A part of this paper was presented at Crypto'90.

Codes for Interactive Authentication

Pete Gemmell[1] and Moni Naor[2]

[1] Sandia National Labs. Part of this work was done while the author was visiting the IBM Almaden Research Center.

[2] Dept. of Applied Math and Computer Science, Weizmann Institute of Science, Rehovot 76100, Israel. Part of this work was done while the author was with the IBM Almaden Research Center.

Abstract. An **authentication protocol** is a procedure by which an **informant** tries to convey n bits of information, which we call an **input message**, to a **recipient**. An **intruder**, I, controls the network over which the informant and the recipient talk and may change any message before it reaches its destination. a If the protocol has security p, then the the recipient must detect this a cheating with probability at least $1-p$. This paper is devoted to characterizing the amount of secret information that the sender and receipient must share in a p-secure protocol. We provide a single-round authentication protocol which requires $\log(n)+5\log(\frac{1}{p})$ bits of secrecy. as well as a single-round protocol which requires $\log(n)+2\log(\frac{1}{p})$ bits of secrecy based on non-constructive random codes. We prove a lower bound of $\log(n)+\log(\frac{1}{p})$ secret bits for single-round protocols.

We introduce authentication protocols with more than one round of communication (multi-round protocols) and present a k-round protocol which reduces the amount of secret information that the two parties need to $\log^{(k)}(n)+5\log(\frac{1}{p})$. When the number of rounds is $\log^*(n)$, our protocol requires $2\log(\frac{1}{p})+O(1)$ bits. Hence interaction helps when $\log(n)>\log(\frac{1}{p})$. We also show a lower bound of $\log^{(k)}(n)$ on the number of shared random bits in a k-round protocol.

1 Introduction

Authentication is one of the major issues in Cryptography. Authentication protocols can take on a variety of forms. The the informant and recipient may or may not rely on complexity assumptions (e.g. that factoring is hard). They may or may not wish to be able to prove to third parties that the message was indeed sent by the informant. For a general survey of authentication issues and results, the reader may refer to [8].

This paper deals with the simple scenario where two parties A and B communicate and want to assure that the message received by B is the one sent by

A. We provide nearly tight bounds for the case of "two party unconditionally secure authentication without secrecy" defined as follows. A protocol is "Without secrecy" if the informant and recipient make no attempt to hide the content of the input message from the intruder. In many cases the intruder may know the input message which the informant is trying to convey and wants only to convince the recipient that the informant is trying to communicate a different message.

If a protocol is "unconditionally secure," with security parameter p, then no intruder, regardless of computational strength, can cheat the communicating parties with probability more than p. An unconditionally secure protocol does not rely on complexity theoretic assumptions such as "There is no polynomial time algorithm to invert function f". Note that unconditionally secure protocols can be used in conjunction with computational hardness based protocols.

If we desire unconditional security then clearly the two parties must share some secret bits. In this paper we try to characterize the number of shared random bits, as a function of p, that the two parties must share in order to assure that any change made to the message will be discovered with probability at least $1-p$. We distinguish between single-round and multi-round protocols. Single-round protocols have been investigated extensively. For this case we provide tight bounds on the number of shared bits up to constant factors: it is $\Theta(\log n + \log 1/p)$, where n is the length of the input message. More precisely, it is between $\log n + \log 1/p$ and $\log n + 2\log 1/p$.

In this paper, we discuss multi-round authentication protocols, a subject which, to our knowledge, has not appeared in the literature. In a multi-round protocol, in order to authenticate an input message, the two parties send messages back and forth for several rounds and at the end if the (original) message has been altered it should be detected. We provide a multi-round protocol that requires $2\log 1/p + O(1)$ bits, i.e. it is independent of the message length. Hence we can conclude that interaction helps, i.e that the number shared secret bits required by a multi-round protocol is smaller than the number required by a single-round, when $\log 1/p < \log n$.

We also investigate the number of rounds required to achieve these bounds. In general, $O(\log^* n)$ round suffice to achieve the $2\log 1/p$ bound, but no constant round protocol can achieve them, since we have a lower bound of $\log^{(k)} n$ for a k-round protocol.

1.1 Previous Work

The one-round case has received a lot of attention in the literature. Gilbert, MacWilliams, and Sloane [4], who were the first to formally consider the problem, provided in 1974 a protocol requiring $2\max\{n, \log 1/p\}$ shared secret bits. Wegman and Carter [14] suggested using *ϵ-almost strongly universal$_2$* hash functions to achieve authentication. They described a protocol that requires $O(\log n \log 1/p)$ secret bits.

Stinson [9] improved upon this result, using *ϵ-almost strongly universal$_2$* hash functions to produce a protocol which requires approximately $(2\log(n) + 3 -$

$2\log\log(\frac{1}{p}))(\log(\frac{1}{p}))$ secret bits.

A fair amount of work has also been devoted to the question of designing protocols where the probability of cheating is exactly inversely proportional to the number of authenticators (the information sent in addition to the message) (see [3], [5], [10], [11], [12], [13]). Adding this constraint makes the task much harder. The number of secret bits required is $\Omega(n)$, and it is only possible to construct such protocols for values of $p = \frac{1}{q} : q$ a prime power.

We apply the idea of *ϵ-almost strongly universal$_2$* hash functions to obtain a near optimal one round protocol. One version of this protocol uses hash functions based on non-constructive codes and requires only $\log(n) + 2\log(\frac{1}{p})$ secret bits.

As for lower bounds, still in the single-round case, Gilbert, MacWilliams, and Sloane [4] showed that the number of secret bits must be at least $2\log(\frac{1}{p})$, a factor of 2 higher than the obvious bound implied by the intruder simply guessing the secret bits. In 1991, Stinson [9] showed that the size of a family of ϵ-almost strongly universal hash functions, mapping a set of size a into a set of size b, is at least $\frac{a(b-1)^2}{b\epsilon(a-1)+b-a}$.

Blum et al. [2] worked on the problem of checking the correctness of (untrusted) memories. They showed that a processor who wishes to store n bits of information in an untrusted (adverserial) memory must have a private, trusted memory of at least $\log(n)$ bits. This lower bound argument can be converted to the authentication scenario considered in this paper.

We use ideas from coding theory to improve this lower bound on single-round protocols to $\log(n) + \log(\frac{1}{p})$.

The multi-round case has not been considered previously in the literature. We show that allowing k rounds of interaction between the sender and recipient enables them to get by with as few as $\log^{(k)}(n) + 5\log(\frac{1}{p})$ bits of secret information, or $\log^{(k)}(n) + 2\log(\frac{1}{p})$ bits using non-constructive codes. When $k = \log^* n$ we have a protocol requiring $2\log(\frac{1}{p}) + 2$ secreet bits. The protocols achieving these bounds reduce in every round the effective length of the message by a logarithmic factor.

We obtain a lower bound of $\log^{(k)}(n)$ for k round protocols by showing that the existence of a k round authentication protocol using l secret bits implies the existence of a $k-1$ round authentication protocol using $l + 2^l$ secret bits.

1.2 Organization of the paper

In the next section we define the model and the parameters involved. Section 3 describes the single-round protocols and Section 4 the multi-round protocols. Section 5 shows the lower bounds on the number of shared random bits, both for the single-rounds and for the multi-round protocols. Section 6 shows a lower bound on the redundancy, i.e on the the length of the authenticator (the parts of the transmissions that are not the input message). The full paper contains bounds for authentication series, i.e. schemes that are designed to authenticate several messages and also a discussion on the issue of the definitions of security.

Some of the lower bounds proofs are not included here and will appear in the full paper.

2 The Model

Definition 1. A **k-round, secrecy l, probability p-authentication scheme for a message of n bits** is a protocol in which informant A and recipient B alternate sending each other k messages (altogether) over an insecure line controlled by an intruder I. A and B share l bits of secret information and each of them has a separate private source of random bits. Their goal is for A to communicate an arbitrary n bit input message m to B. The intruder I, which has unbounded computational power, may intercept any of their communications and replace these communications with whatever I wishes. The intruder does not have to keep A and B synchronized and can feed A with a message before B has sent it.

For all input messages m:

- If there is no interference in the transmissions, then B must output m and both A and B must accept with probability at least $1-p$.
- If B receives a message $m' \neq m$, then with probability at least $1-p$: A or B must output $FAIL$.

In the first round, A sends the input message m and authenticator x_1. In subsequent rounds $i > 1$, only an authenticator x_i is sent. A sends authenticators $x_1, x_3 \ldots$ and B sends $x_2, x_4 \ldots$. The adversary I receives each of these messages x_i and replaces it with x_i'. B receives $m', x_1', x_3' \ldots$ and A receives $x_2', x_4' \ldots$. If $\exists i : x_i \neq x_i'$ then we say I cheats in that round.

If A or B outputs $FAIL$, then either A or B has detected the intruder and knows that the message delivered in the first round may not be valid. For the single-round protocols it is B who detects any intrusion. For the multi-round protocols it may be either A or B who detects the error. Note that if we desire to have both parties alerted in the case of an intrusion, then we could add the stipulation that, at the end of the protocol, they exchange $\log \frac{1}{p}$ bit passwords which are appended to the secret string.

Definition 2. An authentication protocol P is **sound**, if, whenever there is no interference, A and B accept with probability 1.

2.1 Synchronization

For single-round protocols, synchronization is not an issue. The recipient simply waits for some authenticator, message pair to arrive and then either accepts or FAILS. For multi-round protocols, the intruder is able to carry on two separate, possibly asynchronous, conversations, one with the informant and one with the recipient. However, the party that is supposed to send the message in the $i+1$st round always waits until it receives the intruder's ith-round message. Therefore, for each of the two conversations, the protocol forces the intruder to commit to any possible ith-round cheating before soliciting the $i+1$st round message. This is used in the proof of validity for our k-round protocol.

3 Single-Round Protocols

3.1 ϵ-Almost Strongly Universal$_2$ hash functions

For single-round protocols, Wegman and Carter [14] observed that we can view the secret shared information as a hash function, s, secretly chosen by A and B from a publicly known family of hash functions $\mathcal{H}$. If $s \in \mathcal{H}$ then s maps the set M of possible input messages into the set X of authenticators. The requirement for the family of hash functions is that, given the value of a hash function at any one point, it must be impossible to predict the value at any other point with probability greater than p.

Definition 3. We call a hash function family $\mathcal{H}$ an **ϵ-almost strongly universal$_2$** if $\forall m, m' \in M : m \neq m', \forall x, y \in X, Pr_{s\in\mathcal{H}}[s(m') = x | \ s(m) = y] \leq \epsilon$.

The single-round protocols which we present are based on the following idea: A and B choose the secret string s as a description of a member of p-almost-universal$_2$ family of hash functions. In order for A to send B the input message m , it sends the authenticator pair $m, s(m)$. Upon receiving the pair m', x', B checks that $x' = s(m')$.

Claim 4 *The probability that the intruder succeeds in fooling B in the above protocol is at most p.*

Proof: From the definition of ϵ-almost strongly universal$_2$ hash functions, knowing only the value of $s(m)$ for one value of m, I can guess the value of $s(m')$, for $m' \neq m$, with probability at most p. □

3.2 A Single-Round Protocol

Theorem 5. *$\forall p > 0$, there is a sound single-round, secrecy $\lceil \log(n) \rceil + 5\lceil \log(\frac{1}{p}) + 1\rceil)$, probability p authentication scheme.*

Proof: The idea behind the protocol is that A and B share a secret hash function $s : \{0,1\}^n \rightarrow GF[Q] : Q \approx \frac{2}{p}$ chosen uniformly at random from a p-almost strongly universal$_2$ family of hash functions $\mathcal{H}$ such that $|\mathcal{H}| = nQ^5$. $GF[Q]$ refers to the field containing Q elements. Given an input message m, A sends $m, s(m)$ to B. Since $\mathcal{H}$ is p-almost strongly universal$_2$, I has little idea what the value of $s(m')$ is for any m' such that $m' \neq m$.

We now describe the construction of the p-almost strongly universal$_2$ hash function s. Let C be a code $C : \{0,1\}^n \rightarrow GF[Q]^{n'}$ with the properties:

- Q is roughly equal to $\frac{2}{p}$
- n' is roughly equal to nQ^3
- $\forall m_1, m_2$, with $m_1 \neq m_2$, $C(m_1)$ and $C(m_2)$ differ in at least $1 - p$ fraction of their entries.

The best known construction for such a code C is described by Alon et al. in [1]. The shared string $s = (i, a, b)$ consists of three random values where: $i \in_r \{1 \ldots n'\}$, $a \in_r GF[Q] - \{0\}$, $b \in_r GF[Q]$. Using those three values $s(m)$ is evaluated as: $s(m) = aC_i(m) + b$.

The single-round protocol P_1 is:

P_1: A Sound Single-Round, Secrecy $\lceil\log(n)\rceil + 5\lceil\log(\frac{1}{p}) + 1\rceil$, Probability p Authentication Protocol

`A and B share random secret string` $s = (i, a, b)$.

A: `sends to` B `the message, authenticator pair:` $m, s(m)$

B: `receives` m', x' `and accepts` m' `iff` $x' = s(m')$

To see that P_1 is a single-round probability p authentication protocol, we show:

Claim 6 *$\mathcal{H}$ is p-almost strongly universal$_2$.*

Proof: Fix messages m and m', $m \neq m'$. Let $s \in_r H, y = s(m)$. Let $x \in Q$. We will separate the analysis into two cases, $x = y$ and $x \neq y$.

1. Let $x = y$. Since b is chosen uniformly at random, independent of i and a, the distribution on i given $y = s(m) = aC_i(m) + b$ is the same as the original uniform distribution on i. Due to the definition of the code C, we have: $Pr_{i\in_r[1\ldots n']}[C_i(m) = C_i(m')] \leq p$. This implies that

$$Pr_{s\in_r\mathcal{H}}[s(m') = x|s(m) = y]$$
$$= Pr_{s\in_r\mathcal{H}}[aC_i(m') + b = s(m') = s(m) = aC_i(m) + b] \leq p$$

2. Let $x \neq y$. Choose and fix random values for i and b. The distribution on a given the knowledge $y = s(m) = aC_i(m) + b$ is the same as the original uniform distribution on a. Since $m' \neq m$,

$$Pr_{s\in_r\mathcal{H}}[x = s(m')|y = s(m)]$$
$$= Pr_{a\in_r GF[Q]-\{0\}}[x - y = s(m') - s(m)|y = s(m)]$$
$$= Pr_{a\in_r GF[Q]-\{0\}}[x - y = a(C_i(m') - C_i(m))|y = s(m)] \leq \frac{1}{Q} < p$$

We have shown that $\forall m, m' : m \neq m', \forall x, y \in X$, $Pr_{s\in\mathcal{H}}[s(m') = x|\ s(m) = y] \leq p$. Therefore $\mathcal{H}$ is p-almost strongly universal$_2$. □

3.3 Existence of a Single-Round, Secrecy $\log(n) + 2\log(\frac{1}{p})$ Protocol

We note here that $\forall p > 0$, there exists a sound single-round, secrecy $\lceil\log(n)\rceil + 2\lceil\log(\frac{1}{p}) + 1\rceil)$, probability p authentication scheme.

This better upper bound on the number of secret bits is attained by using a smaller family of p-almost strongly universal$_2$ hash functions based on a more powerful family of codes which exist, but are not necessarily constructible.

Using probabilistic arguments one can show, as was done by Roth [7], that there exists a code C^* with the following properties:

- C^* maps $\{0,1\}^n$ into $GF[Q]^{n'}$
- Q is roughly equal to $\frac{2}{p}$
- n' is roughly equal to nQ^2
- $\forall m_1, m_2$, with $m_1 \neq m_2$, and $\forall y_1, y_2 \in GF[Q]$, if $1 \leq i \leq n'$ is chosen at random, then $Pr_i[C_i^*(m_1) = y_1, C_i^*(m_2) = y_2] \leq \frac{2}{Q^2}$.

In this case, we could define $s = i$ where $s(m) = C_i^*(m)$.

4 Multi-Round Protocols

The multi-round protocols which we present in this section are based on the idea that the informant can send the input message in the first round and then the recipient can carry on the authentication by using a $k-1$-round protocol to send back a small, random "fingerprint" of the input message it has received. If the intruder has changed the input message that the informant sent to the recipient, then with a very high probability the random fingerprint computed by the recipient will not match any fingerprint for the message that the informant sent in the first round. If the intruder will not alter any message sent in subsequent rounds, then the informant will be aware of the bad fingerprint sent back by the recipient.

4.1 The k-round protocol

The protocol applies codes similar to those used for the single-round protocol. Let C^k be a code $C^k : \{0,1\}^n \rightarrow GF[Q_k]^{n_k}$ with the properties:

- Q_k is roughly equal to $\frac{2^k}{p}$
- n_k is roughly equal to nQ_k^3
- $\forall m_1, m_2$, with $m_1 \neq m_2$, $C^k(m_1)$ and $C^k(m_2)$ differ in at least $1 - \frac{p}{2^{k-1}}$ fraction of their entries.

P_k: **a k-Round, Secrecy $\lceil log^{(k)}(n)\rceil + 5\lceil \log(\frac{1}{p}) + 1\rceil$,**
Probability $2(1-\frac{1}{2^k})p$ Authentication Protocol

`For` $k = 1$ `the use protocol` P_1 `to authenticate` m.
`Otherwise:`
A `and` B `share the random secret string necessary for`
`a` $(k-1)$`-round protocol on inputs of size` $\log(n) + 4k + 4\log(\frac{1}{p})$.
A: `Send only the input message` m `to` B
B: `(after receiving` m'`)`
  `Choose a random index,` $i_k \in_r \{1 \ldots n_k\}$.
  `Use the protocol` P_{k-1} `to send`
  $i_k, C_i^k(m')$ `to` A
A: `(after the authentication of` $i_k, C_{i_k}^k(m')$ `is complete)`
  `Verify that` $C_{i_k}^k(m') = C_{i_k}^k(m)$
`If neither party finds an inconsistency, then` *ACCEPT*
`Otherwise,` *FAIL*

Theorem 7. *For all n, k and $0 < p \leq 1$ the above protocol is a sound k-round, secrecy $\lceil log^{(k)}(n)\rceil + 5\lceil\log(\frac{1}{p})+1\rceil$), probability $2(1-\frac{1}{2^k})p$ authentication protocol.*

Claim 8 *For all $k \geq 1$, P_k is a k round security $2(1-\frac{1}{2^k})p$ authentication scheme.*

Proof: We have shown previously that P_1 is a single-round security $p = 2(1 - \frac{1}{2^1})p$ authentication scheme. Assume inductively that P_{k-1} is a valid $(k-1)$ round security $2(1 - \frac{1}{2^{k-1}})p$ authentication scheme.

When I commits to message m', I has no idea what the value of index i_k will be. This is true because B chooses i_k uniformly at random only after receiving m'. If $m' \neq m$ then by the definition of the code C^k, we have:

$$Pr_{i_k}[C^k_{i_k}(m) \neq C^k_{i_k}(m')] \geq 1 - \frac{p}{2^{k-1}}$$

If $C^k_{i_k}(m) \neq C^k_{i_k}(m')$ then I must cheat in the second round, i.e. not send to A the message $(i_k, C^k_{i_k}(m))$ or I will surely be caught. If I cheats in the second round, then it is caught with probability at least $1 - 2(1 - \frac{1}{2^{k-1}})p$. Therefore the probability that I can cheat successfully is at most $2(1 - \frac{1}{2^{k-1}})p + \frac{p}{2^{k-1}} = 2(1 - \frac{1}{2^k})p$. □

Claim 9 *P_k uses $\log^{(k)}(n) + 5\log(\frac{1}{p})$ secret bits to authenticate messages of length n.*

Proof: For $k > 1$, the number of secret bits used by P_k to authenticate an n bit message is the same as the number of secret bits P_{k-1} uses to authenticate a message $(i_k, C^k_{i_k}(m'))$ of length $4k + 4\log(\frac{1}{p}) + \log(n)$. So long as $n \geq \left(\frac{1}{p}\right)^4$, the length of the message decreases to roughly $\log(n)$. If $n < \left(\frac{1}{p}\right)^4$ then $5\log(\frac{1}{p})$ dominates other terms in the expression for the number of secret bits used. □

This concludes the proof of theorem 7. □

Corollary 10. *For all n and p there exists a sound $\log^*(n)$ round, secrecy $2\log(\frac{1}{p}) + 2$, probability p authentication protocol.*

Proof: We will use the protocol $P_{\log^*(n)}$ except that we modify the last level of recursion, using the following 1-round authentication protocol instead of P_1.

Consider the following single-round protocol for a message of the form $m = (x, y)$ where $x, y \in GF[Q]$. The secret string is (a, b) where $a, b \in GF[Q]$. To authenticate $m = (x, y)$ send $a^2x + ay + b$. It is not hard to verify that this is a protocol for messages of length $2\log Q$, the security of this protocol is $2/Q$ and it uses a shared secret string of length $2\log Q$.

Set $p' = p/2$ and $k = \log^*(n)$ and run the protocol P_k with security p'. When the length of the message becomes smaller than $2\log(\frac{1}{p'})$ (as it would eventually), use the above one round protocol. □

5 Lower Bounds

We now consider lower bounds on the number of secret bits which A and B require.

Gilbert, Macwilliams and Sloane [4] showed that any single-round probability p authentication protocol requires at least $2\log(\frac{1}{p})$ secret bits. Their argument was based on lower bounding $H(s)$, the entropy of the secret string s. They showed that $H(s) = H(sx|m) = H(x|m) + H(s|mx)$, where $H(X|Y)$ is the conditional entropy of X given Y, averaged over all possible Y's.

Blum et. al. [2] showed that any single-round probability p authentication protocol requires at least $\log(n)$ secret bits for any $p < \frac{1}{2}$. We improve this second lower bound here.

5.1 Lower Bound for Sound Single-round protocols

We now show a lower bound on the number of shared secret bits in single-round protocols. The bound is achieved via a reduction from an authentication scheme to an error-correcting code.

Theorem 11. *There exists a function f such that $f(x) = o(\log(x))$ and such that there is no sound single-round, secrecy $\log(n) + \log(\frac{1}{p}) - f(\frac{n}{p})$, probability p authentication protocol for $p < 1$.*

Proof: Let P be a single-round, probability p authentication protocol. The outline of the proof is:

1. We define one probability distribution $\mathcal{D}_{m,x}$ on the secret strings for each input message, authenticator pair, (m, x).
2. We argue that some large subset of these distributions must be "far apart".
3. We convert this subset of distributions into a set of codewords which forms a code with high minimum distance.
4. We use a lower bound from coding theory to show that the alphabet of the code (which has the same size as the set of possible secret strings) is large. Let L be the number of possible secret strings. We will show that $L \log(L)$ at least $\frac{n}{p}$.

The rest of the proof appears in the full paper. Recently, Noga Alon (private communication) improved the lower bound to $\log(n) + 2\log(\frac{1}{p}) - \log\log\frac{1}{p}$, a better lower bound, using a bound on distances for codes with maximum weight.

5.2 Lower Bound for k Round Protocols

The idea behind our lower bound for k round protocols is to show that the existence of a k round, secrecy l, protocol implies the existence of a k round, secrecy l, protocol whose last authenticator has at most 2^l bits and that the existence of this second protocol implies the existence of a $k-1$ round, secrecy $l + 2^l$ protocol.

Definition 12. Given a conversation consisting of input message m and authenticators $x_1, x_2, \ldots x_k$, let the **characteristic vector** $CV(m, x_1, \ldots x_k)$ be a binary vector of length 2^l such that the sth bit, $CV(m, x_1, \ldots x_k)_s$, is 1 iff the recipient of the last message accepts given that the shared secret string was s and that the conversation which the recipient of the last message saw was $m, x_1 \ldots x_k$.

Note that, for a sound protocol, if the recipient of the last message has any chance of accepting a conversation given a particular secret string, it does so with probability 1 since it must accept all untampered conversations.

Theorem 13. *For $p < 1$, there is no sound k-round, probability p, secrecy $\lfloor \log^{(k)}(n) \rfloor - 1)$ -authentication scheme.*

Proof: We will show that if there is a sound k-round (p, l) -authentication scheme P_k then there is a sound $k-1$-round $(p, l+2^l)$ -authentication scheme P_{k-1}.

Claim 14 *If there is a sound k-round (p, l) -authentication scheme P_k then there is a sound k-round (p, l) -authentication scheme $\hat{P}_k$ such that the length of the last authenticator, x_k, is 2^l.*

Proof: Given P_k we describe a protocol $\hat{P}_k$. $\hat{P}_k$ is identical to P_k except for the last authenticator. The new last authenticator is the characteristic vector of the conversation that the sender of the last authenticator would have seen in P_k:

$$\hat{x}_k = CV(w, \ldots, x'_{k-3}, x_{k-2}, x'_{k-1}, x_k)$$

w is the input message understood by the sender of the last authenticator.
The recipient of $\hat{x}'_k$ accepts iff:

- There exists an authenticator x'_k such that $\hat{x}'_k = CV(w', \ldots, x_{k-1}, x'_k)$. In other words, there is an equivalent authenticator which the sender of the last authenticator could have sent in protocol P_k. Here w' refers to the input that the recipient of the last authenticator understands.
- For the shared secret string s, $(\hat{x}'_k)_s = 1$. The recipient of the last authenticator would have accepted in P_k if s/he received x'_k.

To see that $\hat{P}_k$ is a k-round (p,l)-authentication protocol, we show the following:

1. If $\hat{I}$, the adversary for the second protocol, does not interfere with any of the messages, then both A and B will accept and B will know the input. This is clear since the input m is sent in the first round and since the last message is $CV(m, x_1, \ldots x_k)$ where $x_1 \ldots x_k$ are the authenticators A and B actually send.
2. If $\hat{I}$ is able to cheat A and B in protocol $\hat{P}_k$ then given the same circumstances I could cheat A and B in protocol P_k.
   I's strategy would be to behave exactly as would $\hat{I}$ except that on the last round I replaces x_k with any x'_k such that $\hat{x}'_k = CV(w', \ldots, x_{k-2}, x_{k-1}, x'_k)$

□

The proof of the theorem is now completed by defining a new $k-1$ round protocol, P_{k-1}:

Claim 15 *If there exists a k round (p, l)-authentication protocol P^*_k such that the length of the last authenticator, x^*_k, is $|x^*_k| = 2^l$, then there exists a $k-1$ round $(p, l+2^l)$-authentication protocol, P_{k-1}.*

Proof:

Description of P_{k-1}

- We do away with the kth round completely by adding the advice $\overline{x^*_k}$ to the shared secret string s where $\overline{x^*_k}$ is the last authenticator that would have been sent in the conversation as it would have occurred in P^*_k with no interference from the adversary. The advice for the protocol P_{k-1} consists of the original

l bits of advice from protocol P_k appended to this 2^l bit $\overline{x_k^*}$. We note that, in this situation, the secret string depends on the input message and possibly the random bits of A and B. However this is acceptable since the lower bound of $\log(n)$ presented in [2] applies to such protocols.

- At the end of the $k-1$st round, the party who would have sent the kth authenticator, x_k^*, in protocol P_k^* instead checks to see that $x_k^* = \overline{x_k^*}$.
- The party who would have received the kth message checks to see that they would have accepted $\overline{x_k^*}$ in P_k^* In other words, the party who would have received the kth message in P_k^* looks at $\overline{x_k^*}$ and acts as if s/he received that.

To show that P_{k-1} is a $k-1$-round $(p, l+2^l)$-authentication protocol, we note:

1. If there is been no interference by an intruder, then the party that would have sent the last authenticator in P_k^* will note that $x_k^* = \overline{x_k^*}$. Furthermore, because P_k^* is sound, the other party would accept $\overline{x_k^*}$.
2. If A and B accept an altered input message in protocol P_{k-1}, then the adversary in the protocol P_k^* could convince A and B to accept by acting as s/he would in P_{k-1} and then delivering, unaltered, the last authenticator x_k^*. The recipient of the last authenticator would accept because we have $x_k^* = \overline{x_k^*}$.

□ This concludes the proof of the theorem. □

5.3 Lower bounds for protocols which are not necessarily sound

We now consider lower bounds for protocols which are not necessarily sound: even with no interference from the adversary, they are allowed some probability of failure.

For this section, we modify the definition of characteristic vector:

Definition 16. $CV(m, x_1, \ldots x_k)_s = 1$ iff the recipient of the last message would accept with probability $\geq 1/2$ given the conversation $m, x_1 \ldots x_k$ and secret string s. Otherwise $CV(m, x_1 \ldots x_k) = 0$.

Corollary 17. *There is no single-round, secrecy* $\log(n) - 1$, *probability* p *authentication protocol for* $p < 1/3$.

Proof: Suppose that $l < \log(n)$. If the secret string contains l bits then there are at most 2^{2^l} distinct characteristic vectors. Since $l < \log(n)$ then there are fewer than 2^n characteristic vectors. Therefore, there is some input m such that $\forall x_1 \exists x'_1, m' : m' \neq m$ such that $CV(m', x'_1) = CV(m, x_1)$.

The way we redefined characteristic vectors implies that the probability that B will reject m', x'_1 is at most twice the probability that B will reject m, x_1. Therefore, if an adversary always replaced m, x_1 with m', x'_1, with probability at least $1 - 2p > 1 - 2\frac{1}{3} = 1/3 > p$, B accepts a bad message.

So we must have $p \geq 1/3$. □

Theorem 18. *For* $p < \frac{1}{3}\frac{1}{2^{k-1}}$, *there is no* k*-round, secrecy* $o(\log^{(k)}(n)))$, *probability* p *authentication protocol for any* c *independent of* n.

Proof: The proof is similar to that lower bounding the number of secret bits needed in a k-round sound protocol. As in the previous theorem $CV(m, x_1, \ldots, x_k)$ $= 1$ iff the recipient of the last message would accept with probability $\geq 1/2$ given the conversation has been $m, x_1 \ldots x_k$. This approximation leads to a possible doubling of the error for each conversion of the k round protocol P_k to a k round protocol $\hat{P}_k$ which has a short last message. If the intruder I has interfered in conversation $m, x_1, \ldots, x_k$ and the probability that A and B accept in P_{k-1} is at least $q = \frac{1}{2}$ then $CV(m, x_1, \ldots, x_k) = 1$ and the probability that A and B accept in $\hat{P}_k$ is $1 \leq 2q$. □

6 Redundancy lower bounds

In the previous sections, we showed that multi-round protocols can be used to lessen the number of secret bits that two parties need to share in order to authenticate an n bit message. However, in the protocols we presented, the number of bits exchanged, including the input message and the authenticators, was more than n. Here, we show a lower bound on the **redundancy**, the extra information which they have to share *or* transmit in order to authenticate an n bit input message.

Definition 19. The **redundancy** of an authentication protocol is equal to the sum of the number of authentication bits – the x_i's transmitted between A and B – plus the number of shared secret bits.

Theorem 20. *For any sound k-round authentication protocol P, the redundancy of P is at least* $\log(n)$.

This is significant since it shows that while more rounds may decrease the number of secret bits needed, more rounds cannot decrease the redundancy below $\log(n)$.

Proof: Assume that the protocol P uses: t bits for the authenticators and l bits for the shared secret string. For each input message m and secret string s, define:

- $\mathcal{D}(m, s)$ is the probability distribution on the authenticators that would appear in a conversation between A and B using message m and secret string s.
- Given a probability distribution $\mathcal{D}(m, s)$ on t-bit strings $\overline{x}$, the set of possible authenticator sequences for (m, s) equals

$$N(\mathcal{D}(m, s)) = \{\overline{x} | Pr_{x \in \mathcal{D}(m,s)}[x] > 0\}$$

For each possible input message m, define a vector of sets, $V(m)$, of length 2^l such that $V(m)_s = N(\mathcal{D}(m, s))$.

There are 2^{2^t} possible subsets of all t bit strings and hence at most $(2^{2^t})^{2^l} =$ $2^{2^{t+l}}$ possible vectors $V(m)$. If the redundancy $t + l$ is less than $\log(n)$ then the number of possible vectors $V(m)$ is less than the number of input messages.

By the pigeon hole principle, there would be two input messages, $m, m'; m \neq m'$, which have the same vector, $V(m) = V(m')$. Because the two vectors have the same set of possible authenticator sequences in each entry, for any s, any authenticator sequence $\overline{x}$ which could be generated during a conversation using m and s could also be generated by A and B during a conversation using m' and s. From soundness, we know that such authenticators must also be accepted. Therefore, if $t + l < \log(n)$, an intruder could always substitute m' for m with no chance of being detected. □

7 Acknowledgments

We thank Manuel Blum for many thought-provoking and useful conversations about the defining of the problem. We thank Ronny Roth for helpful conversations about the theory of error correcting codes. Also, we thank Mike Luby for his detailed comments on the drafts of the paper.

References

1. N. Alon, J. Bruck, J. Naor, M. Naor, R. Roth, *Construction of Asymptotically Good Low-Rate Error-Correcting Codes through Pseudo-Random Graphs*, IEEE Transactions on Information Theory, Vol. 38, No. 2, March 1992
2. M. Blum, W. Evans, P. Gemmell, S. Kannan, M. Naor *Checking the Correctness of Memories*, Proc. 31st Symp. on Foundations of Computer Science, October 1990.
3. E. F. Brickell. *A Few Results in Message Authentication* Congressus Numerantium 43 (1984), 141-154.
4. E. Gilbert, F. J. MacWilliams, N. Sloane, *Codes Which Detect Deception*, The Bell System Technical Journal, Vol. 53, No. 3, March 1974
5. M. Jimbo, R. Fuji-hara. *Optimal Authentication Systems and Combinatorial Designs,* IEEE Transactions on Information Theory, vol. 36, no 1, January 1990, pp 54-62.
6. F. J. MacWilliams, N. Sloane. **The Theory of Error Correcting Codes,** North Holland, Amsterdam, 1977.
7. R. Roth. Personal Communication
8. G. Simmons, *A Survey of Information Authentication*, Proceedings of the IEEE, Vol. 76, No. 5, May 1988
9. D. Stinson. *Universal Hashing and Authentication Codes.* Advances in Cryptology: CRYPTO '91, pp 74-85.
10. D. Stinson. *Combinatorial Characterizations of Authentication Codes.* Advances in Cryptology: CRYPTO '91, pp 62-73.
11. D. Stinson. *The Combinatorics of Authentication and Secrecy Codes.* Journal of Cryptology, 1990, vol.2, (no.1):23-49.
12. D. Stinson. *Some Constructions and Bounds for Authentication Codes.* Journal of Cryptology, 1988, vol.1, 37-51.
13. D. Stinson. *A Construction of Authentication/Secrecy Codes from Certain Combinatorial Designs* Journal of Cryptology, 1988, vol.1, (no.2):119-127.
14. Wegman and Carter, *New Hash functions and their use in authentication and set equality* J. Computer and System Sci. **22**, 1981, pp. 265-279.

Hash functions based on block ciphers: a synthetic approach

Bart Preneel*, René Govaerts, and Joos Vandewalle

Katholieke Universiteit Leuven, Laboratorium ESAT-COSIC,
Kardinaal Mercierlaan 94, B–3001 Heverlee, Belgium
bart.preneel@esat.kuleuven.ac.be

Abstract. Constructions for hash functions based on a block cipher are studied where the size of the hashcode is equal to the block length of the block cipher and where the key size is approximately equal to the block length. A general model is presented, and it is shown that this model covers 9 schemes that have appeared in the literature. Within this general model 64 possible schemes exist, and it is shown that 12 of these are secure; they can be reduced to 2 classes based on linear transformations of variables. The properties of these 12 schemes with respect to weaknesses of the underlying block cipher are studied. The same approach can be extended to study keyed hash functions (MAC's) based on block ciphers and hash functions based on modular arithmetic. Finally a new attack is presented on a scheme suggested by R. Merkle.

1 Introduction

Hash functions are functions that compress an input of arbitrary length to a string of fixed length. They are a basic building block for cryptographic applications like integrity protection based on "fingerprinting" and digital signature schemes. The cryptographic requirements that are imposed on hash functions are [4, 5, 19, 27, 28]:

one-wayness: in the sense that given X and $h(X)$, it is "hard" to find a second preimage, i.e., a message $X' \neq X$ such that $h(X') = h(X)$,

collision resistance: it should be "hard" to find a collision, i.e., two distinct arguments that hash to the same result.

The main motivation to construct a hash function based on a block cipher is the minimization of design and implementation effort. Designing secure constructions seems to be a difficult problem; this is illustrated by the large number of schemes that have been broken [23, 27].

The first constructions for hash functions based on a block cipher were one-way hash functions intended for use with the Data Encryption Standard (DES)

* N.F.W.O. postdoctoral researcher, sponsored by the National Fund for Scientific Research (Belgium).

[10]. In this case the size of the hashcode (64 bits) is equal to the block length of the block cipher and the size of the key (56 bits) is approximately equal to the block size. This type of hash functions will be studied in this extended abstract. Later constructions for collision resistant hash functions were developed based on the DES; in that case it is required that the size of the hash code is at least 112...128 bits (because of the birthday attack [33]). Examples of schemes that have not been broken can be found in [20, 22]. Recent work considers the construction of hash functions if the key size is twice the block length [17], and if the key is kept constant [26]. The original constructions are still of interest since for some applications a one-way hash function is sufficient. Moreover, they can yield a collision resistant hash function if a block cipher with sufficiently large block length is available.

2 The General Model

The encryption of plaintext X with key K will be denoted with $E(K,X)$. The corresponding decryption operation applied to ciphertext C will be denoted with $D(K,C)$. Unless stated otherwise, it will be assumed that the block cipher has no weaknesses. The block length, i.e., the size of plaintext and ciphertext in bits is denoted with n and the key size in bits is denoted with k. The argument of the iterated hash function is divided into t blocks X_1 through X_t. If the total length is no multiple of n, the argument has to be padded with an unambiguous padding rule. The hash function h can subsequently be described as follows:

$$H_i = f(X_i, H_{i-1}) \qquad i = 1, 2, \ldots t\,.$$

Here f is the *round function*, H_0 is equal to the initial value (IV), that should be specified together with the scheme, and H_t is the hashcode. The *rate* R of a hash function based on a block cipher is defined as the number of encryptions to process a block of n bits.

The general model for the round function of the hash functions that will be studied in this extended abstract is depicted in Fig. 1. For simplicity it will be assumed that $k = n$. The block cipher has two inputs, namely the key input K and the plaintext input P, and one output C. One can select for the inputs one of the four values: X_i, H_{i-1}, $X_i \oplus H_{i-1}$, and a constant value V. It is also possible to modify with a feedforward FF the output C by addition modulo 2 of one of these four possibilities. This yields in total $4^3 = 64$ different schemes. In the following it will be assumed w.l.o.g. that V is equal to 0.

The exor operation was chosen because it has been used in the proposals that are generalized here; one can show that it can be replaced by any operation that is an easy-to-invert permutation of one of its inputs when the second input is fixed. The main restrictions of this model are that only 1 DES operation is used per round function and that the internal memory of the hash function is restricted to a single n-bit block.

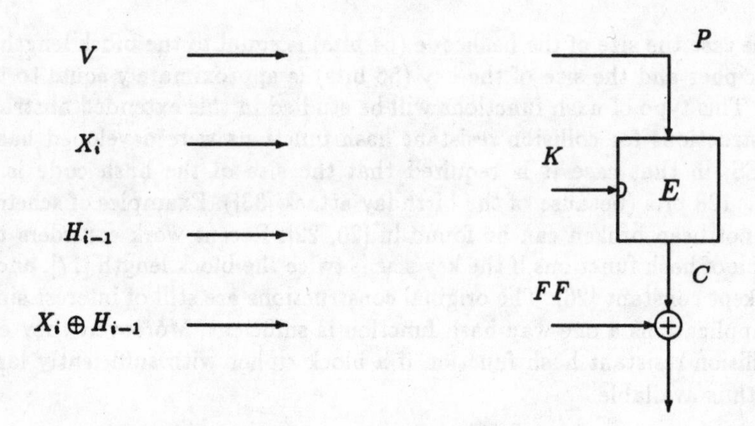

Fig. 1. Configurations where the size of the hashcode is equal to the block length. P, K, and FF can be chosen from the set $\{V, X_i, H_{i-1}, X_i \oplus H_{i-1}\}$.

3 A Taxonomy of Attacks

Five important attacks on the round function $f(X_i, H_{i-1})$ can be identified:

Direct Attack (D): given H_{i-1} and H_i, it is easy to find X_i. All schemes that are vulnerable to a direct attack can in principle be used for encryption, where the encryption of X_i is given by H_i. Of course the CBC and CFB mode belong to this class.

Permutation Attack (P): in this case H_i can be written as $H_{i-1} \oplus f'(X_i)$, where f' is a one-way function: X_i can not be recovered from H_i and H_{i-1}, but the hashcode is independent of the order of the message blocks, which means that a second preimage or collision can be found easily. Moreover one can also insert the same message block twice. These attacks are in fact trivial, as H_i depends only linearly on H_{i-1}.

Forward Attack (F): given H_{i-1}, H'_{i-1}, and X_i (note that this means that H_i is fixed), it is easy to find an X'_i such that $f(X'_i, H'_{i-1}) = f(X_i, H_{i-1}) = H_i$. In this case one can easily construct a second preimage for a given hashcode, but it is not necessarily easy to construct a preimage of a given element in the range.

Backward Attack (B): given H_i, it is easy to find a pair (X_i, H_{i-1}) such that $f(X_i, H_{i-1}) = H_i$. In this case it is trivial to find a preimage (or a second preimage) with a random initial value; a preimage (or a second preimage) can be found with a meet in the middle attack.

Fixed Point Attack (FP): find H_{i-1} and X_i such that $f(X_i, H_{i-1}) = H_{i-1}$. This attack is not very dangerous: if the hash function satisfies the one-way property, it is hard to produce a message yielding this specific value H_{i-1}.

The order of these attacks has some importance: the possibility of a direct attack means that a forward and a backward attack are also feasible, but the converse does not hold. In case of a permutation attack, one can also apply a backward attack by first selecting X_i and subsequently calculating H_{i-1}. It is easy to show that if both a forward and a backward or permutation attack are possible, a direct attack is also feasible. The proof will be given in the full paper.

4 Analysis of the 64 Schemes

Table 1 indicates which attacks are possible for each of the 64 schemes in the general model. The attacks are indicated with their first letter(s), while a "–" means that the round function f is trivially weak as the result is independent of one of the inputs. If none of these five attacks applies, a $\surd$ is put in the corresponding entry.

Table 1. Attacks on the 64 different schemes. The schemes are numbered according to the superscript.

		choice of P			
choice of FF	choice of K	X_i	H_{i-1}	$X_i \oplus H_{i-1}$	V
V	X_i	–	B^{13}	B^{25}	–
	H_{i-1}	D^{1}	–	D^{26}	–
	$X_i \oplus H_{i-1}$	B^{2}	B^{14}	F^{27}	F^{41}
	V	–	–	D^{28}	–
X_i	X_i	–	B^{15}	B^{29}	–
	H_{i-1}	$\surd^{3}$	D^{16}	$\surd^{30}$	D^{42}
	$X_i \oplus H_{i-1}$	FP^{4}	FP^{17}	B^{31}	B^{43}
	V		D^{18}	B^{32}	–
H_{i-1}	X_i	P^{5}	FP^{19}	FP^{33}	P^{44}
	H_{i-1}	D^{6}	–	D^{34}	
	$X_i \oplus H_{i-1}$	FP^{7}	FP^{20}	B^{35}	B^{45}
	V	D^{8}	–	D^{36}	–
$X_i \oplus H_{i-1}$	X_i	P^{9}	FP^{21}	FP^{37}	P^{46}
	H_{i-1}	$\surd^{10}$	D^{22}	$\surd^{38}$	D^{47}
	$X_i \oplus H_{i-1}$	B^{11}	B^{23}	F^{39}	F^{48}
	V	P^{12}	D^{24}	F^{40}	D^{49}

Schemes 18 and 28 correspond to the CFB mode respectively the CBC mode for encryption as specified in [11, 14]; the fact that these modes are useful for keyed hash functions but not sufficient to construct one-way hash functions was pointed out by S. Akl in [1]. This is also connected to the fact that the integrity protection offered by these modes is limited. Scheme 13 was proposed by Rabin in 1978 [28]: however, R. Merkle has shown that a backward attack is possible, from which it follows that one can find a preimage with a meet in the middle attack.

The next proposal was scheme 14, attributed to W. Bitzer in [7, 9]. R. Winternitz [31] has shown that the meet in the middle attack by R. Merkle is applicable to this scheme as well. He also has pointed out that schemes 13 and 14 are vulnerable to a weak key attack: for a weak key K_w the DES is an involution which means that $E(K_w, E(K_w, X)) = X, \forall X$. Inserting twice a weak key as a message block will leave the hashcode unchanged in all the schemes.

The first secure scheme (scheme 3) was proposed by S. Matyas, C. Meyer, and J. Oseas in [18]. Its 'dual', scheme 19, is attributed to D. Davies in [31, 32], and to C. Meyer by D. Davies in [8]. D. Davies has confirmed in a personal communication to the authors that he did not propose the scheme. Nevertheless, this scheme is widely known as the Davies-Meyer scheme (see e.g., [23]). The fact that this scheme is vulnerable to a fixed point attack was pointed out in [25]. Scheme 10 was proposed by the authors and studied in [30]. It appeared independently in [24] as a mode for N-hash. In 1990 the same scheme was proposed by Japan to ISO/IEC [15]. The international standard ISO/IEC 10118 Part 2 [16] specifying hash functions based on block ciphers contains scheme 3. Scheme 20 was proposed as a mode of use for the block cipher LOKI in [3]. Finally it should be remarked that scheme 40 (together with its vulnerability to a forward attack) was described in [18].

It is the merit of this approach that all schemes based on the general model have been classified once and for all. The second advantage is that the properties of the 12 'secure' schemes can now be compared. First a further classification will be made based on an equivalence transformation.

5 Equivalence Classes

This large number of schemes can be classified further by considering linear transformations of the inputs. A class of schemes that is derived from a single scheme by linear transformation of variables will be called an equivalence class.

- In 7 equivalence classes the round function depends on two independent inputs (X_i and H_{i-1}), and 6 transformations are possible, as there are 6 invertible 2×2 matrices over $GF(2)$. It can be shown that in 2 cases the round function is secure or is vulnerable to a fixed point attack, and in 5 cases the round function is vulnerable to a direct attack, a permutation attack, or a backward attack.
- In 7 equivalence classes the round function depends on a single independent input. Hence one has three possible inputs, namely X_i, H_{i-1}, and $X_i \oplus H_{i-1}$, corresponding to the 3 nonzero vectors of length 2 over $GF(2)$. If the round function depends on the sum of the two inputs, it is not trivially weak. However, it is vulnerable to a direct attack (2 cases out of 7) or to a forward attack (5 cases out of 7).
- In 1 equivalence class the round function is simply constant.

Table 2 describes the equivalence classes. A further classification is made based on the number CI of constants in the choices. To characterize a class, a relation is given between plaintext P, key K, and feedforward FF.

Table 2. Overview of the 15 variants, sorted according to the number CI of constant inputs.

CI	characterization	class size	-	D	P	B	F	FP	$\surd$
0	$FF = P, (P \neq K)$	6						4	2
	$FF = P \oplus K, (P \neq K)$	6						4	2
	$FF = K, (P \neq K)$	6		2		4			
	$P = K, (FF \neq P)$	6		2	2	2			
	$FF = P = K$	3	2				1		
1	$FF = V, (P \neq K)$	6		2		4			
	$P = V, (FF \neq K)$	6		2	2	2			
	$K = V, (FF \neq P)$	6		4	1	1			
	$FF = V, (P = K)$	3	2				1		
	$P = V, (FF = K)$	3	2				1		
	$K = V, (P = FF)$	3	2				1		
2	$FF = P = V$	3	2				1		
	$FF = K = V$	3	2	1					
	$P = K = V$	3	2	1					
3	$FF = P = K = V$	1	1						
Total		64	15	14	5	13	5	8	4

One can conclude that only 4 schemes of the 64 are secure, and that 8 insecure schemes are only vulnerable to a fixed-point attack. These 12 schemes are listed (and re-numbered) in Table 3 and graphically presented in Fig. 2. The roles of X_i and H_{i-1} in the input can be arbitrarily chosen, and the dotted arrow is optional (if it is included, the key is added modulo 2 to the ciphertext). For the dash line, there are three possibilities: it can be omitted or it can point from key to plaintext or from plaintext to key. There are two equivalence classes that are secure, and their simplest representatives are the scheme by Matyas et al. (number 1) and the scheme by Miyaguchi and the authors (number 3). For each of these schemes it is possible to write a 'security proof' based on a black box model of the encryption algorithm, as was done for the Davies-Meyer scheme (number 5) in [32]. The basic idea is that finding a (pseudo)-preimage for a given hash value is at least as hard as solving the equation $H_i = f(X_i, H_{i-1})$ for a given value of H_i. The expected number of evaluations of $f()$ is shown to be 2^{n-1}.

In the full paper these 12 schemes will be compared in more detail based on their vulnerability to fixed point attacks, to attacks based on weaknesses of the underlying block cipher (in this case the DES), and to differential attacks [2]. Also their efficiency will be compared. The main results are summarized in Table 4: column 5 indicates what the output of the round function is if the key is a weak key and the plaintext is one of the corresponding fixed points, column 6 has a $\surd$ if one can exploit the complementation property, and the last column indicates which variables have to be modified if a differential attack

Table 3. A list of the 12 secure schemes for a one-way hash function based on a block cipher and a feedforward.

no.	function expression
1	$E(H_{i-1}, X_i) \oplus X_i$
2	$E(H_{i-1}, X_i \oplus H_{i-1}) \oplus X_i \oplus H_{i-1}$
3	$E(H_{i-1}, X_i) \oplus X_i \oplus H_{i-1}$
4	$E(H_{i-1}, X_i \oplus H_{i-1}) \oplus X_i$
5	$E(X_i, H_{i-1}) \oplus H_{i-1}$
6	$E(X_i, X_i \oplus H_{i-1}) \oplus X_i \oplus H_{i-1}$
7	$E(X_i, H_{i-1}) \oplus X_i \oplus H_{i-1}$
8	$E(X_i, X_i \oplus H_{i-1}) \oplus H_{i-1}$
9	$E(X_i \oplus H_{i-1}, X_i) \oplus X_i$
10	$E(X_i \oplus H_{i-1}, H_{i-1}) \oplus H_{i-1}$
11	$E(X_i \oplus H_{i-1}, X_i) \oplus H_{i-1}$
12	$E(X_i \oplus H_{i-1}, H_{i-1}) \oplus X_i$

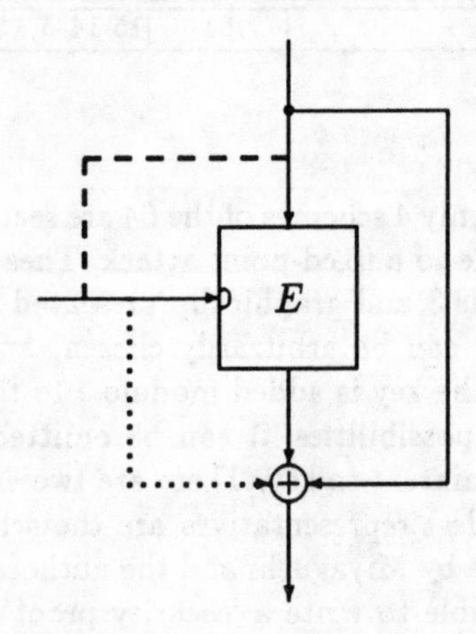

Fig. 2. Secure configuration for a one-way hash function based on an block cipher and a feedforward.

with a fixed key is used. (One could also think of a dual differential attack, where the plaintext is fixed and a given key difference is applied; this has not been considered here.)

If $k \neq n$, the question arises whether it is possible to use variables X_i and H_{i-1} of length $max(n, k)$, in order to maximize the security level. The idea is that bits that are not used in the key or as plaintext might influence the output through some exors. The following proposition shows that this is not possible

Table 4. Properties of the 12 secure schemes: fixed points, properties if the DES is used as the underlying block cipher, and variables to be modified in case of a differential attack.

no.	fixed points		properties if E = DES			differential
	X_i	H_{i-1}	rate R	K_w	compl.	attack
1	–	–	1	0	√	X_i
2	–	–	1	0	√	X_i
3	–	–	1	K_w	-	X_i
4	–	–	1	K_w	-	X_i
5	K	$D(K,0)$	n/k	0	√	H_{i-1}
6	K	$D(K,K) \oplus K$	n/k	0	√	H_{i-1}
7	K	$D(K,K)$	n/k	K_w	-	H_{i-1}
8	K	$D(K,0) \oplus K$	n/k	K_w	-	H_{i-1}
9	$D(K,K)$	$D(K,K) \oplus K$	1	0	√	X_i, H_{i-1}
10	$D(K,0) \oplus K$	$D(K,0)$	n/k	0	√	H_{i-1}
11	$D(K,0)$	$D(K,0) \oplus K$	1	K_w	-	X, H_{i-1}
12	$D(K,K) \oplus K$	$D(K,K)$	n/k	K_w	-	X_i, H_{i-1}

for the hash functions which follow our general model; it will be proven in the full paper.

Proposition 1 *The security level of the one-way hash function is determined by the minimum of k and n, with k the size of the key and n the block length.*

6 Merkle's Improvement to Rabin's Scheme

In order to avoid the backward attack in case of Rabin's scheme (Scheme 13), R. Merkle proposed to encrypt the message in CBC or CFB mode (with a random non-secret key K and initial value IV) before applying the hash function [6]. This implies a reduced performance: the rate equals 2. The idea is to introduce a dependency between the blocks that enter the hash function. It will be shown how the meet in the middle attack can be modified to take into account this extension.

- Generate a set of r messages for which the last ciphertext block of the CBC encryption with initial value IV and key K is equal to IV' ; this can be done easily with an appropriate selection of the last plaintext block (or with a meet in the middle attack).
- Generate a second set of r messages and encrypt these messages in CBC mode with initial value IV' and key K.
- As the two message parts are now independent, one can use the set of two 'encrypted' messages in a simple meet in the middle attack.

This shows that finding a preimage requires only $O(2^{n/2})$ encryptions.

7 Extensions

The same approach can be applied to keyed hash functions (or Message Authentication Code) based on a block cipher. The main result in this case is that the round function

$$f = E(K, X_i \oplus H_{i-1}) \oplus X_i$$

is preferable over the CBC or CFB mode that are specified in most standards (e.g., [13]). Note that an additional protection is required at the end. More details will be given in the full paper.

The design of hash functions based on modular squaring can also benefit from this synthetic approach. Since there is no key, only 16 cases have to be studied. Most hash functions in this class base their security on the fact that taking modular square roots is hard for someone who does not know the factorization of the modulus. Seven of these schemes are trivially weak, and 4 (from the remaining 9) have appeared in the literature. The main conclusion is that the round function

$$f = (X_i \oplus H_{i-1})^2 \bmod N \oplus X_i$$

is the most promising. In order to obtain a secure hash function a redundancy scheme has to be specified. The redundancy is necessary to thwart the exploitation of the algebraic structure like fixed points of the modular squaring and 'small' numbers for which no reduction occurs [12, 27].

Finally it should be noted that a similar structure has been employed in dedicated hash functions like the MD4 family [29] and Snefru [21]. The analysis with respect to differential attacks and fixed points can also be transferred to these schemes.

8 Conclusion

The construction of cryptographic hash functions based on block ciphers is apparently a difficult problem. In this extended abstract a general treatment has been developed for the simplest case, namely size of hashcode equal to block length and key size. This approach allows to identify the secure schemes and to compare these with respect to several criteria. It is also useful to study keyed hash functions, hash functions based on modular arithmetic, and dedicated hash functions.

References

1. S.G. Akl, "On the security of compressed encodings," *Advances in Cryptology, Proc. Crypto'83*, D. Chaum, Ed., Plenum Press, New York, 1984, pp. 209–230.
2. E. Biham and A. Shamir, "Differential cryptanalysis of DES-like cryptosystems," *Journal of Cryptology*, Vol. 4, No. 1, 1991, pp. 3–72.
3. L. Brown, J. Pieprzyk, and J. Seberry, "LOKI – a cryptographic primitive for authentication and secrecy applications," *Advances in Cryptology, Proc. Auscrypt'90, LNCS 453*, J. Seberry and J. Pieprzyk, Eds., Springer-Verlag, 1990, pp. 229–236.

4. I.B. Damgård, "Collision free hash functions and public key signature schemes," *Advances in Cryptology, Proc. Eurocrypt'87, LNCS 304*, D. Chaum and W.L. Price, Eds., Springer-Verlag, 1988, pp. 203–216.
5. I.B. Damgård, "A design principle for hash functions," *Advances in Cryptology, Proc. Crypto'89, LNCS 435*, G. Brassard, Ed., Springer-Verlag, 1990, pp. 416–427.
6. D. Davies and W. L. Price, "The application of digital signatures based on public key cryptosystems," *NPL Report* DNACS 39/80, December 1980.
7. D. Davies, "Applying the RSA digital signature to electronic mail," *IEEE Computer*, Vol. 16, February 1983, pp. 55–62.
8. D. Davies and W. L. Price, "Digital signatures, an update," *Proc. 5th International Conference on Computer Communication*, October 1984, pp. 845–849.
9. D. Denning, "Digital signatures with RSA and other public-key cryptosystems," *Communications ACM*, Vol. 27, April 1984, pp. 388–392.
10. FIPS 46, *"Data Encryption Standard,"* Federal Information Processing Standard, National Bureau of Standards, U.S. Department of Commerce, Washington D.C., January 1977.
11. FIPS 81, *"DES Modes of operation,"* Federal Information Processing Standard, National Bureau of Standards, US Department of Commerce, Washington D.C., December 1980.
12. M. Girault, "Hash-functions using modulo-n operations," *Advances in Cryptology, Proc. Eurocrypt'87, LNCS 304*, D. Chaum and W.L. Price, Eds., Springer-Verlag, 1988, pp. 217–226.
13. ISO/IEC 9797, *"Information technology - Data cryptographic techniques - Data integrity mechanisms using a cryptographic check function employing a block cipher algorithm,"* 1993.
14. ISO/IEC 10116, *"Information technology - Security techniques - Modes of operation of an n-bit block cipher algorithm,"* 1991.
15. *"Hash functions using a pseudo random algorithm,"* ISO-IEC/JTC1/SC27/WG2 N98, Japanese contribution, 1991.
16. ISO/IEC 10118, *"Information technology - Security techniques - Hash-functions - Part 1: General and Part 2: Hash-functions using an n-bit block cipher algorithm,"* 1993.
17. X. Lai and J.L. Massey "Hash functions based on block ciphers," *Advances in Cryptology, Proc. Eurocrypt'92, LNCS 658*, R.A. Rueppel, Ed., Springer-Verlag, 1993, pp. 55–70.
18. S.M. Matyas, C.H. Meyer, and J. Oseas, "Generating strong one-way functions with cryptographic algorithm," *IBM Techn. Disclosure Bull.*, Vol. 27, No. 10A, 1985, pp. 5658–5659.
19. R. Merkle, *"Secrecy, Authentication, and Public Key Systems,"* UMI Research Press, 1979.
20. R. Merkle, "One way hash functions and DES," *Advances in Cryptology, Proc. Crypto'89, LNCS 435*, G. Brassard, Ed., Springer-Verlag, 1990, pp. 428–446.
21. R. Merkle, "A fast software one-way hash function," *Journal of Cryptology*, Vol. 3, No. 1, 1990, pp. 43–58.
22. C.H. Meyer and M. Schilling, "Secure program load with Manipulation Detection Code," *Proc. Securicom 1988*, pp. 111–130.
23. C. Mitchell, F. Piper, and P. Wild, "Digital signatures," in *"Contemporary Cryptology: The Science of Information Integrity,"* G.J. Simmons, Ed., IEEE Press, 1991, pp. 325–378.

24. S. Miyaguchi, M. Iwata, and K. Ohta, "New 128-bit hash function," *Proc. 4th International Joint Workshop on Computer Communications*, Tokyo, Japan, July 13–15, 1989, pp. 279–288.
25. S. Miyaguchi, K. Ohta, and M. Iwata, "Confirmation that some hash functions are not collision free," *Advances in Cryptology, Proc. Eurocrypt'90, LNCS 473*, I.B. Damgård, Ed., Springer-Verlag, 1991, pp. 326–343.
26. B. Preneel, R. Govaerts, and J. Vandewalle, "On the power of memory in the design of collision resistant hash functions," *Advances in Cryptology, Proc. Auscrypt'92, LNCS 718*, J. Seberry and Y. Zheng, Eds., Springer-Verlag, 1993, pp. 105–121.
27. B. Preneel, *"Cryptographic hash functions,"* Kluwer Academic Publishers, 1994.
28. M.O. Rabin, "Digitalized signatures," in *"Foundations of Secure Computation,"* R. Lipton and R. DeMillo, Eds., Academic Press, New York, 1978, pp. 155–166.
29. R.L. Rivest, "The MD4 message digest algorithm," *Advances in Cryptology, Proc. Crypto'90, LNCS 537*, S. Vanstone, Ed., Springer-Verlag, 1991, pp. 303–311.
30. K. Van Espen and J. Van Mieghem, *"Evaluatie en Implementatie van Authentiseringsalgoritmen (Evaluation and Implementation of Authentication Algorithms – in Dutch),"* ESAT Laboratorium, Katholieke Universiteit Leuven, Thesis grad. eng., 1989.
31. R.S. Winternitz, "Producing a one-way hash function from DES," *Advances in Cryptology, Proc. Crypto'83*, D. Chaum, Ed., Plenum Press, New York, 1984, pp. 203–207.
32. R.S. Winternitz, "A secure one–way hash function built from DES," *Proc. IEEE Symposium on Information Security and Privacy 1984*, 1984, pp. 88–90.
33. G. Yuval, "How to swindle Rabin," *Cryptologia*, Vol. 3, 1979, pp. 187–189.

Security of Iterated Hash Functions Based on Block Ciphers

Walter Hohl Xuejia Lai Thomas Meier Christian Waldvogel

Signal and Information Processing Laboratory
Swiss Federal Institute of Technology
CH-8092 Zurich, Switzerland

Abstract. Cryptographic hash functions obtained by iterating a round function constructed from a block cipher and for which the hash-code length is twice the block length m of the underlying block cipher are considered. The computational security of such hash functions against two particular attacks, namely, the free-start target and free-start collision attacks, is investigated; these two attacks differentiate themselves from the "usual" target and collision attacks by not specifying the initial value of the iterations. The motivation is that computationally secure iterated hash functions against these two particular attacks implies computationally secure iterated hash functions against the "usual" target and collision attacks. For a general class of such $2m$-bit iterated hash functions, tighter upper bounds than the one yet published in the literature on the complexity of free-start target and free-start collision attacks are derived. A proposal for a $2m$-bit iterated hash function achieving these upper bounds is made; this new proposal is shown to be computationally more secure against free-start target and free-start collision attacks than some of the already proposed schemes falling into this general class. It is also shown that our proposal is better than the present proposal for an ISO standard in the sense that both schemes achieve these upper bounds but one encryption is required in our proposal for hashing one m-bit message block as opposed to two encryptions in the ISO proposal. Finally, two new attacks on the LOKI Double-Block-Hash function are presented with lower complexities than the known ones.

1 Introduction

A *hash function* is an easily computable mapping from the set of all binary sequences of some minimum length or greater to the set of binary sequences of some fixed length. In cryptography, hash functions are used to provide data integrity and to produce short digital signatures.

One well-known method to obtain hash function is the use of iteration. Let H_0 denote an m-bit initial value ($m > 0$) and let M denote a binary message whose length is a positive multiple of m, i.e., $M = (M_1, M_2, \ldots, M_n)$ for some positive n with M_i representing an m-bit block. [Note that one can extend the message length to a multiple of m by applying deterministic "padding" techniques, cf. [ISO 91, Merkle 90].] Then, we write $hash(\cdot\ ,\ \cdot)$ to denote the *m-bit iterated hash function* which, given H_0 and M, computes the hash value $hash(H_0, M) = H_n$ according to the recursive equation

$$H_i \ = \ round(H_{i-1}, M_i) \qquad\qquad i = 1, 2, \ldots, n \qquad\qquad (1)$$

where $round(\cdot\ ,\ \cdot)$ is an easily computable function from two m-bit inputs to an m-bit output. The function $round(\cdot\ ,\ \cdot)$ will be called the ***m*-bit round function.**

Let H_0 and $\hat{H}_0$ be two m-bit initial values and, let $M = (M_1, \ldots, M_n)$ and $\hat{M} = (\hat{M}_1, \ldots, \hat{M}_{\hat{n}})$ be two binary messages with M_i $(1 \leq i \leq n)$ and $\hat{M}_i$ $(1 \leq i \leq \hat{n})$ denoting m-bit blocks. For the iterated hash function $hash(\cdot\ ,\ \cdot)$ one can distinguish between the following five attacks, cf. [Lai 92]:

- **target attack:** Given H_0 and M, find $\hat{M}$ such that $\hat{M} \neq M$ but $hash(H_0, \hat{M}) = hash(H_0, M)$;
- **free-start target attack:** Given H_0 and M, find $\hat{H}_0$ and $\hat{M}$ such that $(\hat{H}_0, \hat{M}) \neq (H_0, M)$ but $hash(\hat{H}_0, \hat{M}) = hash(H_0, M)$;
- **collision attack:** Given H_0, find M and $\hat{M}$ such that $\hat{M} \neq M$ but $hash(H_0, \hat{M}) = hash(H_0, M)$;
- **semi-free-start collision attack:** Find H_0, M and $\hat{M}$ such that $\hat{M} \neq M$ but $hash(H_0, \hat{M}) = hash(H_0, M)$;
- **free-start collision attack:** Find H_0, $\hat{H}_0$, M and $\hat{M}$ such that $(\hat{H}_0, \hat{M}) \neq (H_0, M)$ but $hash(\hat{H}_0, \hat{M}) = hash(H_0, M)$.

When the messages M and $\hat{M}$ contain only one block, i.e., $n = \hat{n} = 1$, we have $hash(H_0, M) = round(H_0, M)$, from which it follows that each above attack reduces to an attack of the same type on the m-bit round function. For example, a target attack reads: given H_0 and M, find $\hat{M}$ such that $\hat{M} \neq M$ but $round(H_0, \hat{M}) = round(H_0, M)$.

We will consider iterated hash functions based on (m, k) block ciphers, where an (m, k) *block cipher* defines, for each k-bit key, a reversible mapping from the set of all m-bit plaintexts onto the set of all m-bit ciphertexts. Given an (m, k) block cipher, we write $\mathbf{E}_Z(X)$ to denote the encryption of the m-bit plaintext X under the k-bit key Z, and $\mathbf{D}_Z(Y)$ to denote the decryption of the m-bit ciphertext Y under the k-bit key Z. In our discussion, we will always assume that the (m, k) block cipher has no known weaknesses. We define the *rate* of such an iterated hash function (or equivalently, of an round function) as the number of m-bit message blocks processed per encryption or decryption.

Given that the m-bit round function is based on an (m, k) block cipher, we define the *complexity* of an attack as the total number of encryptions or decryptions of the (m, k) block cipher required for this attack, e.g., an attack requiring 2^s encryptions or decryptions is said to have complexity 2^s. Because an attack on the m-bit round function implies an attack of the same type on the corresponding m-bit iterated hash function with roughly the same complexity, the design of computationally secure round functions is a necessary (but not sufficient) condition for the design of computationally secure iterated hash functions. Moreover, under certain conditions (cf. [Merkle 90, Damgaard 90, Naor 89, Lai 92]), a computationally secure round function implies a computationally secure iterated hash function. We will therefore concentrate our attention to the design of computationally secure round functions.

In Section 2 we will consider a general class of $2m$-bit iterated hash functions of rate 1 based on an (m, m) block cipher. Several previously proposed schemes [Preneel 89, Quisquater 89, Brown 90] are shown to be in this class. For this class, we derive upper bounds on the complexities of free-start target and free-start collision attacks, by describing attacks that are better than the brute-force attacks. In Section 3, we propose a $2m$-bit iterated hash function which will be proven, under plausible assumptions, to achieve these upper bounds. Section 4 contains a new free-start collision attack using two encryptions on the LOKI Double-Block-Hash scheme [Brown 90] and a new semi-free-start collision attack requiring about $2^{m/2}$ encryptions. In Section 5 we investigate a class of $2m$-bit iterated hash functions with rate 1/2. It is shown that the upper bounds derived in Section 2 also hold for this class of rate 1/2 hash functions. It then follows that both our proposal and the Meyer-Schilling scheme [Meyer 88] (which is presently under consideration for an ISO standard [ISO 91]) achieve the same computational security against free-start attacks; however, our proposal is more efficient in the sense that one encryption is required for hashing one m-bit message block as opposed to two encryptions in the Meyer-Schilling scheme.

2 $2m$-bit round functions with rate 1

In this section, we consider $2m$-bit round functions with rate 1 based on (m, m) block ciphers, i.e., block ciphers with m-bit ciphertext-plaintext and m-bit keys. Let the $2m$-bit hash values H_i be written as the concatenation (denoted by the symbol :) of two m-bit vectors H_i^1 and H_i^2 such that $H_i = H_i^1 : H_i^2$; similarly, let the $2m$-bit message block M_i be written as $M_i = M_i^1 : M_i^2$ with M_i^1 and M_i^2 denoting two m-bit vectors. Using this new notation, we can rewrite (1) as

$$H_i^1 : H_i^2 \quad = \quad round\left(H_{i-1}^1 : H_{i-1}^2 \ , \ M_i^1 : M_i^2\right) \qquad . \tag{2}$$

One of the proposals of $2m$-bit iterated hash functions based on the $2m$-bit round function with rate 1 defined in (2) is the following:

LOKI Double Block Hash (DBH) scheme: For the $2m$-bit iterated hash function proposed in [Brown 90], the $2m$-bit round function is given by

$$\begin{cases} H_i^1 & = \quad \mathbf{E}_{H_{i-1}^1 \oplus M_i^1}(H_{i-1}^1 \oplus M_i^2) \qquad \oplus \ H_{i-1}^1 \oplus H_{i-1}^2 \oplus M_i^2 \\ H_i^2 & = \quad \mathbf{E}_{H_{i-1}^2 \oplus M_i^2}(H_{i-1}^1 \oplus M_i^1 \oplus H_i^1) \ \oplus \ H_{i-1}^1 \oplus H_{i-1}^2 \oplus M_i^1 \end{cases} \tag{3}$$

with the symbol $\oplus$ denoting bitwise modulo-2 addition.

Other similar proposals are the Preneel-Bosselaers-Govaerts-Vandewalle (PBGV) scheme proposed in [Preneel 89] and the Quisquater-Girault (QG) scheme proposed in [Quisquater 89]. The purpose of these schemes was to obtain secure $2m$-bit round functions by modifying the apparently secure m-bit round function proposed by Davies and by Meyer [Davies 85, Matyas 85, Winternitz 84]. However, several recent works (cf. [Quisquater 89, Miyaguchi 91, Lai 92, Preneel 93]) and the attacks on the LOKI-DBH scheme that will be presented in this paper show that these proposals of $2m$-bit round functions are in fact weaker than the underlying m-bit round function against free-start attacks. In order to give a systematic solution to this problem, we will consider the following general form of such $2m$-bit round functions:

General form of the $2m$-bit round function with rate 1:

$$\begin{cases} H_i^1 &= \mathbf{E}_A(B) \oplus C \\ H_i^2 &= \mathbf{E}_R(S) \oplus T \end{cases} \tag{4}$$

where A, B and C are binary linear combinations of the m-bit vectors H_{i-1}^1, H_{i-1}^2, M_i^1 and M_i^2, and where R, S and T are some (not necessarily binary linear) combinations of the vectors H_{i-1}^1, H_{i-1}^2, M_i^1, M_i^2 and H_i^1. We can therefore write A, B and C in matrix-form as

$$\begin{bmatrix} A \\ B \\ C \end{bmatrix} = \begin{bmatrix} a_1 \; a_2 \; a_3 \; a_4 \\ b_1 \; b_2 \; b_3 \; b_4 \\ c_1 \; c_2 \; c_3 \; c_4 \end{bmatrix} \begin{bmatrix} H_{i-1}^1 \\ H_{i-1}^2 \\ M_i^1 \\ M_i^2 \end{bmatrix} \tag{5}$$

for some binary values a_i, b_i and c_i $(1 \leq i \leq 4)$.

One can easily see that for the LOKI-DBH scheme we have

$$\begin{bmatrix} A \\ B \\ C \end{bmatrix} = \begin{bmatrix} 1\,0\,1\,0 \\ 1\,0\,0\,1 \\ 1\,1\,0\,1 \end{bmatrix} \begin{bmatrix} H_{i-1}^1 \\ H_{i-1}^2 \\ M_i^1 \\ M_i^2 \end{bmatrix} \quad \text{and} \quad \begin{bmatrix} R \\ S \\ T \end{bmatrix} = \begin{bmatrix} 0\,1\,0\,1\,0 \\ 1\,0\,1\,0\,1 \\ 1\,1\,1\,0\,0 \end{bmatrix} \begin{bmatrix} H_{i-1}^1 \\ H_{i-1}^2 \\ M_i^1 \\ M_i^2 \\ H_i^1 \end{bmatrix}.$$

The PBGV and QG schemes can also be represented in a similar way.

We now show upper bounds on the complexity of a free-start target and free-start collision attacks on $2m$-bit iterated hash functions whose round function is of type (4).

Proposition 1: For the $2m$-bit iterated hash function with rate 1 whose $2m$-bit round function is of type (4), the complexity of a free-start target attack is upper-bounded by about 2^m encryptions, and the complexity of a free-start collision attack is upper-bounded by about $2^{m/2}$.

Proof: Since there are 2^{4m} possible values for the $4m$-bit vector $(H_{i-1}^1, H_{i-1}^2, M_i^1, M_i^2)$ and 2^{3m} possible values for the $3m$-bit vector (A, B, C), it follows that,

for each possible value of (A, B, C), there are at least $\frac{2^{4m}}{2^{3m}} = 2^m$ values of $(H_{i-1}^1, H_{i-1}^2, M_i^1, M_i^2)$ satisfying (5). The problem of finding these 2^m values is computationally negligible, because from (5) we see that this problem is equivalent to solving a system of three linear equations with four unknowns.
Free-start target attack: For a given value of $(H_{i-1}^1, H_{i-1}^2, M_i^1, M_i^2)$, we will find a different value of $(H_{i-1}^1, H_{i-1}^2, M_i^1, M_i^2)$ yielding the same value of (H_i^1, H_i^2) according to (4). We proceed as followed: for the given value of $(H_{i-1}^1, H_{i-1}^2, M_i^1, M_i^2)$ we compute the value of (A, B, C) according to (5) and, using the above argument we produce 2^m different values of $(H_{i-1}^1, H_{i-1}^2, M_i^1, M_i^2)$ yielding the same value for (A, B, C) as the given value. We then compute the value of H_i^2 for the given value and for each of the 2^m newly produced values of $(H_{i-1}^1, H_{i-1}^2, M_i^1, M_i^2)$. Because there are 2^m possible values of the m-bit block H_i^2, it follows that one must compute H_i^2 for about 2^m different values of $(H_{i-1}^1, H_{i-1}^2, M_i^1, M_i^2)$ to have an 0.63 probability of finding a value of $(H_{i-1}^1, H_{i-1}^2, M_i^1, M_i^2)$ yielding the same value for H_i^2 as the given value of $(H_{i-1}^1, H_{i-1}^2, M_i^1, M_i^2)$. Such an attack requires therefore about 2^m encryptions, which gives an upper bound of about 2^m for the complexity of a free-start target attack on the $2m$-bit iterated hash function.
Free-start collision attack: We will find two different values for $(H_{i-1}^1, H_{i-1}^2, M_i^1, M_i^2)$ yielding the same value of (H_i^1, H_i^2) according to (4). We proceed as follows: we first produce $2^{m/2}$ values for $(H_{i-1}^1, H_{i-1}^2, M_i^1, M_i^2)$ yielding the same value for H_i^1. Because there are 2^m possible values for the m-bit block H_i^2, it follows from the usual "birthday argument" that one must compute H_i^2 for about $2^{m/2}$ values of $(H_{i-1}^1, H_{i-1}^2, M_i^1, M_i^2)$ to have an 0.63 probability of finding two values of $(H_{i-1}^1, H_{i-1}^2, M_i^1, M_i^2)$ yielding the same hash value H_i^2. Thus, such an attack requires about $2^{m/2}$ encryptions, which gives an upper bound of about $2^{m/2}$ for the complexity of a free-start collision attack on the $2m$-bit iterated hash function. □

Remark. The basic idea behind the attacks in Proposition 1 is to attack the two equations in (4) separately. If one can find many values for $(H_{i-1}^1, H_{i-1}^2, M_i^1, M_i^2)$ yielding the same value for (A, B, C), then the attack on the $2m$-bit round function of type (4) is reduced to an attack on one m-bit round function. Thus, similar attacks as the ones described in the proof of Proposition 1 will also work, even if the mapping from $(H_{i-1}^1, H_{i-1}^2, M_i^1, M_i^2)$ to (A, B, C) in (4) is not a binary linear combination. Therefore, Proposition 1 also holds if it is easy to find 2^m different values of $(H_{i-1}^1, H_{i-1}^2, M_i^1, M_i^2)$ for a given value of (A, B, C).

Given the $2m$-bit iterated hash function $hash(\cdot\ ,\ \cdot)$ whose $2m$-bit round function is of type (4), we say that $hash(\cdot\ ,\ \cdot)$ is *optimum against a free-start target attack* when the best possible free-start target attack has complexity about 2^m; similarly, $hash(\cdot\ ,\ \cdot)$ is said to be *optimum against a free-start collision attack* when the best possible free-start collision attack has complexity about $2^{m/2}$.

3 Proposal for a 2*m*-bit hash function with rate 1

In this section, we propose a new $2m$-bit iterated hash function whose $2m$-bit round function is of type (4). We will prove that this new proposal is optimum against a free-start target and free-start collision attacks.

Before introducing our proposal, we describe the m-bit iterated hash function which was proposed independently by Davies and Meyer, cf. [Davies 85, Matyas 85, Winternitz 84].

Davies-Meyer (DM) scheme: This scheme consists of an m-bit iterated hash function as defined in (1) whose m-bit round function is based on an (m,k) block cipher. For our purpose, we will assume that $k = m$, i.e., the plaintext-ciphertext length and the key length are the same. Letting H_{i-1} and M_i denote two m-bit blocks, the m-bit output H_i of the DM round function for the input pair (H_{i-1}, M_i) is given by

$$H_i = \mathbf{E}_{M_i}(H_{i-1}) \oplus H_{i-1} \quad . \tag{6}$$

The DM scheme is generally considered to be secure, i.e., a free-start target and free-start collision attacks on (6) need about 2^m and $2^{m/2}$ encryptions, respectively. Our proposal is based on the DM scheme:

Parallel Davies-Meyer (Parallel-DM) scheme: For the $2m$-bit iterated hash function defined in (1), we propose the following $2m$-bit round function with rate 1 based on an $(m, m,)$ block cipher:

$$\begin{cases} H_i^1 &= \mathbf{E}_{M_i^1 \oplus M_i^2}(H_{i-1}^1 \oplus M_i^1) \oplus H_{i-1}^1 \oplus M_i^1 \\ H_i^2 &= \mathbf{E}_{M_i^1}(H_{i-1}^2 \oplus M_i^2) \oplus H_{i-1}^2 \oplus M_i^2 \end{cases} \quad . \tag{7}$$

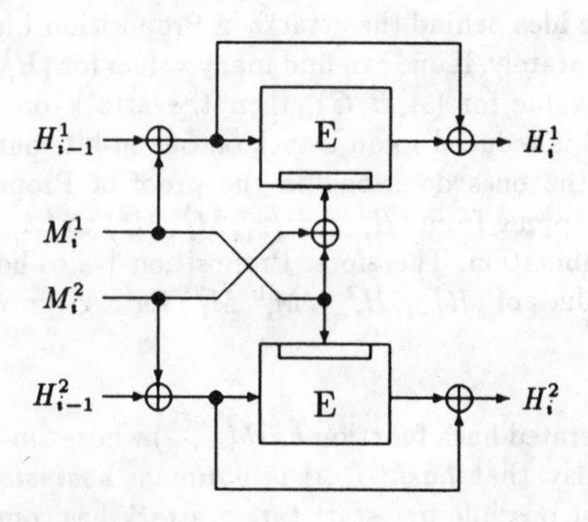

Fig. 1. *The 2m-bit round function of the proposed Parallel-DM scheme (the small boxes indicate the key input of the block cipher).*

In order to avoid trivial attacks based on the fact that a falsified message can have a different length from that of the genuine message, we defined the following strengthening on the iterated hash functions, which was proposed independently by Merkle and Damgaard, cf. [Merkle 90, Damgaard 90]:

Merkle-Damgaard (MD) Strengthening: For the iterated hash function defined by (1), the MD-strengthening consists of specifying that the last block M_n of the binary message $M = (M_1, M_2, \ldots, M_n)$ to be hashed must represent the length of M (in binary form), i.e., the total length of $(M_1, M_2, \ldots, M_{n-1})$.

Proposition 2: Assuming that the DM scheme is secure and that the $2m$-bit iterated hash function is used with MD-strengthening, it follows that the Parallel-DM scheme is optimum against free-start target and free-start collision attacks.

Proof: By applying the invertible transformation
$(H^1_{i-1}, H^2_{i-1}, M^1_i, M^2_i) \rightarrow (h^1_{i-1}, h^2_{i-1}, m^1_i, m^2_i)$
such that

$$\begin{array}{llll} h^1_{i-1} = H^1_{i-1} \oplus M^1_i, & & h^2_{i-1} = H^2_{i-1} \oplus M^2_i, \\ m^1_i = M^1_i, & & m^2_i = M^1_i \oplus M^2_i \end{array}$$

to the Parallel-DM round function defined in (7), we obtain the new $2m$-bit round function

$$\begin{cases} h^1_i = \mathbf{E}_{m^2_i}(h^1_{i-1}) \oplus h^1_{i-1} \\ h^2_i = \mathbf{E}_{m^1_i}(h^2_{i-1}) \oplus h^2_{i-1} \end{cases} \quad . \tag{8}$$

From (8) we see that h^1_i depends only on m^2_i and h^1_{i-1} and, h^2_i depends only on m^1_i and h^2_{i-1}, which implies that h^1_i and h^2_i can be attacked separately. Moreover, we see from (8) that the equations for h^1_i and h^2_i correspond each to the m-bit round function of the DM scheme defined in (6). By the assumption that the DM scheme is secure, it follows that the best possible free-start target and collision attacks on (8) require about 2^m and $2^{m/2}$ encryptions, respectively. From the *Transformation principle*, viz. applying any simple (in both directions) invertible transformation to the input and to the output of the round function produces a new round function with the same computational security as the original one against free start attacks (cf. [Lai 92]), it follows that the best possible free-start target and collision attacks on the Parallel-DM round function defined by (7) require about 2^m and $2^{m/2}$ encryptions, respectively. Because a free-start attack on an iterated hash function with MD-strengthening is roughly as hard as an attack of the same type on its round function (cf.[Merkle 90, Damgaard 90, Naor 89, Lai 92]), we have completed the proof. □

4 New attacks on the LOKI-DBH scheme

We describe here a new free-start collision attack on the LOKI-DBH scheme which requires two encryptions and a semi-free-start collision attack which uses about $2^{m/2}$ encryptions. These attacks can be applied to the LOKI-DBH scheme with any underlying (m, m) block cipher. Such low attacking complexities on the LOKI-DBH scheme have not yet being reported in the literature. Moreover, the low complexity of this new free-start collision shows that the LOKI-DBH scheme is *not* optimum against a free-start collision attack.

By applying the invertible transformation
$(H_{i-1}^1, H_{i-1}^2, M_i^1, M_i^2) \rightarrow (h_{i-1}^1, h_{i-1}^2, m_i^1, m_i^2)$
such that

$$\begin{array}{llll} h_{i-1}^1 = H_{i-1}^1, & & h_{i-1}^2 = H_{i-1}^2, \\ m_i^1 = M_i^1 \oplus H_{i-1}^1, & & m_i^2 = M_i^2 \oplus H_{i-1}^2 \end{array} \tag{9}$$

to the LOKI-DBH round function defined in (3), we obtain the new $2m$-bit round function

$$\begin{cases} h_i^1 = \mathbf{E}_{m_i^1}(h_{i-1}^1 \oplus h_{i-1}^2 \oplus m_i^2) \oplus h_{i-1}^1 \oplus m_i^2 \\ h_i^2 = \mathbf{E}_{m_i^2}(m_i^1 \oplus h_i^1) \oplus h_{i-1}^2 \oplus m_i^1 \end{cases} . \tag{10}$$

Free-start collision attack: We will find two different values for $(H_{i-1}^1, H_{i-1}^2, M_i^1, M_i^2)$ yielding the same value for (H_i^1, H_i^2) according to (3). We proceed as followed:

Step 1: We randomly choose an m-bit value x and two distinct m-bit values y and $\hat{y}$.
Step 2: Computing z and $\hat{z}$ such that

$$z = \mathbf{E}_y(x \oplus y) \oplus x \oplus y \tag{11}$$

$$\hat{z} = \mathbf{E}_{\hat{y}}(x \oplus \hat{y}) \oplus x \oplus \hat{y} \tag{12}$$

we obtain two different values $(z, x \oplus z, y, y)$ and $(\hat{z}, x \oplus \hat{z}, \hat{y}, \hat{y})$ for $(h_{i-1}^1, h_{i-1}^2, m_i^1, m_i^2)$.
Step 3: By applying the inverse transformation of (9), we obtain two different values $(z, x \oplus z, y \oplus z, x \oplus y \oplus z)$ and $(\hat{z}, x \oplus \hat{z}, \hat{y} \oplus \hat{z}, x \oplus \hat{y} \oplus \hat{z})$ for $(H_{i-1}^1, H_{i-1}^2, M_i^1, M_i^2)$.

Note that substituting $(z, x \oplus z, y, y)$ for $(H_{i-1}^1, H_{i-1}^2, M_i^1, M_i^2)$ in (3) gives

$$\begin{cases} H_i^1 = \mathbf{E}_y(x \oplus y) \oplus y \oplus z \\ H_i^2 = \mathbf{E}_y(y \oplus H_i^1) \oplus x \oplus y \oplus z \end{cases}$$

which, using the expression for z defined in (11), gives

$$\begin{cases} H_i^1 = x \\ H_i^2 = \mathbf{E}_y(H_i^1 \oplus y) \oplus \mathbf{E}_y(x \oplus y) \end{cases} .$$

Replacing H_i^1 by x in the right side of the second equation gives $(H_i^1, H_i^2) = (x, 0)$. In exactly the same manner, it can be shown that the substitution of $(\hat{z}, x \oplus \hat{z}, \hat{y} \oplus \hat{z}, x \oplus \hat{y} \oplus \hat{z})$ for $(H_{i-1}^1, H_{i-1}^2, M_i^1, M_i^2)$ in (3) also yields $(H_i^1, H_i^2) = (x, 0)$, which proves the correctness of our attack. Because an attack on the round function implies an attack of the same type on the iterated hash function with the same complexity, we conclude that the above attack implies a free-start collision attack on the LOKI-DBH iterated hash function requiring two encryptions.

Semi-free-start collision attack: We will find a value for (H_{i-1}^1, H_{i-1}^2) and two different values for (M_i^1, M_i^2) yielding the same value (H_i^1, H_i^2) according to (3). We proceed as followed:

Step 1: We randomly choose an m-bit value x.
Step 2: Let z and $\hat{z}$ be defined by (11) and (12), respectively. Given x, we randomly choose a pair $(y, \hat{y})$ of two distinct m-bit values until we find a pair yielding matching z and $\hat{z}$, i.e., $z = \hat{z}$. By the usual "birthday argument", it takes about $2^{m/2}$ encryptions to have an 0.63 probability of finding such a pair $(y, \hat{y})$.
Step 3: By applying the inverse transformation of (9) we obtain a value $(z, x \oplus z)$ for (H_{i-1}^1, H_{i-1}^2) and two different values $(y \oplus z, x \oplus y \oplus z)$ and $(\hat{y} \oplus z, x \oplus \hat{y} \oplus z)$ for (M_i^1, M_i^2).

By applying similar substitutions as for the free-start collision attack, one can easily prove the correctness of this attack. We then conclude that the just described attack implies a semi-free-start collision attack on the LOKI-DBH $2m$-bit iterated hash function using about $2^{m/2}$.

5 $2m$-bit round functions with rate $1/2$

In this section, we consider $2m$-bit round functions with rate 1/2 based on (m, m) block ciphers. Letting the $2m$-bit hash values H_i be written as the concatenation (denoted by the symbol :) of two m-bit vectors H_i^1 and H_i^2 such that $H_i = H_i^1 : H_i^2$, we can rewrite (1) as

$$H_i^1 : H_i^2 \quad = \quad round\left(H_{i-1}^1 : H_{i-1}^2, \; M_i\right) \qquad (13)$$

where M_i denotes an m-bit message block.

General form of the $2m$-bit round function with rate $1/2$:

$$\begin{cases} H_i^1 & = & \mathbf{E}_A(B) \; \oplus \; C \\ H_i^2 & = & \mathbf{E}_R(S) \; \oplus \; T \end{cases} \qquad (14)$$

where A, B and C are binary linear combinations of the m-bit vectors H_{i-1}^1, H_{i-1}^2 and M_i, and where R, S and T are some (not necessarily binary linear) combinations of the vectors H_{i-1}^1, H_{i-1}^2, M_i and H_i^1. We can therefore write A, B and C in matrix-form as

$$\begin{bmatrix} A \\ B \\ C \end{bmatrix} = \prod \begin{bmatrix} H_{i-1}^1 \\ H_{i-1}^2 \\ M_i \end{bmatrix} \qquad (15)$$

where $\prod$ denotes a 3×3 matrix whose entries are 0's and 1's.

From Section 2, we know that the complexity of a $2m$-bit iterated hash function with *rate* 1 whose round function is of type (4) is upper-bounded by about 2^m for a free-start target attack and by about $2^{m/2}$ for a free-start collision attack. We now show that the same upper-bounds hold for $2m$-bit iterated hash function with *rate* 1/2 whose round function is of type (14).

Proposition 3: For the $2m$-bit iterated hash function with rate 1/2 whose $2m$-bit round function is of type (14), the complexity of a free-start target attack is upper-bounded by about 2^m and the complexity of a free-start collision attack is upper-bounded by about $2^{m/2}$.

Proof: We first consider the **free-start target attack**, i.e., for a given value of $(H^1_{i-1}, H^2_{i-1}, M_i)$, we will find a different value for $(H^1_{i-1}, H^2_{i-1}, M_i)$ yielding the same value for (H^1_i, H^2_i) according to (14). When the matrix $\prod$ defined in (15) is non-singular, let D be the value of H^1_i for the given value of $(H^1_{i-1}, H^2_{i-1}, M_i)$. We then generate 2^m different values of $(H^1_{i-1}, H^2_{i-1}, M_i)$ yielding the same value D by first computing $C = D \oplus \mathbf{E}_A(B)$ for 2^m randomly chosen values of (A, B) and then, for each value of (A, B, C), by computing $(H^1_{i-1}, H^2_{i-1}, M_i)$ according to

$$\begin{bmatrix} H^1_{i-1} \\ H^2_{i-1} \\ M_i \end{bmatrix} = \prod^{-1} \begin{bmatrix} A \\ B \\ C \end{bmatrix},$$

where $\prod^{-1}$ denotes the inverse of the non-singular matrix $\prod$. When the matrix $\prod$ is singular, there exist, for the value of (A, B, C) obtained from the given value of $(H^1_{i-1}, H^2_{i-1}, M_i)$, at least 2^m different values of $(H^1_{i-1}, H^2_{i-1}, M_i)$ yielding the same value for (A, B, C), i.e., the same value for H^1_i. For the given and the 2^m newly generated values of $(H^1_{i-1}, H^2_{i-1}, M_i)$, we compute the value of H^2_i according to (14). Because there are 2^m possible values of the m-bit block H^2_i, it follows that one must compute H^2_i for about 2^m different values of $(H^1_{i-1}, H^2_{i-1}, M_i)$ to have an 0.63 probability of finding a value of $(H^1_{i-1}, H^2_{i-1}, M_i)$ yielding the same value for H^2_i as the given value of $(H^1_{i-1}, H^2_{i-1}, M_i)$. Such an attack requires therefore about 2^m encryptions.
We now consider the **free-start collision attack**, i.e. we will find two different values of $(H^1_{i-1}, H^2_{i-1}, M_i)$ yielding the same value for (H^1_i, H^2_i) according to (14). This attack is similar to the free-start target attack just described, except that here, one only generates $2^{m/2}$ values of $(H^1_{i-1}, H^2_{i-1}, M_i)$ yielding the same value of H^1_i. This follows from the usual "birthday paradox" which says that one only needs to try $2^{m/2}$ randomly chosen values of $(H^1_{i-1}, H^2_{i-1}, M_i)$ to have an 0.63 probability of finding two values of $(H^1_{i-1}, H^2_{i-1}, M_i)$ yielding the same value for H^2_i. □

Example: Meyer-Schilling scheme (modified)
In [Meyer 88], Meyer and Schilling proposed a $2m$-bit hash function based on the (m=64, k=56) block cipher DES, which was later named as MDC-2 in [Matyas 91] and which is presently under consideration as an ISO standard [ISO 91]. The Meyer-Schilling scheme, after some minor modifications, namely,

using an (m, m) block cipher instead of DES and not considering the additional "cut-and-paste" and swapping operations, can be written as

$$H_i^1 = \mathbf{E}_{H_{i-1}^1}(M_i) \oplus M_i$$
$$H_i^2 = \mathbf{E}_{H_{i-1}^2}(M_i) \oplus M_i$$

or, under its general form (14), as

$$\begin{bmatrix} A \\ B \\ C \end{bmatrix} = \begin{bmatrix} 1\,0\,0 \\ 0\,0\,1 \\ 0\,0\,1 \end{bmatrix} \begin{bmatrix} H_{i-1}^1 \\ H_{i-1}^2 \\ M_i \end{bmatrix} \quad \text{and} \quad \begin{bmatrix} R \\ S \\ T \end{bmatrix} = \begin{bmatrix} 0\,1\,0 \\ 0\,0\,1 \\ 0\,0\,1 \end{bmatrix} \begin{bmatrix} H_{i-1}^1 \\ H_{i-1}^2 \\ M_i \end{bmatrix}.$$

Thus, the upper bounds of Proposition 3 also hold for the Meyer-Schilling scheme. By applying the similar approach as in the proof of Proposition 2, we can show that the Meyer-Schilling scheme indeed achieves these upper bound if the underlying cipher has no weaknesses.

6 Conclusion

In this paper, we have derived upper bounds for the complexities of free-start target and free-start collision attacks on a large class of $2m$-bit iterated hash functions based on an (m, m) block cipher. Eventhough these free-start target and free-start collision attacks are "non-real" attacks (because the initial value of the iterated hash function is usually fixed), their complexities give a lower bound for the complexities of "real" target and collision attacks, respectively.

We have also proposed a $2m$-bit iterated hash function with rate 1 which, under the assumption that the DM scheme is secure and that MD-strengthening is applied, was proven to be optimum against free-start target and free-start collision attacks. Moreover, our new free-start collision attack on the LOKI-DBH scheme shows that this scheme is not optimum against a free-start target collision attack. Finally, even though the Meyer-Schilling $2m$-bit iterated hash function is also optimum with respect to these two attacks, it only achieves a rate of 1/2, while our proposal achieves a rate of 1, i.e., two encryptions are needed in the Meyer-Schilling scheme to hash one m-bit message block as opposed to one encryption for our proposal.

Since the m-bit Davies-Meyer scheme appears to be secure, it has been an open question [Preneel 93, Lai 92] whether one can modify the DM scheme to construct a $2m$-bit iterated hash function that is more secure than the original m-bit DM scheme. This problem is partly solved by the results of Propositions 1 and 3, which show that, given a $2m$-bit round function which hashes at least one m-bit message block by using twice an m-bit block cipher with an m-bit key, there always exist free-start attacks that are better than the brute-force attacks. That is, by using any (m, m) block cipher twice plus some "simple operations", it is impossible to obtain a $2m$-bit round function of rate 1/2 or greater that is more secure than the m-bit DM round function against free-start attacks. Thus, to obtain such a secure $2m$-bit round function, one has to apply an (m, m) block cipher at least three times in each round, or, to use twice an m-bit block cipher with key length greater than m. In fact, it appears that [Lai 92] secure $2m$-bit round functions of rate 1/2 can be constructed by using twice an m-bit block cipher with a $2m$-bit key.

References

[Brown 90] L. Brown, J. Pieprzyk and J. Seberry, *"LOKI - A Cryptographic Primitive for Authentication and Secrecy Applications"*, Advances in Cryptology - AUSCRYPT'90 Proceedings, pp. 229-236, Springer-Verlag, 1990.

[Damgaard 90] I.B. Damgaard, *"A Design Principle for Hash Functions"*, Advances in Cryptology - CRYPT0'89 Proceedings, pp. 416-427, Springer-Verlag, 1990.

[Davies 85] R.W. Davies and W.L. Price, *"Digital Signature - an Update"*, Proc. International Conference on Computer Communications, Sydney, Oct. 1984, Elsevier, North Holland, pp. 843-847, 1985.

[ISO 91] ISO/IEC CD 10118, *Information technology - Security techniques - Hash-functions*, I.S.O., 1991.

[Lai 92] X. Lai and J.L. Massey, *"Hash Functions Based on Block Ciphers"*, Advances in Cryptology - EUROCRYPT'92 Proceedings, pp. 55-70, LNCS 658, Springer-Verlag, 1993.

[Matyas 85] S.M. Matyas, C.H. Meyer and J. Oseas, *"Generating Strong One-Way Functions with Cryptographic Algorithm"*, IBM Technical Disclosure Bulletin, Vol. 27, No. 10A, pp. 5658-5659, March 1985.

[Matyas 91] S.M. Matyas, *"Key Processing with Control Vectors"*, Journal of Cryptology, Vol.3, No.2, pp. 113-136, 1991.

[Merkle 90] R.C. Merkle, *"One-Way Hash Functions and DES"*, Advances in Cryptology - CRYPTO'89 Proceedings, pp. 428-446, Springer-Verlag, 1990.

[Meyer 88] C. H. Meyer and M. Schilling, *"Secure Program Code with Modification Detection Code"*, Proceedings of SECURICOM 88, pp. 111-130, SEDEP.8, Rue de la Michodies, 75002, Paris, France.

[Miyaguchi 91] S. Miyaguchi, K. Ohta and M. Iwata, *Confirmation that Some Hash Functions Are Not Collision Free*, Advances in Cryptology-EUROCRYPT '90, Proceedings, LNCS 473, pp. 326-343, Springer Verlag, Berlin, 1991.

[Naor 89] M. Naor and M. Yung, *"Universal One-way Hash Functions and Their Cryptographic Applications"*, Proc. 21 Annual ACM Symposium on Theory of Computing, Seattle, Washington, May 15-17, 1989, pp. 33-43.

[Preneel 89] B. Preneel, A. Bosselaers, R. Govaerts and J. Vandewalle, *"Collision-Free Hashfunctions Based on Blockcipher Algorithm"*, Proceedings of 1989 International Carnahan Conference on Security Technology, pp.203-210, 1989.

[Preneel 93] B. Preneel, *Analysis and Design of Cryptographic Hash Hashfunctions* , Ph.D thesis, Katholieke Universiteit Leuven, Belgium, January 1993.

[Quisquater 89] J.J. Quisquater and M. Girault, *"2n-bit Hash Functions Using n-bit Symmetric Block Cipher Algorithms"*, Abstracts of EUROCRYPT'89.

[Winternitz 84] R.S. Winternitz, *"Producing One-Way Hash Function from DES"*, Advances in Cryptology - CRYPTO'83 Proceedings, pp. 203-207, Plenum Press, New York, 1984.

Improved Algorithms for the Permuted Kernel Problem

Jaques Patarin
Bull CP8
BP 45
68 Route de Versailles
78430 Louveciennes
France

Pascal Chauvaud
CNET-France Telecom
PAA-TSA-SRC
38-40 Rue du Général Leclerc
92131 Issy-les-Moulineaux
France

Abstract

In 1989, Adi Shamir published a new asymmetric identification scheme, based on the intractability of the Permuted Kernel Problem (PKP) [3]. In 1992, an algorithm to solve the PKP problem was suggested by J. Georgiades [2], and also in 1992 T. Baritaud, M. Campana, P. Chauvaud and H. Gilbert [1] have independently found another algorithm for this problem. These algorithms still need huge amount of time and/or memory in order to solve the PKP problem with the values suggested by A. Shamir.

In this paper, we will see that it is possible to solve the PKP problem using less time that which was needed in [1] and [2], and much less memory than that needed in [1].

First we will investigate how the ideas of [1] and [2] can be combined. This will enable us to obtain a little reduction in the time needed. Then, some new ideas will enable us to obtain a considerable reduction in the memory required, and another small reduction in time.

Since our new algorithms are quicker and more practical than previous algorithms they confirm the idea stated in [1] that for strong security requirements, the smallest values ($n = 32$, $m = 16$, $p = 251$) mentioned in [3] are not recommended.

1 Recall of the algorithms of [1].

In this section, we will briefly recall the attack given in [1] for the PKP Problem (see [1] for more details). Then in the next sections, we will study how to improve these algorithms.
The PKP problem is the following :

<u>Given</u> : a prime number p,
a $m \times n$ matrix $A = (a_{ij})$, $i = 1 \dots m$, $j = 1 \dots n$, over $\mathbf{Z_p}$,
a n-vector $V = (V_j)$, $j = 1 \dots n$, over $\mathbf{Z_p}$,
<u>Find</u> : a permutation π over $(1, \dots, n)$ such that $A * V_\pi = 0$, where $V_\pi = (V_{\pi(j)})$, $j = 1, \dots, n$.

We will assume that A is of rank m and is generated under the form $[\ A'|I\ ]$ where A' is a fixed $m*(m-n)$ matrix and I the $m*m$ identity matrix. As mentioned in [3] this is not restrictive because both the prover and the verifier can apply Gaussian elimination. We will use, as far as possible, the notations of [1], and we will denote by $(x_1,\ldots,x_n)$ the components of the V_π vector. So

$$A * V_\pi = 0 \Longleftrightarrow \begin{cases} \sum_{i=1}^{n-m} a'_{1i} x_i + x_{n-m+1} & = & 0 \quad (1) \\ \vdots & & \vdots \\ \sum_{i=1}^{n-m} a'_{mi} x_i + x_n & = & 0 \quad (m) \end{cases}$$

In the algorithms described in [1], one first tries to solve the equations (1) to (k), where k is a parameter of the algorithms. In these k equations, there are $n-m+k$ distinct variables x_i. These variables are divided into two groups : one group of l variables (l is another parameter of the algorithm) will be written on the right-hand side of the equations, and the other $n-m+k-l$ variables will be on the left-hand side of the equations. (We will denote by $(x_1,\ldots,x_l)$ the l variables on the right-hand side).

So the equations (1) to (k) will be represented in a scheme like this :

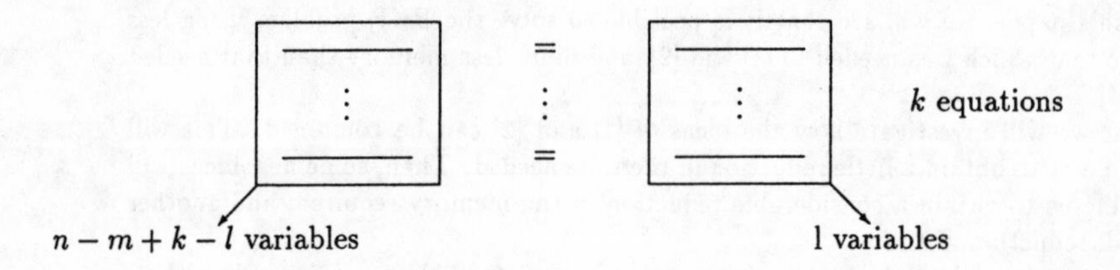

Then the algorithm of [1] proceeds in the following way :

<u>Step 1</u> : **Precomputation.** For each of the $\frac{n!}{(n-l)!}$ possible values for the l variables on the right-hand side, the right-hand side value (of k equations) is calculated. These values are stored in a file $F1$ in such a way that for each of the p^k possible values for the right-hand side, the list of the corresponding $(x_1,\ldots,x_l)$ can be accessed by very few elementary operations.

<u>Step 2</u> : **Generation and test of candidates.** For each of the $\frac{n!}{(n-k+l)!}$ possible values for the $(n-k+m-l)$ variables on the left-hand side, the left-hand side value (of k equations) is calculated. Then the file $F1$ of step 1 is used to obtain a list of possible $(x_1,\ldots,x_l)$ such that the value on the left-hand side is the same as the value on the right-hand side. Then these candidates are tested, by using also the equations $(k+1)$ to (n). (One also has to check that the variables used for the right-hand side are not used on the left-hand side).

Example 1 : Algorithm $A0$.
For a PKP (16,32), that is to say with $n = 32$, $m = 16$, $p = 251$, let us choose (as suggested in [1]) $k = 6$ and $l = 10$. Then step 1 needs $\frac{32!}{22!} \approx 2^{47.7}$ 10-uples of memory. Since $2^{47.7} < 251^6$, in step 2 on average less than one candidate for the right-hand side will have to be checked for each candidate for the left-hand side. But there are 12 variables on the left-hand side, so the time for the step 2 will be about $\frac{32!}{20!} \approx 2^{56.6}$ multiplications by A. We will call this algorithm $A0$. More generally, in this paper we will call by Ai, i=0,1,.. some algorithms with $k = 6$. $A0$ was the quickest algorithm described in [1] but it needs a huge amount of memory(about 1300 Terabytes).
Remarks.
1. To memorize a 10-uple one can use 10 bytes (since each value is modulo 251) or 10 5-bits (since there are only 32 possible values for each x_i).
2. The unit of time is, in first evaluation, the time for a multiplication by A. But it is possible to show that on average it will be appreciably less. This is because when the value of only one x_i changes (or a few x_i), the new value of $A * V_\pi$ can be quickly calculated from the old value of $A * V_\pi$.
These two remarks will be true for all the algorithms we are studying.

Example 2 : Algorithm $B0$.
If we choose $k = 3$ and $l = 5$ then we will need about 2^{24} 5-uples to memorize and $2^{65.1}$ in time for calculation. We will call this algorithm $B0$. More generally, in this paper we will call by Bi, i=0, 1,.. some algorithms with $k = 3$.
In the following paragraphs we will show how to modify $A0$ and $B0$ in order to improve these algorithms.

2 How to combine the ideas of [1] and [2].

In [2], J. Georgiades pointed out that we can add some equations in the system (1) to (n) of equations such that $A * V_\pi = 0$. As a matter of fact, $V = (v_1, \dots, v_n)$ is known, and $V_\pi = (x_1, \dots, x_n)$ is a permutation of V, so every symmetrical function of the x_i is known.
So we can add :

$$\left\{ \begin{array}{lcl} \sum_{i=1}^{n} x_i & = & \alpha_1 \quad (G1) \\ \sum_{i=1}^{n} x_i^2 & = & \alpha_2 \quad (G2) \\ \sum_{i=1}^{n} x_i^3 & = & \alpha_3 \quad (G3) \quad etc\dots \end{array} \right.$$

where $\alpha_1 = \sum_{i=1}^{n} v_i \bmod p$, $\alpha_2 = \sum_{i=1}^{n} v_i^2 \bmod p$ etc ...

For us, equation $(G1)$ will be very useful due to the fact that $(G1)$ is of the first degree. So with $(G1)$ we can apply Gaussian elimination with one more equation. Then we will be able to use the algorithms of [1] but with one variable less. (We will obtain in this way new algorithms which combine the ideas of [1] and [2] in a straightforward way!) Let us call "Algorithm A1" the modified version of algorithm $A0$ where the Gaussian reduction takes $(G1)$ into account, and "Algorithm B1" the modified version of algorithm $B0$ where the Gaussian reduction takes $(G1)$ into account. It is easy to see that algorithm $A1$ $(k = 6, l = 10)$, needs $2^{52.2}$ in time (instead of 2^{56} for A0), for the same memory as A0 ($2^{47.7}$ 10-uples). The algorithm $B1$ $(k = 3, l = 5)$, needs $2^{60.9}$ in time (instead of $2^{65.1}$) for the same memory as B0 (2^{24} 5-uples).

So we have obtained a small reduction of the time needed.

Remark. In order to defend PKP, one may consider to choose the A matrix and the vector V in such a way that the equation $(G1)$ will be just a consequence of the equations (1) to (m). But this idea doesn't work because the number m of equations has been chosen in such a way that there is about one solution π for $A * V_\pi = 0$. So m is chosen such that $n! \approx p^m$ as explained in [3]. But for all the permutations π, (G1) is satisfied. So (G1) does not restrict the permutations solutions. So (G1) is really a "free" equation for the attack.

How can we use equation (G2)? In [2], J Georgiades was able to use the equation (G2) in his algorithm. Nevertheless we didn't see how to use it in a combined algorithm with the algorithms of [1]. This is because (G2) uses all the x_i and we eliminate some of them with the other equations (as we did for$(G1)$). Thus we will obtain, in general, an equation of the second degree which seems impossible to split in two groups of distinct variables (some variables on the right-hand side of the equation and the others on the left-hand side), because we will have all the double products. So we do not use $(G2)$, $(G3)$ etc.., in our algorithms.

3 Introducing a set E on the left (or right) hand side.

In this section, we describe a new algorithm, which we will call : algorithm "B2". Like $B1$ and $B0$, our algorithm B2 will first be used for k=3 equations. But here, the new idea is that we will take into account the fact that the variables on the left-hand side cannot be used on the right-hand side.

After the Gaussian reduction with (G1) (as described in Section 2), there are 15+3=18 variables left in our equations. Here we split these variables with l=6.

This is represented schematically as follows :

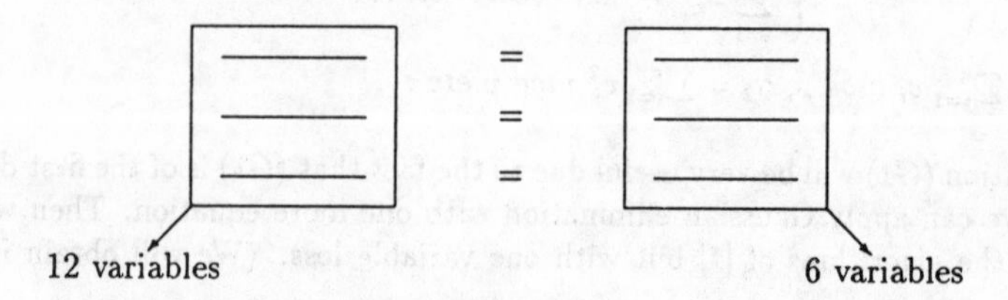

The new idea is that we will distinguish C_{32}^{12} cases : one for each possible value for the set E of the 12 variables on the left-hand side. For every such case we do :

<u>Step 1</u> : Precomputation compatible with E.

For each of the $\frac{20!}{14!} \approx 2^{24.7}$ possible values for the 6 variables on the right-hand side (when E is given), the right-hand side value is calculated and stored in a file $F1$ with the corresponding 6 variables.

<u>Step 2</u> : Generation and test of a candidate.

For each of the $12! \approx 2^{28.8}$ possible values for the 12 variables on the left-hand side (when E is given) the left-hand side value is calculated. Then the file $F1$ of step 1 is used to obtain a list of possible $(x_1, ..., x_6)$ values. On average 1.76 such candidates will be given by $F1$ (because $\frac{20!}{14!} \approx 1.76 * (251)^3$). All these candidates will be immmediately tested (as usual) with more equations.

Total time for this algorithm B2 : $C_{32}^{12} * (2^{24.7} + 2^{28.8} * 1.76) \approx 2^{57.4}$.
Total memory : $2^{24.7}$ 6-uples.

For each new set E the new file $F1$ can use the memory of the old file $F1$ because we can forget the old file $F1$. So this algorithm B2 is quicker than B1 and needs about the same amount of memory.

<u>Remark</u> : This idea to fix the set E of the variables used on the left-hand side can be generalized for values of the parameter other than the $k = 3$ and $l = 6$ that we have given. By convention, in our diagrams, the values that we will store during the precomputation phase are the values of the right-hand side. However, we can choose to fix the set E of these variables (with a higher value of l), or we can choose to fix the set E of the variables of the left-hand side (as we did in algorithm $B2$). In this paper, we will only describe some of the best algorithms that we have obtained.

4 Introducing some "middle values".

The idea given in paragraph 3 (fixing a set E in order to take into account the fact that the variables on the left-hand side cannot be used on the right-hand side) was useful to improve algorithm $B1$, because we were able to split the 18 variables in a group with much more variables (12) than in the other group (6). However, it is not possible to do this in order to improve algorithm $A1$ because here to minimize the time we have to split the 21 variables in two groups of about the same number of variables. So the idea of paragraph 3 does not work for algorithm $A1$.

Nevertheless, we will describe here another idea that will allow us to reduce the memory needed in $A1$ by $(251)^3$ without significative loss of time! We will call this algorithm "A2". Here k=6 and l=10 (as for A1), we have 15+6=21 variables after reduction with G1, but we will introduce the middle values c_1, c_2, c_3 of the first three equations.

This is represented schematically as follows :

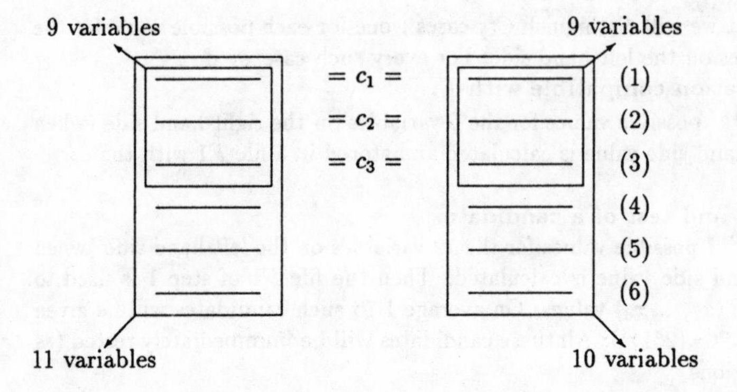

Here the new idea is that, in order to need less memory, we will try to find the solutions with a given (c_1, c_2, c_3). So, our algorithm A2 will proceed in $(251)^3$ cases : one for each possible value for (c_1, c_2, c_3). For each such cases we do :

Step 1 : Precomputation compatible with (c_1, c_2, c_3).

The aim of this step is to calculate and store in a file F1 all the possible values for the right-hand side of equations (4), (5), (6), (with the corresponding 10 variables of the right-hand side) when the right-hand side of equations (1), (2), (3) is (c_1, c_2, c_3). The problem is that we do not want to do that in time $\frac{32!}{22!}$ but in time $\frac{32!}{22! * (251)^3}$ (because we will have to do that $(251)^3$ times). For that purpose, we first notice that the right-hand side of (1), (2), (3) gives us three equations with 9 variables, when (c_1, c_2, c_3) is fixed. We will write these equations with 5 variables on the left and 4 variables on the right. We denote by (A) this system of three equations. (A) depends on c_1, c_2, c_3.
This is represented schematically as follows :

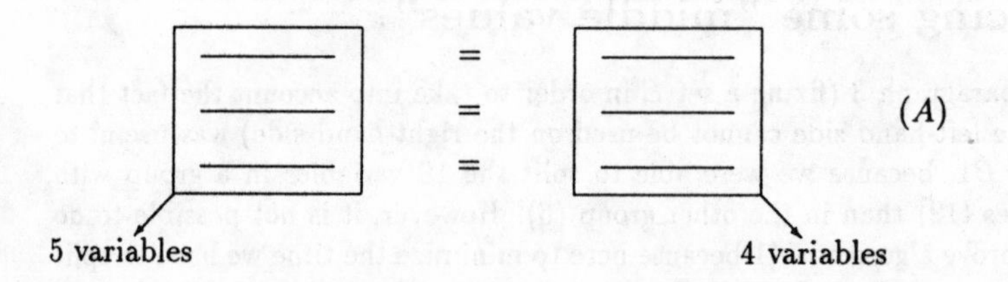

All the possible values for the right-hand side of (A) are calculated and stored in a file $F0$. So $F0$ contains $32 * 31 * 30 * 29$ values $\approx 0.055 * (251)^3$. We can notice that if c_1, c_2, c_3 are put on the left-hand side, then the file $F0$ can be precomputed once and for all : it does not depend on (c_1, c_2, c_3). Then for each of the 5 possible variables on the left-hand side of (A), the value of the left-hand side of (A) is calculated (this value depends on c_1, c_2, c_3). Then file $F0$ is used to obtain suitable values of the right-hand side variables (if any). If values are formed, and if all the 9 variables are distinct then the variable number 10 is introduced (there are 32-9=23 possible values for this variable) and the 23 corresponding values for the right-hand side of (4), (5), (6) are stored in $F1$ with the corresponding 10 variables. So, the time to generate $F1$ will be about $\frac{32!}{27!} \approx 2^{24.5}$ plus the number of values stored in $F1$, that is to say about $2^{24.5} + \frac{32!}{22! * (251)^3} \approx 2^{25.2}$. This will be done $(251)^3$ times. The memory needed is about $2^{19.7}$ 4-uples for $F0$ and $2^{23.8}$ 10-uples for $F1$.

Step 2 : Generation of a candidate compatible with (c_1, c_2, c_3).

When (c_1, c_2, c_3) is given, the left-hand side of equations (1), (2), (3) gives three equations with 9 variables. We will denote by (B) this system of equations. We will write (as in step 1) these equations with 5 variables on the left and 4 on the right.
This is represented schematically as follows :

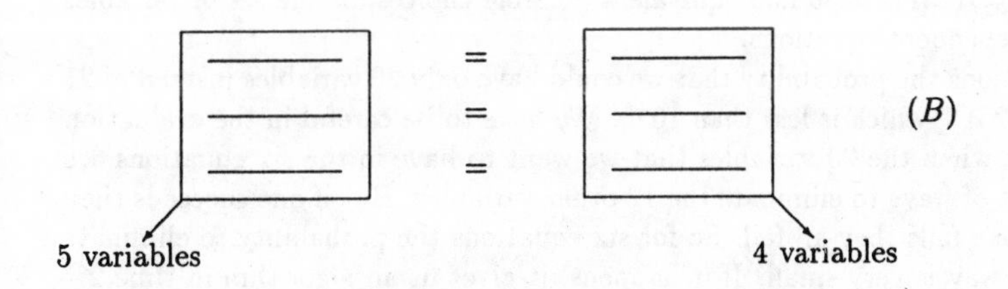

All the possible values for the right-hand side of (B) are calculated and stored in a file $F'0$. Then a candidate compatible with (c_1, c_2, c_3) values is quickly generated as follows :
(i). Choose values for the 5 variables on the left-hand side of (B).
(ii). Then look in the file $F'0$ to see whether there are any 4 variables compatible with these 5 variables.
(iii). If (ii) provides 9 variables which are all distinct, then introduce variables 10 and 11 of the left-hand side of equations (4), (5), (6) and for each possible value calculate the left-hand side of (4), (5), (6).
(iv). Then look in the file $F1$ to see whether there are any 10 variables compatible with these 11 variables.
(v). If (iv) provides 21 variables which are all distincts then quickly test this candidate as usual with extra equations.
(vi). If the test proves negative then retry (i) with a new choice for the 5 variables.

Finally, there are $\frac{32!}{27!} \approx 2^{24.5}$ values for the 5 variables on the left-hand side of (B).
We will generate about $\frac{32!}{21! * (251)^3} \approx 2^{28.3}$ candidates for the 11 variables compatible with (c_1, c_2, c_3). So the total time for step 2 is about $2^{24.5} + 2^{28.4}$. The memory needed for the file $F'0$ is $\approx 2^{19.7}$ 4-uples.

Conclusion : The total time for this algorithm A2 is : $(251)^3.(2^{25.2} + 2^{28.4}) \approx 2^{52.4}$. The total memory is about 2^{24} 10-uples (files $F0 + F'0 + F1$). So A2 is a great improvement on A1: with about the same time we need $(251)^3$ times less memory.

This idea to fix some "middle values" is very useful in order to improve the algorithms of [1]. We can use it also to improve the algorithm B2 of section 3 : it is possible to reduce the memory needed by making $251.C_{32}^{12}$ cases ; one for each value of c_1 and E. We will call this algorithm $B3$. We will not go into details for $B3$ because we will now introduce another idea to improve $B3$ and we will explain in details the algorithm $B4$ obtained.

5 Introducing precomputation on the A matrix.

In algorithms $B2$ and $B3$ we had $k = 3$, and in the first three equations we had $15+3 = 18$ variables (after Gaussian reduction with G1). We will now see that if the variables are carefully chosen we can do even better and have only 17 variables in three independent equations. For this, the idea is that we can choose a particularly convenient Gaussian reduction. When the 17 variables that we want to keep in the equations are chosen, we will try to eliminate the other variables by Gaussian reduction. For three equations there are on average about $C_{32}^{17}/(251)^3 \approx 36$ non equivalent possible choices for the set of variables providing three independent equations.
Note : For six equations the probability that we could have only 20 variables instead of 21 is only about $C_{32}^{20}/(251)^6$ which is less than 10^{-6}. We have to be careful in the evaluation of this probability : when the 20 variables that we want to have in the six equations are fixed, there are a lot of ways to eliminate the 12 other variables. But if one succeeds they all succeed, and if one fails they all fail. So for six equations the probability to eliminate one variable in this way is very small. If it happens, it gives us an algorithm in time 2^{48} instead of 2^{52}. However for $k = 3$ equations we can easily eliminate one variable. The probability of success is close to 100% , and we can quickly find a good system of three equations : we will have to do at most $C_{32}^{17} \approx 2^{29}$ Gaussian reductions (this is negligible as compared with the time of the algorithm). With a program of Gaussian elimination on a PC we have made simulations, and we were able to find such equations quite easily. We will now describe an "algorithm B4", which will use such equations.
Algorithm B4 :

Step 0 : Find three independent equations with only 17 variables. (This is done as described above with Gaussian reduction with G1 and a good choice of the 17 variables). Then these are written with l=6, that is to say with 11 variables on the left and 6 on the right-hand side. Now the algorithm proceeds in $C_{32}^{11} \approx 2^{26.9}$ cases, one case for each possible value in the set E of the 11 variables on the left-hand side. Each cases is separated in 251 subcases : one for each possible value of the "middle value" c_1, as described in the following diagram :

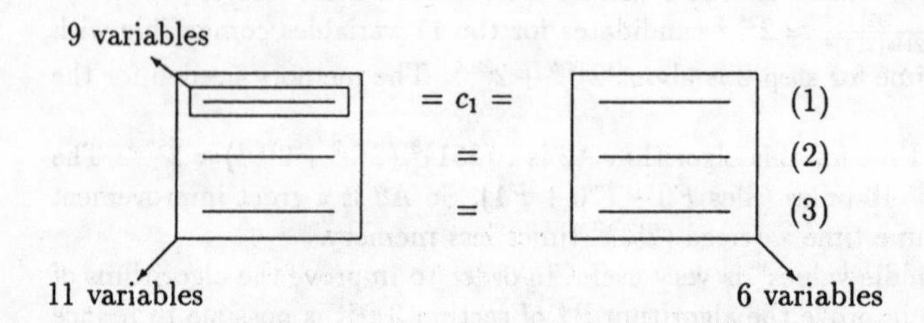

For each of the $251 * C_{32}^{11} \approx 2^{34.9}$ subcases we perform the following steps 1, 2 and 3.

Step 1 : Precomputation compatible with E and c_1. The aim of this step 1 is to calculate and store in a file $F1$ all the possible values for the right-hand side of equations (2) and (3) (with the corresponding 6 variables) when the right-hand side of equation (1) is c_1 and when the set E of the 11 variables on the left-hand side is known. The problem is that we want to do that not in $\frac{21!}{15!}$ time but in about $\frac{21!}{15! * 251}$ time. For that purpose, we first notice that the right-hand side of (1) gives us one equation with six variables when c_1 is fixed. We will call this equation (A) and we will write (A) with 4 variables on the left and 2 variables on the right. A file F0 of the possible value for the right-hand side of (A) is calculated, so F0 contains $21 * 20 \approx 1.67 * 251$ values. Now for all possible values for the 4 left variables of (A), this file $F0$ will be used in order to find the corresponding 2 variables and to store in $F1$ all the solutions such that the 6 variables are distinct. The time to generate $F1$ will be then about $1.67 * 21 * 20 * 19 * 18 \approx 2^{17.9}$. $F1$ will contain about $\frac{21!}{15! * 251} \approx 2.47 * 251^2 \approx 2^{17.3}$ values.

Step 2 : Generation of a candidate compatible with E and c_1.
The left-hand side of (1) gives us one equation with nine variables when c_1 is fixed. We will call this equation (B) and we will write (B) with 5 variables on the right-hand side and 4 variables on the left-hand side. Now we will distinguish $C_{11}^5 = 462$ more subcases (for each cases for E and c_1), one for each possible value for the 5 variables on the right-hand side of (B). As usual, a file $F'0$ of the 5! possible values for the right-hand side of (B) is introduced. Then for all possible values for the 4 left variables of (B) this file $F'0$ will be used in order to find whether there is a corresponding set of 5 variables (the probability is about $5!/251 \approx 0.48$ for each attempt). If this is the case, then variables number 10 and 11 of the left-hand side are introduced (E is fixed so we will have only 2 possible values for these variables when the first 9 variables are known). Then the left-hand side of equations (2) and (3) are calculated for this 11-variable candidate. Then the file $F1$ of step 1 is used in order to add 6 variables to our candidate. (In average, we will have 2.47 solutions for these 6 more variables). Then these 17-variable candidate is quickly tested as we will describe in step 3.

Step 3 : test of a 17-variable candidate.
At the end of step 2 we have obtained a 17-variable solution of equations (1), (2), (3). As usual we will now test this candidate with more equations. But the problem here is that when introducing one more equation, we introduce two more variables (because of the special equations (1), (2), (3) chosen with 17 variables instead of 18 variables). We want to test a candidate very quickly. In order to do this, we will first add only two equations that we call equations (4) and(5). These equations are written with the variables of (1), (2), (3) on the left and the three new variables on the right.

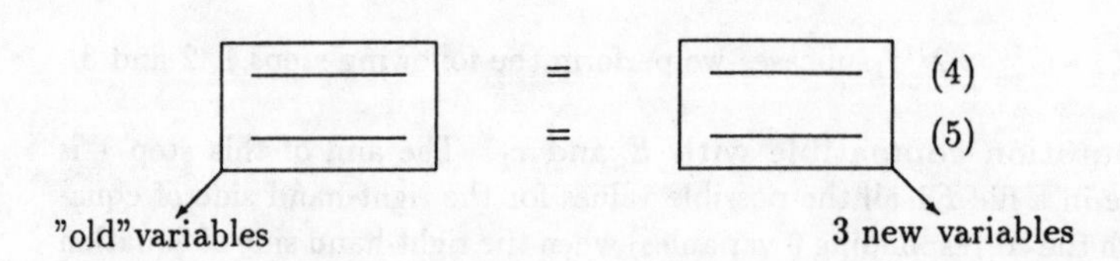

Once and for all at the beginning of this algorithm, a file $F"0$ with the possible values for the right-hand side of (4) and (5) has been stored. So $F"0$ contains $32 * 31 * 30 \approx 0.47(251)^2$ values. Now with $F"0$ a 17-variable candidate is immediately transformed to a 20-variable candidate which will be tested as usual with more equations (each extra equation needs only one additional variable).

Conclusion : The total time of this algorithm $B4$ is about $C_{32}^{11} * 251 * (2^{17.9} + 462 * (5! + 6 * 5 * 4 * 3 * 0.48 * 2 * 2.47)) \approx 2^{54}$. The total memory is about 2^{17} 6-uples (files $F'0$ and $F"0$ requiring negligible memory in comparison with $F1$).

Note : As an algorithm with $k = 3$, the total time of $B4$ is near from optimal because there are about $\frac{32!}{15!} \cdot \frac{1}{251^3} \approx 2^{53.5}$ solutions for 3 equations and 17 variables. So an algorithm with $k = 3$ will need at least this time.

6 Introducing a special equation (1).

In the first note of paragraph **5** we have seen that with a small probability (less than 10^{-6}) we can have an algorithm in time 2^{48} instead of 2^{52} and it is possible to know if this algorithm will be convenient after only $C_{32}^{20} \approx 2^{28}$ gaussian reductions. Here we will see that there is also an algorithm in time 2^{48} with a probability of about 0.057 of success (or about 1/18). It is possible to know if this algorithm will be convenient after only $C_{32}^{12} \approx 2^{28}$ gaussian reductions.

The algorithm.

Step 0 : First, we have to find an equation with only 12 variables (instead of 16). The probability to find such an equation is about $C_{32}^{12} \cdot \frac{1}{251^4} \approx 0.0569$. If it exists, then we will find it after at most C_{32}^{12} gaussian reductions. If we find such an equation, call it equation (1). Then we introduce 6 more equations. The 12 variables of (1) are written on the left-hand side, the 10 more variables are written on the right-hand side. This is represented schematically as follows.

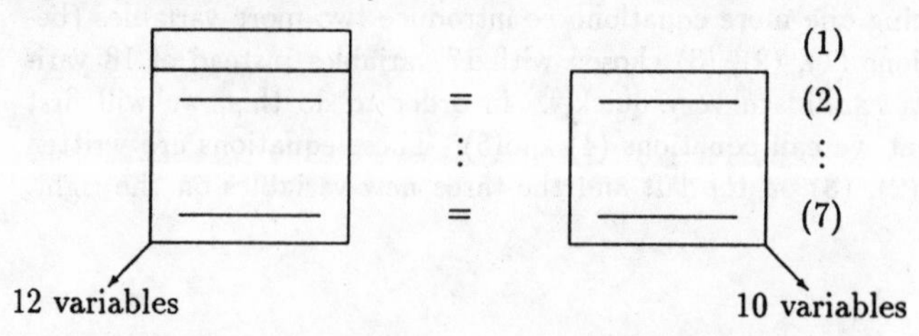

Step1 : precomputation. For each possible values for the 10 variables of the right hand side, the right-hand side value is calculated. By introducing middle values, like in paragraph 4, it is possible to reduce the memory needed. We do not give details since we are mainly interested in the time here.

Step2 : Equation (1) is written with 10 variables on the left-hand side, and 2 variables on the right-hand side. Now the algorithm proceeds in $C_{32}^{10} \approx 2^{26}$ steps : one for each possible value for the set E of the 10 variables on the left side of (1). At the beginning of each such step, a file F2 of the possible values for the right-hand side of (1) is stored. So there is 22.21 = 462 values in $F2$. Then, each possible value for the left-hand 10 variables is tested, $F2$ gives then in average 462/251 = 1.84 solutions for variables 11 and 12. Then $F1$ gives variables 13 to 20 and this 20-variable candidate is tested as usual with more equation.

Conclusion: The total time is only $C_{32}^{10}.10!.(1.84) \approx 2^{48.6}$. But the probability that this algorithm works is only about 0.057.

For PKP (37,64) a similar algorithm with $k = 15$ and $l = 21$ works. In this case the probability to find an equation (1) with 21 variables instead of 27 is closed to 100 per cent, because $C_{64}^{21}/251^6 \approx 164$. It gives the fastest algorithm we found for PKP(37,64) with 2^{116} in time.

7 Conclusion

We have investigated different ideas in order to improve the algorithms for the PKP problem. Let us recall the numerical results that we have obtained for the PKP (16,32), with a probability of success of about 100 per cent.

Previous algorithms for PKP(16,32) :

Memory	Time	Name
Negligible	2^{65}	J. Georgiades (cf[2])
2^{24} (5 – *uples*)	2^{65}	see [1] with k=3
2^{38} (8 – *uples*)	2^{60}	see [1] with k=5
2^{48} (10 – *uples*)	2^{56}	see [1] with k=6

Our algorithms for PKP(16,32) :

Memory	Time	Name
2^{17} (6 – *uples*)	2^{54}	Algorithm B4
2^{24} (10 – *uples*)	2^{52}	Algorithm A2

<u>Note</u> : Numeric values show that algorithm B4 needs only about 600 kbytes of memory and algorithm A2 about 100 Megabytes of memory, instead of about 1300 Terabytes for [1] with k=6, for example.

Now we summarize some results for PKP(37,64) (with a probability of success of about 100 per cent).

Previous algorithms for PKP(37,64) :

Memory	Time	Name
Negligible	2^{142}	J. Georgiades (cf[2])
2^{64}	2^{137}	see [1] with k=8 and l=11
2^{120}	2^{122}	see [1] with k=15 and l=21

Our algorithms for PKP(37,64) :

Memory	Time	Name
2^{27}	2^{123}	k=7, l=19, one variable eliminated, E on the right, c_1 to c_4 fixed
2^{52}	2^{119}	k=14, l=20, c_1 to c_8 fixed
2^{65}	2^{116}	k=15, l=21, equation (1) with 21 variables, c_2 to c_7 fixed

Note : The times given above are the times for finding all the solutions. If there is only one solution, this solution will be found in average after half the time given. Moreover, for PKP(37,64), there will be about three solutions: the secret solution plus about two solutions because $64! \approx 2 * 251^{37}$. So, for PKP(37,64), the average time to find one solution will really be a little less than the time given.

As we can see, our algorithms are still impracticable for PKP (37,64). However, although they need a lot of time, they are not completely unrealistic for PKP(16,32). Furthermore, the times given above are for sequential algorithms. Nevertheless, all our algorithms are very easy to implement in parallel, since they are not only designed in independent cases, but the memory needed in each of these cases is moderate.

Acknowledgements
We want to thank Thierry Baritaud for doing the Latex version of this paper.

References

[1] T. Baritaud, M. Campana, P. Chauvaud and H. Gilbert, "On the security of the Permuted Kernel Identification Scheme", Crypto'92, Springer Verlag.

[2] J. Georgiades, "Some Remarks on the security of the Identification Scheme Based on Permuted Kernels", Journal of Cryptology Vol. 5, n 2, 1992.

[3] A. Shamir, "An Efficient Identification Scheme Based on Permuted Kernels", Crypto'89, Springer Verlag.

On the Distribution of Characteristics in Composite Permutations

Luke O'Connor

Distributed Systems Technology Center
Brisbane, Australia
email: oconnor@fitmail.fit.qut.edu.au

Abstract. Differential cryptanalysis is a method of attacking iterated mappings which has been applied with varying success to a number of product ciphers and hash functions [1, 2]. Let $\rho : Z_2^c \times Z_2^m \rightarrow Z_2^m$ be a mapping that consists of c 'control' bits and m 'data' bits. The mapping ρ mapping contains 2^c m-bit permutations $\pi_i : Z_2^m \rightarrow Z_2^m$, $0 \leq i \leq 2^c - 1$, one of which is selected (multiplexed) by the control bits, and a substitution is then performed on the data bits using the selected permutation. Such mappings will be called *composite permutations*. The S-boxes of DES are composite permutations of the form $Si : Z_2^2 \times Z_2^4 \rightarrow Z_2^4$ with 2 control bits and 4 data bits.
In differential cryptanalysis the attacker is interested in the largest entry in a given XOR table, and the fraction of the XOR table that is zero. In this paper we determine the distribution of characteristics in the XOR tables of composite permutations, which leads to approximations for the largest entry in the XOR table and the density of zero entries.

Keywords: Differential cryptanalysis, iterated mapping, product cipher.

1 Introduction and Results

Differential cryptanalysis is a statistical attack popularized by Biham and Shamir [1, 2] that has been applied to a wide range of iterated mappings including LUCIFER, DES, FEAL, REDOC, Kahfre [3, 4, 8, 9, 12, 13]. As explained below, the attack is based on a quantity Ω called a *characteristic*, which has some probability p^{Ω} of giving information about the secret key used in the mapping. The attack is universal in that characteristics Ω will always exist for any iterated mapping, though p^{Ω} may be very small, and possibly less likely than the probability of guessing the secret key at random.

An r-round characteristic Ω is an $(r+1)$-tuple of differences $\Delta X, \Delta Y_1, \ldots, \Delta Y_r$; the probability p^{Ω} is defined as the fraction of plaintext pairs X, X' for which $X + X' = \Delta X$ and ΔY_i is the difference[1] of the encryption of X and X' after i rounds for all $1 \leq i \leq r$. If Ω incorrectly predicts an intermediate difference ΔY_i for a plaintext pair X, X' satisfying $X + X' = \Delta X$, then X, X' is said to be a *wrong pair* with respect to the characteristic. The probability of the differences ΔY_i being correctly predicted will typically depend on the distribution of differences in the auxiliary tables used by the iterated mapping. The *XOR table* of a mapping $\rho : Z_2^n \to Z_2^m$ shows the number of input pairs of difference $\Delta X \in Z_2^n$ that map to outputs of difference $\Delta Y \in Z_2^m$. In the case of DES, these auxiliary tables are known as S-boxes, and a study of the corresponding XOR tables by Biham and Shamir [1] yielded the following information: (*i*) the most likely input/output difference pair for any one S-box occurs with probability $\frac{1}{4}$; (*ii*) approximately 70–80% of the entries in the XOR tables are zero ; (*iii*) it is conjectured that if Ω is an r-round characteristic then extending Ω by an additional 2 rounds will decrease p^{Ω} by a factor of at least β where $\beta \approx 1/234$.

Each S-box in DES is a union of 4 permutations of the integers $\{0, 1, \ldots, 15\}$ where 2 of the 6 input bits select the permutation that will be used as the current substitution. We will call such mappings *composite permutations.* The design criteria of employing composite permutations in DES has not been adequately explained, but the experimental work of Dawson and Tavares [5] indicates that composite permutations have better XOR table distributions than single permutations $\pi : Z_2^m \to Z_2^m$ (as well as being close to optimal with respect to several other design criteria for S-boxes). In this paper we complement this experimental work by showing that for large m composite permutations yield XOR tables that are optimized against at least two properties that facilitate differential attacks.

Example 1. The S-boxes of DES are mappings of the form $Si : Z_2^2 \times Z_2^4 \to Z_2^4$ with 2 'control' bits and 4 'data' bits. For example, $S3$ may be written as the following 4×16 *table, where the control bits select the row, and the data bits select the column.*

[1] In this paper the difference operator + will refer to addition in the vector space Z_2^m, though it is possible to define other difference operators.

$$S3 = \begin{bmatrix} 15 & 1 & 8 & 14 & 6 & 11 & 3 & 4 & 9 & 7 & 2 & 13 & 12 & 0 & 5 & 10 \\ 3 & 13 & 4 & 7 & 15 & 2 & 8 & 14 & 12 & 0 & 1 & 10 & 6 & 9 & 11 & 5 \\ 0 & 14 & 7 & 11 & 10 & 4 & 13 & 1 & 5 & 8 & 12 & 6 & 9 & 3 & 2 & 15 \\ 13 & 8 & 10 & 1 & 3 & 15 & 4 & 2 & 11 & 6 & 7 & 12 & 0 & 5 & 14 & 9 \end{bmatrix}.$$

□

Let $\rho : Z_2^c \times Z_2^m \rightarrow Z_2^m$ be a mapping that consists of c control bits and m data bits. The mapping ρ mapping contains 2^c m-bit permutations $\pi_i : Z_2^m \rightarrow Z_2^m$, $0 \leq i \leq 2^c - 1$, one of which is selected (or basically multiplexed) by the control bits, and a substitution is then performed on the data bits using the selected permutation[2]. Let c be bound as $1 \leq c \leq m$. Assume that for an input $X \in Z_2^{c+m}$ the first c bits (the c most significant bits) are the control bits, which we will refer to as the *control prefix* of X. Then for $X \in Z_2^{c+m}$ the expression $\lfloor \frac{X}{2^m} \rfloor$ will extract the control prefix of X. A zero control prefix is one for which $\lfloor \frac{X}{2^m} \rfloor = 0$; all other control prefixes will be called *nonzero*.

In the XOR table for ρ, let $\Lambda_\rho(\Delta X^*, \Delta Y)$ be the entry for the input difference ΔX^* and the output difference ΔY. We will show that the distribution of the random variable $\Lambda_\rho(\Delta X^*, \Delta Y)$ for the composite mapping $\rho : Z_2^{c+m} \rightarrow Z_2^m$ when ΔX^* has a nonzero control prefix takes the form

$$\Pr(\Lambda_\rho(\Delta X^*, \Delta Y) = 2k) = \sum_{\substack{p_1 + p_2 + \cdots + p_{2^c - 1} = k \\ p_i \geq 0}} \prod_{i=1}^{2^{c-1}} \Pr(\lambda_i(m) = p_i). \tag{1}$$

We will prove that the $\lambda_i(m)$ are independent identically distributed (i.i.d.) random variables, described by the following probability distribution

$$\Pr(\lambda(m) = k) = \frac{1}{2^m!} \cdot \binom{2^m}{k} \cdot \frac{(2^m - k)!}{e} \cdot \left(1 + O\left(\frac{1}{(2^m - k)!}\right)\right). \tag{2}$$

This distribution is derived from the number of fixed points in an m-bit permutation (see Theorem 1). On the other hand, when ΔX^* has a zero control prefix it will be shown that

$$\Pr(\Lambda_\rho(\Delta X^*, \Delta Y) = 2k) = \sum_{\substack{p_1 + p_2 + \cdots + p_{2^c} = k \\ p_i \geq 0}} \prod_{i=1}^{2^c} \Pr(\Lambda_{\pi_i}(\Delta X, \Delta Y) = 2p_i) \tag{3}$$

where $\Lambda_{\pi_i}(\Delta X, \Delta Y)$ is the XOR table entry for $\Delta X, \Delta Y$ in π_i and ΔX are the m data bits of ΔX^*. Again these random variables are i.i.d. In this case, O'Connor [10, 11] has shown that $\Lambda_{\pi_i}(\Delta X, \Delta Y)$ is described by the following probability distribution

$$\Pr(\Lambda_{\pi_i}(\Delta X, \Delta Y) = 2k) = \binom{2^{m-1}}{k}^2 \cdot \frac{k! \cdot 2^k \cdot \Phi(2^{m-1} - k)}{2^m!} \tag{4}$$

[2] In this paper let ρ denote a composite permutation and let π denote a permutation.

where

$$\Phi(d) = \sum_{i=0}^{d} (-1)^i \cdot \binom{d}{i}^2 \cdot 2^i \cdot i! \cdot (2d-2i)!.$$

Examining the work of Biham and Shamir [1] on DES indicates that the differential cryptanalyst is interested in two properties related to the individual XOR tables: (*a*) the largest entry in the XOR table; (*b*) the fraction of the XOR table that is zero. The value of (*a*) will influence the probability of the most likely characteristic, while the value (*b*) will influence the signal-to-noise ratio in the experiments to determine the key (see [1] for definitions and details). The system designer should then attempt to minimize the quantity in (*a*) and maximize the quantity in (*b*). Our basic result is that composite permutations are well-suited to this min-max problem.

We will model an XOR table by assuming that each entry of the table is distributed according to either eq. (1) or eq. (3), and further assume that the entries are independent. Using this model we are able to show that the number of zero entries in an XOR table for a composite permutation is well-approximated by the expression

$$(2^m - 1) \cdot \left[2 + e^{-2^{c-1}} \cdot (2^m - 1) \right] + \frac{2^{2m}(2^c - 1)}{e^{2^{c-1}}}. \tag{5}$$

By considering the cases where $\Delta X = 0$ or $\Delta Y = 0$ it is easily shown that every XOR table will have at least $2^{m+1} - 2$ entries that are zero. From eq. (5) we see that as c approaches m, the fraction of zero entries in the XOR table approaches $2^{m+1} - 2$, the least number possible (see the computational results in Table 1).

From eqs. (1) and (3) we see that $\Lambda_\rho(\Delta X, \Delta Y)$ is a sum of i.i.d. random variables. We will use the law of large numbers to show that as c is increased the expected value of $\Lambda_\rho(\Delta X, \Delta Y)$ approaches 2^c. It then follows that the probability of a characteristic for which both ΔX and ΔY are not equal to zero is approximately 2^{-m}.

2 Some notation

If a difference $X + X' \in Z_2^{c+m}$ is written as $b\Delta X$, then let b be the control prefix, ie. $\lfloor (b\Delta X)/2^m \rfloor = b$. The set of all bijective mappings $\pi : Z_2^m \to Z_2^m$ is known as the symmetric group on 2^m objects and is denoted as S_{2^m}. Let $[\cdot]$ be a boolean predicate that evaluates to 0 or 1 such as $[n \text{ is prime}]$.

3 Characteristics in Composite Permutations

Initially consider the case where $c = 1$ and ρ consists of two permutations π_0 and π_1, from which we will directly generalize to the cases where $c > 1$. Consider determining the pairs XOR table distribution for ρ. Let $\Delta X^* = 0\Delta X \in Z_2^{m+1}$

where $\Delta X \in Z_2^m$ such that ΔX^* has a zero control prefix. Let $\Delta X^*, \Delta Y \in Z_2^m$ be a characteristic for ρ. We then have that

$$\begin{aligned}
\Lambda_\rho(\Delta X^*, \Delta Y) &= \sum_{\substack{X, X' \in Z_2^{m+1} \\ X + X' = \Delta X^*}} [\rho(X) + \rho(X') = \Delta Y] \\
&= \sum_{\substack{X, X' \in Z_2^{m} \\ X + X' = \Delta X}} [\pi_0(X) + \pi_0(X') = \Delta Y] + [\pi_1(X) + \pi_1(X') = \Delta Y] \\
&= \Lambda_{\pi_0}(\Delta X, \Delta Y) + \Lambda_{\pi_1}(\Delta X, \Delta Y).
\end{aligned}$$

Then for a zero control prefix, the probability of the characteristic $\Delta X^*, \Delta Y$ will be the average of the probabilities for the characteristic $\Delta X, \Delta Y$ in π_0 and π_1. On the other hand, consider the case where $\Delta X^* = 1\Delta X \in Z_2^m$. Then we have that

$$\begin{aligned}
\Lambda_\rho(\Delta X^*, \Delta Y) &= \sum_{\substack{X, X' \in Z_2^{m+1} \\ X + X' = \Delta X^*}} [\rho(X) + \rho(X') = \Delta Y] \\
&= \sum_{\substack{X < X',\ X, X' \in Z_2^{m+1} \\ X + X' = \Delta X^*}} 2 \cdot [\rho(X) + \rho(X') = \Delta Y] \\
&= \sum_{\substack{X < X',\ X, X' \in Z_2^{m+1} \\ X + X' = \Delta X^*}} 2 \cdot [\pi_0(X) + \pi_1(X') = \Delta Y] \\
&\stackrel{\text{def}}{=} 2 \cdot \lambda(m).
\end{aligned} \tag{6}$$

In the theorem below we prove that the expected value of $\lambda(m)$ approaches unity.

Theorem 1. Assuming that π_0 and π_1 are selected independently and uniformly from S_{2^m}

$$\mathbf{E}[\lambda(m)] = 1 + O\left(\frac{e^{2^m}}{2^{(m-1)\cdot 2^m}}\right).$$

Proof. From eq. (6) we have that

$$\begin{aligned}
\lambda(m) &= \sum_{\substack{X < X',\ X, X' \in Z_2^{m+1} \\ X + X' = \Delta X}} [\pi_0(X) + \pi_1(X') = \Delta Y] \\
&= \sum_{\substack{X < X',\ X, X' \in Z_2^{m+1} \\ X + X' = \Delta X}} \Pr(\pi_0(X) = \alpha \mid \pi_1(X') + \Delta Y = \alpha)
\end{aligned} \tag{7}$$

since this conditional probability in eq. (7) is either 1 or 0 for all X, X'. Notice that all choices for π_1 are equivalent in that there is a unique solution to $\pi_1(X') + \Delta Y = \alpha$ for any fixed X' and ΔY. Then without loss of generality assume that π_1 is the identity permutation, $\pi_1(a) = a$, $a \in Z_2^m$. Also since π_0 and π_1 are independent, then without loss of generality assume that $\Delta X = \Delta Y = 0$. It then follows that

$$\begin{aligned}
\lambda(m) &= \sum_{\substack{X<X',\, X,X'\in Z_2^{m+1} \\ X+X'=\Delta X}} \Pr(\pi_0(X)=\alpha \mid \pi_1(X')=\alpha) \\
&= \sum_{\substack{X<X',\, X,X'\in Z_2^{m+1} \\ X+X'=\Delta X}} \Pr(\pi_0(X)=X' \mid \pi_1(X')=X') \\
&= \sum_{\substack{X<X',\, X,X'\in Z_2^{m+1} \\ X+X'=\Delta X}} [\pi_0(X)=X'] \\
&= \sum_{X\in Z_2^m} [\pi_0(X)=X]
\end{aligned}$$

where the last two simplifications follow from the fact that $X = X'$ since $\Delta X = 0$, and π_1 is the identity permutation. It then follows that $\lambda(m)$ is equivalent to the number of fixed points $\pi_0(a) = a$ for $a \in Z_2^m$. A permutation that has no fixed points is called a *derangement*. It is well-known [7] that the number of m-bit permutations π that are derangements D_n is given as

$$D_n = 2^m! \cdot \sum_{i=0}^{2^m} \frac{(-1)^i}{i!} = \frac{2^m!}{e} \cdot \left(1 + O\left(\frac{1}{2^m!}\right)\right).$$

It then follows that for large m

$$\mathbf{E}[\lambda(m)] = \frac{1}{2^m!} \cdot \sum_{k=0}^{2^m} k \cdot \binom{2^m}{k} \cdot \frac{(2^m-k)!}{e} \cdot \left(1 + O\left(\frac{1}{(2^m-k)!}\right)\right)$$

which simplifies to

$$\begin{aligned}
\mathbf{E}[\lambda(m)] &= \frac{1}{e} \cdot \sum_{k=0}^{2^m-1} \frac{1}{k!} \cdot \left(1 + O\left(\frac{1}{(2^m-k)!}\right)\right) \\
&= \frac{1}{e} \cdot \left[\sum_{k=0}^{2^m-1} \frac{1}{k!} + \sum_{k=0}^{2^m-1} O\left(\frac{1}{(2^m-k)! \cdot k!}\right) \right] \\
&= \frac{1}{e} \cdot \left[e + O\left(\frac{1}{2^m!}\right) + O\left(\frac{2^m}{(2^{m-1}!)^2}\right) \right] \qquad (8) \\
&= 1 + O\left(\frac{e^{2^m}}{2^{(m-1)\cdot 2^m}}\right).
\end{aligned}$$

The simplification in eq. (8) follows from the fact that the summands $\sum_{k=0}^{2^m} \frac{1}{k!\cdot(2^m-k)!}$ are unimodal and symmetric. □

Corollary 3.1 $\Pr(\lambda(m) = 0) = e^{-1} + o(1)$.

Proof. From the previous theorem, the probability that $\lambda(m)$ is zero is equal to the probability that an m-bit permutation is a derangement, which is $e^{-1} + O\left(\frac{1}{2^m!}\right)$. □

We have now computed the exact distribution of an entry in the XOR table corresponding to a characteristic with a nonzero control prefix when $c = 1$. It follows that

$$\mathbf{E}[\Lambda_\rho(\Delta X^*, \Delta Y)] = 2 \cdot \mathbf{E}[\lambda(m)]$$
$$\Pr(\Lambda_\rho(\Delta X^*, \Delta Y) = 0) = e^{-1} + o(1)$$

since $\Pr(2 \cdot \lambda(m) = 0) = \Pr(\lambda(m) = 0/2) = \Pr(\lambda(m) = 0)$.

We are able to generalize our results for $c > 1$. Let ρ consist of 2^c permutations selected independently and uniformly from S_{2^m}. Let $\Delta X^* = b\Delta X$, $b \neq 0 \in Z_2^c$. By definition we have that

$$\begin{aligned}
\Lambda_\rho(\Delta X^*, \Delta Y) &= \sum_{\substack{X, X' \in Z_2^{c+m} \\ X + X' = \Delta X^*}} [\rho(X) + \rho(X') = \Delta Y] \\
&= \sum_{\substack{X, X' \in Z_2^m \\ X + X' = \Delta X}} \sum_{\substack{a + a' = b \in Z_2^c \\ b\Delta X = \Delta X^*}} [\pi_a(X) + \pi_{a'}(X') = \Delta Y] \\
&= \sum_{\substack{a + a' = b \in Z_2^c \\ b\Delta X = \Delta X^*}} \sum_{\substack{X < X' \\ X, X' \in Z_2^m \\ X + X' = \Delta X}} 2 \cdot [\pi_a(X) + \pi_{a'}(X') = \Delta Y] \\
&\stackrel{\text{def}}{=} 2 \cdot \sum_{k=1}^{2^{c-1}} \lambda_{i,j,k}(m)
\end{aligned} \tag{9}$$

where $\Delta X^* = i$ and $\Delta Y = j$. For fixed i, j in the range $1 \leq i, j \leq 2^m - 1$ and $1 \leq k \leq 2^{c-1}$, the $\lambda_{i,j,k}(m)$ are independent and are distributed identically as $\lambda(m)$ from eq. (6). When ΔX^* has a zero control prefix it follows that

$$\begin{aligned}
\Lambda_\rho(\Delta X^*, \Delta Y) &= \sum_{\substack{X, X' \in Z_2^m \\ X + X' = \Delta X}} \sum_{a \in Z_2^c} [\pi_a(X) + \pi_a(X') = \Delta Y] \\
&= \sum_{i=0}^{2^c - 1} \Lambda_{\pi_i}(\Delta X, \Delta Y).
\end{aligned}$$

4 Joint distributions

Let the XOR table for ρ be denoted as $\Lambda_\rho(i, j)$ for $0 \leq i < 2^{c+m}$, $0 \leq j < 2^m$, where i and j are interpreted as binary strings of length $c + m$ and m

respectively. We may then define the XOR table in terms of its characteristics $\Delta X^* = i, \Delta Y = j$ as follows:

$$\Lambda_\rho(i,j) = 2 \cdot \sum_{k=1}^{2^{c-1}} \lambda_{i,j,k}(m) \qquad \left\lfloor \frac{i}{2^m} \right\rfloor > 0 \tag{10}$$

$$\Lambda_\rho(i,j) = \sum_{k=1}^{2^c} \Lambda_{\pi_k}(i,j) \qquad \left\lfloor \frac{i}{2^m} \right\rfloor = 0 \tag{11}$$

where the distribution for $\Lambda_\pi(i,j)$ is given in eq. (4) and $\sum_{k=1}^{2^{c-1}} \lambda_{i,j,k}(m)$ is defined in eq. (9). In analyzing properties of the XOR table for a composite permutation, we are concerned with the joint distribution of the $\Lambda_\rho(i,j)$, which in turn, is the joint distribution of the $\lambda_{i,j,k}(m)$. Observe that for fixed i,j and varying k the $\lambda_{i,j,k}(m)$ are *independent*, but for varying i,j,k the $\lambda_{i,j,k}(m)$ are *dependent*. However, for sake of analysis, *we will assume the $\lambda_{i,j,k}(m)$ to be independently distributed.* That is, we will assume that the individual XOR table entries are independently distributed. This assumption allows the probability of events for the XOR table, such as the size of the largest entry, to be cast in terms of events for the individual table entries. Results obtained by experimentation presented below show that this assumption leads to only a small deviation from the actual value of an XOR entry (see Table 1).

Theorem 2. Let the composite permutation $\rho : Z_2^{c+m} \to Z_2^m$ consist of 2^c independently and uniformly selected m-bit permutations. Let $\Lambda_{\rho,0}$ be the number of entries in the XOR table that are expected to be zero. Then

$$\Lambda_{\rho,0} \sim (2^m - 1) \cdot \left[2 + e^{-2^{c-1}} \cdot (2^m - 1) \right] + \frac{2^{2m}(2^c - 1)}{e^{2^{c-1}}}. \tag{12}$$

□

Comparisons between $\Lambda_{\rho,0}$ as derived in Theorem 2 and the observed fraction $\overline{\Lambda_{\rho,0}}$ of zero entries in the XOR table of m_p random composite permutations ρ are given in Table 1. The table indicates that the estimates $\Lambda_{\rho,0}$ from Theorem 2 are very accurate, which validates the assumption that the distribution of the $\lambda_{i,j,k}(m)$ is independent.

Biham and Shamir [1] report that 20%–30% of the entries in the S-boxes of DES are zero. Theorem 2 yields that approximately 16% of the XOR table entries in a mapping with 4 data bits and 2 control bits are expected to be zero; further, a random sample of 10,000 such mappings has yielded an average of 15.7% zero entries in the corresponding XOR tables. This suggests that the set of design criteria for the S-boxes has increased the expected density of zeroes in the XOR table.

The (strong) law of large number states that for random variables $Y_1, Y_2, \ldots, Y_N$ which are i.i.d. with mean $\mathbf{E}[Y]$

$$\left| \frac{Y_1 + Y_2 + \cdots + Y_N}{N} - \mathbf{E}[Y] \right| < \epsilon \tag{13}$$

m	c	$\Lambda_{\rho,0}/2^{c+2m}$	$\overline{\Lambda_{\rho,0}}$	m_p
4	1	0.40419	0.39433	10000
4	2	0.16053	0.15707	10000
4	3	0.03268	0.03221	10000
4	4	0.00765	0.00765	10000
5	1	0.38683	0.38160	1000
5	2	0.14839	0.14666	1000
5	3	0.02574	0.02544	1000
5	4	0.00411	0.00410	1000
5	5	0.00189	0.00189	1000
6	1	0.37755	0.37486	1000
6	2	0.14197	0.14106	1000
6	3	0.02208	0.02196	1000
6	4	0.00225	0.00225	1000

Table 1. Estimates of $\Lambda_{\rho,0}$ for composite permutations.

for any $\epsilon > 0$ as N becomes large. That is, the sample mean approaches the expectation of the random variable Y_i. Observe that an individual entry of the XOR table for a multiple permutation is a sum of i.i.d. random variables. If $\Delta X^* = i, \Delta Y = j$ then for characteristics with zero control prefix, the XOR entry is a sum of 2^c random variables $\Lambda_{\pi_k}(i,j)$ as defined in eq. (11), and when the control prefix is nonzero, the XOR entry is a sum of 2^{c-1} random variables $\lambda_{i,j,k}(m)$ as defined in eq. (10). The mean of $\lambda_{i,j,k}(m)$ was proven to be $1+o(1)$ in Theorem 1. It can also be shown [11] from eq. (4) that the mean of $\Lambda_{\pi_k}(i,j)$ is $1+o(1)$. Both these $o(1)$ terms dominate $2^c \leq 2^m$ and the largest value in the table for large m will be $2^c + o(1)$ given our independence assumption.

5 Conclusion

Our analysis has shown that for sufficiently large c, a very small fraction of the XOR table will be zero, and that the largest entry will be close to $2^c + o(1)$. We note that no special algorithms are required to construct composite permutation ρ that have these properties as they are a consequence of the law of large numbers. For this reason it appears that composite permutations provide are more resistant to differential attacks than most other mappings, including single permutations π.

References

1. E. Biham and A. Shamir. Differential cryptanalysis of DES-like cryptosystems. *Journal of Cryptology*, 4(1):3–72, 1991.

2. E. Biham and A. Shamir. Differential cryptanalysis of Snefru, Khafre, REDOC-II, LOKI and LUCIFER. *Advances in Cryptology, CRYPTO 91, Lecture Notes in Computer Science, vol. 576, J. Feigenbaum ed., Springer-Verlag*, pages 156–171, 1992.
3. L. P. Brown, J. Pieprzyk, and J. Seberry. LOKI - a cryptographic primitive for authentication and secrecy applications. *Advances in Cryptology, AUSCRYPT 90, Lecture Notes in Computer Science, vol. 453, J. Seberry and J. Pieprzyk eds., Springer-Verlag*, pages 229–236, 1990.
4. T. Cusick and M. Wood. The REDOC-II cryptosystem. *Advances in Cryptology, CRYPTO 90, Lecture Notes in Computer Science, vol. 537, A. J. Menezes and S. A. Vanstone ed., Springer-Verlag*, pages 545–563, 1991.
5. M. H. Dawson and S. E. Tavares. An expanded set of S-box design criteria based on information theory and its relation to differential-like attacks. *Advances in Cryptology, EUROCRYPT 91, Lecture Notes in Computer Science, vol. 547, D. W. Davies ed., Springer-Verlag*, pages 352–367, 1991.
6. W. Feller. *An Introduction to Probability Theory with Applications.* John Wiley and Sons, 3rd edition, Volume 1, 1968.
7. R. L. Graham, D. E. Knuth, and O. Patshnik. *Concrete Mathematics, A Foundation for Computer Science.* Addison Wesley, 1989.
8. X. Lai and J. L. Massey. A proposal for a new block encryption standard. In *Advances in Cryptology, EUROCRYPT 90, Lecture Notes in Computer Science, vol. 473, I. B. Damgård ed., Springer-Verlag*, pages 389–404, 1991.
9. R. Merkle. Fast software encryption functions. *Advances in Cryptology, CRYPTO 90, Lecture Notes in Computer Science, vol. 537, A. J. Menezes and S. A. Vanstone ed., Springer-Verlag*, pages 476–501, 1991.
10. L. J. O'Connor. On the distribution of characteristics in bijective mappings. presented at Eurocrypt 93, Norway, May 1993. Also accepted for publication in the *Journal of Cryptology.*
11. L. J. O'Connor. *An analysis of product ciphers using boolean functions.* PhD thesis, Department of Computer Science, University of Waterloo, 1992.
12. A. Shimizu and S. Miyaguchi. Fast data encipherment algorithm FEAL. *Advances in Cryptology, EUROCRYPT 87, Lecture Notes in Computer Science, vol. 304, D. Chaum and W. L. Price eds., Springer-Verlag*, pages 267–278, 1988.
13. A. Sorkin. LUCIFER: a cryptographic algorithm. *Cryptologia*, 8(1):22–35, 1984.

Remark on the Threshold RSA Signature Scheme

Chuan-Ming Li, Tzonelih Hwang, Narn-Yih Lee
Institute of Information Engineering
National Cheng-Kung University
Tainan, Taiwan, R.O.C.

Abstract

Shared generation of secure signatures, called the threshold signatures, was introduced by Desmedt and Frankel in 1991. A threshold signature scheme is not only a threshold scheme, but also a signature scheme. Therefore, it should possess the properties of both threshold scheme and digital signature scheme. In this paper, we investigate conspiracy attacks on the Desmedt and Frankel's threshold RSA signature scheme. We also discuss the requirements of secure threshold signature scheme.

1 Introduction

A digital signature is some information which is dependent on the message and on data known only to the signer. Thus, a secure signature scheme should be able to protect a signature from being forged by the receiver or another person, and the sender cannot deny having signed the message.

The signer of the conventional digital signature schemes is usually a single user. However, the responsibility of signing messages needs to be shared from time to time. For example, a company may require that any policy decision must be signed by k directors before it is issued. It may also be a company's policy that checks should be signed by k individuals rather than by one person. Shared generation of secure signatures, called the *threshold signatures*, are used to solve these problems [1].

It is very important that a (k, l) threshold signature scheme is not only a threshold scheme in the sense that less than k users cannot generate a valid signature, but also a signature scheme such that the signature of a particular set of k users should not be forged by another set of k users. In general, there are several requirements that (k, l) threshold signatures should satisfy:

1. Similar to the (k, l) threshold secret sharing scheme, the group secret key K can be divided into l different "*secretshares*", K_1 K_2, ..., K_l, such that

   (a) the group signature can be easily produced with knowledge of any k secret shares ($k \leq l$);
   (b) with knowledge of any $k-1$ or fewer secret shares, it is impossible to generate a group signature;
   (c) the group secret key cannot be derived from the released group signature and all partial signatures; and
   (d) it is impossible to derive any secret share from the released group signature and all partial signatures.

2. It is better that the size of the group signature is equivalent to the size of an individual signature.

3. The group signature can be verified by any outsider and the verification process should be as simple as possible.

4. The signing group holds the responsibility to the signed message. That is, each signer in this group cannot deny having signed the message.

5. The partial signatures and the group signature cannot be forged by malicious users.

In 1991, Desmedt and Frankel [1] proposed the first (k, l) threshold digital signature scheme based on the RSA assumption. In this paper we shall show that if k or more shareholders conspire, then the system secret of their scheme will be revealed with a high probability. Once the system secret is revealed by these shareholders, they can impersonate another set of shareholders to sign messages without holding the responsibility to the signatures, and a malicious group of k signers can deny having signed a message though in fact they have signed the message. However, this paper does not weaken the security of the threshold signature scheme in the sense that our remarks do not give more power to $k-1$ or less shareholders than estimated before.

The structure of this paper is as follows. In Section 2, we review the Desmedt et al.'s threshold RSA signature scheme. In Section 3, we shall show how $k+1$ users can conspire to derive the system's secret with a high probability. Then, we show that the system secret can also be revealed by the conspiracy of k users. Finally, we conclude this paper in Section 5.

2 Desmedt et al.'s Threshold RSA Signature Scheme

Let $n = pq$ where p, q are large primes. For p and q to be safe primes, let $p = 2p' + 1$ and $q = 2q' + 1$ where p', q' are primes [2][3]. Define $\lambda(n) = 2p'q'$ (λ is the Carmichael function, i.e., the exponent of $Z_n^*(\,\cdot\,)$.) The secret key is d which was chosen at random such that $\gcd(d, \lambda(n)) = 1$ (so d is odd) and the public key is e such that $de \equiv 1 \mod \lambda(n)$. All shareholders in the Desmedt et al.'s scheme form a set A ($|A| = l$) such that any subset B ($|B| = k$) in A can collectively generate a signature for a message.

Basically, the Desmedt et al.'s threshold RSA signature scheme is based on interpolation polynomials over the integers. As in the Lagrange interpolation scheme of [4], let $f(x)$ be a polynomial of degree $k-1$ such that $f(0) = d-1$. In the scheme, there is a share distribution center (SDC) which would choose $p, q, d, f(x)$ and distribute to each shareholder i a public integer x_i and a secret share K_i:

$$K_i = \frac{f(x_i)/2}{\left(\prod_{\substack{j\in A\\ j\neq i}}(x_i - x_j)\right)/2} \pmod{p'q'}, \tag{1}$$

where all the x_i's are odd and all $f(x_i)$'s are even.

To create a signature S_m of a message m, each shareholder $i \in B$ will generate a modified share $a_{i,B}$:

$$a_{i,B} = K_i \cdot \Big(\prod_{\substack{j\in A\\ j\notin B}}(x_i - x_j)\prod_{\substack{j\in B\\ j\neq i}}(0 - x_j)\Big) \tag{2}$$

and will calculate the partial result $s_{m,i,B} \equiv m^{a_{i,B}} \bmod n$. Since $f(0) = d-1$ and

$$f(x) = \sum_{i\in B} K_i \prod_{\substack{j\in A\\ j\notin B}}(x_i - x_j)\prod_{\substack{j\in B\\ j\neq i}}(x - x_j) \pmod{\lambda(n)},$$

we can have that

$$\sum_{i\in B} a_{i,B} \equiv d-1 \mod \lambda(n). \tag{3}$$

Each $i \in B$ sends the partial result $s_{m,i,B}$ to a Combiner C. To create the signature S_m, C calculates

$$S_m \equiv m \cdot \prod_{i\in B} s_{m,i,B} \equiv m \cdot m^{d-1} \equiv m^d \bmod \ n. \tag{4}$$

The receiver of the S_m can check the correctness of S_m by verifying

$$m \overset{?}{\equiv} (S_m)^e \bmod n, \quad \text{where } e \text{ is the public key.}$$

The following example is used to illustrate Desmedt et al.'s threshold RSA signature scheme.

Example 1 : We assume that there are five shareholders in the system and any two of them can generate the signature S_m for a message m, i.e., $|A| = l = 5$, $|B| = k = 2$. First of all, the SDC chooses

$$\begin{aligned} p &= 2p' + 1 = 2 \cdot 23 + 1 = 47, \\ q &= 2q' + 1 = 2 \cdot 29 + 1 = 59, \\ d &= 221, \\ f(x) &= 6 \cdot x^{k-1} + (d-1) = 6 \cdot x + 220, \\ \text{and } x_1 &= 3,\ x_2 = 7,\ x_3 = 13,\ x_4 = 17,\ x_5 = 19. \end{aligned}$$

Thus,

$$\begin{aligned} n &= pq = 47 \cdot 59 = 2773, \\ \lambda(n) &= 2p'q' = 2 \cdot 23 \cdot 29 = 1334\ , \\ p'q' &= 23 \cdot 29 = 667\ , \\ e &= 833, \text{where } de \equiv 1 \mod \lambda(n)\ , \\ \text{and } f(x_1) &= 238,\ f(x_2) = 262,\ f(x_3) = 298,\ f(x_4) = 322,\ f(x_5) = 334. \end{aligned}$$

By using Eq. (1), the SDC can calculate the secret share K_1 as follows:

$$\begin{aligned} K_1 &= \frac{f(x_1)/2}{(x_1 - x_2)(x_1 - x_3)(x_1 - x_4)(x_1 - x_5)/2} \mod p'q' \\ &= \frac{238/2}{-4 \cdot -10 \cdot -14 \cdot -16/2} (\text{mod } 667) = 197. \end{aligned}$$

Similarly, the SDC calculates $K_2 = 219$, $K_3 = 179$, $K_4 = 575$ and $K_5 = 219$. Then, the SDC sends x_i and K_i to the Shareholder i. We assume that the Shareholder 1 and the Shareholder 4, i.e., $B = \{1, 4\}$, would like to generate a signature S_m for a message m. The Shareholder 1 uses Eq. (2) to generate the modified share $a_{1,B}$,

$$\begin{aligned} a_{1,B} &= K_1 \cdot (x_1 - x_2)(x_1 - x_3)(x_1 - x_5)(0 - x_4) \\ &= 197 \cdot\ -4 \cdot\ -10 \cdot\ -16 \cdot\ -17 = 2143360, \end{aligned}$$

and sends the partial result $s_{m,1,B}$,

$$s_{m,1,B} \equiv m^{a_{1,B}} \mod n = m^{2143360} \mod 2773$$

to the Combiner C. Similarly, The Shareholder 4 also calculates the modified share $a_{4,B} = 138000$ and sends the partial result

$$s_{m,4,B} = m^{138000} \mod 2773$$

to C. Finally, the C creates the signature as:

$$S_m \equiv m \cdot s_{m,1,B} \cdot s_{m,4,B} = m^{2281361} \equiv m^{221} \mod 2773.$$

The signature receiver can check

$$m \stackrel{?}{\equiv} (S_m)^e \mod n, \stackrel{?}{\equiv} (S_m)^{833} \mod 2773,$$

to verify whether the signature S_m is correct or not.

3 Conspiracy Attack by k+1 Users

It is clear that $k-1$ or less shareholders cannot generate a valid signature, and once a signature is generated by k shareholders, no one can perform an impersonation or substitution attack. However, we are going to show that arbitrary set of $k+1$ shareholders can conspire to compute the system's secret $\lambda(n)$ or $p'q'$ with a high probability. This attack refers to the observation of Davida et al. in [5].
Let

$$\begin{aligned} w_1 &\equiv y \bmod \lambda(n), \\ w_2 &\equiv y \bmod \lambda(n), \\ w_3 &\equiv y \bmod \lambda(n). \end{aligned}$$

If all the values of w_1, w_2, and w_3 are known, then one can have

$$\begin{aligned} w_3 - w_2 &= z \cdot \lambda(n), \\ w_2 - w_1 &= z' \cdot \lambda(n), \\ &\quad \text{for some integers } z \text{ and } z'. \end{aligned}$$

Therefore, $\lambda(n)$ will be revealed with a high probability by finding the greatest common divisor (GCD) of $(w_3 - w_2)$ and $(w_2 - w_1)$.

According to the combination theorem, there are $k+1$ possible ways to choose any k shareholders from $k+1$ shareholders. Each choice forms a subset, B_j, of k users. If $k+1$ (*i.e.*, $k \geq 2$) shareholders act in collusion in the Desmedt et al.'s scheme, then they may compute w_j, where $w_j = \sum_{i \in B_j} a_{i,B_j}$, $1 \leq j \leq k+1$. By the Davida et al.'s observation, we have

$$\begin{aligned} w_1 &\equiv (d-1) \bmod 2p'q', \\ w_2 &\equiv (d-1) \bmod 2p'q', \\ w_3 &\equiv (d-1) \bmod 2p'q'. \\ &\cdots \end{aligned}$$

Thus, the system secret $2p'q'$ or $p'q'$ will be revealed with a high probability by finding the GCD of $(w_3 - w_2)$ and $(w_2 - w_1)$. Once the system secret is revealed by these $k+1$ shareholders, they are able to compute all of the secret shares. Then, these malicious shareholders can impersonate other shareholders to sign messages without holding the responsibility to the signatures. The following example is used to illustrate this attack.

Example 2 : Let us consider the Example 1 again. We assume that the Shareholders 1, 3 and 4 conspire to compute the system secret $p'q'$. They constitute three subsets $B_{(1)} = \{1,3\}$, $B_{(2)} = \{1,4\}$, $B_{(3)} = \{3,4\}$ and compute $\sum a_{i,B}$, for each subset.

$$\begin{aligned}
\sum_{i \in B_{(1)}} a_{i,B_{(1)}} &= a_{1,B_{(1)}} + a_{3,B_{(1)}} \\
&= 2294656 + -77328 = 2217328, & \text{(a)} \\
\sum_{i \in B_{(2)}} a_{i,B_{(2)}} &= a_{1,B_{(2)}} + a_{4,B_{(2)}} \\
&= 2143360 + 138000 = 2281360, & \text{(b)} \\
\sum_{i \in B_{(3)}} a_{i,B_{(3)}} &= a_{3,B_{(3)}} + a_{4,B_{(3)}} \\
&= 1095480 + 2093000 = 3188480, & \text{(c)}
\end{aligned}$$

Then, they compute :

$$\begin{aligned}
\text{(c) - (b)} &= 907120 = z \cdot \lambda(n) = z \cdot 2p'q', & \text{(e)} \\
\text{(b) - (a)} &= 60432 = z' \cdot \lambda(n) = z' \cdot 2p'q', & \text{(f)}
\end{aligned}$$

for some integer z and z'.

Because p' and q' are odd primes, the product of $p' \cdot q'$ must be an odd number. By removing the factor 2 in (e) and (f) (i.e., $(e) = 11339$ and $(f) = 2001$), they will obtain

$$GCD(11339, 2001) = 667 = p' \cdot q'$$

Thus, $\lambda(n) = 2p'q' = 1334$.

4 Conspiracy Attack by k Users

In this section, we will show that if only k shareholders act in collusion, they can also compute the system secret with a high probability. The attack is described as follows.

As the conspiracy attack by $k+1$ shareholders, any subset B of k malicious shareholders can have

$$w = \sum_{i \in B} a_{i,B} \equiv (d-1) \mod \lambda(n), \tag{5}$$

Multipling Eq. (5) by the public key e, we can have

$$\begin{aligned}
e \cdot w &\equiv e \cdot (d-1) && \mod \lambda(n), \\
e \cdot w &\equiv ed - e && \mod \lambda(n),
\end{aligned}$$

Because $ed \equiv 1 \mod \lambda(n)$, we can compute

$$\begin{aligned} e \cdot w &\equiv 1 - e && \mod \lambda(n), \\ e \cdot w + e - 1 &\equiv 0 \equiv z \cdot \lambda(n) && \mod \lambda(n), \end{aligned} \qquad (6)$$

for some integer z.

The result of Eq. (6) is a multiple of $\lambda(n)$. Therefore, n can be factored in polynomial time by [6].

5 Conclusions

In this paper, we have showed that the system secret of Desmedt et al.'s threshold RSA signature scheme can be revealed with a high probability by the conspiracy of $k+1$ or k shareholders. It will be very challenging to devise a threshold signature scheme that satisfies the requirements proposed in this paper.

Acknowledgement

The authors wish to thank G. R. Blakley, Y. Desmedt, Y. Frankel, M. Yung and G. I. Davida for their valuable suggestions to this paper at Crypto '93. Dr. Y. Desmedt pointed out that the system secret can also be revealed by the conspiracy of k users as shown in Section 4. This paper is supported by the National Science Council of R. O. C. under the contract NSC 82 - 0408 - E - 006 - 026.

Reference

[1] Y. Desmedt and Y. Frankel: "Shared Generation of Authenticators and Signatures", Advances in Cryptology - Cryto '91, Proceedings, pp.457-469, Springer Verlag, 1991.

[2] B. Blakley and G. R. Blakley: "Security of Number Theoretic Public Key Cryptosystems Against Random Attack", Cryptologia, 1978. In three parts: Part 1: 2(4), pp.305-321, Oct 1978; Part 2: 3(1), pp.29-42, Jan 1979; Part 3: 3(2), pp.105-118, Apr 1979;

[3] G. R. Blakley and I. Borosh: "RSA Public Key Cryptosystems Do Not Always Conceal Messages", Computers & Mathematics with Applications, 5(3): 169-178, 1979.

[4] A. Shamir: "How to share a secret", Commun. ACM, 22:612-613, 1979.

[5] G. I. Davida, D. L. Wells, and J. B. Kam: "A Database Encryption System with Subkeys", ACM Trans. Database System, 6, (2), pp.312-328, 1981.

[6] Gary L. Miller: "Riemann's hypothesis and tests for primality", J. Computer Systems Sci., 13, pp.300-317, 1976.

Another Method for Attaining Security Against Adaptively Chosen Ciphertext Attacks

Chae Hoon Lim and Pil Joong Lee

Department of Electrical Engineering,

Pohang University of Science and Technology (POSTECH)

Pohang, 790-784, KOREA

ABSTRACT Practical approaches to constructing public key cryptosystems secure against chosen ciphertext attacks were first initiated by Damgard and further extended by Zheng and Seberry. In this paper we first point out that in some cryptosystems proposed by Zheng and Seberry the method for adding authentication capability may fail just under known plaintext attacks. Next, we present a new method for immunizing public key cryptosystems against adaptively chosen ciphertext attacks. In the proposed immunization method, the deciphering algorithm first checks that the ciphertext is legitimate and then outputs the matching plaintext only when the check is successful. This is in contrast with the Zheng and Seberry's methods, where the deciphering algorithm first recovers the plaintext and then outputs it only when the checking condition on it is satisfied. Such a ciphertext-based validity check will be particularly useful for an application to group-oriented cryptosystems, where almost all deciphering operations are performed by third parties, not by the actual receiver.

1 Introduction

Recently much attention has been devoted to constructing public key cryptosystems secure against chosen ciphertext attacks, from the theoretical and practical points of view. Theoretically, non-interactive zero-knowledge proof was shown to be a nice tool for this purpose [3] [9] and several such concrete public key cryptosystems have been proposed [16] [18]. However, due to the enormous data expansion during the enciphering transformation, the resulting schemes are highly inefficient and thus no one would try to implement them in practice.

Practical approaches to this field were initiated by Damgard [7] and further extended by Zheng and Seberry [25]. The key idea of Damgard's approach is to construct a public key cryptosystem in such a way that an attacker cannot produce legitimate ciphertexts (i.e., the ciphertexts whose plaintexts he can get from the deciphering oracle) without knowing the plaintext. This makes useless the attacker's ability to gain access to the deciphering oracle under chosen ciphertext attacks. Based on this idea, Damgard [7] proposed simple methods for modifying any deterministic public key cryptosystems and the ElGamal/Diffie-Hellman cryptosystem so that the resulting cryptosystems may be more secure under chosen ciphertext attacks. Later, by refining Damgard's idea and combining the probabilistic encryption technique [10], Zheng and Seberry [25] presented three practical methods for immunizing public key cryptosystems and proved that their cryptosystems are semantically secure against adaptively chosen ciphertext attacks under reasonable assumptions.

In this paper, we first point out that in some cryptosystems presented by Zheng and Seberry the method for adding authentication capability may fail just under known plaintext attacks. Next we propose a new method for immunizing public key cryptosystems, which is illustrated by constructing cryptosystems based on the Diffie-Hellman/ElGamal scheme and the RSA scheme. In the modified cryptosystems, the deciphering algorithm first checks that ciphertexts are properly constructed according to the enciphering algorithm and only when the check is successful, does it output the matching plaintexts. A main difference of our approach from that of Zheng and Seberry is that it is determined based on the ciphertext, not on the recovered plaintext, whether or not the deciphering algorithm outputs the result.

Such a ciphertext-based validity check is especially useful for an application to group-oriented cryptosystems [8]. In group-oriented cryptosystems, ciphertexts are usually accompanied by the indicator indicating the nature of the ciphertexts and all or substantial part of deciphering operations are performed independently of the actual receiver(s). Then the partial computations are distributed to the legitimate receiver(s) according to the indicator and the security policy of the receiving company. Thus, the main threat is an illegal modification of the indicator by an inside group member who violates the security policy and tries to read the ciphertexts. This requires a concrete scheme for combining the ciphertexts and the indicator so that no one can produce a legal ciphertext by modifying the intercepted ciphertext, especially by changing the indicator. The proposed cryptosystem is well suited for this application.

This rest of this paper is organized as follows. Section 2 briefly mentions probabilistic encryption and pseudorandom number generators. Section 3 discusses the previous works of Damgard and Zheng-Seberry in this field. Here, we also point out some weakness of Zheng and Seberry's method for adding authentication capability to their cryptosystems. In section 4, applying the proposed immunization method, we present two cryptosystems based on the Diffie-Hellman/ElGamal scheme and the RSA scheme and analyze their security. Finally we conclude in section 5.

2 Random Number Generators and Probabilistic Encryption

Goldwasser and Micali [10] presented a general scheme for constructing public key probabilistic encryption schemes which hide all partial information, in the sense that whatever is efficiently computable about the plaintext given the ciphertext is also efficiently computable without the ciphertext. (This is an informal definition of semantic security which can be thought of as a polynomially bounded version of Shannon's perfect secrecy. See [13] for other equivalent notions of security for public key cryptosystems.) These encryption schemes can be thought of as the best we are seeking for, as far as passive attacks are concerned, since a polynomially bounded passive attacker can extract no information on the plaintexts from the ciphertexts. They also gave a concrete implementation under the intractability assumption of deciding quadratic residuosity modulo a large composite number. However, their scheme expands each plaintext bit into a ciphertext block of length of the composite modulus and thus is highly inefficient.

Cryptographically strong pseudorandom number generators whose notion was first introduced by Blum and Micali [5] and extended by Yao [24] is one of the most powerful tools in many cryptographic applications. The output sequences produced by such a generator cannot be distinguished by a polynomial-time algorithm from truly random sequences of the same length (such a generator is said to be *perfect*). Thus these generators can be used for constructing more efficient probabilistic encryption schemes, as first illustrated by Blum and Goldwasser [4] : "Send the exclusive-or of a message sequence with an output sequence of the same length of a pseudorandom number generator, together with a public key encryption of a random seed used." Consequently, cryptosystems constructed like this can be proved to be secure against any passive attacks (e.g., see [4] for detailed proof) and thus as far as passive attacks are concerned, the problem of constructing a secure public key cryptosystem is settled. Furthermore, the plaintext is only expanded by a constant factor in this case, the portion of public key encryption of a random seed used.

Several perfect pseudorandom number generators have been established. Long and Wigderson [12] generalized the Blum-Micali's generator [5] based on the discrete logarithm problem and showed that O(log k) bits can be securely produced per each exponentiation where k is the bit-length of a modulus. The same result was obtained by Peralta [17] with different technique. Alexi et al. [1] showed that RSA/Rabin function can hide O(log k) bits under the intractability assumption of RSA encryption and factoring. Vazirani and Vazirani [22] showed that O(log k) bits can be securely extracted from the x^2 mod N generator of Blum, Blum and Shub [2] as well as from the RSA/Rabin functions. Recently Micali and Schnorr [14] developed a very efficient polynomial random number generator which can be based on an arbitrary prime modulus as well as on RSA modulus. This generator can produce more than k/2 bits per iteration at a cost of about one full modular multiplication, though it is open whether the generator with this efficiency is perfect. This, if perfect, will lead to very efficient probabilistic encryption schemes which hide all partial information.

3 Overview and Discussion of Previous Works

3.1 Damgard's Approach

A main drawback of probabilistic encryption schemes is that while being provably secure against any passive attacks, they can be completely broken under chosen ciphertext attacks. In a chosen ciphertext attack, an attacker can query the deciphering oracle with polynomially many ciphertexts, and use the information obtained from the answers to extract any useful information for the target ciphertext.

Recently, Damgard [7] made a first step into the research of practical public key cryptosystems secure against chosen ciphertext attacks. His key idea is to modify a public key cryptosystem in such a way that an attacker cannot produce ciphertexts whose plaintexts he can get from the deciphering oracle unless he starts by first choosing the plaintext. This nullifies the ability to have access to the deciphering oracle under chosen ciphertext attacks. Based on this idea, he presented two concrete examples of public key cryptosystems which appear to be secure against chosen ciphertext attacks, one using any deterministic public key cryptosystems and the other using the Diffie-Hellman /ElGamal public key cryptosystem.

To get a better insight into the Damgard's approach, we briefly describe his second scheme based on the Diffie-Hellman/ElGamal public key cryptosystem. Let a user A's secret key be a pair (x_{A1}, x_{A2}) of elements chosen at random over [1, p-1] and the corresponding public key be a pair (y_{A1}, y_{A2}), where $y_{A1} \equiv_p g^{x_{A1}}$ and $y_{A2} \equiv_p g^{x_{A2}}$. (Here, "$Y \equiv_p X$" denotes "Y is congruent to X in mod p".) Then the ciphertext for a message m to be sent to user A consists of a triple (c_1, c_2, c_3) :

$$c_1 \equiv_p g^r,\ c_2 \equiv_p y_{A1}^{\ r} \text{ and } c_3 = m \oplus (y_{A2}^{\ r} \bmod p),$$

where r is uniform in [1, p-1] and the symbol $\oplus$ denotes the bit-wise exclusive-or operation. The deciphering algorithm by user A who has the secret key (x_{A1}, x_{A2}) is as follows :

$$m = c_3 \oplus (c_1^{\ x_{A2}} \bmod p) \text{ if } c_2 \equiv_p c_1^{\ x_{A1}},\ \text{NULL otherwise.}$$

Here, NULL is a special symbol used for meaning "no plaintext output".

The intuitive reason of the security against chosen ciphertext attacks is that given g and y_{A1}, it seems hard to generate a pair (g^r mod p, $y_{A1}{}^r$ mod p), unless one starts by simply choosing r, which in turn implies that it is hard for an attacker to generate a legitimate ciphertext (on which the deciphering algorithm produces a non-null output), unless he already knows the plaintext. Therefore, this modified ElGamal cryptosystem will be secure against chosen ciphertext attacks, if we assume that the original ElGamal is secure against passive attacks and that there is no other way to produce ciphertexts than to first choose r. This approach suggests a method of gaining more security against chosen ciphertext attacks : "Modify a public key cryptosystem in such a way that it is infeasible to generate a legitimate ciphertext without first choosing the plaintext. Then the modified cryptosystem will be as secure under a chosen ciphertext attack as under a passive attack."

Here, we have to note that Damgard considered a restricted model for chosen ciphertext attacks, known as indifferently chosen ciphertext attacks (also called a lunchtime attack or a midnight attack), where an attacker has access to the deciphering oracle only *before* seeing the ciphertext he attempts to decrypt itself. This is a less satisfying model of attacks inherent in real-life applications of cryptosystems, as illustrated by Zheng and Seberry [25]. One of the most severe type of attack against a public key cryptosystem is an adaptively chosen ciphertext attack, where an attacker is allowed to have access to the deciphering algorithm even *after* seeing the target ciphertext. Note that, in an adaptively chosen ciphertext attack, an attacker can feed the deciphering algorithm with the ciphertexts correlated to the target ciphertext and obtain the matching plaintexts. Consequently, the Damgard's scheme described above can be completely broken under this model of attacks as follows : For a given ciphertext $c = (c_1, c_2, c_3)$, an attacker feeds the deciphering algorithm with the modified ciphertext $c' = (c_1, c_2, c_3')$ where $c_3' = c_3 \oplus r$ with a random message r. Then he will get a message $m' = m \oplus r$ as an answer and thus can obtain the desired message m by computing $m = m' \oplus r$.

3.2 Zheng and Seberry's Extension

Zheng and Seberry [25] further extended and generalized the Damgard's approach in order to attain security against adaptively chosen ciphertext attacks. Extending Damgard's idea, they introduced the notion of sole-samplability, defined informally as follows : The space induced by function $f : D \to R$ is said to be *sole-samplable* if there is no other way to generate an element y in R than to first choose an element x in D and then to evaluate the function at the point x. This notion is very similar to the assumption used by Damgard in proving the security of his modified ElGamal cryptosystem under chosen ciphertext attacks. However, according to this notion, for the enciphering transformation to be a sole-samplable function, the whole ciphertext should be hard to generate without knowing the plaintext. Note that the space induced by the enciphering algorithm of Damgard's modified ElGamal cryptosystem is not sole-samplable due to the last part c_3 of the ciphertext.

Using different techniques in order to approximate sole-samplability of the enciphering transformation, they presented three methods for immunizing public key cryptosystems against adaptively chosen ciphertext attacks. Thanks to the generation of ciphertexts in a sole-samplable way along with probabilistic encryption, they could attain semantic security of their cryptosystems against adaptively chosen ciphertext attacks, under reasonable assumptions. More generally, they proved the following : "Assume that the space induced by the enciphering algorithm of a public key cryptosystem is sole-samplable. Then the cryptosystem is semantically secure against adaptively chosen ciphertext attacks, if it is semantically secure against chosen plaintext attacks." This is a quite obvious consequence resulting from the sole-samplability assumption on the enciphering transformation.

The main point of Zheng and Seberry's immunization methods is to make the enciphering transformation into a sole-samplable function by appending to each ciphertext a tag computed as a function of the message to be enciphered, much as a manipulation detection code (MDC) under encipherment or a message authentication code (MAC) in clear is used for message authentication. Note that a MDC is computed solely as a function of the message and transmitted under encipherment, while a MAC is computed from the message using a secret key and transmitted in clear [11]. The three immunization methods proposed by Zheng and Seberry differ only in the ways of generating tags. That is, they applied three basic tools that can be used to realize message authentication : one-way hash functions, universal classes of hash functions [6] [23] and digital signature schemes. Among them, the method based on one-way hash function (the resulting cryptosystem is denoted by C_{owh}) is first explained, since it is very simple, but most reflects the approach taken by them.

Assume h hashes arbitrary input strings into t-bit output strings. In this method, the ciphertext for an n-bit message m consists of a pair (c_1, c_2), where c_2 is the exclusive-or of the (n+t)-bit concatenated message m || h(m) with an (n+t)-bit output sequence of a pseudorandom number generator on a secret seed s and c_1 is a public key encryption of the seed s. In the decryption process, the deciphering algorithm first recovers a message m' || h(m)' from the ciphertext c_2 and outputs m' as the matching plaintext only when h(m') = h(m)'. Due to the involvement of a tag h(m), it is reasonable to assume that a polynomially bounded adaptively chosen ciphertext attacker cannot produce a ciphertext whose plaintext passes the check of the deciphering algorithm. This justifies the sole-samplability assumption and thus the ability to have access to the deciphering oracle gives no advantage to the attacker. Therefore, the cryptosystem is as secure under adaptively chosen ciphertext attacks as under chosen plaintext attacks. Now its semantic security against adaptively chosen ciphertext attacks follows immediately from the fact that the cryptosystem can be proved to be semantically secure under chosen plaintext attacks as in the Blum and Goldwasser's scheme [4].

Next we describe in more details the cryptosystem C_{sig} which is based on an adaptation of digital signature schemes, since it will be a basis of our proposed method. Let p be a large prime and let g be a generator of the multiplicative group $GF(p)^*$. A user A possesses a secret key $x_A \in_R [1, p-1]$ and the corresponding public key is computed as $y_A \equiv_p g^{x_A}$. Let G(n, s) be an n-bit output sequence produced on seed s by a pseudorandom number generator based on the intractability of computing discrete logarithms in finite fields [5] [12] [17] [14]. Assume that a user B wants to send in secret an n-bit message m to A. Then the enciphering and deciphering algorithms are as follows.

Enciphering Algorithm (user B) :

i) Choose $r_1, r_2 \in_R [1, p-1]$, where $\gcd(r_2, p-1) = 1$.

ii) Compute $s \equiv_p y_A^{r_1+r_2}$ and $z = G(n, s)$.

iii) Compute $c_1 \equiv_p g^{r_1}$, $c_2 \equiv_p g^{r_2}$,

$c_3 \equiv_{p-1} (h(m) - sr_1)/r_2$, and $c_4 = z \oplus m$.

iv) Send (c_1, c_2, c_3, c_4) to user A.

Deciphering Algorithm (user A) :

i) Compute $s' \equiv_p (c_1 c_2)^{x_A}$ and $z' = G(n, s')$ where $n = |c_4|$.

ii) Recover a plaintext m' by $m' = z' \oplus c_4$.

iii) Check that $g^{h(m')} \equiv_p c_1^{s'} c_2^{c_3}$.

If ok, output m' ; Else, output NULL.

The first three parts (c_1, c_2, c_3) of the ciphertext correspond to an adaptation of the ElGamal's signature scheme. Assuming that the hash function h produces output with almost uniform distribution, the ciphertext also leaks no partial information on the message m. Note that a tag is generated like a MDC in C_{owh} whereas in C_{sig} it is generated much like a MAC. Also note that in all three methods of Zheng and Seberry the validity check is based on the recovered plaintexts. In section 4, we will present a new immunization method using a ciphertext-based validity check.

3.3 Problem of Zheng-Seberry's Authentication Method

Zheng and Seberry also presented the method for adding authentication capability to their cryptosystems. Here, we point out that in their cryptosystems C_{owh} and C_{uhf} (based on universal class of hash functions), their method may fail to provide this capability under known plaintext attacks. First note that authentication scheme fails when a user different from the sender can create a message which the receiver will accept as being originated from the sender and that its security may be independent of the security of the cryptosystem used for secrecy. The reason of authentication failure in these two schemes is that tags are computed just as a function of the message to be enciphered and/or a pseudorandom sequence used for encryption, both of which are available under known plaintext attacks. This makes it possible for an attacker to reuse a pseudorandom sequence obtained from a ciphertext-plaintext pair to encrypt and falsely authenticate his chosen message.

Let's first consider the cryptosystem C_{owh} in which the ciphertext for an n-bit message m consists of (c_1, c_2) where $c_1 \equiv_p g^r$, $c_2 = G(n+t, s) \oplus (m \| h(m))$ and the secret seed s is computed by $s \equiv_p y_A^r$. As suggested by Zheng and Seberry, authentication capability may be added to this cryptosystem by the sender B computing a seed s as $s \equiv_p y_A^{r+x_B}$ where x_B is the secret key of user B. But in this case an attacker knowing the plaintext m for this ciphertext (mounting known plaintext attacks, for example) can obtain the (n+t)-bit pseudorandom sequence $G(n+t, s)$ by simply exclusive-oring $m \| h(m)$ with c_2 (i.e., $G(n+t, s) = c_2 \oplus (m \| h(m))$). Then he will be able to generate a legal ciphertext for his chosen message m' of length $n' \leq n$ as (c_1, c_2') where $c_2' = G(n'+t, s) \oplus (m' \| h(m'))$. Here, the pseudorandom sequence $G(n'+t, s)$ is just the first (n'+t) bits of $G(n+t, s)$. Clearly this ciphertext will be correctly deciphered and the receiver A will accept the plaintext m' as a valid message sent by user B.

The same attack can be applied to the cryptosystem C_{uhf} as well. To send an n-bit message m to user A using the cryptosystem C_{uhf}, a user B generates the ciphertext $c = (c_1, c_2, c_3)$ as : $c_1 \equiv_p g^r$, $c_2 = h_u(m)$, and $c_3 = z \oplus m$. Here, z denotes the first n bits of a pseudorandom sequence $G(n+k, s)$ of length (n+k) produced on seed $s \equiv_p y_A^r$ and u denotes the remaining k bits of $G(n+k, s)$. The function h_u denotes a hash function specified by a string u in a universal class of hash functions mapping n-bit input into t-bit output [6] [21] [23]. On receiving the ciphertext c, user A (the deciphering algorithm) computes $s' \equiv_p c_1^{x_A}$, generates $z' \| u' = G(n+k, s')$, computes $m' = z' \oplus c_3$ and finally checks that $h_{u'}(m') = c_2$. Only when the check is successful, does it output m' as the matching plaintext. Now consider the case where user B computes a secret seed s as $s \equiv_p y_A^{r+x_B}$ to provide authentication in addition, as Zheng and Seberry suggested, and suppose that an attacker obtained the ciphertext-plaintext pair (c, m). Then the attacker can extract from this pair the pseudorandom sequence z of length n. Therefore, as in C_{owh}, he can generate a legitimate ciphertext for a message of length n' ≤ n-k' where k' is the key length needed to specify a hash function in a universal class of hash functions mapping n'-bit input into t-bit output.

Finally, consider possible countermeasures against the described attack. First note that tags are computed as a function of the message alone in C_{owh} or as a function of the message and pseudorandom sequence in C_{uhf}, while in C_{sig} the random numbers used to generate a seed are also involved in generating a tag. Therefore, we can see that to defeat the attack, either a secret seed itself or random numbers used to compute a seed should also be involved in generating a tag. Simple countermeasure may be such that h(m) (resp. $h_u(m)$) is replaced by h(s || m) (resp. $h_u(s \| m)$) where in C_{owh} the secret seed s may be used as an initialization variable of the hash function h.

4 Proposed Immunization method

4.1 Motivation

Before presenting our immunization method, we first give the motivation that drives us to devise such a system that the decision on deciphering can be made solely based on the ciphertext. First recall that in an adaptively chosen ciphertext attack, an attacker can query the deciphering algorithm with any ciphertexts, *except* for the target ciphertext to decipher. However, in some applications, the attacker may feed the target ciphertext itself into the deciphering oracle and directly obtain the corresponding plaintext. This is seemingly a meaningless attack, but such a case may arise in group-oriented cryptosystems [8].

In group-oriented cryptosystems, the name of the destined receiver (or an indicator denoting the nature of the ciphertext, according to which the receiving group processes the ciphertext and distributes the partial results so that the legitimate group member can read the message) is usually accompanied by the ciphertext and in particular all or substantial part of the deciphering operations are carried out apart from the actual receiver. This separation of deciphering process from the actual receiver may make inside attacks easy, unless the ciphertext and the receiver's name are inalterably combined. This is because anyone inside the group can intercept the ciphertext not directed to him and then can change the receiver's name into his, if necessary in collusion with an outside colleague. Then all partial computations for decryption will be sent to him and thus he can decipher it. In fact, this inside attack may be mounted independently of the security of cryptosystems used. Ordinary authentication or digital signature schemes do not help to prevent an illegal modification of the receiver's name, since the modification should be detected, before decryption, by third parties not knowing the message. All that is required is to adapt the cryptosystem in such a way that any change in ciphertexts including the receiver's name can be detected before decryption by any third parties.

Motivated by the case considered above, we present a ciphertext-based immunization method and illustrate it by examples of the Diffie-Hellman /ElGamal scheme and the RSA scheme in the following two subsections. In the proposed schemes, any attempt to illegally modifying ciphertexts can be detected at the start of the deciphering process, which makes useless adaptively chosen ciphertext attacks.

4.2 Immunizing Diffie-Hellman/ElGamal Cryptosystem

Let p be a large prime (say, $\geq$ 512 bits) such that t-bit prime q divides p-1 (for example, t = 160) and let α be a generator of the unique subgroup $GF(q)^*$ of the multiplicative group $GF(p)^*$. Let h be a one-way hash function hashing arbitrary input strings into t-bit output strings (for example, secure hash standard [20]). Denote by G(n, s) an n-bit output sequence produced on a secret random seed s by a cryptographically strong pseudorandom number generator such as [12], [14] and [17]. As before, each user A possesses a secret key $x_A \in_R GF(q)^*$ and let $y_A \equiv_p \alpha^{x_A}$ be the corresponding public key. Assume user B wants to send in secret an n-bit message m to user A. Then they can proceed as follows.

Enciphering Algorithm (user B) :

i) Choose r_0, $r_1 \in_R [1, q\text{-}1]$.

ii) Compute $c_0 \equiv_p \alpha^{r_0}$, $c_1 \equiv_p \alpha^{r_1}$ and $s \equiv_p {y_A}^{r_1} c_0$.

iii) Compute $z = G(n, s)$, $c_2 = z \oplus m$, $c_3 = h(c_0 \,\|\, c_2)$ and $c_4 \equiv_q r_0 + c_3 r_1$.

iv) Send $c = (c_1, c_2, c_3, c_4)$ to user A.

Deciphering Algorithm (user A) :

i) Check that $c_3 = h(c_0' \,\|\, c_2)$ where $c_0' \equiv_p \alpha^{c_4} c_1^{-c_3}$.

If ok, continue ; Else, stop and output NULL.

ii) Compute $s \equiv_p c_1^{x_A} c_0'$ and $z = G(n, s)$ where $n = |c_2|$.

iii) Output m such that $m = z \oplus c_2$.

For efficiency reason, we applied the Schnorr's signature scheme [19], with a secret key chosen at random. If the pseudorandom number generator requires a generator g of $GF(p)^*$, such a g can be published in addition.

Security against chosen plaintext attacks : Since among the ciphertext the only message embedding part is $c_2 = z \oplus m$, we know that as far as it is computationally infeasible to compute the seed s from the ciphertext, no partial information on m will be released to a polynomial-time chosen plaintext attacker. First an algorithm for computing $s \equiv_p y_A^{r_1}\alpha^{r_0}$ from $y_A \equiv_p \alpha^{x_A}$, $c_1 \equiv_p \alpha^{r_1}$, $c_3 \in_R [1, 2^t\text{-}1]$ and $c_4 \equiv_q r_0 + c_3 r_1$ can be shown to be used to solve the Diffie-Hellman problem of computing $v \equiv_p \alpha^{x_1 x_2}$ from $v_1 \equiv_p \alpha^{x_1}$ and $v_2 \equiv_p \alpha^{x_2}$: On inputs $y_A = v_1$, $c_1 = v_2$, $c_3 = k_1 \in_R [1, 2^t\text{-}1]$ and $c_4 = k_2 \in_R GF(q)^*$, the algorithm will output $s \equiv_p v_1^{x_2}\alpha^r$ where r is an element of $GF(q)^*$ uniquely determined (by the algorithm) from the equation $k_2 \equiv_q r + k_1 x_2$. Therefore, one can compute $(v_1^{k_2} s^{-1})^{(k_1 - 1)^{-1}} \equiv_p v_1^{x_2} \equiv_p \alpha^{x_1 x_2}$, the desired result. Here we note that the relation $c_3 = h(c_0 \parallel c_2)$ where $c_0 \equiv_p \alpha^{c_4} c_1^{-c_3}$ does not affect the difficulty of computing a seed from the ciphertext. This shows that under the Diffie-Hellman assumption it is computationally infeasible to compute the seed s from the ciphertext and thus the cryptosystem is semantically secure against chosen plaintext attacks.

Security against adaptively chosen ciphertext attacks : In the following, we show that a polynomial-time attacker can extract no additional information on the plaintext, even if he is given the ability to have access to the deciphering algorithm under adaptively chosen ciphertext attacks. Then, the proposed cryptosystem will be as secure under adaptively chosen ciphertext attacks as under chosen plaintext attacks.

First we note that, in order to obtain any useful information on the plaintext corresponding to the ciphertext $c = (c_1, c_2, c_3, c_4)$ from the accessibility to the deciphering oracle, an attacker must generate a ciphertext $c' = (c_1', c_2', c_3', c_4')$ so that the following two conditions are satisfied at the same time : First, the ciphertext c' must satisfy the checking condition of step i) of the deciphering algorithm to get a non-null output. Second, the seed computed from the modified ciphertext by the deciphering algorithm must be equal to the seed used to produce the original ciphertext, since otherwise the output will be just another pseudorandom string indistinguishable from the truly random string. The second condition requires that no change in c_0 and c_1 should be made. Consequently, the attacker must solve the congruence equation $c_0 \equiv_p \alpha^{c_4} c_1^{-c_3} \equiv_p \alpha^{c_4'} c_1^{-c_3'}$ where $c_3' = h(c_0 \parallel c_2')$. Now solving this equation can be easily shown to be as difficult as solving the discrete logarithm problem : To solve $y \equiv_p \alpha^x$ in x using an algorithm for solving the equation, we provide as inputs $c_0 = c_1 = y$. Then from the outputs c_2' and c_4', we can compute the desired logarithm $x \equiv_q (1 + h(c_0 \parallel c_2'))^{-1} c_4'$.

In the above, we have shown that a polynomially bounded attacker gains no advantage from the accessibility to the deciphering oracle under adaptively chosen ciphertext attacks. This, together with semantic security under chosen plaintext attacks, shows that the proposed cryptosystem is semantically secure against adaptively chosen ciphertext attacks, under the Diffie-Hellman assumption.

The main difference of our immunization method from that of Zheng and Seberry is that a signature is generated based on the ciphertext, not on the message m. Consequently, everyone can check that ciphertexts are properly constructed according to the enciphering algorithm. This property may be useful in many applications, as exemplified by an application to group-oriented cryptosystems in subsection 4.1.

Adding authentication capability : Authentication capability is easily incorporated into the system : just replace (r_1, c_1) by the secret key/public key pair (x_B, y_B) of the sender (user B). In this case, the last two parts (c_3, c_4) of the ciphertext constitute a signature for the message embedding part c_2 of the ciphertext. Though the signature is generated on the ciphertext, not on the message m, generally considered as a bad practice for signing, it does not raise any problem as a signature for the message m (for example, the authorship problem raised in the CCITT X.509 token structure [15]). This is because message encryption and signature generation are tightly combined through the same random number r_0, which ensures that no one can produce a signature for the ciphertext c_2 without knowing the plaintext m.

Some additional cleartexts may be included as arguments of the hash function, e.g., the name of the sender and receiver, time information, and especially the indicator of the ciphertext when used for a group-oriented cryptosystem. As shown above, these cleartexts cannot be modified by an attacker. In particular, time information may be used to prevent the playback attack within the (predetermined) clock skew limit. We think that adding these additional cleartexts can provide much convenience in most communications, since the source and destination and the timeliness of a ciphertext can be easily verified by using only the ciphertext.

4.3 Immunizing RSA Cryptosystem

Let $N_A = p_A q_A$ be the modulus of user A in the RSA scheme, where p_A and q_A are large primes of the same size. Each user A chooses the public exponent e_A as a t-bit prime (t = 64, for example) and keeps secret d_A such that $e_A d_A = 1 \bmod \phi(N_A)$ where ϕ denotes the Euler phi function. Let h be a one-way hash function hashing arbitrary input strings into output values less than e_A. Let G(n, s) be the same as before. But it can be based on the modulus N_A of the receiver, such as the RSA/Rabin scheme based generators [1] [14] or the x^2 mod N generator [2] [22]. Of course, a common, possibly standardized, pseudorandom number generator may be used independently of the individual modulus. Assume that user B wants to send user A an n-bit message

m. Then the enciphering and deciphering algorithms are as follows.

Enciphering Algorithm (user B) :

i) Choose $s \in_R [1, N_A-1]$.

ii) Compute $c_1 = s^{3e_A} \bmod N_A$ and $z = G(n, s)$.

iii) Compute $c_2 = z \oplus m$, $c_3 = h(c_1 \| c_2)$ and $c_0 = s^{3c_3} \bmod N_A$.

iv) Send $c = (c_0, c_1, c_2)$ to user A.

Deciphering Algorithm (user A) :

i) Check that $c_0^{e_A} = c_1^{c_3'} \bmod N_A$ where $c_3' = h(c_1 \| c_2)$.

If ok, continue ; Else, stop and output NULL.

ii) Compute $s = c_1^{d_A'} \bmod N_A$ and $z = G(n, s)$,

where $d_A' = 3^{-1} d_A \bmod \phi(N_A)$ and $n = |c_2|$.

iii) Output m such that $m = z \oplus c_2$.

With the same argument as before, under the assumption that RSA is secure, the ciphertext (c_0, c_1, c_2) leaks no partial information on m and thus the cryptosystem is semantically secure against chosen plaintext attacks. Under adaptively chosen ciphertext attacks, an attacker should not change the second part c_1 of the ciphertext in order to get a useful output from the accessibility to the deciphering algorithm. Therefore the attacker is faced with the problem of solving $c_0^{e_A} = c_1^{h(c_1 \| c_2)} \bmod N_A$ in c_0 and c_2 for a fixed c_1. Solving this equation can be shown to be at least as difficult as inverting the RSA function : To compute x such that $y = x^{e_A} \bmod N_A$, one provides as inputs e_A, N_A and y to an algorithm for solving the equation, which then will output c_0 and c_2 such that $c_0 = x^{h(y \| c_2)} \bmod N_A$. Now one can easily compute the desired number x from two equations $y = x^{e_A} \bmod N_A$ and $c_0 = x^{h(y \| c_2)} \bmod N_A$ using the extended Euclidean algorithm, since e_A and $h(y \| c_2)$ is relatively prime. This shows that an algorithm for computing c_0 and c_2 from $c_0^{e_A} = c_1^{h(c_1 \| c_2)} \bmod N_A$ for a fixed c_1 can be used to invert the RSA function. Therefore, we have shown that under the assumption that RSA is secure, the above cryptosystem can be proved to be secure against adaptively chosen ciphertext attacks.

The cryptosystem can also provide authentication capability. This can be done by the sender generating c_0 as follows. Assume user B also has the RSA keys, public key (N_B, e_B) and secret key d_B. Then he can compute c_0 by $c_0 = (s^{3c_3} \bmod N_A)^{d_B} \bmod N_B$ where we assume that $N_B > N_A$. The checking condition can be changed accordingly : check that $(c_0^{e_B} \bmod N_B)^{e_A} = c_1^{c_3'} \bmod N_A$ with $c_3' = h(c_1 \| c_2)$.

5 Conclusion

This paper presented a new method for immunizing public key cryptosystems against adaptively chosen ciphertext attacks, which is illustrated by examples of the Diffie-Hellman/ElGamal cryptosystem and the RSA cryptosystem. In the proposed immunization method, everyone can check that ciphertexts are properly constructed according to the enciphering algorithm. This property is particularly useful for an application to group-oriented cryptosystems, where deciphering operations are separated from the actual receiver and the main threat is an illegal modification of the indicator by an inside group member. We also pointed out that in some of the cryptosystems presented by Zheng and Seberry the method for adding authentication capability may fail just under known plaintext attacks and presented a simple countermeasure.

References

[1] W.Alexi, B.Chor, O.Goldreich and C.P.Schnorr, *RSA and Rabin functions : certain parts are as hard as the whole*, SIAM J. Computing vol.17 no.2 (1988), 194-208.

[2] L.Blum, M.Blum and M.Shub, *A simple unpredictable pseudo-random number generator*, SIAM J. Computing vol.15 no. 2 (1986), 364-383.

[3] M.Blum, P.Feldman and S.Micali, *Non-interactive zero-knowledge proof systems and applications*, Proc. 20th Annual ACM Symposium on Theory of Computing (STOC) (1988), 103-112.

[4] M.Blum and S.Goldwasser, *An efficient probabilistic public key encryption scheme which hides all partial information*, Advances in Cryptology - Crypto'84, Lecture Notes in Computer Science vol.196, Springer-Verlag (1985), 289-299.

[5] M.Blum and S.Micali, *How to generate cryptographically strong sequences of pseudo-random bits*, SIAM J. Computing vol.13 no.4 (1984), 850-864.

[6] J.Carter and M.Wegman, *Universal classes of hash functions*, J. Computer and System Sciences vol.18 (1979), 143-154.

[7] I.Damgard, *Towards practical public key systems secure against chosen ciphertext attacks*, Advances in Cryptology - Crypto'91, LNCS vol.576, Springer-Verlag (1992), 445-456.

[8] Y.Desmedt, *Society and group oriented cryptography : a new concept*, Advances in Cryptology - Crypto'87, LNCS vol.293, Springer-Verlag (1988), 120-127.

[9] Z.Galil, S.Haber and M.Yung, *Symmetric public key cryptosystems*, submitted to J. Cryptology.

[10] S.Goldwasser and S.Micali, *Probabilistic encryption*, J. Computer and System Sciences vol.28 no.2 (1984), 270-299.

[11] R.R.Jueneman, S.M.Matyas and C.H.Meyer, *Message authentication with manipulation detection codes*, Proc. 1983 Symposium on Security and Privacy, IEEE Computer Society Press, 33-54.

[12] D.L.Long and A.Wigderson, *The discrete logarithm hides O(log n) bits*, SIAM J. Computing vol.17 no.2 (1988), 363-372.

[13] S.Micali, C.Rackoff and B.Sloan, *The notion of security for probabilistic cryptosystems*, SIAM J. Computing vol.17 no.2 (1988), 412-426.

[14] S.Micali and C.P.Schnorr, *Efficient, perfect polynomial random number generators*, J. Cryptology vol.3 no.3 (1991), 157-172.

[15] C.Mitchell, M.Walker and D.Rush, *CCITT/ISO standard for secure message handling*, IEEE J. Selected Areas on Commun. vol.7 no.4 (1989), 517-524.

[16] M.Naor and M.Yung, *Public key cryptosystems provably secure against chosen ciphertext attacks*, Proc. 22th Annual ACM Symposium on Theory of Computing (STOC) (1990), 427-437.

[17] R.Peralta, *Simultaneous security of bits in the discrete log*, Advances in Cryptology - Eurocrypt'85, LNCS vol.219, Springer-Verlag (1986), 62-72.

[18] C.Rackoff and D.Simon, *Non-interactive zero-knowledge proof of knowledge and chosen ciphertext attacks*, Advances in Cryptology - Crypto'91, LNCS vol.576, Springer-Verlag (1992), 433-444.

[19] C.P.Schnorr, *Efficient signature generation by smart cards*, J. Cryptology vol.4 no.3 (1991), 161-174.

[20] *A proposed federal information processing standard for secure hash standard*, Federal Register Announcement (Jan. 31, 1992), 3747-3749.

[21] D.R.Stinson, *Combinatorial techniques for universal hashing*, submitted to J. Computer and System Sciences.

[22] U.V.Vazirani and V.V.Vazirani, *Efficient and secure pseudo-random number generation*, Advances in Cryptology - Crypto'84, LNCS vol.196, Springer-Verlag (1985), 193-202.

[23] M.Wegman and J.Carter, *New hash functions and their use in authentication and set equality*, J. Computer and System Sciences vol.22 (1981), 265-279.

[24] A.C.Yao, *Theory and applications of trapdoor functions*, Proc. 23rd Annual IEEE Symposium on Foundations of Computer Science (FOCS), IEEE Computer Society Press (1982), 80-91.

[25] Y.Zheng and J.Seberry, *Practical approaches to attaining security against adaptively chosen ciphertext attacks*, Proc. Crypto'92.

Attacks on the Birational Permutation Signature Schemes

Don Coppersmith
IBM Research
T. J. Watson Research Center
Yorktown Heights, NY 10598

Jacques Stern, Serge Vaudenay
Laboratoire d'Informatique
Ecole Normale Supérieure
75230 Paris, France

Abstract. Shamir presents in [3] a family of cryptographic signature schemes based on birational permutations of the integers modulo a large integer N of unknown factorization. These schemes are attractive because of the low computational requirements, both for signature generation and signature verification. However, the two schemes presented in Shamir's paper are weak. We show here how to break the first scheme, by first reducing it algebraically to the earlier Ong-Schnorr-Shamir signature scheme, and then applying the Pollard solution to that scheme. We then show some attacks on the second scheme. These attacks give ideas which can be applied to schemes in this general family.

1 The first scheme

The public information in Shamir's first scheme consists of a large integer N of unknown factorization (even the legitimate users need not know its factorization), and the coefficients of $k-1$ quadratic forms $f_2, \cdots, f_k$ in k variables $x_1, \cdots, x_k$ each. Each of these quadratic forms can be written as

$$f_i = \sum_{j,\ell} \alpha_{ij\ell} x_j x_\ell \tag{1}$$

where i ranges from 2 to k and the matrix $\alpha_{ij\ell}$ is symmetric i.e. $\alpha_{ij\ell} = \alpha_{i\ell j}$.

The secret information is a pair of linear transformations. One linear transformation B relates the quadratic forms $f_2, \cdots, f_k$ to another sequence of quadratic forms $g_2, \cdots, g_k$. The second linear transformation A is a change of coordinates that relates the variables $(x_1, \cdots, x_k)$ to a set of "original" variables $(y_1, \cdots, y_k)$. Denoting by Y the column vector of the original variables and by X the column vector of the new variables, we can simply write $Y = AX$.

Of course, the coefficients of A and B are known only to the legitimate user. The trap-door requirements are twofold: when expressed in terms of the original variables $y_1, \cdots, y_k$, the quadratic form g_2 is computed as:

$$g_2 = y_1 y_2 \tag{2}$$

and the subsequent g_i's, $3 \leq i \leq k$ are *sequentially linearized*, i.e. can be written

$$g_i(y_1, \cdots, y_k) = l_i(y_1, \cdots, y_{i-1}) \times y_i + q_i(y_1, \cdots, y_{i-1}) \tag{3}$$

where l_i is a linear function of its inputs and q_i a quadratic form.

To sign a message M, one hashes M to a $k-1$-tuple $(f_2, \cdots, f_k)$ of integers modulo N, then finds a sequence $(x_1, \cdots, x_k)$ of integers modulo N satisfying (1). This is easy from the trap-door.

We let A_i, $2 \leq i \leq k$ denote the $k \times k$ symmetric matrix of the quadratic form f_i, namely:

$$A_i = (\alpha_{ij\ell})_{1\leq j\leq k, 1\leq \ell\leq k} \tag{4}$$

The kernel K_i of g_i is the kernel of the linear mapping whose matrix is A_i. It consists of vectors which are orthogonal to all vectors with respect to g_i. The rank of the quadratic form g_i is the rank of A_i. It is the dimension of K_i as well as the unique integer r such that g_i can be written as a sum of squares of r independent linear functionals. Actually, all this is not completely accurate as N is not a prime number and therefore $\mathbf{Z}/N$ is not a field. This question is addressed at the end the paper and, meanwhile, we ignore the problem.

An easy computation shows that K_i is the subspace defined in terms of the original variables by the equations

$$y_1 = \cdots = y_i = 0 \tag{5}$$

It follows from this that
i) K_i is decreasing
ii) the dimension of K_i is $k - i$
iii) any element of K_{i-1} not in K_i is an isotropic element wrt g_i, which means that the value of g_i is zero at this element.

We will construct a basis b_i of the k-dimensional space, such that $b_{i+1}, \cdots, b_k$ spans K_i for $i = 2, \cdots, k-1$. The main problem we face is the fact that the g_i's and therefore the K_i's are unknown. In place, we know the f_i's. We concentrate on the (unknown) coefficient δ_i of g_k in the expression of f_i, i.e. we write

$$f_i = \delta_i g_k + \sum_{j=2}^{k-1} \beta_{ij} g_j \tag{6}$$

As coefficients have been chosen randomly, we may assume that δ_k is not zero. Let $i < k$. Consider the quadratic form $Q_i(\lambda) = f_i - \lambda f_k$. When $\lambda = \delta_i/\delta_k$, this form has a non-trivial kernel and therefore δ_i/δ_k is a root of the polynomial $P_i(\lambda) = det(Q_i(\lambda))$. This is not enough to recover the correct value of λ. Computing the matrix of $Q_i(\lambda)$ for $\lambda_i = \delta_i/\delta_k$ in the basis corresponding to the original coordinates $y_1, \cdots, y_k$ yields the following

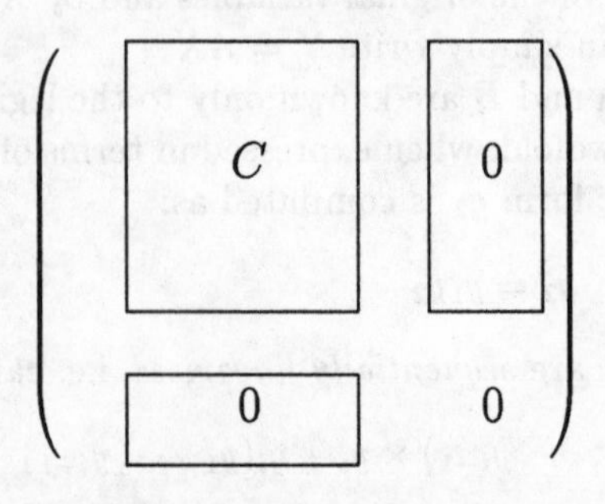

In the same basis, the matrix of $Q_i(\lambda)$ for any λ, can be written as

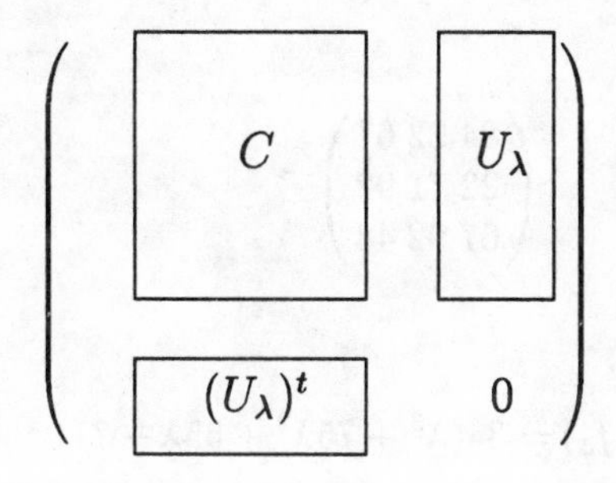

We observe that U_λ is linear in λ and vanishes at λ_i. Since determinants can be computed up to a multiplicative constant in any basis, it follows that $(\lambda-\lambda_i)^2$ factors out in $P_i(\lambda)$. Thus the correct value of λ_i can be found by observing that it is a double root of the polynomial equation $P_i(\lambda) = 0$. This double root is disclosed by taking the g.c.d. $\pmod N$ of P_i and P_i' with respect to λ. We find a linear equation in λ, from which we easily compute λ_i.

Once all coefficients λ_i have been recovered, we set for $i = 2, \cdots, k-1$

$$\tilde{f}_i = f_i - \lambda_i f_k \qquad i < k \tag{7}$$

and $\tilde{f}_k = f_k$. We note that all quadratic forms $\tilde{f}_i$ have kernel K_{k-1}. This allows to pick a non-zero vector b_k in K_{k-1}. The construction can then go on inductively in the quotient space of the k-dimensional space by the vector spanned by $\{b_k\}$ with $\tilde{f}_2, \cdots, \tilde{f}_{k-1}$ in place of $f_2, \cdots, f_k$.

At the end of the recursive construction, we obtain a sequence b_i, $3 \leq i \leq k$ such that $b_{i+1}, \cdots, b_k$ spans K_i for $i = 2, \cdots, k-1$ and a sequence of quadratic forms $\tilde{f}_2, \cdots, \tilde{f}_k$ such that

i) $\tilde{f}_i$ has kernel K_i

ii) b_i is an isotropic element wrt $\tilde{f}_i$

Choosing b_1, b_2 at random, we get another set of coordinates $z_1, \cdots, z_k$ such that

i) $\tilde{f}_2$ is a quadratic form in the coordinates z_1, z_2

ii) $\tilde{f}_3, \cdots, \tilde{f}_k$ is sequentially linearized

The rest is easy. From a sequence of prescribed values for $f_2, \cdots, f_k$, we can compute the corresponding values of $\tilde{f}_2, \cdots, \tilde{f}_k$. Next, we can find values of $\{z_1, z_2\}$ achieving a given value of $\tilde{f}_2 \pmod N$ in exactly the same way as the Pollard solution of the Ong-Schnorr-Shamir scheme [2]. Then, values for $z_3, \cdots, z_k$ achieving given values of $\tilde{f}_3, \cdots, \tilde{f}_k$ are found by successively solving $k-2$ linear equations. Finally, the values of $z_1, \cdots, z_k$ can be translated into values of $x_1, \cdots, x_k$.

Example. In Shamir's paper [3], an example is given with $N = 101$. (We use 101 to maintain consistency with Shamir's paper, even though 101 is prime, while N should be composite. We treat 101 as a number of unknown factorization; in particular we never solve nonlinear equations mod 101.)

$$v_2 = 78x_1^2 + 37x_2^2 + 6x_3^2 + 54x_1x_2 + 19x_1x_3 + 11x_2x_3 \pmod{101}$$

$$v_3 = 84x_1^2 + 71x_2^2 + 48x_3^2 + 44x_1x_2 + 33x_1x_3 + 83x_2x_3 \pmod{101}$$

Matrices of f_2, f_3 are as follows

$$\begin{pmatrix} 78 & 27 & 60 \\ 27 & 37 & 56 \\ 60 & 56 & 6 \end{pmatrix} \qquad \begin{pmatrix} 84 & 22 & 67 \\ 22 & 71 & 92 \\ 67 & 92 & 48 \end{pmatrix}$$

We get:

$$P(\lambda) = det(f_2 - \lambda f_3) = 34(\lambda^3 + 75\lambda^2 + 55\lambda + 71) \tag{8}$$

$$P'(\lambda) = \lambda^2 + 50\lambda + 52 \tag{9}$$

$$\gcd(P, P') = \lambda - 63 \tag{10}$$

We let

$$\tilde{f}_2 = f_2 - 63f_3 \quad ; \quad \tilde{f}_3 = f_3 \tag{11}$$

The kernel of $\tilde{f}_2$ is spanned by vector $b_3 = (31, 12, 1)^t$. We pick $b_2 = (0, 1, 0)^t$ and $b_1 = (1, 31, 0)^t$. We get, in the corresponding coordinates z_1, z_2, z_3:

$$\tilde{f}_2 = 26z_1^2 + 8z_2^2 \quad ; \quad \tilde{f}_3 = z_3(26z_1 + 20z_2) + 90z_1^2 + 2z_1z_2 + 71z_2^2 \tag{12}$$

2 The second scheme

We now treat Shamir's [3] second scheme. The ideas developed in this section will have general applicability.

Throughout, we will pretend we are working in $\mathbf{Z}/p$ rather than $\mathbf{Z}/N$.

We treat first the case $s = 1$. We begin with k variables $y_1, y_2, \ldots, y_k$, with k odd. These are subjected to a secret linear change of variables which gives $u_i = \sum_j a_{ij}y_j, i = 1, 2, \ldots, k$, with the matrix $A = (a_{ij})$ secret. The products u_iu_{i+1}, including u_ku_1, are subjected to a second secret linear transformation $B = (b_{ij})$, so that $v_i = \sum_j b_{ij}u_ju_{j+1}, i = 1, 2, \ldots, k-1$. The public key is the set of coefficients $(c_{ij\ell})$ expressing v_i in terms of pairwise products y_jy_ℓ, for $1 \leq i \leq k-1$,

$$v_i = \sum_{j,\ell} c_{ij\ell}y_jy_\ell, 1 \leq i \leq k-1, c_{ij\ell} = c_{i\ell j} \tag{13}$$

(Here i is ranging to $k-1$, so we have discarded $s = 1$ of the v_i.)

The first step in our solution: linear combinations of the v_i are linear combinations of the u_iu_{i+1}, but they form only a subspace of dimension $k-1$. Some linear combinations of the v_i,

$$v_1 + \delta v_2 + \sum_{3 \leq j \leq k-1} \epsilon_j v_j \tag{14}$$

will be quadratic forms in the y_i of rank 2. A computation shows that the only linear combinations of the products u_iu_{i+1} of rank 2 are of the form

$$\alpha_i u_{i-1}u_i + \beta_i u_i u_{i+1} = u_i(\alpha_i u_{i-1} + \beta_i u_{i+1}), \tag{15}$$

for any values of α_i, β_i, i. Because the v_j span a subspace of codimension 1, and because we are further restricting to one lower dimension by the choice of the multiplier 1 for v_1 in the linear combination, we find that for each i there will be one pair (α_i, β_i) and one set of coefficients (δ, ϵ_j) such that

$$\alpha_i u_{i-1}u_i + \beta_i u_i u_{i+1} = u_i(\alpha_i u_{i-1} + \beta_i u_{i+1}) = v_1 + \delta v_2 + \sum_{3\le j\le k-1} \epsilon_j v_j. \tag{16}$$

The condition of being rank 2 is an algebraic condition: setting

$$v_1 + \delta v_2 + \sum_{3\le j\le k-1} \epsilon_j v_j = \sum_{ij} \tau_{ij} y_i y_j, \tag{17}$$

with $\tau_{ij} = \tau_{ji}$, we find that each 3×3 submatrix of the (τ_{ij}) has vanishing determinant. Each of these determinants is a polynomial equation in δ, ϵ_j. Use resultants to eliminate ϵ_j from this family of polynomial equations (in the ring $\mathbf{Z}/N$) and find a single polynomial F of degree k satisfied by δ. We also find ϵ_j as polynomials in δ, by returning to the original equations and eliminating the variables $\epsilon_i, i \neq j$.

Thus each solution δ to $F(\delta) = 0$ gives rise to a linear combination of v_j which is of rank 2. The root δ corresponds to that index i for which

$$v_1 + \delta v_2 + \sum_{3\le j\le k-1} \epsilon_j v_j = u_i(\alpha_i u_{i-1} + \beta_i u_{i+1}). \tag{18}$$

We will indicate this correspondence by writing $\delta = \delta_i$.

For each solution $\delta = \delta_i$, the rows of the resulting matrix (τ_{ij}) span a subspace $Y(\delta_i) = Y_i$ of $\mathbf{Z}_p^k$ of rank 2; namely, Y_i is spanned by u_i and $\alpha_i u_{i-1} + \beta_i u_{i+1}$.

Observe that u_i, u_{i+2}, and $(\alpha_{i+1}u_i + \beta_{i+1}u_{i+2})$ are linearly related, as are u_i, u_{i-2}, and $(\alpha_{i-1}u_{i-2} + \beta_{i-1}u_i)$. So

$$u_i \in Y_i \cap (Y_{i+1} + Y_{i+2}) \cap (Y_{i-1} + Y_{i-2}) \tag{19}$$

This is an algebraic relation among δ_{i-2}, δ_{i-1}, δ_i, δ_{i+1}, and δ_{i+2}.

We formulate the relation as the vanishing of several determinants, and reduce the resulting ideal by factoring out any occurrences of $(\delta_i - \delta_j), i \neq j$ to assure that δ_i, δ_j are really two different solutions. That is, we consider the ideal formed by $F(\delta_i)$, $(F(\delta_i) - F(\delta_j))/(\delta_i - \delta_j)$, etc., and the various determinants. We apply the Groebner basis and the Euclidean algorithm to this ideal to find a basis.

Only multiples of some u_i satisfy such a relation (19) over $\mathbf{Z}/p$, namely, two different linear relations. We fix a multiple of each u_i by normalizing u_i to have first coordinate 1. The linear relations serve to define u_i in terms of δ_i.

By similar argument, there is a quadratic equation expressing δ_{i+1} in terms of δ_i, whose two solutions are δ_{i+1} and δ_{i-1}. The algebraic condition is that the corresponding spaces Y_i, Y_{i+1} are in two different triples of subspaces enjoying linear relations:

$$rank(Y_i + Y_{i+1} + Y_{i+2}) = rank(Y_i + Y_{i+1} + Y_{i-1}) = 5 \tag{20}$$

We represent the solution of the quadratic equation by τ, and say that (δ, τ) generates a pair of 'adjacent' elements (u_i, u_{i+1}) (elements which are multiplied together in the original signature). We think of δ as generating an extension of degree k over $\mathbf{Z}/N$, and τ as generating an extension of degree 2 over $\mathbf{Z}/N[\delta]/F(\delta)$. The ability to distinguish the unordered pairs of 'adjacent' roots $\{\delta_i, \delta_{i+1}\}$ makes the system similar, in spirit, to a Galois extension of $\mathbf{Q}$ whose Galois group is the dihedral group on k elements. We will call on this analogy later. (Remark: it is only an analogy, because δ and τ really are elements of the ground fields.)

We can get the missing kth equation

$$v'_k = \sum_i u_i u_{i+1}. \tag{21}$$

The coefficients of v'_k in terms of $y_j y_\ell$ ostensibly depend on δ_i and on the pairings (δ_i, δ_{i+1}), or equivalently on (δ, τ). But the coefficients would come out the same no matter which solution (δ, τ) were chosen, that is, no matter whether we assigned the ordering $(1, 2, 3, \ldots, k)$ or $(3, 2, 1, k, k-1, \ldots, 4)$ to the solutions u_i. This means that the coefficients will be in fact independent of (δ, τ). They will be expressible in terms of only the coefficients of the original v_i, $1 \le i \le k$. This is because they are symmetric (up to dihedral symmetry) in the solutions δ_i.

The arguments here are analogous to those of Galois theory. Each coefficient c of v'_k is expressed as

$$c = \sum_{0 \le i < k, 0 \le j \le 1} w_{ij} \delta^i \tau^j \tag{22}$$

For each of $2k$ different choices of (δ, τ) the value of c comes out the same. Treating (22) as $2k$ linear equations in the $2k$ unknowns w_{ij}, with coefficients given by $\delta^i \tau^j$ for various choices of (δ, τ), we must find (if the matrix has full rank) that $w_{00} = c$, and $w_{ij} = 0$ for $(i, j) \neq (0, 0)$.

Now we wish to solve a particular signature. We are given the values $v_1, \ldots, v_{k-1}$, and we assign an arbitrary value to v'_k. We have the equations relating v_i to $u_j u_{j+1}$:

$$v_i = \sum_j b'_{ij} u_j u_{j+1}, \tag{23}$$

where b'_{ij} depends on δ_j. Select (symbolically) one pair (δ, τ) to fix the first two solutions (u_1, u_2), and compute the others in terms of (δ, τ). Then we have $b'_{ij} u_j u_{j+1}$ depending only on (δ, τ).

Invert this matrix b' to solve for $u_j u_{j+1}$ in terms of the given v_i and (δ, τ).

Now assign

$$u_1 = \xi, \tag{24}$$

where ξ is an unknown. Compute

$$u_2 = \frac{(u_1u_2)}{\xi}, u_3 = \frac{\xi(u_2u_3)}{(u_1u_2)}, u_4 = \frac{(u_1u_2)(u_3u_4)}{\xi(u_2u_3)}, \ldots, u_1 = \frac{(u_1u_2)(u_3u_4)\ldots(u_ku_1)}{\xi(u_2u_3)\ldots(u_{k-1}u_k)} \tag{25}$$

The last equation gives a quadratic equation which ξ must satisfy:

$$(u_1u_2)(u_3u_4)\ldots(u_ku_1) = \xi^2(u_2u_3)\ldots(u_{k-1}u_k) \tag{26}$$

We do not solve for ξ (we cannot). So now we have three algebraic unknowns: δ, τ, ξ, of successive degrees $k, 2, 2$.

These equations give u_i in terms of δ, τ, ξ. Notice that each u_i is an odd function of ξ: either ξ times a function of (δ, τ) or ξ^{-1} times a function of (δ, τ). We also have u_i as linear combinations of y_j with coefficients depending on (δ, τ). Solve for y_j in terms of (δ, τ, ξ), and note that y_j is again an odd function of ξ.

Now each product y_jy_ℓ will be a function only on (δ, τ), since it will be an even function of ξ, and we know ξ^2 in terms of (δ, τ). But again the value y_jy_ℓ will be independent of the dihedral ordering $(1, 2, 3, \ldots, k)$ versus $(3, 2, 1, k, k-1, \ldots, 4)$, and thus independent of the choice of solutions (δ, τ). That means, by standard Galois theory arguments, that (δ, τ) will not appear in the expressions of y_jy_ℓ.

So we have found the products y_jy_ℓ in terms of the given coefficients, the given values $v_1, v_2, \ldots, v_{k-1}$, and the assumed value v'_k. We have given a valid signature.

3 Comments and extensions

3.1 Working mod N versus working mod p

Some justification is needed to go from calculations mod$p$ to calculations mod N. In section 1, we basically use tools from linear algebra such as Gaussian elimination or determinants. Thus all computations go through regardless the fact that N is composite. The situation is a bit more subtle in section 2. For instance, F has k solutions mod$p$ but k^2 solutions mod$N$, each obtained by mixing some solution mod$p$, with some solution mod$q$. But if we consider only the image, mod p, of our calculations mod N, things are all right: the symmetric functions of the k roots of a polynomial are expressible in terms of the coefficients of the polynomial, and the expressions of the products y_jy_ℓ in terms of the coefficients of the public key are valid mod p. They are also valid mod q, and the Chinese remainder theorem suffices to make them valid mod$N$. This in spite of the fact that a solution δ of F mod N might well mix different solutions δ_i mod p and δ_j mod q. Since we never explicitly solve for δ, but only work with it symbolically and use the fact that $F(\delta) = 0$ mod N, we never are in danger of factoring N.

3.2 Extension to the case $s > 1$ (Sketch)

The case $s > 1$ is more complicated. Suppose again that we have k variables $y_1, y_2, \ldots, y_k$, with k odd, whose pairwise products constitute the signature, and that the hashed message has $k - s$ quantities $v_1, v_2, \ldots, v_{k-s}$, together with coefficients $c_{ij\ell}$ expressing v_i in terms of $y_j y_\ell$. Suppose for simplicity that $s > 1$ is odd, so that $k - s$ is even.

Some linear combinations of the $k - s$ quadratic forms v_i will have rank $s + 1$. Namely, for each index set $I \subseteq \{1, 2, \ldots, k\}$ of size $(s+1)/2$ such that $\forall i, j \in I$: $|i - j| \geq 2$, there is such a linear combination of the form

$$\sum_{i \in I} u_i(\alpha_{iI} u_{i-1} + \beta_{iI} u_{i+1}) \tag{27}$$

The number of such index sets I is

$$\frac{k}{\frac{s+1}{2}} \binom{k - \frac{s+3}{2}}{\frac{s-1}{2}} \tag{28}$$

There are more than k linear combinations, leading to increased complication. The space Y_I, spanned by rows of the corresponding quadratic form, contains u_i for each index $i \in I$. So each u_i is in the intersection of a large number of subspaces Y_I, and hopefully only multiples of u_i will be in such an intersection. This algebraic condition should distinguish the u_i, hopefully indexing them by the roots δ of some polynomial $F(\delta)$ of degree k. Pairs $\{u_i, u_{i+2}\}$ of solutions with index differing by 2 should be distinguished by appearing together in many different subspaces Y_I. Using this we would be able to distinguish pairs $\{u_i, u_{i+1}\}$. We would fabricate the missing equations: for $j = k - s + 1, \ldots, k$, let $u'_{i(j)}$ be a multiple of u_i, normalized to have a 1 in position j, and set $v'_j = \sum_i u'_{i(j)} u'_{i+1(j)}$.

3.3 The case k=3, s=1

In the special case $k = 3$, $s = 1$, where we must satisfy two quadratic equations in three variables, we can employ an *ad hoc* method, since the methods outlined above don't work. Take a linear transformation of the two quadratic equations so that the right-hand side of one equation vanishes; that is, if the given values are v_1 and v_2, take v_2 times the first equation minus v_1 times the second. This gives a homogeneous quadratic equation in three variables y_1, y_2, y_3:

$$\sum_{ij} c_{ij} y_i y_j = 0 \tag{29}$$

The second equation is inhomogeneous:

$$\sum_{ij} d_{ij} y_i y_j = d_0 \tag{30}$$

By setting $z_1 = y_1/y_3$, $z_2 = y_2/y_3$ in (29), we obtain an inhomogeneous quadratic equation in two variables z_1, z_2. We can easily find an affine change of basis from z_1, z_2 to z_1', z_2' which transforms the equation to the form

$$c_{11}' {z_1'}^2 + c_{12}' z_1' z_2' + c_{22}' {z_2'}^2 = c_0' \bmod N \tag{31}$$

and a further linear change of variables to z_1'', z_2'' yielding

$$c_{11}'' {z_1''}^2 + c_{22}'' {z_2''}^2 = c_0'' \bmod N \tag{32}$$

which can be solved by the Pollard [2] attack on the Ong-Schnorr-Shamir [1] scheme. We find from this a set of ratios y_j/y_3, and, by extension, a set of ratios $y_i y_j / y_3^2$, satisfying (29). Setting $y_3^2 = \lambda$, the second equation (30) becomes a linear equation in λ. Thus we find a consistent set of pairwise products $y_i y_j$ satisfying the desired equations (29), (30).

3.4 Open questions

The birational permutation signature scheme has many instances, of which we have attacked only the first few examples. For a more complex instance of the scheme, the ideas of the present paper will still apply: the trap door conditions lead to algebraic equations on the coefficients of the transformations, and we hope to gather enough such equations to make it possible to solve them by g.c.d. or Groebner basis methods. But, for any specific instance, it remains to see whether the ideas of the present paper would be sufficient to mount an attack.

One general theme is that when solutions of the algebraic equations enjoy a symmetry, it makes the equations harder to solve, but we don't need to solve them, since the final solution will enjoy the same symmetry, and quantities symmetric in the roots of the equation can be expressed in terms of the coefficients of the equation alone, not in terms of the roots. When the roots fail to enjoy a symmetry, they can be distinguished by algebraic conditions, which yield further algebraic equations, and the Groebner basis methods have more to work with. This gives us hope that the methods outlined in this paper will apply with some generality to many instances of the birational permutation signature scheme.

References

1. H. Ong, C. P. Schnorr, and A. Shamir: A fast signature scheme based on quadratic equations. Proc. 16th ACM Symp. Theory of Computing, pp.208-216; 1984.
2. J. M. Pollard and C. P. Schnorr: An efficient solution of the congruence $x^2 + ky^2 = m \pmod{n}$. IEEE Trans. Inform. Theory vol IT-33 no 5, pp.702-709; Sept., 1987.
3. A. Shamir: Efficient signature schemes based on birational permutations. Manuscript March 1993. To appear, *Crypto 93*.

Interaction in Key Distribution Schemes* (Extended Abstract)

Amos Beimel** and Benny Chor ***

Department of Computer Science
Technion, Haifa 32000, Israel

Abstract. A (g, b) *key distribution scheme* allows conferences of g users to generate secret keys, so that disjoint coalitions of b users cannot gain any information on the key (in the information theoretic sense). In this work we study the relationships between interaction and space efficiency of key distribution schemes. We prove that interaction does not help in the context of *unrestricted schemes*. On the other hand, we show that for restricted schemes, which are secure for a limited number of conferences, interaction can substantially improve the space efficiency.

1 Introduction

A non-interactive key distribution scheme for conferences of size g which is secure against b "bad" users (denoted (g, b)-scheme) is a method in which an off-line server initially distributes private individual pieces of information to n users such that:

1. The pieces of every "good" conference G of g users determine a key, such that every user in G can reconstruct the key from his piece. This reconstruction requires no interaction (either among users or with the server).
2. Every "bad" coalition B of b users does not gain any information on the key of any disjoint conference G.

It is clear that non-interactive schemes require initial distribution of pieces of information to the users. The (space) efficiency of the scheme is measured by the cardinality of the domain of pieces. The cardinality is a function of the cardinality of the domain of possible keys, $|S|$, of the number of users, n, of the size of conferences g, and of the size of coalitions b.

Blom [1] was the first to consider non-interactive schemes for conference of size 2 and coalitions of size b. He presented an efficient $(2, b)$ scheme, based on MDS codes. Other works dealing with non-interactive schemes in our setting are [7, 9]. Matsumoto and Imai [8] suggest the use of symmetric linear functions for (g, b) schemes. Blundo, De Santis, Herzberg, Kutten, Vaccaro and Yung [2]

---

* Supported by the Fund for the Promotion of Research at the Technion.

** email: beimel@cs.technion.ac.il

*** email: benny@cs.technion.ac.il

present (g, b) schemes, based on symmetric multinomials. Their multinomials have g variables and degree at most b in each variable. The pieces in their scheme are taken from a domain of cardinality $|S|^{\binom{g+b-1}{g-1}}$ (where S is the domain of keys). For large values of g and b, this expression is quite large. However, using entropy arguments, Blundo et. al. [2] prove a tight lower bound on the cardinality of the domain of pieces. Therefore, their scheme is space-optimal. We use direct arguments (no entropy) to prove the same lower bound. Our proof has two advantages. First, it seems more intuitive and less technical. Second, it actually applies to a weaker notion of security, thereby providing a stronger result. This stronger result is used in proving our lower bound on interactive schemes which is described in the next paragraph.

The large lower bound (for big conferences or coalitions) raises the question whether interaction could be of help in reducing the size of pieces. Interaction has some subtle implications on the security requirement (see section 5 for details). Just like the non-interactive schemes, we require that even if all conferences interact in order to generate keys, these keys remain secure with respect to disjoint coalitions of size b. Since no secure channels among users can be assumed, interaction takes place via a broadcast media. One problem which arises is that the communication of one conference could leak information on the keys of *other* conferences. Therefore, we require that even if a "bad" coalition heard the communication of all the conferences, the coalition does not gain any information on keys of disjoint conferences. We argue that this is the right security requirement for interactive schemes. We prove that, regrettably, such unrestricted interactive schemes require pieces from a domain as large as non-interactive schemes.

This negative result motivates the introduction of ***restricted*** interactive schemes. These schemes can be used only for a limited number of conferences, whose identity is not known beforehand. We construct an efficient one-time secure scheme, where the size of the domain of pieces is of cardinality $|S|^{2+2\cdot(b-1)/g}$. This is a substantial improvement over the $|S|^{g+b-1}$ cardinality in the one-time secure interactive scheme of [2]. (The fact that this scheme is only one-time secure was not mentioned in [2]). Other, less efficient, one time secure interactive schemes are presented in [5, 6].

We contrast our results with known results in the computational model, where users are restricted to probabilistic polynomial time computations. Diffie and Hellman [3], in their pioneering work on public key cryptography, introduced an interactive scheme of key generation for conferences of size two[4]. This interactive scheme requires no server and no pieces. In this scheme a given communication *uniquely* determines the key, but it is (presumably) intractable for a third party to compute the key from the communication (of course, in our setting this information enables other users to find the conference key). On the other hand, even in the computational model, a non-interactive scheme requires

[4] Let p be a prime number, and let α be a primitive element in the field GF(p). User i (respectively j) chooses a random number $r_i \in \text{GF}(p)$ (respectively r_j) and sends the message $m_i = \alpha^{r_i}$ (respectively $m_j = \alpha^{r_j}$). The joint key of users i and j is $\alpha^{r_i \cdot r_j}$, which i easily computes from m_j and r_i using the equality $m_j^{r_i} = \alpha^{r_i \cdot r_j}$.

pieces taken from a domain which is at least as large as the domain of keys. So in the computational model, interaction does reduce the size of pieces, up to complete elimination. Fiat and Naor [4] present a non-interactive $(n, 1)$-scheme in the computational model. In their scheme, which is based on the assumed intractability of extracting root modulo composites, the domain of pieces has the same cardinality as the domain of keys. Recall that in the computationally unbounded model a non interactive $(n, 1)$-scheme requires that the domain of pieces is at least of cardinality $|S|^n$.

The remaining of this extended abstract is organized as follows. In section 2 we give formal definition of interactive and non-interactive schemes. Section 3 contains our proof of the lower bound for weak non-interactive schemes. In section 4 we use this result to prove a lower bound to unrestricted interactive schemes. In section 5 we introduce restricted interactive schemes, present an efficient construction, and prove some weak lower bounds.

2 Definition of Key Distribution Schemes

In this section, we present formal definition of *interactive* and *non-interactive* key distribution schemes. We start with the interactive schemes.

Definition 1. Let $\{1, \ldots, n\}$ be a set of users, g and b be positive integers such that $g + b \leq n$, S be a set of keys, and $\mathcal{P}$ be an a-priori probability distribution on S. Let R be a set of random inputs. For every i $(1 \leq i \leq n)$ let U_i denote a domain of pieces for user i. An *unrestricted interactive* (g, b) *key distribution scheme* (later denoted by (g, b) scheme) with n users and domain of keys S is a function $\mathcal{U} : R \rightarrow U_1 \times U_2 \times \ldots \times U_n$. A server distributes the vectors $\{\mathcal{U}(r) : r \in R\}$ to the users according to some a-priori probability distribution on the random inputs. We denote by $\mathcal{U}_i(r)$ the i-th coordinate of $\mathcal{U}(r)$. This coordinate is the piece of user i. When users of a set G (of cardinality g) wish to generate a conference key, each user in G chooses a local random input for this conference. Then, the users communicate among themselves over a broadcast channel. We denote the resulting communication by C_G (it is a function of the g pieces and the local random input). As the messages are sent over a broadcast channel, they can be heard by all the users (including the users *not* in G). The key distribution function $\mathcal{U}$ and the conversations satisfy the following requirements:

reconstruction requirement At the end of the conversation, each member of G can reconstruct a key from the conversation and his piece. The key that every member of G reconstructs is the same, and is denoted by $s_G(r, \vec{r_G})$, where r is the random input of the server, and $\vec{r_G}$ is the vector of random inputs of the users in G.

unrestricted security requirement Every coalition B of b (bad) users, having their pieces and knowing the conversations of *all* possible conferences, does not gain any information on the key of every subset G_0 such that $G_0 \cap B = \emptyset$. That is, for every vector of pieces $\langle u_1, \ldots, u_n \rangle$ which is distributed with positive probability, every set of random inputs $\vec{r_B}$ to coalition members, every possible key $s \in S$, and every possible consistent conversations $C_1, \ldots, C_{\binom{n}{g}}$ of all sets of cardinality g:

$$\Pr [\, s_{G_0}(r, \vec{r_{G_0}}) = s \,|\vec{r_B} \wedge \bigwedge_{j \in B} \mathcal{U}_j(r) = u_j \wedge \bigwedge_{|G|=g} C_G(\mathcal{U}_G(r), \vec{r_G}) = C_G \,] = \mathcal{P}(s)$$

Where the probability is taken over r – the random input of the server, and over $\vec{r}$ – the random inputs of all the users for all conferences. We denote by $\vec{r_A}$ the restriction of $\vec{r}$ to a set $A \subseteq \{1, \ldots, n\}$.

The security property implies that for every conference G of cardinality g, it holds that $\Pr [\ s_G(r, \vec{r_G}) = s \ ] = \mathcal{P}(s)$, where the probability is taken over r, the random input of the server,r, and $\vec{r_G}$ the random inputs of the users of G. In other words, the conference key of G is a random variable, which is distributed according to the a-priori probability distribution on the keys. It is *not* guaranteed that keys of different conferences are *independent* random variables. The security requirement does imply some independence between the keys. For example, it is possible to prove that every $b+1$ keys are independent. In the rest of this paper we assume that the a-priori probability of each key is positive. That is, for every key $s \in S$ it holds that $\mathcal{P}(s) > 0$.

We now define non-interactive schemes, which are a special case of interactive schemes.

Definition 2. A *non interactive* (g, b) *key distribution scheme* with n users and domain of keys S is a (g, b) scheme, in which every set G of cardinality g has a key which depends only on the vector of pieces (and not on any communication), and every user $i \in G$ can reconstruct G's key from his piece. In this case the random input of the server determines the key of every set G. That is, s_G is only a function of r.

We now consider a weakening of the security requirement. Instead of requiring that the conditional probability, given any pieces of a bad set B, of every key equals the a-priory probability, we will only require that this conditional probability is positive. We claim that this security requirement is *not* reasonable, since every bad set B could gain a lot of information. The reason we do define weak schemes is because we show that the lower bounds on the size of the pieces hold even for these weak schemes. To simplify this discussion, we will only consider non-interactive weak schemes.

Definition 3. A *weak non-interactive* (g, b) *key distribution scheme* is a non-interactive (g, b) scheme in which the security property is relaxed:

weak security property Let B be a coalition of b (bad) users, and let G be a conference of g (good) users, such that $G \cap B = \emptyset$. Then the users in B, having their pieces, can not rule out any key of G. That is, for every vector of pieces $\vec{u} = \langle u_1, \ldots, u_n \rangle$ that is dealt with positive probability, and every possible key $s \in S$, there exists a vector of pieces $\vec{u'}$ that agrees with $\vec{u}$ on the pieces of B, but the key of the set G according to the vector $\vec{u'}$ is s. Formally,

$$\Pr [\ s_G(r) = s \mid \bigwedge_{j \in B} \mathcal{U}_j(r) = u_j] > 0$$

where the probability is taken over the random input of the server.

It is obvious that unrestricted non-interactive schemes are a special case of weak non-interactive schemes. Therefore, every lower bound for weak schemes, implies the same lower bound for unrestricted non-interactive schemes.

3 Lower Bound for Non-Interactive Schemes

Blundo et. al. [2] prove a tight lower bound on the size of the pieces in every non-interactive key distribution scheme. Their proof is based on the entropy function, and does not seems to reveal the intuition behind this lower bound. We present a simpler proof of this lower bound, which is not based on entropy. Furthermore, this proof gives a stronger result, which we use in the sequel.

Theorem 4. [2] *Let $\mathcal{U}$ be a weak non-interactive (g, b) scheme with n users and domain of keys S. Let U_i be the domain of pieces of user i in $\mathcal{U}$. Then for every i $(1 \le i \le n)$:*

$$|U_i| \ge |S|^{\binom{g+b-1}{g-1}}$$

Proof. Consider a (g, b) scheme with a domain of keys S. Without loss of generality, we assume that there are *exactly* $g+b$ users, which we denote by $\{1, \ldots, g+b\}$. We prove the lower bound on the domain of pieces of user 1. Let $G_1, \ldots, G_\ell$ be all the sets of cardinality g that contain user 1, where $\ell = \binom{g+b-1}{g-1}$. Let $\vec{s} = \langle s_1, s_2, \ldots, s_\ell \rangle$ be any vector in S^ℓ. We claim that there exists a vector of pieces $\vec{u} = \langle u_1, \ldots, u_{g+b} \rangle$ (that is dealt with positive probability), such that for every $1 \le i \le \ell$ the key of the set G_i reconstructed from $\vec{u}$ equals s_i. Otherwise, let i be a maximal index such that there exist keys $s'_i, \ldots, s'_\ell \in S$ for which the vector $\vec{s'} = \langle s_1, \ldots, s_{i-1}, s'_i, \ldots, s'_\ell \rangle$ is the vector of keys for some possible vector of pieces $\vec{u}$. Such index $i \ge 1$ exists, since given any b pieces, each key is distributed according to the a-priori distribution. Consider the set $B = \{1, \ldots, g+b\} \setminus G_i$, which contains exactly b users. Since the set B intersects every G_j for $j \ne i$, then the users in B can compute the keys of the sets $G_1, \ldots, G_{i-1}, G_{i+1}, \ldots, G_\ell$. Therefore, the pieces from $\vec{u}$ of the users of B determine that the keys of $G_1, \ldots, G_{i-1}$ are $s_1, \ldots, s_{i-1}$ respectively. By the maximality of i it follows that:

$$\Pr[\ s_{G_i}(r) = s_i \mid \bigwedge_{j \in B} \mathcal{U}_j(r) = u_j\ ] = 0$$

But this violates the weak security property of the (g, b) scheme, a contradiction to our assumption.

Hence for every $\vec{s} \in S^\ell$, there is a vector of pieces for the users, in which the vector of reconstructed keys for the sets $G_1, \ldots, G_\ell$ is $\vec{s}$. Since user 1 computes the keys of the sets $G_1, \ldots, G_\ell$ from his piece, it follows that his piece must be different for every pair of different vectors of keys for the sets $G_1, \ldots, G_\ell$. There are $|S|^\ell$ possible vectors of keys, therefore there are at least $|S|^\ell$ different pieces for user 1. That is, $|U_1| \geq |S|^\ell = |S|^{\binom{g+b-1}{g-1}}$, as claimed. □

In this proof we use the weak security requirement. That is, even weak (g, b) schemes must have large domain of pieces. Thus, our proof yields a stronger result than the lower bound of [2]. We remark that if the keys of all sets were independent random variables, then using the same ideas of this proof, we can prove a lower bound of $|S|^{\binom{n}{g-1}}$. Another observation is that we can consider a key distribution scheme in which only some pre-defined subsets of size g can reconstruct a key. Our proof actually supplies a lower bound for this setting as well.

Lemma 5. *Let $\mathcal{U}$ be a (weak) non-interactive (g, b) scheme with exactly $g + b$ users and domain of keys S, in which user i is a member of at least ℓ sets that can reconstruct a key. Let U_i be the domain of pieces of user i in $\mathcal{U}$. Then:*

$$|U_i| \geq |S|^\ell$$

Notice that ℓ can be at most $\binom{g+b-1}{g-1}$.

Using symmetric degree b multinomials with g variables, Blundo et. al [2] have constructed an unrestricted non-interactive (g, b) scheme with domains of pieces $|U_i| = |S|^{\binom{g+b-1}{g-1}}$. provided that $|S| \geq n$ and $|S|$ is a prime power. So the lower bound is tight (except for small domains of keys).

4 Removing Interaction from Unrestricted Schemes

In this section we show how to transform an unrestricted interactive scheme into a unrestricted (weak) non-interactive key distribution scheme, without changing the domain of pieces. This means that the lower bound on the cardinality of the domain of pieces applies to unrestricted interactive schemes.

Theorem 6. *Let $\mathcal{U}$ be an interactive (g, b)-KDS with $n \geq g+b$ users and domain of keys S. Let $U_1, \ldots, U_n$ be the domains of pieces of the users in u. Then for every user i:*

$$|U_i| \geq |S|^{\binom{g+b-1}{g-1}}$$

Proof. The high level idea of the proof is to fix, for every set G of g users, a possible communication C_G (i.e. one that is exchanged with positive probability when G interacts in order to generate a conference key). Now the server deals only vectors of pieces that are consistent with all the communications C_G's. When a member of a set G wishes to determine a conference key, he applies the reconstruction function to his piece and the fixed communication C_G. This way, no interaction is required. In the proof, we show first how to choose communications for different conferences such that they are consistent among themselves. Therefore there are vectors of pieces that are consistent with all the communications. Once this is done, it is clear that the non-interactive scheme has the reconstruction property. We then prove that the resulting non-interactive scheme has the weak security property. Therefore it is a weak non-interactive (g, b) scheme.[5] By Theorem 4 the cardinality of the domain of pieces of every user in the resulting non-interactive scheme is at least $|S|^{\binom{g+b-1}{g-1}}$. But the domain of the pieces in the non-interactive scheme is not larger than that of the interactive scheme. Therefore, the lower bound on the size of the pieces applies to the original interactive scheme as well.

To complete this proof we first show how to choose a set of communications C_G (for all G's) in a consistent way. To do this, we first fix an arbitrary vector of pieces $\vec{u}$, that the server deals with positive probability. We also fix the local random input of each user. Each communication C_G is the one determined when the users of G hold pieces from $\vec{u}$, and have the fixed random inputs. It is clear that $\vec{u}$ is consistent with all these conversations. The server chooses at random a vector of pieces that is consistent with the communications. That is, the server chooses from all the vectors of pieces $\vec{v}$ for which there exists a vector of random inputs $\vec{r}$, such that every set G of g users, holding the pieces of $\vec{v}$, and having the random inputs $\vec{r_G}$, communicate C_G.

We next prove the weak security property of the non-interactive scheme. Let G be any set of cardinality g, and B be a disjoint set of cardinality b. By the security property of the interactive scheme, it follows that for every vector of pieces that is consistent with the fixed conversations, and every key $s \in S$, there exists a vector of pieces in which the pieces of the users in B are the same, but the key of the conference G is s. That is, the non-interactive scheme has the weak security property, as claimed. □

We can define the notion of weak security for unrestricted interactive schemes as well. The lower bound of Theorem 6 is also applicable to such weak unrestricted interactive schemes.

5 Restricted Interactive Key Distribution Schemes

5.1 Motivation and Definition

By Theorem 6, interaction cannot decrease the size of the pieces of information given to the users in key distribution schemes. In order to decrease the size of the pieces of information, we relax the security requirement. We require that the key

[5] In this proof we do not define the probability distribution under which the server distributes the consistent vectors of pieces. We only require that every consistent vector is distributed with positive probability. It is possible to define a probability distribution on the consistent vectors, such that the induced (g, b) scheme will have the unrestricted security property.

distribution schemes should be secure only for a limited number of conferences. Which conference will generate a key is not known a-priori, so the distributed pieces should accommodate any combination of conferences (up to the limit on their number). We will show that if this limit is relatively small, then the size of the pieces can be substantially reduced. For example, if the scheme is only required to be secure for a single conference, then for $g = b = n/2$, we present a scheme whose domain of pieces is of cardinality $|S|^4$, regardless of n. Recall that for unrestricted schemes with these parameters, the cardinality of the domain of pieces is $|S|^{2^{\Omega(n)}}$ (Theorem 6). First, we state the exact definition of τ-restricted key distribution scheme, and then prove upper and lower bounds on the size of the pieces in such schemes. There is still a gap between our upper and lower bounds.

Before going any further, we remark that the notion of key distribution schemes restricted to a limited number of conferences is meaningful only with respect to interactive schemes. For non-interactive schemes, the generation of a conference key does not add any information with respect to any user (either in the conference, or not in the conference). Therefore a one-time secure non-interactive scheme would also be secure in the unrestricted sense, and no saving can be expected. On the other hand, in interactive schemes the interaction, heard by all users (not only conference members), could reduce the secrecy of the remaining pieces. Finally, after sufficiently many interactions take place, no uncertainty is left, and the pieces become useless for additional conferences. This means that the amount of initial secrecy in restricted interactive schemes can be smaller than in unrestricted schemes. The proof that unrestricted interactive schemes can not be more space efficient than unrestricted non-interactive schemes (Theorem 6) can not be used for restricted schemes. For example, one could transform a one-time secure interactive scheme into a non-interactive scheme, using the technique of Theorem 6. However, this would yield a non-interactive scheme which is secure with respect to a single *fixed* conference, depending one initiating the interaction.

Definition 7. A τ-restricted (g, b)-scheme is an interactive (g, b)-scheme in which the security property is replaced by the following one:

τ-restricted security property Let B be a subset of b (bad) users. Then the users in B, having their pieces and knowing the conversations sent in any τ conferences, do not have any information on the key of any disjoint set G_0. That is, for every vector of pieces $\langle u_1, \dots, u_n \rangle$ which is dealt with positive probability, every combination of τ sets of users of cardinality g, denoted by $G_0, \dots, G_{\tau-1}$, every coalition B of b users, such that $G_0 \cap B = \emptyset$, every set of random inputs $\vec{r_B}$ to coalition members, every possible key $s \in S$, and every possible consistent conversations $C_0, \dots, C_{\tau-1}$ sent by the users of $G_0, \dots, G_{\tau-1}$ respectably:

$$\Pr[\, s_{G_0}(r, \vec{r}) = s \mid \vec{r_B} \wedge \bigwedge_{j \in B} \mathcal{U}_j(r) = u_j \wedge \bigwedge_{0 \le j \le \tau-1} C_{G_j}(\mathcal{U}_{G_j}(r), \vec{r}) = C_j \,] = \mathcal{P}(s)$$

Where the probability is taken over r – the random input of the server, and $\vec{r}$ – the random inputs of all the users for all conferences, where the restriction of $\vec{r}$ to the coalition B equals $\vec{r_B}$.

We denote 1-restricted scheme by *one-time scheme.*

5.2 Upper Bound

Blundo et. al. [2] present a one-time (g,b)-scheme in which the domain of pieces of each user is of cardinality $|S|^{g+b-1}$. We improve their one-time scheme, and present a one-time key distribution scheme in which the domain of pieces of each user is of cardinality $|S|^{2+2(b-1)/g}$. (In [2] it is not mentioned that this scheme is only one-time secure). To construct τ-restricted schemes, we use τ copies of our one-time scheme.

Lemma 8. *Let S be a domain of keys of cardinality q^g, such that q is a prime-power which is greater or equal to $\sqrt{n}$. There exists a one-time (g,b) scheme with n users and domain of keys $|S|$ in which the cardinality of the domain of pieces of every user is $|S|^{2+2\cdot(b-1)/g}$.*

Proof Sketch. We construct our interactive one-time (g,b)-scheme as following: The server deals vectors of pieces according to the non-interactive $(2, g+b-2)$ scheme of Blom [1] for n users, with keys taken from a domain of cardinality $|S|^{2/g} = q^2$ (this is where we need $q \geq \sqrt{n}$). When the users of a set G want to generate a conference key, every user $i \in G$ picks at random $s_i \in \{0,\ldots,q-1\}$. The conference key s of the set G is the concatenation of these random s_i's. That is

$$s = s_1 \circ s_2 \circ \ldots \circ s_g$$

We will show how every user $i \in G$ sends a message on a broadcast channel, such that every user in G will be able to reconstruct s_i, and every user not in G does not learn anything from these messages. Every user $i \in G$ will send a message to every user $j \in G$, that will be meaningful only to user j. The idea is to use the keys of the non-interactive scheme as a one-time pad. More formally, every pair of users $i,j \in G$ reconstruct the joint key $s_{i,j} \in \{0,\ldots,q^2-1\}$ according to the pieces from the non-interactive scheme. Now we view this joint key as consisting of two sub-keys $s'_{i,j}, s''_{i,j}$, both in $\{0,\ldots,q-1\}$. In order to inform user j of s_i, user i broadcasts $s_i + s'_{i,j} \pmod q$, in the case $i < j$, and $s_i + s''_{i,j} \pmod q$, in the case $i > j$. Notice that every sub-key is used only once.

To prove that the interactive scheme has the 1-restricted security property, it is enough to show that the messages sent are all uniformly distributed and independent of the conference key of G and the pieces of any coalition B with b users (provided $G \cap B = \emptyset$). This fact, in turn, follows the next claim from [2] about unrestricted non-interactive $(2, g+b-2)$ schemes. The claim states that the vector of keys of all pairs of users in G in the non-interactive scheme is uniformly distributed and independent of the pieces of the users in B. Formally,

Lemma 9. *(Lemma 4.1 of* [2]*) Let $\mathcal{U}$ be a non-interactive unrestricted $(2, g+b-2)$-scheme with the uniform a-priori distribution on the key domain S. Let G and B be sets of g and b users respectably, such that $G \cap B = \emptyset$. Let $G_1, G_2, \ldots, G_{\binom{g}{2}}$ be all the subsets of G of cardinality 2. Let $s_1, \ldots, s_{\binom{g}{2}}$ be any combination of $\binom{g}{2}$ keys from S, and $\langle u_1, \ldots, u_n \rangle$ be a vector of pieces that is distributed with positive probability. Then:*

$$\Pr[\bigwedge_{1 \le i \le \binom{g}{2}} s_{G_i}(r) = s_i | \bigwedge_{j \in B} \mathcal{U}_j(r) = u_j] = \frac{1}{|S|^{\binom{g}{2}}}$$

We used the non-interactive $(2, g+b-2)$-scheme with domain of keys of cardinality $|S|^{2/g}$. So the cardinality of the domain of pieces of each user is

$$(|S|^{2/g})^{g+b-1} = |S|^{2+2\cdot(b-1)/g}$$

as claimed. □

Notice that the conference key of G ($s = s_1 \circ s_2 \circ \ldots \circ s_j$) is distributed uniformly in S. It is possible to change this probability distribution on the keys. One way to achieve this goal is to first generate a key s as in the previous way. Then user 1 chooses the real key k for the conference according to any desired distribution. User 1 sends the message $(k+s) \bmod q$.

One property of our interactive scheme is that it uses only one-way interaction. The messages of different members of G do not depend on other messages. Another property of our scheme is that for a fixed b, the cardinality of the domain of pieces of each user is a monotonically *decreasing* function of g. This feature stands in contrast to unrestricted (g, b) schemes, where the cardinality of the domain of pieces of each user is a monotonically *increasing* function of g.

We remark that the scheme cannot be reused. For example, if users $\{1, 2\}$ are members of two conferences G_1, G_2, then the part of the keys generated by them in the two conferences will be known to all the users in $G_1 \cup G_2$. We use τ independent copies of the one time scheme in order to extend our scheme to a τ-secure one. Since the copies are independent, each conference does not add any information on other conferences. Hence the security of the one-time scheme, implies the security of the τ-restricted scheme.

Theorem 10. *Let S be a domain of keys, such that $|S| = q^g$ for some prime-power $q \ge \sqrt{n}$. There exists a τ-restricted (g, b)-scheme with n users and domain of keys S, in which the domain of pieces is of each user is of cardinality $|S|^{2\tau(1+(b-1)/g)}$.*

This τ-restricted interactive schemes requires that the users hold a counter, which is incremented each time a conference key is generated. Given such a reliable counter, *active attack* by users sending messages deviating from the protocol, do not reveal information on different conferences. Such attack could prevent the generation of the present conference key. Our scheme does not work in the absence of a reliable counter.

5.3 Lower Bound

The cardinality of the domain of pieces in the τ-restricted scheme depends on τ. We show that if $\tau \leq \binom{g+b-1}{g-1}$, then this dependency on τ cannot be avoided. We conclude, that for $\tau \geq \binom{g+b-1}{g-1}$, the unrestricted scheme of [2] are space optimal even for τ-restricted schemes.

Theorem 11. *Let* $\ell = \min\left\{\tau, \binom{g+b-1}{g-1}\right\}$. *In every τ-restricted (g,b)-scheme, with n users and domain of keys S, the cardinality of the domain of pieces of every user is at least $|S|^{\ell}$.*

Proof Sketch. Again, we limit the number of users to $g+b$. Using the same ideas as in the proof of Theorem 6, we transform a τ-restricted (g,b)-scheme into a a non-interactive (g,b)-scheme in which ℓ pre-defined sets can reconstruct a key. That is, we fix consistent conversations of the ℓ sets, and the server generates vectors of pieces consistent with these conversations. The original scheme is secure for τ conferences, therefore by fixing $\ell \leq \tau$ conversations,we get a secure scheme in which these ℓ sets can reconstruct a key without any interaction. Since $\ell \leq \binom{g+b-1}{g-1}$, then there are ℓ sets that contain user i. Choosing ℓ such sets, we can apply Lemma 5 to the transformed scheme. So, by Lemma 5. the cardinality of the domain of pieces of user i in the transformed scheme at least $|S|^{\ell}$. By the transformation, the cardinality of the domain of pieces in the transformed scheme is at most the cardinality of the domain of pieces in the τ-restricted secure scheme. Therefore, the cardinality of the domain of pieces of every user in the τ-restricted scheme is at least $|S|^{\ell}$. □

This lower bound is not tight. For example, we can prove that for a one-time $(2,1)$-scheme, the domain of pieces has to be bigger than $|S|$. We believe that for $(2,1)$-schemes the lower bound can be improved to $|S|^2$ (which is the upper bound).

Acknowledgement Thanks to Amir Herzberg for helpful discussions on these topics.

References

1. R. Blom. An Optimal Class of Symmetric Key Generation Systems. In T. Beth, N. Cot, and I. Ingemarsson, editors, *Advances in Cryptology – proceeding of Eurocrypt 84*, volume 209 of *Lecture notes in computer Science*, pages 335–338. Springer-Verlag, 1984.
2. C. Blundo, A. De Santis, A. Herzberg, S. Kutten, U. Vaccaro, and M. Yung. Perfectly-Secure Key Distribution for Dynamic Conferences. In *Advances in Cryptology - CRYPTO '92 proceeding*, 1992.
3. W. Diffie and M. E. Hellman. New Directions in Cryptography. *IEEE Transactions on Information Theory*, 22(6):644–654, 1976.
4. A. Fiat and M. Naor. Broadcast Encryption. In *Advances in Cryptology - CRYPTO '93 proceeding*, 1993.

5. M. J. Fischer, M. S. Paterson, and C. Rackoff. Secure Bit Exchange Using a Random Deal of Cards. In *Distributed Computing and Cryptography*, pages 173–181. AMS, 1991.
6. M. J. Fischer and R. N. Wright. Multiparty Secret Key Exchange Using a Random Deal of Cards. In J. Feigenbaum, editor, *Advances in Cryptology – proceeding of CRYPTO 91*, volume 576 of *Lecture notes in computer Science*, pages 141–155. Springer-Verlag, 1992.
7. L. Gong and D. J. Wheeler. A matrix Key-Distribution Scheme. *Journal of Cryptology*, 2:51–59, 1990.
8. T. Matsumoto and I. Imai. On the Key Predistribution Systems: A Practical Solution to the key Distribution Problem. In C. Pomerance, editor, *Advances in Cryptology – proceeding of CRYPTO 87*, volume 293 of *Lecture notes in computer Science*, pages 185–193. Springer-Verlag, 1988.
9. E. Okamoto and K. Tanaka. Key Distribution System Based on Identification Information. *IEEE Journal on Selected Areas in Communications*, 7(4):481–485, 1989.

Secret-Key Agreement without Public-Key Cryptography (Extended Abstract)

Tom Leighton[1,2] and Silvio Micali[2]

[1] Mathematics Department and
[2] Laboratory for Computer Science
Massachusetts Institute of Technology
Cambridge, Massachusetts 02139

Abstract. In this paper, we describe novel approaches to secret-key agreement. Our schemes are not based on public-key cryptography nor number theory. They are extremely efficient implemented in software or make use of very simple unexpensive hardware. Our technology is particularly well-suited for use in cryptographic scenarios like those of the Clipper Chip, the recent encryption proposal put forward by the Clinton Administration.

1 Introduction

1.1 The Problem of Secret-Key Agreement

Private-key cryptosystems are the most common type of cryptosystems; indeed, they are also refered to as "conventional systems." Their goal is to allow two parties A and B, who have agreed on a common and secret key K_{AB}, to exchange private messages via a network whose communication lines are easy to tap by an adversary. If properly designed, conventional cryptosystems are extremely fast, and believed to be very secure in practice. They are also very attractive from a theoretical point of view. In fact, provably-secure conventional cryptosystems can be build based on a very mild complexity assumption: the existence of one-way functions. (Roughly, these are functions that are easy to evaluate, but for which finding pre-images is hard.) Among so many advantages, these systems have a major drawback: agreeing beforehand on a common secret key with every one with which we wish to talk in private is not trivial. Certainly, meeting in a secure physical location is not a practical approach to obtain such an agreement. It is the goal of this paper to forward new, secure, and practical approaches to secret-key agreement. Prior to duscussing our ideas, let us briefly review the main approaches that have been considered so far.

1.2 Prior Approaches

Most of the protocols currently being used for key agreement are either classified or company-confidential. In the public domain, the most popular key-agreement protocols fall into two wide categories: (1) those based on public-key cryptography, and practically implemented with number theory (e.g., the Diffie-Hellman [4] and the RSA algorithms), and (2) those based on symmetric-key generation and a trusted agent, and practically implemented with some form of polynomial or integer arithmetic (e.g., those of Blom [2] and Blundo, De Santis, Herzberg, Kutten, Vaccaro, and Yung [3]). Unfortunately, these approaches are somewaht wanting both with respect to security and efficiency.

DRAWBACKS OF THE PUBLIC-KEY APPROACH. Although very elegant, the Diffie-Hellman and RSA algorithms require that number theoretic problems such as factoring or discrete-log be computationally intractable. In particular, the RSA scheme would become insecure if someone discovered a much improved algorithm for factoring large integers, and the Diffie-Hellman algorithm would suffer a similar fate if an improved algorithm were found for computing discrete logs. If either scheme is used to select the session keys for all government traffic, then all this traffic would be decipherable to anyone who found improved algorithms for these problems. (Although it might be the case that these problems are truly intractable, it would be nice if there were a key agreement protocol that was secure even if there are algorithmic advances made in number theory in future decades.)

Most likely, similar drawbacks will be suffered by any other proposed solution based on the framework of public-key cryptography. This framework is very "distributed" in nature; namely, every user A *individually chooses* a pair of matching encryption and decryption keys (E_A, D_A), publicizes E_D and keeps secret D_A. Any message encrypted via E_A can be easily (and, hopefully, solely) decrypted via the corresponding key D_A. Since the E_A is made public, any one can send A a private message, because any one can encrypt via key E_A. In this setting, it is conceptually very easy for two parties A and B to agree on a common key K_{AB}. For instance, B may individually choose K_{AB} at random and, since E_A is public, send K_{AB} to A encrypted via E_A. This key is common because A can decrypt it thanks to his private knowledge of D_A. For K_{AB} to be secret for everyone else, however, several conditions must be met; in particular, D_A must not be easily computable from E_A. Indeed, public-key cryptography requires very strong complexity assumptions: the existence of *one-way trap-door predicates* or that of *one-way trap-door functions*. (The latter, roughly, are functions that not only are easy to evaluate and hard to invert, but *also* possess an associated secret whose knowledge allows one to easily invert them.) Such assumptions appear to be much more demanding than existence of a one-way function. Indeed, while the existence of one-way functions is widely believed (and plenty of candidates are available), trap-door one-way functions may not exist or may be very hard to find (indeed, only a handful of them have been proposed without being immediately dismissed). In sum, therefore, while one-way functions are sufficient to communicate securely in a conventional cryptosystem once a common secret key has been established, the process of establishing such a key based on public-key cryptography appears to require a much stronger type of assumption, thereby creating a weaker link in the overall security.

In addition to the above concerns about security, and perhaps of more immediate importance, key agreement protocols based on public-key cryptography and number theory tend to be very expensive to implement. Indeed, the cost of building hardware that can quickly perform modular exponentiation is far greater than the cost of building encryption devices based on one-way functions. Indeed, when providing encryption devices to a country of the size of the United States, the cost of the key-agreement hardware is far from being insignificant.

SECURITY DRAWBACKS OF THE TRUSTED-AGENT APPROACH. In this approach there are three main parameters: N, the total number of users in the system, k the length of a common secret key, and B, a bound on the number of (collaborating) malicious users. The algorithms of [2] and [3] make use of polynomial and integer arithmetic for implementing a *Symmetric Key-Generation System.* This consists of computing (from a single, secret, system value K) a set of n secret values, $K_1, \ldots, K_n$, which in turn yield n^2 quantities, K_{ij}, satisfying the following properties. For each i and j the quantity K_{ij} can be computed easily either on inputs j and K_i, or on inputs i and K_j. Moreover, given any B individual values, $K_{a_1}, \ldots, K_{a_B}$, one has no information about the quantity K_{ij} whenever $i, j \neq a_i, \ldots, a_B$.

A symmetric key-generation system can be effectively used by trusted agent to enable the users to provide an elegant solution to the secret-key agreement problem. The trusted agent (after choosing the system master secret) simply computes n secret values $K_1, \ldots, K_n$ and assigns to user i value K_i as his individual secret key. The common key between two users i and j will then be K_{ij}, which can be easily computed by either one of the two users. This key is secret in that no coalition of less than B users, no matter how much computation they perform, can infer the common of two other users. While this is a very attractive property, relying on a trusted agent for assigning the proper, individual secret keys is in itself from a security point of view, a serious drawback.

From an efficiency point of view, while the algorithms of [2] and [3] require small individual keys (i.e., $\Theta(Nk)$ bits per user), they do require a fair amount of algebraic computation in oreder to compute common secret keys from individual keys. Even if this computation could be sped up in a hardware implementation, the cost of this circuitry may not be trivial with respect to that required for conventionally encrypting messages via a one-way function, after common secret keys have been established. Moreoever, for both security and efficiency reasons, conventional encryption schemes are not algebraic; thus the "algebraic hardware" necessary for their secret-key agreement represents pure additional cost, since it will have very little to share with the "encrypting hardware."

1.3 Our Contribution

In this paper, we advocate two new approaches to secret-key exchange, none of which is based on public-key cryptography or polynomial/integer arithmetic.

In the first we introduce a new class of symmetric key-generation systems requiring longer individual keys than [2] or [3], but guaranteeing a that computing common secret keys is absolutely trivial. To enhance the security of these algorithms further we recommend using them in conjunction with special hardware and a set of moderatly trusted agents. This whole approach is described in Section 2.

The second approach requires a simple interaction with moderatly trusted agents, and can be impleneted in software with great efficiency and security (and, of course, wit great savings). This approach is described in Section 3.

Both approaches are information-theoretically secure, though they can be more convenently impemented if *ordinary* one-way functions are used. (In any case, therefore, they will be immune to attacks based on advances in number theory.)

Finally, both of our approaches are capable, *if so wanted by society*, of making very secure encryption "compatible with law-enforcement;" that is, in case a court authorizes tapping the communication lines of users suspected of illegal activities (and only in case of these legitimate authorizations), the relevant secret keys can be reconstructed by the Police. This is indeed the major feature of two recent encryption proposals, *Fair cryptosystems*, as put forward by the second author, and *the Clipper Chip* as put forward be the Clinton administration. To illustrate both this additional important point and the cryptographic use of tamper-proof hardware that may enhance the security of our first scheme, let us provide a brief introduction to the government proposal.

1.4 The Clipper Chip Project

In April, 1993, the Clinton Administration announced its intention to develop a cryptographic scheme for widespread use within the government. The scheme is centered around a device known as the *Clipper Chip* which is expected to become standardized for encryption and decryption of telephone, fax, email, and modem traffic. The Clipper Chip does not offer a solution to the secret-key agreement problem; rather, it *assumes* that such a solution exists. Its goal is making conventional cryptosystems compatible with law-enforcement.

The Clipper Chip will be made using a special VLSI process which is designed to prevent reverse engineering. In particular, the conventional encryption and decryption algorithms used on the Clipper Chip will be classified, but the chip itself will not be classified. The Clipper Chip will also contain a protected memory for secret keys. The protected memory is designed to prevent anyone (even the legitimate user of the chip) from gaining access to the keys contained therein.

According to the government press release, each Clipper Chip will be equipped with a unique secret key K_i that is formed by an irreversible process from two pieces of the secret key $K_i^{(1)}$ and $K_i^{(2)}$. The pieces of the secret keys will be held by system-wide trusted agents $\mathcal{T}_1$ and $\mathcal{T}_2$. (Actually, only one of the agents needs to be trusted since $\mathcal{T}_1$ will hold only the first piece of each secret key and $\mathcal{T}_2$ will hold only the second piece.) When two parties wish to communicate using the new system, they first agree on a session key S and they enter this key into their respective Clipper Chips. This key is used by the Clipper Chips as an encryption/decryption key for the message traffic. In other words, once the session key is selected, the Clipper Chips function as a private-key cryptosystem.

There is a major difference between the Clipper Chips and a conventional private-key cryptosystem, however. That is, the Clipper Chips also transmit the session key S being used in encrypted form using the secret key for the chip, thereby allowing the trusted agents to eavesdrop on the conversation. The reason for transmitting the session key in this fashion is so that law enforcement can (upon obtaining the relevant court order) obtain the secret key of the user from the trusted agents and then decrypt the conversation (thereby preserving current wiretapping capabilities) but no other unauthorized person can eavesdrop (thereby providing greater privacy than exists currently for most individuals).

1.5 Our Contribution to the Clipper Chip Project

In the proposed Clipper Chip project, it is assumed that every pair of users has already agreed on a common, secret, (session) key. In practice, therefore, devices that incorporate the Clipper Chip will use a specific key-agreement protocol, or a crucial link would be missing. This, however, has the potential of introducing a host of new difficulties. For instance, if public-key cryptography is used for providing this missing link, the system might become more vulnerable (since it now must rely on stronger —and possibly false— comlexity assumptions) and much more costly.

By contrast, as we shall see, our two schemes can greatly enhance the security and the economicity of the Clipper Chip project. In fact, not only can we guarantee compatibility of law-enforcement with strong encryption, but, within the same framework (without additional costs or loss of security), also the necessary secret-key agreement that was missing in the Administration proposal.

2 A Hardware-Based Approach

For simplicity of exposition, the secret-key agreement of this section is described in two phases. The first phase consists of a special class of symmetric key-generation systems —and thus of a basic scheme for secret-key agreement based on a trusted agent.

In the second phase, we show how to enhance the security of our basic schemes assuming the availability of tamper-proof hardware and the existence of a group of "only moderately trusted" agents rather than a single, totally trusted one.

2.1 The First Basic Scheme

The symmetric key-generation scheme described in this subsection is both very efficient and quite natural. We thus expect that it might have beed already discussed in the literature, but we have been unable to find such reference. We of course be very grateful to anyone who can provide us with such a piece of information.

Recall that in a symmetric key-generation system there are three relevant parameters: N, the total number of possible users in the system; B, an upperbound on the size of a coalition of dishonest users, and k, the number of unpredictable bits contained in each common secret key of two honest users (e.g., $k = 100$). For didactic purposes, however, we shall make use of an auxiliary parameter M, and then show that M should be about $B^3 \ln N$.

COMPUTING INDIVIDUAL KEYS. On input N, B, and k, the trusted agent performs the following steps:

- First, he randomly and secretly selects M, k-bit long, *system secret keys*: $X_1, \ldots, X_M$, where $M = \Theta(B^3 \ln N)$ (the precise constant will be worked out in the final paper).
- Then, he constructs (in $poly(N, M)$ time) a $N \times M$ $0-1$ matrix $A = \{a_{i,n}\}$ with the properties that:
  1) any pair of rows have 0's in at least $F = \Theta(M^{1/3} \ln^{2/3} N)$ common columns, and
  2) any triple of rows all have 0's in at most $G = \Theta(\ln N)$ common columns.
  1. $B < \lceil F/G \rceil$.

  (We will show how to construct such a matrix shortly.)
- He then assignes to player i, where i is a $\log N$-bit identifier, the individual secret key consisting of the vector $(v_{i,1}, \ldots, v_{i,M})$, where $v_{i,n}$ equals the secret system key M_n if $a_{i,n} = 0$, and the empty word otherwise.

COMPUTING COMMON SECRET KEYS. The common secret key of users i and j, $K_{i,j}$, consists of the M-vector whose nth component equals M_n if $a_{i,n} = b_{j,n} = 1$, and the empty word otherwise. Thus, each user trivially computes his common secret key with another user by taking a subset of the system secrets in his possession.

(If so wanted, the size of these common secret keys can be reduced; for instance, by evaluating a one-way hash function on them. Such reductions are not, however, a concern of this paper.)

SECURITY. Assume, for now, that a matrix A as above has been constructed. Then, given the individual secret keys of a set of B users, the common secret key of two users contains at least k random and unpredictable bits. The proof is very simple. By Property 1, above, every common secret key $K_{i,j}$ must contain at least F system secrets. By Property 2, however, the individual secret key of a user other than i or j contains at most G of these secrets. Hence, at least $\lceil F/G \rceil$ other individual secret keys are necessary in order to recover all of the system secrets in a common secret key.

CONSTRUCTING THE MATRIX. Next, let us show that a matrix A satisfying Properties 1 and 2 above exists. We do this by showing that such an A can be constructed with probability > 0 by the following probabilistic algorithm.

Set each entry of A to 0 with probability $p = \Theta\left(\left(\frac{\ln N}{M}\right)^{1/3}\right)$. The probability that Property 1 is not satisfied is then at most

$$\sum_{h=0}^{F-1} \binom{N}{2} \binom{M}{h} (1-p^2)^{M-h} (p^2)^h \leq \sum_{h=0}^{F-1} \frac{N^2}{2} \left(\frac{Mep^2}{h}\right)^h e^{-p^2(M-h)}$$

$$\leq \sum_{h=0}^{F-1} \frac{N^2}{2} \left(\frac{Me^{1+p^2}p^2}{h}\right)^h e^{-p^2 M}$$

$$\leq \frac{FN^2}{2} \left(\frac{Me^{1+p^2}p^2}{F}\right)^F e^{-p^2 M}$$

provided that $F \leq Mp^2$. Similarly, the probability that Property 2 is not satisfied is at most

$$\binom{N}{3} \binom{M}{G+1} (p^3)^{G+1} \leq \frac{N^3}{6} \left(\frac{Mep^3}{G}\right)^G.$$

Hence, the probability that A fails to satisfy either property is at most

$$\frac{FN^2}{2} \left(\frac{Me^{1+p^2}p^2}{F}\right)^F e^{-p^2 M} + \frac{N^3}{6} \left(\frac{Mep^3}{G}\right)^G. \tag{1}$$

By setting $p = 2(\frac{\ln N}{M})^{1/3}$, $F = \frac{1}{2} M^{1/3} \ln^{2/3} N$, and $G = 16e \ln N$, we find that if $M \geq 8 \ln N$, then this probability is strictly less than 1, as desired. (It is worth noting that substantially better constants can be derived for particular values of N and M with a more careful analysis.)

Finally, let us show that such a matrix A not only exists, but is also easy to compute deterministically. This is so thanks to the *method of conditional probabilities* [1]. In fact, it should be noted that the task of evaluating a simple upper bound on the probability that either Property 1 or Property 2 is violated (as in Equation 1) conditioned on some of the values in A being fixed is easily accomplished in polynomial time.

(Indeed, matrix A needs not to be computed "at once," but can be easily computed row-by-row. This way, the cost of a row can be incurred only when one more user —of a budgeted total of Nusers— joins the system.)

In the case when N is small, even better constructions for A exist. For example, when $N = M^{2/3}$, we can construct an A for which $F = M^{1/3}$ and $G = 1$ by letting each row of A correspond to a plane through an $M^{1/3} \times M^{1/3} \times M^{1/3}$ lattice of points mod$M$. By choosing non-parallel planes for the rows of A, each pair of planes will intersect in a line, and each triple of planes will intersect in exactly one point, thereby achieving the desired bounds for F and G. In this example, it suffices to have $B = M^{1/3} - 1$.

EFFICIENCY. Given what we have said so far, the bit-length of an individual key is about $kB^3 \log N$. An individual key, in fact, consists of a subset of the

$M = \Theta(B^3 \log N)$ system secrets. One must observe, however, that the subset tends to be quite sparse. Indeed, we can construct matrices A enjoying also the following additional property:

0) Every row has at most $O(M^{2/3} \log^{1/3} N)$ 0's.

The number of system secrets entering an individual key is thus about $B^2 \log N$.

Though this is longer than the kB bits of individual key (roughly) needed by an individual key in the Blom' algorithm with the same parameters, in our case the construction of a common secret key is trivial, since it only consists of taking a subset of stored secrets. It is unquestionable that, whether in software or in hardware, this operation is much preferable to computing with B-degree polynomials over fields with size-k elements.

2.2 The Second Basic Scheme.

Also this scheme should have, in our opinion, been discussed in the litterature. Nonetheless, we have not been able to find it.

In this scheme the trusted agent distributes to the users longer individual secret keys. As we shall see in the subsection after next, however, the scheme offers additional advantages if some special hardware is available. This scheme too has a perfectly-secure version, but one-way functions can be used in order to shorten the length of the individual secret keys.

COMPUTING INDIVIDUAL SECRET KEYS. Parameters N, B, k, and M are as in the first scheme. The secod basic scheme has, however, an additional and independent parameter $L \geq 2$.

In the perfectly secure version, the trusted agent computes individual keys as follows.

- He chooses ML, k-bit long, system secrets, $\{X_{i,n}\}$, where i ranges between 1 and M, and n between 0 and $L-1$.
- He assigns to each user in the system an identifier consisting of M random values between 1 and $L-1$, $\alpha = (\alpha_1, \ldots, \alpha_M)$. (In practice, the α_i values can be generated by applying a proper cryptographic function to the "real name" of the user.)
- Then, he assigns to user $(\alpha_1, \ldots, \alpha_M)$ the secret key consisting of the (properly ordered) subset of system secrets $\{X_{i,n} : i \in [1, M]\ \alpha_i \in [0, L-1]\}$.

COMPUTING COMMON SECRET KEYS. The common secret key of two users with identifiers $\alpha = (\alpha_1, \ldots, \alpha_M)$ and $\beta = (\beta, \ldots, \beta)$ consists of the (properly ordered) set of system values $\{X_{i,max_i}, \ldots, X_{i,L-1-max_i} : i \in [1, M]\ max_i = max(\alpha_i, \beta_i)$.

(Again, such a key can be substantially reduced in size, if so wanted, but this is not a concern of this paper.)

THE COMPUTATIONALLY-SECURE VERSION. The above scheme requires quite long individual keys. This drawback can be eliminated by using one-way functions. This, of course, will turn the perfect security of the above scheme into computational security, but this is quite tolerable, since the conventional encryption that follows the key-agreement can be at most be computationally secure.

The most straightforward way to use one-way functions to shorten individual key-length in our second basic scheme consists of having the trusted agent set $X_{i,n} = h^n(X_i)$ where X_i is the ith system secret (randomly selected as before) and h is a one-way function. The user identifiers are chosen like in the perfectly-secure version, but the individual key of user $\alpha = (\alpha_1, \ldots, \alpha_M$ now is $(h^{\alpha_1}(X_1), \ldots, h^{\alpha_M}(X_M))$, since user α can reconstruct all other values by evaluating h in the easy direction. This version of our scheme approximates the perfectely-secure one in that the value $X_{i,n}$ is computationally hard to predict from the values $X_{i,m}$ for $m > n$. Nonetheless, if the correct value $X_{i,n}$ is supplied, one can verify its correctness by evaluating h sufficiently many times on it.

In most cases, the above approximation can be deemed sufficient. But, though this lies beyond the scope of this paper, let us mention that a better way to approximate computationally the perfectly-secure, second basic scheme, can be obtained by using a slight variation of the pseudo-random function construction of Goldreich, Goldwasser, and Micali [5]. In particular, one can guarantee that, to someone who is given $X_{i,n+1}$, $X_{i,n}$ is undistinguishable from a random value of the same length. The lenth of an individual key, however, becomes $kM \log L$.

SECURITY. In order for an adversary to compute the common secret key of two users with identifiers $\alpha = (\alpha_1, \ldots, \alpha_M)$ and $\beta = (\beta_1, \ldots, \beta_M)$, he will need to obtain enough individual keys so as to have all system values X_{m,ρ_m} ($Z_m = h^{\rho_m}(X_m)$ in the computationally secure scenario) where $\rho_m \leq max(\alpha_m, \beta_m)$ for $1 \leq m \leq M$. In order for the system to be secure, then, we would like it to be the case that no matter for what set of B users, if the adversary enters in possession of the individual keys of those users, he still does not have enough information to recover a common secret key for any other two users in the system.

To this end, we need to select the identifier $K_i = \{\alpha_{i,1}, \alpha_{i,2}, \ldots, \alpha_{i,M}\}$ for each user i so that for any $i_1, i_2, \ldots, i_B \in [1, N]$ and any $j_1, j_2 \in [1, N]$ such that $i_r \neq j_s$ for $1 \leq r \leq B$ and $1 \leq s \leq 2$, there exists an integer t $(1 \leq t \leq M)$ such that

$$\min_{1 \leq r \leq B} \{\alpha_{i_r,t}\} > \max_{1 \leq s \leq 2} \{\alpha_{j_s,t}\}.$$

In other words, for any pair of users j_1 and j_2 and any set of B users $i_1, i_2, \ldots, i_B$ whose individual keys are learned by the adversary, there needs to be at least one X_t such that none of the B learned keys contains $X_{t,\omega}$ ($h^{\omega}(X_t)$ in the computationally-secure scenario) for any $\omega \leq \max(\alpha_{j_1,t}, \alpha_{j_2,t})$.

Fortunately, it is not too difficult to find identifiers for the users that satisfy the preceding property. For example, if L is a multiple of B and B is about $(M/e \ln N)^{1/3}$, then this property is likely to hold if each $\alpha_{i,m}$ is chosen uniformly at random in $[0, L-1]$. This is because the probability that

$$\min_{1 \le r \le B} \{\alpha_{i_r,t}\} \le \max_{1 \le s \le 2} \{\alpha_{j_s,t}\}. \tag{2}$$

for some particular values of $i_1, i_2, \ldots, i_B, j_1, j_2$, and t is at most

$$1 - \left(\frac{1}{B}\right)^2 \left(1 - \frac{1}{B}\right)^B \tag{3}$$

since there is always a $\frac{1}{B^2}$ chance that $\alpha_{j_1,t}, \alpha_{j_2,t} < \frac{L}{B}$ and a $\left(1 - \frac{1}{B}\right)^B$ chance that $\alpha_{i_1,t}, \ldots, \alpha_{i_B,t} \ge \frac{L}{B}$. For $B \ge 2$, $(1 - \frac{1}{B}) \ge e^{-\frac{1}{B} - \frac{1}{B^2}}$ and thus

$$\begin{aligned} 1 - \left(\frac{1}{B}\right)^2 \left(1 - \frac{1}{B}\right)^B &\le 1 - \frac{e^{-1-\frac{1}{B}}}{B^2} \\ &\le 1 - \frac{e^{-1}(1-\frac{1}{B})}{B^2} \\ &\le 1 - \frac{e^{-1}(B-1)}{B^3} \end{aligned}$$

Hence, the probability that Equation 2 holds for all t, $1 \le t \le M$ is at most

$$\left(1 - \frac{(B-1)e^{-1}}{B^3}\right)^M \le e^{-(B-1)e^{-1}M/B^3}.$$

Summing this probability over the

$$\binom{N}{B+2} \le \left(\frac{Ne}{B+2}\right)^{B+2}$$

possible choices for $i_1, i_2, \ldots, i_B, j_1, j_2$, we find that the probability that any pair key can be recovered by opening any B other chips is at most

$$\left(\frac{Ne}{B+2}\right)^{B+2} e^{-(B-1)e^{-1}M/B^3} = e^{(B+2)[\ln N + 1 - \ln(B+2)] - (B-1)e^{-1}M/B^3}$$

which becomes small when B is about $(M/e \ln N)^{1/3}$.

If L is larger than B^2, then the preceding sort of analysis can also be used to show that the probability that any pair key can be recovered by opening B other chips is small when B is about $(2M/\ln N)^{1/3}$. The key difference is that when L is large compared to B^2, the probability in Equation 3 can be improved to

$$1 - \frac{(1-o(1))2B!}{(B+2)!} = 1 - \frac{2-o(1)}{B^2},$$

which is smaller than $1 - \frac{e^{-1}(B-1)}{B^3}$ for large B.

Efficiency. Given what we have said so far, in the computationally secure version, the length of an individual key equals that of the first basic scheme:

$kB^3 \ln N$ bits. The time needed to compute a common secret key from an individual one is however slightly longer, consisting of $M(L-1)$ one-way function evaluations. Since one-way functions are particularly fast to evalaute, however, this operation, implemented in software, may still be negligible with respect to the equivalent one of [2] and [3]. Our common-secret-key computation also is more convenient than that of [2] and [3] with respect to hardware implementations. Not only because the circuitry needed for one-way function evaluation tends to be simpler than that of polynomial arithmetic, but also because the circuitry necessary to evaluate one-way functions needs to be present *any way*, in order to encrypt messages (or session keys) once a common secret key has been computed.

Let us just mention in passing, however, that, like for the first basic scheme, the length of an individual key can be reduced to about $B^2 \log N$, but the initial (omitted) constant can be better than that of the first scheme. (Details will be given in the final paper.)

The real advantage of the second scheme, however, comes in if a special type of hardware is available. This advantage is discussed in section 2.5.

2.3 Optimality of our Basic Schemes

We now show that the schemes for assigning public keys that were just described are all optimal to within a logarithmic factor among those where common secret keys are constructed as "subsets of common secrets." In particular, we will show that no matter how large L is and no matter how public keys are assigned, then there is always a set of $B = \Theta(M^{1/3} \log^{2/3} M)$ chips which, when opened, will lead to the recovery of a pair key for some other pair of users. Because of space limitations, we will only sketch the proof in this extended abstract, and we will not worry about constant factors. (In fact, the constant hidden behind the Θ is quite small.)

Given any set of N public keys $\{K_i | 1 \leq i \leq N\}$ where $K_i = \{\alpha_{i,m} | 1 \leq m \leq M\}$, define an $N \times M$ matrix $A = \{a_{i,j}\}$ as follows:

$$a_{i,j} = \begin{cases} -1 \text{ if } \alpha_{i,j} \text{ is the smallest item in } \{\alpha_{r,j} | 1 \leq r \leq N\}, \\ 0 \text{ if } \alpha_{i,j} \text{ is among the next smallest } \frac{N \log^{1/3} M}{M^{1/3}} \text{ items in } \{\alpha_{r,j} | 1 \leq r \leq N\}, \\ 1 \text{ otherwise.} \end{cases}$$

Ties can be broken arbitrarily. Thus, each column of A will have one -1, $\frac{N \log^{1/3} M}{M^{1/3}}$ 0's, and $N - 1 - \frac{N \log^{1/3} M}{M^{1/3}}$ 1's.

We next use the pigeonhole principle to show that there are two rows u and v in A such that

1) neither row u nor row v contains any -1's, and
2) there are at most $F = \Theta(M^{1/3} \log^{2/3} M)$ columns for which both rows u and v contain a 0.

This is because at most M rows contain a -1, and at most

$$\binom{\frac{N\log^{1/3}M}{M^{1/3}}}{2} \leq \frac{N^2\log^{2/3}M}{2M^{2/3}}$$

pairs of rows can both contain a 0 in column j for any j. Hence, there exists a pair of rows satisfying Properties 1 and 2 above if

$$\binom{N-M}{2}F > \frac{MN^2\log^{2/3}M}{2M^{2/3}}. \tag{4}$$

Equation 4 is satisfied for $F = \Theta(M^{1/3}\log^{2/3}M)$ provided that $N \geq 2M$.

We next show how to find $\Theta(M^{1/3}\log^{2/3}M)$ chips which, when opened, can be used to recover the pair key for u and v. The first step towards this goal is to find $\Theta(M^{1/3}\log^{2/3}M)$ rows of A (other than u and v) such that for each column t, one of these rows contains a 0 in column t. We can find such a set of rows because every column of A has $\frac{N\log^{1/3}M}{M^{1/3}}$ 0's and at most 2 of these zeros can be in rows u and v. In particular, there is always a row in A (besides u and v) with at least

$$M\left(\frac{N\log^{1/3}M}{M^{1/3}} - 2\right)/N = \Theta(M^{2/3}\log^{1/3}M)$$

0's. When we remove this row (call it i_1) and any columns j in A for which $a_{i_1,j} = 0$, there is still a row with $\Omega(M^{2/3}\log^{1/3}M)$ 0's in the remaining columns. Continuing in this fashion, we can easily find a collection of $O(M^{1/3}/\log^{1/3}M)$ rows which collectively have at least one 0 in half of the columns. Continuing further, we can find a collection of $O(M^{1/3}\log^{2/3}M)$ rows which collectively have at least one 0 in each column. By opening the chips corresponding to these $O(M^{1/3}\log^{2/3}M)$ rows, we can then recover the portions of the secret pair key for u and v corresponding to columns t for which $\max\{a_{u,t}, a_{v,t}\} = 1$.

All that remains is to open the $F = O(M^{1/3}\log^{2/3}M)$ chips corresponding to the rows ω for which $a_{\omega,t} = -1$ where t is one of the $F = O(M^{1/3}\log^{2/3}M)$ columns for which $a_{u,t} = a_{v,t} = 0$. These chips contain the remaining secrets necessary to reproduce the pair key for users u and v. Hence B can be at most $O(M^{1/3}\log^{2/3}M)$ if the system is to be secure.

2.4 Enhancing the Security of the Basic Scheme.

MULTIPLE TRUSTEES. Let us now modify the basic schemes described above so that they no longer need such a strong assumption as the existence of a trusted agent selecting and distributing individual secret keys to the users. At the simplest level, as envisaged in [6] and in the Clipper Chip scenario, this modification consists of replacing the trusted agent with a group of agents that are moderatly trusted. That is, trusted collectively, but not necessarily individually.

For instance, there may be a set of t independent trustees, each one of which acts as the trusted agent of Sections 2.1 and 2.2. Thus, denoting by $K^n_{i,j}$ the secret common key between users i and j relative to the nth trustee, the *overall* secret common key between i and j, $K_{i,j}$, will be the XOR of the $K^n_{i,j}$'s. It is easily seen that this does not effect the security and the all the relevant properties of the trusted agent scenario. In addition, if society so wants, this way of operating also allows one to make cryptography compatible with law enforcement. In fact, the two trusted agents can give, when presented with a court order, the two individual secret keys of a suspected user i, K^1_i and K^2_i, to the Police. Thus the Police will be allowed, in case of a court order, to understand any message sent or received by i, because it can easily compute all possible common secret keys between i and other users.

SECURE CHIPS. A second modification that may greatly enhance the security of our basic schemes depends on the availability of secure chips. Recall that these are chips whose memory (or parts of it) cannot be read from the outside, and that cannot be tampered with without destroying their content. Given such chips, each of the trustees can then store the individual key of a user into a secure chip. Once a chip has all the required individual keys (i.e., t if there are t trustees), is given to its proper user. We further recommend that he should not see any of the common secret keys enabling him to communicate privately with other users. In fact, we recommend that the only way for one to use his own chip consists of entering the chip a message and a recipient identity and that the only action taken in response by the chip consists of (1) internally (and thus secretly) compute the right common secret key, and then (2) outputting the encryption of that message with that key. (In particular, if the input to the chip is not of the proper form, the chip will take no action, but it will never output part of a secret key, nor will it answer in any way to any "input request" about a secret key.)

This step makes it harder for a coalition of malicious users to compute the common secret key of two honest users. In fact, malicious users no longer know their own individual keys. Even if one assumes that, with a lot of effort and with a substantial investment in sophisticated equipment, one may succeed in reading the content of at most A chips, by selecting the individual keys with parameter $B > A$, one can guarantee that no common secret key between two uncompromised chips can be computed.

The inability of computing common secret keys does not, however, quite coincide with the inability of eavesdropping. Indeed, Yuliang Zheng [8] has observed that, if $B = 0$, then there may exist a *single* chip ω whose individual secret key allows one to compute the common secret key, $K_{\alpha,\beta}$, of two other chips α and β, and that, if this is the case, chip ω can be used (without opening it) in order to eavesdrop communications between α and β. (He also shows, however, that by having the common secret key of two users also depend from their identifiers makes eavesdropping impossible even in that special case. It would be nice to know, by the way, where are such special cases exist, or to have a formal proof that no other sich a case exist.)

ENHANCING THE SECURITY OF PRIOR KEY-DISTRIBUTION SCHEMES. These enhancements can also be used to improve the security of the algebraic schemes of Blom [2] and Blundo et al. [3]. Namely, assume that there are two (for simplicity) trustees, τ_1 and τ_2, at least one of which is honest. Then, each trustee τ_i can produce up to n secret values K_i^x's, and store K_i^x in the protected memory of chip of user i as his secret individual key, without revealing its value to him. Then, the common secret key of two users i and j can be easily computed (without interaction) by either of their tamper-proof chips as a fixed combination of K_{ij}^1 (i.e., their common secret key as if τ_1 were the only trusted agent) and K_{ij}^2 (i.e., their common secret key as if τ_2 were the only trusted agent. This way, gathering together more than B secret individual keys is harder (because all such keys are stored in protected hardware and are not known to the users) and, at the same time, one no longer relies on a single trusted agent. In addition, also these so modified schemes may make, if so wanted, encryption and law enforcement compatible.

These security-enhanced schemes, however, will be less efficient and more expensive than our enhanced ones.

2.5 Partially-Openable Chips

So far we have assumed that an adversary who succeeds in opening the protected memory of a chip, succeeds in reading all of this memory. It may be more realistic, however, to assume that, by tampering with a properly protected chip, an adversary can obtain at most a few bits from the protected memory of each chip before he destroys the rest of the bits. In this case, our second basic scheme may, by using a large parameter L, make the adversary's work dramatically difficult. This is so because learning —say— 5 bits of $h^n(X_m)$ for different values of n will be of little help to the adversary. Thus the ability of opening random chips no longer is very useful: the adversary must open many chips with exactly the same portion of the secret key if he wants to obtain "useful information." But, when L is large, even the simple step of *getting hold* of a large number of chips with such identifiers (and thus such secret keys) may be overwhelmingly hard. In fact, in the cases of the first basic scheme and the second one with $L = 2$, the adversary needs to partially open about $B(B^2 \log N)k/5$ chips (if at most 5 bits can be recovered from such an opened chip) in order to compute the individual keys of B chips. By contrast, in the second basic scheme, when the parameters L increases (and all other parameters remain fixed), the number of chips that need to be opened tends to $LB(B^2 \log N)k/5$. It then becomes clear that even for reasonable values of L, in most applications there will not be that many chips in the system. Thus no adversary has a realistic chance of computing the common secret key of two uncompromised chips even if he gets hold and partially opens all the other ones. This is even more true if one assumes (like it may be realistic to do) that before succeeding in partially opening a chip the adversary is expected to destroy a few chips in the process. In fact, many more than $LB(B^2 \log N)k/5$ chips are needed in this case.

Additional protection can be obtained if chips are initialized with users identifiers that are chosen as an unpredictable function of the names of their users.

Security Hierarchies The previous scheme can be easily modified for use in a scenario where there are various gradations of security. For example, assume that the users are categorized into S security levels $1, 2, \ldots, S$, where level 1 is the highest level of security, and level S is the lowest level of security. Then, we can use the same scheme as before except that the public key for a user at security level q is selected so that α_m is a random integer in the range $[1 + (q-1)L, qL]$ for $1 \leq m \leq qM/S$.

The modification increases the time to compute a pair key by a factor of at most S, but it has the nice feature that conversations between users will always take place at the highest (i.e., most secure) common level of security. In particular, in order for an adversary to recover the pair key for a conversation between users at security level q or better, the adversary will need to open at least B chips of security level q or better where B is as defined above (with M decreased by S). Since there are likely to be fewer chips at the higher security level, and since they are likely to be guarded more closely, it will be much more difficult for an adversary to obtain such chips, and he will have to open more of them before being able to recover a pair key (since we can replace N by the number of users with that level of security, which is smaller).

3 A Software-Based Approach

Let us now describe a scheme for exchanging keys that does not rely on any protected hardware at all. The scheme is very simple and, in its basic version, only relies on a trusted agent. (We shall describe how to implement the scheme with multiple trustees later.)

Again, it seems to us that this very scheme should have been considered before, but we have not been able to find it in the literature. Our software-based approach shares, however, some similarities with that of Needham and Schroeder, and it is actually useful to to contrast the two of them.

3.1 The Needham-Schroeder Approach Versus Ours

THE NEEDHAM-SCHROEDER PARADIGM. Outside the public-key cryptography framework, the most popular approaches to secret-key agreement follow a paradigm put forward by Needham and Schroeder [7], whose essential features are summarized below.

A trusted agent T assigns to each user i in the system an individual secret key K_i. This key is thus a common secret between the user and T. Whenever a user i wishes to talk in private to another user j, i sends T a message (in the clear) specifying his own identity and that of the recipient, j. The trusted agent answers this request by selecting a new session key S and sending i a global message, encrypted with key K_i, consisting of two values: S and E, where E is an encryption of S with key K_j. User i, after recovering S and E, sends j two values: his intended message encrypted with S, and E. User j, thanks to his knowledge of K_j, recovers S from E, and then recovers the message.

(After this basic step, i and j also engage in another protocol to check that indeed S is a common key to both of them. We have decided to ignore these additional protocols in this paper.)

CRITIQUE OF THE NEEDHAM-SCHROEDER PARADIGM. Many objections can be and have been moved to the above paradigm for secret-key exchange; in particular:

1. *It requires that the trusted agent be continuously available.*
   Indeed, if the communication link between a user and the trusted agent is down, then that user is deprived of the possibility of communicate privately.
2. *It exposes arbitrarily many cleartext-ciphertext pairs.* Indeed, whenever user i requests to speak to user j he will receive, for free, a session key S and its encryption (with j's secret key).
3. *It requires encryption to provide authentication.*
   Assume that the encryption between i and T consists of exclusive-oring messages with a one-time pseudo-random pad generated on input K_i. Also assume that it is an enemy (instead of T) who send i the global message, and that he chooses it to consist of two random values: α and β. Then, i will compute the decryptions, S and R, of, respectively, α and β. Of course, it is very unlikely that R will be an encryption of S with key K_j. Thus a message sent by i to j encrypted with S will not be understood by j. But i has no way to realize this without engaging in an additional protocol with j. Thus, even if a good encryption scheme (such as exclusive-oring with a pseudo-random pad) is used in the Needham-Schroeder approach, an additional interactive protocol between users becomes more of a necessity than an option. This is a pity, because now their approach stops from being very simple, and is also made less efficient. Moreover, such additional protocols need to be secure, something that it is not trivial to achieve —at least if one wishes to keep them very simple.
   (Indeed, as we have already mentioned, such an additional protocol was recommended in the original paper, though it did not quite achieve its intended goal. Things, in fact, are not that simple. Not only should i and j verify that they share a common key, but they also better verify that this key is the one chosen by T for this particular session —a task further complicated by the fact that an enemy may impersonate both T and j in order to fool i.)
   Do such complications arise because xoring with a pseudo-random pad is not a secure encryption algorithm? The answer is no: xoring with a pseudo-random pad is a very secure encryption scheme. But the Needham-Schroeder paradigm derives its attractive simplicity from the assumption that encryption schemes can guarantee both privacy and authentication. When T sends i the two values S and E, it would be desirable both that (a) only i will

understand them, and that (*b*) user i can be guaranteed that it is T who is sending them. Now, encryption schemes are traditionally designed so as to guarantee property (*a*), but not property (*b*). Indeed, it should be realized that the problems just discussed, which are evident in the case of one-time-pad encryption, may become harder to see, but not necessarily disappear, if more complex block-ciphers are used. Property (*b*) can instead be guaranteed —say— by digital signature schemes, but requiring their explicit use would deprive the Needham-Schroeder approach of its attractive simplicity, and would substantially increase its efficiency and the cost of its hardware implementations. (In fact, each user should then be given the additional circuitry necessary for digital signature verification.)

ADVANTAGES OF OUR APPROACH. Even before relaxing the assumption of a trusted agent, our approach offers significant advantages.
First, it is very simple, efficient, and easy to implement. In particular, it does not require that i and j engage in additional protocols for guaranteeing i that he and j share the same secret key.
Second, the security of our scheme does not depend from complicated assumptions (such as the existence of —hopefully efficient— encryption schemes that possess additional and little-understood properties); rather, our scheme solely requires that *ordinary* one-way functions exist.
Finally, for talking in private to an user j, user i needs to access the trusted party only the first time that such a conversation is desired; the responsibility of choosing session keys is the users'; and no undue stress (such as revealing an arbitrary number of cleartext-ciphertext pairs for free) is imposed on the secret encryption keys.

3.2 Our Software-Based, Trusted-Agent Scheme.

To analize this scheme we only consider two parameters: N, the total number of users, and k the length of the common secret keys. (As we shall see, in fact, a big advantage of this approach is that an adversary cannot eavesdrop conversations between two honest users no matter how many other users he may compromise.)

The high-level mechanics of the scheme are as follows. The trusted agent, T, gives to each user i two secret individual keys: an *exchange key* K_i and an *authentication key* K_i'. (Using two keys considerably simplifies the description and the analysis of the scheme.) In addition, upon request, T can quckly compute and make available a *pair key* $\mathcal{P}_{i,j}$ for each pair of users i and j. To send an encrypted message to user j, i uses K_i and the pair key $\mathcal{P}_{i,j}$. To decrypt the resulting ciphertext, j only uses his own individual key K_j. The scheme possesses two important properties:

1. Without knowledge of K_i or K_j, the pair key $\mathcal{P}_{i,j}$ (and any other pair key as well) is useless for deciphering the encrypted communications between i and j; and
2. Learning the individual keys of any number of users does not help in eavesdropping the conversations between any other two users.

The first property guarantees that there is no need to protect the pair keys, something that greatly accounts for the simplicity of our scheme. Indeed, though this would be quite inefficient, all N^2 of them could be made public, in which case there would be no need to call the trusted party.

The second property implies that, differently from the hardware-based approach of Section 2, there is nothing to be gained by preventing a user from knowing his own key. Of course, each user should not loose or divulge his individual key (since the secrecy of his communications depend on it), but the scheme is such that compromising a user's individual key is only useful to understand the communications relative to that user alone. This greatly simplifies the logistic requirements of our scheme, and allows it to be implemented with a very unexpensive apparatus.

PSEUDO-RANDOM FUNCTIONS. Also the present secret-key agreement has a perfectly-secure version, but given that it is both rather inefficient and easily derivable from the computationally secure one, it is only the latter version that we shall describe below.

A critical building block for our computationally-secure version is the notion of a pseudo-random function generator, as defined by Goldreich, Goldwasser, and Micali [5]. Roughly said, such a generator is a easy to compute function $h(\cdot,\cdot)$ mapping pairs of —say— k-bit strings into k-bit strings. When the first argument is fixed to a randomly and secretly chosen value K, then the resulting single-argument pseudo-random function $f_K(x) = h(K,x)$ passesses all polynomial-time statistical tests for functions. That is, any $poly(k)$-time observer who asks for and receives function values at inputs of his choice (but does not know the initial value K) cannot distinguish the case when his questions are consistently answered by means of the specially-constructed pseudo-random function $h(K,\cdot)$, or by means of a function $f(\cdot)$ randomly selected among all possible ones mapping k-bit strings to k-bit strings. We stress that only the random initial value K needs to be kept secret; the program for h is instead assumed to be public knowledge. The authors of [5] also show that (good) pseudo-random *function* generators can be constructed given any (good) pseudo-random *number* generator. (Thus pseudo-random functions in their sense exist if a and only if *ordinary* one-way functions exist.) The cost of one evaluation of such a pseudo-random function f_K equals that of generating $2k^2$ pseudo-random bits (with a k-bit seed), and is thus essentially negligible.

In practice, one might prefer to use a one-way hash function

$$H : \{0,1\}^{2k} \rightarrow \{0,1\}^k$$

as his pseudo-random function generator. In this case, $h(K,x) = H(K|x)$ will be his chosen pseudo-random function, where the symbol "|" denotes concatenation. Such pseudo-random function generators will be even faster to evaluate.

In any case, in describing our scheme with inputs N and k, we assume that a pseudo-random function generator $h : \{0,1\}^k \times \{0,1\}^k \rightarrow \{0,1\}^k$ has been publically agreed-upon. We also assume that the number of possible users is upperbounded by a small polynomial in k.

COMPUTING INDIVIDUAL KEYS AND PAIR KEYS. The trusted agent randomly and secretly selects two k-bit *master secret keys* K and K'. He then privately gives to user i, as his individual exchange key, the value $K_i = h(K, i)$, and as his individual authentication key the value $K'_i = h(K', i)$. Whenever user i asks him for the pair key of two users i and j, the trusted agent computes and sends him the k-bit value value

$$\mathcal{P}_{i,j} = h(K_j, i) \oplus h(K_i, j)$$

together with the k-bit authentication value

$$\mathcal{A}_{i,j} = h(K'_i, h(K_j, i)).$$

COMPUTING COMMON SECRET KEYS. The common secret key used by user i to send a private message to user j is $K_{i,j} = h(K_j, i)$. User i computes this key by retrieving the pair key $\mathcal{P}_{i,j}$ from his personal directory (if he has spoken privately to j in the past) or by asking the trusted agent for it (if it is the first time he wishes to speak privately to j). Then i computes $V = \mathcal{P}_{i,j} \oplus h(K_i, j)$ (which is easy for user i because he knows his individual key and the name of the intended recipient) and checks whether $h(K'_i, V) = \mathcal{A}_{ij}$. If so, he then randomly selects a session key S and sends to j an encryption of S with key V.

To read the session key sent by i, j simply computes the common secret key $V = h(K_j, i)$, from the sender's name and his own individual key (no table lookup is needed for receiving) and then decrypts i's ciphertext with V so as to obtain S.

(Notice that the common secret key by which i sends private messages to j does not coincide with that by which j sends messages to i, but this does not cause any problems with respect to their ability of communicating with one another. Also notice that while we guarantee i that

ENCRYPTING MESSAGES. While we recommend the use of session keys, recommendation of a particular conventional encryption scheme is beyond the goals of this paper. Whether the common secret key computation occurs within a secure chip or an ordinary personal computer, we insist that it should not become "known" to the user. That is, in the case of a sender i, the key $K_{i,j}$ should reside within i's computer. User i can access only indirectly, by giving his own computing device a message string m that needs to be encrypted with $K_{i,j}$.

EXTENDING OUR SCHEME. In the scheme above, in order to provide a fairer comparison between our approach and that of Needham and Schroeder, we have addressed, with better results, (after all, technical advances should occur after 15 years) essentially the same security concerns as in the summarized Needham-Schroeder scenario. However, it should be realized (and we will explicitly show it in the final paper) that our techniques easily extend to include much tighter security requirements. For instance, we can accomplish that whenever user i asks the trusted party for the pair key with user j, the trusted agent can check that this request indeed comes from i. As for another example, we can accomplish that once both i and j have computed a common secret (session) key, the recipient of a message encrypted with it can be sure that that particular message was indeed sent by i. Details about this will be given in the final paper. We would like to stress from now, however, that like in the simpler setting below, no authenticating step needs to be interactive (like in a challenge-response mechanism), which makes also these secure protocols simple and efficient.

SECURITY. Let us briefly sketch the security of the above scheme in an itemized manner.

- *Unpredictability of individual keys.* Because the k-bit string K is random and secret, and because $h(\cdot,\cdot)$ is a pseudo-random function generator, the individual key of a user i, $h(K,i)$, is unpredictable to an adversary. In fact, $f_K(\cdot) = h(K,\cdot)$ is poly(k)-time undistinguishable from a truly random, secret function f from k-bit strings to k-bit strings. Thus, $f_K(i)$ is polynomial-time unpredictable because $f(i)$ would be totally unpredictable. (This argument actually shows that individual keys are more than unpredictable, they are undistinguishable from random numbers even when their values are actually revealed —provided that the secret master key is still secret. These are important differences, but we shall not discuss them in detail in this abstract.) Further, because $f(i)$ would be totally umpredictable even given the value of f at any other number of inputs, $K_i = f_K(i)$ remains poly(k)-time unpredictable even if all other individual keys become known.
- *Unpredictability of common secret keys.* When user i wishes to talk to user j in private, he does so by means of the common secret key $h(K_j,i)$. User i can compute this key thanks to the special help given him by the trusted agent (i.e., thanks to the particular way in which the trusted agent chooses the pair key $\mathcal{P}_{i,j}$ based on i's individual key). But $K_{i,j}$ is unpredictable to any number of other users, that is, unpredictable given all possible individual and pair keys in the system except the individual keys of i and j. The unpredictability of $h(K_j,i)$ can be proved in two steps. First, $f_{K_j}(\cdot) = h(K_j,\cdot)$ passes all poly(k)-time statistical tests for functions in the sense of [5] whenever K_i is random and secret. This is not sufficient, however, because key K_j, though secret, is not random, but poly(k)-time undistingushable from a truly random number. In fact, $K_j = f_K(j)$ and thus we are not exactly within the hypothesis of [5]. But given that K is random and secret to all users, we can apply a kind of "transitivity property" for poly(k) undistinguishability. Namely, if $h(K_j,i)$ were predictable, being K_j secret and h a pseudo-random function generator, then this would imply the predictability of $K_j = f_K(j)$. But, since K is random and secret, this would contradict the undistinguishability of f_K from a random function. It should be noticed that the this basic argument for the unpredictability of $h(K_j,i)$ keeps on holding when the adversary also knows additional information, such as the individual keys of users z other than i and j. This is so because (a) for a truly random and secretly selected function $f(\cdot)$, the value of f on input i is un-

predictable no matter for how many other inputs z the value of f becomes known, and (*b*) the undistinguishability (in poly(k)-time) of $h(K_j, \cdot)$ from such an f. This argument can be extended so as to take in consideration the case in which *all* possible pair keys in the system and their authenticating values are also known to the adversary (e.g., because he has eavesdropped all possible requests for pair keys). Full details will be given in the final paper. In addition, the authenticating value $\mathcal{A}_{i,j}$ provided by the trusted agent does not help an adversary who has not compromised the individual keys of i and j to predict $K_{i,j}$. In fact, even if he knew the value of $K_{i,j}$, the authenticating value would undistinguishable from a truly random number to him. Indeed, $\mathcal{A}_{i,j} = h(K'_i, K_{i,j})$, and the key K'_i is secret and undistinguishable from a truly random number. Thus, the function $h(K'_i, \cdot)$ is poly(k)-time undistinguishable from a function $f(\cdot)$ truly randomly and secretly selected among those mapping k-bit strings to k-bit strings. But because for such a function the value $f(K_{i,j})$ cannot possibly betray $K_{i,j}$, the same is true for $h(K'_i, K_{i,j})$. (To be precise, for any users $a, b, \ldots$ for which i asks and obtains pair keys, the adversary also sees the authenticating values $h(K'_i, K_{i,a}), h(K'_i, K_{i,b}), \ldots$. But these additional values are practically useless. In fact, in the case of a truly random and secretly selected function f, the values $f(K_{i,a}), f(K_{i,b}), \ldots$ would be useless because random and independent of $f(K_{i,j})$, and $h(K'_i, \cdot)$ is poly(k)-time undistinguishable from such a function.)

- *Requesting pair keys.* In requesting the pair key $\mathcal{P}_{i,j}$, a user i does not need to autheticate himself to the trusted agent. Indeed, there is no need for the trusted agent to ensure that the request of $\mathcal{P}_{i,j}$ comes from i. This is so because, as we have already said, even if all the pair keys were to be made public, no group of malicious users can compute the common secret key of two honest users.
- *Authenticating pair keys.* We have already argued that the authenticating values produced by the trusted agent do not betray the common secret keys. We must now argue that they prevent an adversary z, who cuts off the communication line between i and the trusted agent and tries to impersonate the latter, from finding a false pair key $\mathcal{FP}_{i,j}$ and a false authenticating value $\mathcal{FA}_{i,j}$ such that honest user i accepts as valid

$$FK_{i,j} = \mathcal{FP}_{i,j} \oplus h(K_i, j)$$

as his common secret key with j. What is immediate to argue is that z cannot mislead i into accepting a false secret key $FK_{i,j}$ known to z. This is so because K'_i is unkown and unpredictable to z and h is a pseudo-random function generator; thus, the function $h(K'_i, \cdot)$ is undistinguishable from a truly-random secret function, and the necessary authenticating value $h(K'_i, FV)$ is unpredictable to z. We must also argue, however, that z does not have a realistic chance of authenticating a false common secret key that he does not know. The reason for this is that the pseudo-random functions $h(K_i, \cdot)$ and $h(K'_i, \cdot)$ not only "behave randomly" (because *each* of K_i and K'_i is random and secret), but also behave as *independent* functions, because the values K_i and K'_i are independent random values. Details will be given in the final paper.

REMARK. Notice that this is not a public-key approach not only because each user does not choose his own keys, but also because no user has a public key. Pair keys not only are associated to pairs of users and are chosen by an external party, but do not need to be made public for the scheme to work. Rather, the scheme remains secure *even if* they become public.

EFFICIENCY AND OTHER CONSIDERATIONS. The scheme above described is most efficient.

For the users, computing common secret keys from pair keys is quite trivial. Moreover, user i needs to ask the trusted agent for pair key $\mathcal{P}_{i,j}$ only when he wants to talk to user j in private for the first time. In fact, $K_{i,j}$ is then stored by i for future use. It is important to notice that if also its associated authenticating value is stored alongside with it, this pair key needs not to be stored in a protected memory. Indeed, it can be stored outside the user's own computer. Indeed, the user needs to keep secret only two k-bit values: K_i and K'_i. Since no particular precaution needs to be applied to the pair keys and authenticating values, the user can easily store them all. Thus, if the link to the trusted party is down, users still can (very much as in the public-key scenario) talk in private with every other user with which they did so in the past.

Moreover, since our scheme does not depend on any interactive authentication protocols, the initial effort of calling up the trusted party is negligible. Indeed, obtaining the necessary two k-bit values from the trusted party can be handled much as we currently handle a call to 411 (information) in the phone system today. In fact, the whole process can be easily automated—the caller dials in his own identity and that of j and then receive two 10-byte values in response.

Our scheme is also most convenient from the trusted agent point of view. Indeed, computing an individual key consists of a single pseudo-random function evaluation, which is trivial todo whether or not one uses a one-way hash function in practice. Also handling a request for a pair key is trivial. In fact, even if decides to securely store only his k-bit master secret key K, the trusted party can satisfy a pair-key request by making 5 pseudo-random function evaluations and one sum modulo 2.

3.3 The Multiple-Trustee Scenario

Also our software-based scheme can be easily adapted for use with multiple trusted agents, only one of whom needs to be honest. For example, if there are two trustees, we can make 2 copies of the preceding scheme (one for each trustee) thus there will be two individual secret keys, two individual secret authentication keys, and two pair keys for each pair of users. There will be only one common secret key for each pair of user, however, set to be the sum modulo 2 of the two common secret keys relative to each trustee.

Once again, not only does this decrease the amount of trust required, but allows our scheme to make strong encryption compatible with law enforcement.

It should also be realized that, though not needed, secure chips can be useful in this approach too. First, storing individual keys in secure chips cannot but be useful. Second, in the multiple-trustee scenario, each of the trustees can inbed his own k-bit master secret key in a number of secure chips, and then gives these chips to the phone company. Thus, the phone companies need not to be trusted with respect to user privacy (or law enforcement) but only to deliver efficiently pair keys on request in a 411-like manner, something that they are set up to do quite well. In fact, once a pair key internally computed by a secure chip is output, it can be handled without other privacy concerns.

Thus, like our first approach, this one too succeeds in simultaneously accomplishing two tasks: (1) making law enforcement compatible with encryption like in the Clipper Chip scenario, and (2) providing the secret-key agreement missing in the Clipper Chip. Moreover, the present approach succeeds in achieving an additional important goal; namely, (3) *being very economical.* Our second approach, in fact, can be totally and securely implemented in software. Indeed, the only operation required from a user consists in summing two numbers modulo 2 —which, rather than in software, can actually be done by hand.

4 Conclusions

In this paper, we have described two simple schemes for key agreement which offer significant advantages in terms of cost and (potentially) security over traditional number-theoretic schemes such as Diffie-Hellman and RSA. The new schemes are also particularly well-suited for use with the emerging Clipper Chip technology proposed by the Clinton Administration, and with Kerborous.

5 Acknowledgment

Many thanks to Mihir Bellare, Joe Kilian, and Phil Rogaway for their helpful comments.

References

1. N. Alon, P. Erdos, and T. Spencer. The Probabilistic Method. Wiley Interscience Series in Discrete Mathematics and Optimization. John Wiley & Sons, NY, 1992.
2. R. Blom. An Optimal Class of Symmetric Key Generation Systems. In *Advances in Cryptology: Proceedings of Eurocrypt 84, Lecture Notes in Computer Science*, vol,. 209, Springer-Verlag, Berlin, 1987, pp. 335–338.
3. C. Blundo, A. De Santis, A. Herzberg, S. Kutten, U. Vaccaro, and M. Yung. Perfectly Secure Key Distribution for Dynamic Conferences. In *Advances in Cryptology: Proceedings of CRYPTO '92, Lecture Notes in Computer Science*, Springer Verlag, 1992.

4. W. Diffie and M.E. Hellman New Direction in Cryptography. In *IEEE Transaction on Information Theory*, vol. 22, no. 6, December 1976, pp. 644–654.
5. O. Goldreich, S. Goldwasser, and S. Micali How To Construct Random Functions. In *J. of the ACM*, vol. 33, no. 4, October 1986, pp. 792-807.
6. S. Micali. Fair Public-Key Cryptosystems. In *Advances in Cryptology: Proceedings of CRYPTO '92, Lecture Notes in Computer Science*, Springer Verlag, 1992.
7. R. M. Needham and M.D. Schroeder. Using Encryption for Authentication in Large Networks of Computers. In *Comm. ACM*, vol. 21, no. 12, December 1978, pp. 993-999.
8. Y. Zheng. Personal Communication, September 1993.

Broadcast Encryption

Amos Fiat[1] and Moni Naor[2]

[1] Department of Computer Science, School of Mathematics, Tel Aviv University, Tel Aviv, Israel, and Algorithmic Research Ltd.

[2] Department of Computer Science and Applied Math, Weizmann Institute, Rehovot, Israel.

Abstract. We introduce new theoretical measures for the qualitative and quantitative assessment of encryption schemes designed for broadcast transmissions. The goal is to allow a central broadcast site to broadcast secure transmissions to an arbitrary set of recipients while minimizing key management related transmissions. We present several schemes that allow a center to broadcast a secret to any subset of privileged users out of a universe of size n so that coalitions of k users not in the privileged set cannot learn the secret. The most interesting scheme requires every user to store $O(k \log k \log n)$ keys and the center to broadcast $O(k^2 \log^2 k \log n)$ messages regardless of the size of the privileged set. This scheme is resilient to *any* coalition of k users. We also present a scheme that is resilient with probability p against a random subset of k users. This scheme requires every user to store $O(\log k \log(1/p))$ keys and the center to broadcast $O(k \log^2 k \log(1/p))$ messages.

1 Introduction

We deal with broadcast encryption. We consider a scenario where there is a center and a set of users. The center provides the users with prearranged keys when they join the system. At some point the center wishes to broadcast a message (e.g. a key to decipher a video clip) to a *dynamically* changing privileged subset of the users in such a way that non-members of the privileged class cannot learn the message. Naturally, the non-members are curious about the contents of the message that is being broadcast, and may try to learn it.

The obvious solution is: give every user its own key and transmit an individually encrypted message to every member of the privileged class. This requires a very long transmission (the number of members in the class times the length of the message). Another simple solution is to provide every possible subset of users with a key, i.e. give every user the keys corresponding to the subsets it belongs to. This requires every user to store a huge number of keys.

The goal of this paper is to provide solutions which are efficient in both measures, i.e. transmission length and storage at the user's end. We also aim that the schemes should be computationally efficient.

To achieve our goal we add a new parameter to the problem. This parameter represents the number of users that have to collude so as to break the scheme. The scheme is considered broken if a user that does not belong to the privileged class can read the transmission. For a given parameter k, our schemes should be resilient to any subset of k users that collude and any (disjoint) subset (of any size) of privileged users.

We also consider another scheme parameter, the random-resiliency of a scheme which refers to the expected number of users, chosen uniformly at random, that have to collide so as to break the scheme.

In many applications, it suffices to consider only the (weaker) random-resiliency measure. For example, if decryption devices are captured from random users, (or were assigned at random to users), it is the random resiliency that determines how many devices need be captured so as to break the scheme. We discuss a number of different scenarios with differing assumptions on the adversary strength. We show that even powerful and adaptive adversaries are incapable of circumventing the protection afforded by our schemes.

The final goal of the broadcast encryption scheme is to securely transmit a message to all members of the privileged subset. If cryptographic tools such as one-way functions exist then this problem can be translated into the problem of obtaining a common key. Let the security parameter be defined to be the length of this key.

1.1 Definitions

A broadcast scheme allocates keys to users so that given a subset T of U, the center can broadcast messages to all users following which all members of T have a common key.

A broadcast scheme is called *resilient* to a set S if for every subset T that does not intersect with S, no eavesdropper, that has all secrets associated with members of S, can obtain the secret common to T. We can relax the requirement that no adversary can obtain the secret to one that says that no adversary that is computationally bounded by probabilistic polynomial time can obtain the key with non-negligible probability (i.e. greater than inverse polynomial).

A scheme is called *k-resilient* if it is resilient to *any* set $S \subset U$ of size k. We also deal with random coalitions: a scheme is called *(k,p)-random-resilient* if with probability at least $1-p$ the scheme is resilient to a set S of size k, chosen at random from U. Let $|U| = n$, we use n and $|U|$ interchangeably hereinafter.

The relevant "resources" which we attempt to optimize are the number of keys associated with each user, the number of transmissions used by the center, and the computation effort involved in retrieving the common key by the members of the privileged class.

1.2 Results

As a function of the resiliency required, we provide a large set of schemes that offer a tradeoff between the two relevant resources: memory per user and transmission length.

If nothing is known about the privileged subset T, any broadcast scheme requires that the transmission be sufficiently long to uniquely identify the privilege subset T. Otherwise, by a simple counting argument, there would be two non-identical sets, T and T', both of which somehow manage to obtain the same common key.

Thus, in general, simply representing a subset $T \subset U$ requires $|U|$ bits. Using our schemes, transmitting an additional $o(|U|)$ bits guarantees security against all coalitions of size $\tilde{O}(\sqrt{|U|})$ users and randomly chosen coalitions of $\tilde{O}(|U|)$ users. The computational and memory requirements for these schemes are $\tilde{O}(\sqrt{U})$. Thus, in some sense, security is available for "free".

In fact, in many contexts the privileged set may be identified by sending a relatively short transmission. *E.g.*, if the set can somehow be computed from an old privileged set or the set representation can be compressed. Thus, we distinguish between the *set identification* transmission and the *broadcast encryption* transmission. Our goal is the study of broadcast encryption transmissions and their requirements. In general, the center will identify every user with a unique identification number, and thus the set representation can be a bit vector. There are distinct advantages that the identification numbers be assigned at random to new users, we discuss this hereinafter in the context of random resiliency.

We distinguish between zero-message schemes and more general schemes. Zero-message schemes (Section 2) have the property that knowing the privileged subset T suffices for all users $x \in T$ to compute a common key with the center without any transmission. Of course, to actually use a zero-message scheme to transmit information implies using this key to encrypt the data transmitted.

More general schemes (Section 3) may require that the center transmit many messages. All the schemes we describe require that the length of the center generated messages be equal in length to the security parameter. Thus, when counting messages transmitted by the center, each messages is s bits in length.

Our general approach to constructing schemes is to use a two stage approach. First, we construct low resiliency zero-message schemes and then use these to construct higher resiliency schemes. The latter are not zero-message type schemes.

For low resiliency schemes, we describe assumption-free constructions, that are based upon no cryptographic assumption (the equivalent of a one-time pad). Then, we describe more efficient schemes based upon a some cryptographic assumptions, either the existence of a one way function or the more explicit assumption that RSA is secure. These results are described in Theorems 1, 2, 3.

We then deal with the more general case, and describe schemes of high resiliency (Section 3). For clarity of exposition, we describe our constructions in terms of the number of "levels" involved in the scheme construction. Informally, the levels refer to a sets of hash functions that partition and group users in a variety of ways. Our proofs are all based upon applications of the probabilistic method [1].

To obtain a resiliency of k, it suffices to store $k \log k \log n$ keys per user, while the number of messages transmitted by the center is $O(k^2 \log^2 k \log n)$ (Theorem 8). To obtain a random resiliency of k, with probability p, it suffices to store $\log k \log(1/p)$ keys per user, while the number of messages transmitted by the center is $O(k \log k \log(1/p))$ (Corollary 9). Other points along the tradeoff between memory and transmission length are given in Theorem 5.

2 Zero Message Schemes

In this section we present several schemes that do not require the center to broadcast any message in order for the member of the privileged class to generate a common key. The main significance of the schemes presented in this section is their application as building blocks for the schemes presented in Section 3.

2.1 The Basic Scheme

The basic scheme we define allows users to determine a common key for every subset, resilient to any set S of size $\leq k$. The idea is very simple.

For every set $B \subset U$, $0 \leq |B| \leq k$, define a key K_B and give K_B to every user $x \in U - B$. The common key to the privileged set T is simply the exclusive or of all keys K_B, $B \subset U - T$. Clearly, every coalition of $S \leq k$ users will all be missing key K_S and will therefore be unable to compute the common key for any privileged set T such that $S \cap T$ is empty.

The memory requirements for this scheme are that every user is assigned $\sum_{i=0}^{k} \binom{n}{k}$ keys. With these requirements we need make no assumptions whatsoever. We therefore have

Theorem 1. *There exists a k-resilient scheme that requires each user to store $\sum_{i=0}^{k}\binom{n}{k}$ keys and the center need not broadcast any message in order to generate a common key to the privileged class.*

2.2 1-Resilient Schemes using Cryptographic Assumptions

We now see how to improve the memory requirements of the scheme described above using cryptographic assumptions such as "one-way functions exist" and that extracting prime roots modulo a composite is hard. The improvements are applicable to any k, however they are the most dramatic for $k = 1$.

A 1-resilient scheme based on one-way functions.

Consider the 1-resilient version of the scheme described above. It requires every user to store $n+1$ different keys. However, this can be reduced to $O(\log n)$ keys per user if the keys are pseudo-randomly generated from a common seed where the pseudo-random function f output is twice the length of the input, as described below.

Assume that one-way functions exist and hence pseudo-random generators exist (see [6])). We first explain how the key distribution is done. Associate the n users with the leaves of a balanced binary tree on n nodes. The root is labeled with the common seed and other vertices are labeled recursively as follows: apply the function f to the root label and taking the left half of the function value to be the label of the right subtree while the right half of the function value is the label of the left subtree. (This is similar to the construction of the tree in the generation of a pseudo-random function in [5].)

By the scheme of Section 2.1, every user x should get all the keys except the one associated with the singleton set $B = \{x\}$. To meet this goal remove the path from the leaf associated with the user x to the root. We are left with a forest of $O(\log n)$ trees. Give the user x the labels associated with the roots of these trees. The user can compute the all leaf labels (except K_B) without additional help. Therefore we have

Theorem 2. *If one-way functions exist, then there exists a* 1*-resilient scheme that requires each user to store* $\log n$ *keys and the center need not broadcast any message in order to generate a common key to the privileged class.*

A 1-resilient scheme based on Computational Number Theoretic Assumptions

A specific number theoretic scheme, cryptographically equivalent to the problem of root extraction modulo a composite, can further reduce the memory requirements for $1-$*resilient* schemes. This scheme is cryptographically equivalent to the RSA scheme [8] and motivated by the Diffie-Hellman key exchange mechanism, and the original Shamir cryptographically secure pseudo-random sequence. [3, 9].

The center chooses a random hard to factor composite $N = P \cdot Q$ where P and Q are primes. It also chooses a secret value g of high index. User i is assigned key $g_i = g^{p_i}$, where p_i, p_j are relatively prime for all $i, j \in U$. (All users know what user index refers to what p_i). A common key for a privileged subset of users

T is taken as the value $g_T = g^{\prod_{i \in T} p_i} \bmod N$ Every user $i \in T$ can compute g_T by evaluating

$$g_i^{\prod_{j \in T-\{i\}} p_j} \bmod N$$

Suppose that for some $T \subset U$ and some $j \notin T$ user j could compute the common key for T. We claim that it implies that the user could also compute g: given $a^x \bmod N$ and $a^y \bmod N$ and x and y one can compute $a^{GCD(x,y)} \bmod N$ by performing a sequence of modular exponentiations/divisions on a^x and a^y (see [9]). As the GCD of p_j and $\prod_{h \in T} p_h$ is 1, it follows that g can be computed by user j in this manner. Thus, the user could compute the p_j'th root of g^{p_j} while knowing only the composite N. Therefore if this is assumed to be hard, then the scheme is 1-resilient and we have

Theorem 3. *If extracting root modulo composites is hard, then there exists a* 1-*resilient scheme that requires each user to store one key (of length proportional to the composite) and the center need not broadcast any message in order to generate a common key to the privileged class.*

Clearly, this scheme is not 2-resilient since any two user can collude and compute g.

3 Low Memory k-Resilient Schemes

The zero message k-resilient schemes described in the proceeding section require for $k > 1$ a great deal of memory, exponential in k. In this section we provide several efficient constructions of k-resilient schemes for $k > 1$. Our schemes are based on a method of converting 1-resilient schemes into k-resilient schemes. Throughout this section we assume the existence of a 1-resilient scheme for any number of users. This can be taken as the no-assumption scheme, or any of the cryptographic assumption variants.

Let w denote the number of keys that a user is required to store in the 1-resilient scheme. I.e $w = n + 1$ if no cryptographic assumptions are made, $w = \log n$ if we assume that one-way functions exists and $w = 1$ if we assume that it is hard to extract roots modulo a composite. The efficiency of our schemes will be measured by how many w's they require.

3.1 One Level Schemes

Consider a family of functions $f_1, \ldots, f_l$, $f_i : U \mapsto \{1, \ldots, m\}$, with the following property: For every subset $S \subset U$ of size k, there exists some $1 \leq i \leq l$ such that for all $x, y \in S$: $f_i(x) \neq f_i(y)$. This is equivalent to the statement that the family of functions $\{f_i\}$ contains a perfect hash function for all size k subsets of U when mapped to the range $\{1, \ldots, m\}$. (See [7] or [4] for more information on perfect hash functions.)

Such a family can be used to obtain a k-resilient scheme from a 1-resilient scheme. For every $1 \leq i \leq l$ and $1 \leq j \leq m$ use an independent 1-resilient scheme

$R(i,j)$. Every user $x \in U$ receives the keys associated with schemes $R(i, f_i(x))$ for all $1 \le i \le \ell$. In order to send a secret message M to a subset $T \subset U$ the center generates random strings $M^1, \ldots, M^\ell$ such that $\bigoplus_{i=1}^{l} M^i = M$. The center broadcasts for all $1 \le i \le \ell$ and $1 \le j \le m$ the message M^i to the privileged subset $\{x \in T | f_i(x) = j\}$ using scheme $R(i,j)$. Every user $x \in T$ can obtain all the messages $M^1, \ldots M^\ell$ and by Xoring them get M.

The number of keys each user must store is m time the number needed in the 1-resilient scheme. The length of the transmission is $\ell \cdot m$ times the length of the transmission for a zero message 1-resilient scheme, equal to the security parameter.

Claim 4 *The scheme described above is a k-resilient scheme*

We now see what values can m and ℓ take. It turns out that setting $m = 2k^2$ and $\ell = k \log n$ is sufficient. This can be seen via a probabilistic construction. Fix $S \subset U$ of size k. The probability that a random f_i is 1-1 on S is at least $1 - \sum_{i=1}^{k-1} \frac{i}{2k^2} \ge \frac{3}{4}$. Therefore the probability that for no i we have that f_i is 1-1 on S is at most $1/4^\ell = 1/n^{2k}$. Hence the probability that for all subsets $S \subset U$ of size k there is a 1-1 f_i is at least $1 - \binom{n}{k} \cdot \frac{1}{n^{2k}} \ge 1 - \frac{1}{n^k}$. We therefore conclude

Theorem 5. *There exists a k-resilient scheme that requires each user to store $O(k \log n \cdot w)$ keys and the center to broadcast $O(k^3 \log n)$ messages. Moreover, the scheme can be constructed effectively with arbitrarily high probability by increasing the scheme parameters appropriately.*

The proof implies that against a randomly chosen subset $S \subset U$ of size k we can have a much more efficient scheme:

Corollary 6. *For any $1 \le k \le n$ and $0 \le p \le 1$ there exists a (k,p)-random-resilient scheme that requires each user to store $O(\log(1/p) \cdot w)$ keys and the center to broadcast $O(k^2 \log(1/p))$ messages. Simply choose $m = k^2$ and $\ell = \log p$. Moreover, the scheme can be constructed effectively with arbitrarily high probability by increasing the scheme parameters appropriately.*

As for explicit constructions for the family $f_1, \ldots f_\ell$, they seem to be at least a factor of k more expensive. Consider the family

$$F = \{f_p(x) = x \bmod p | p \le k^2 \log n \text{ and is a prime}\}$$

F satisfies the above requirement.

The number of keys stored per user in this explicit construction is $O(k^2 \log n / \log\log n)$ and the number of messages that the center broadcasts is $O(k^4 \log^2 n / \log\log n)$.

3.2 Remarks

After having seen the single-level schemes above, we wish to clarify certain points that can be discussed only after seeing an example of the types of schemes we deal with. We will continue with more efficient multi-level schemes in the next section, the remarks of this section are clearly applicable to both single and multi level schemes.

Representing the Functions. In some applications using probabilistic constructions is problematic because of representation problem, i.e that storing the resulting structure may be prohibitively expensive. However, as described above, our schemes do not absolutely require that the f_i functions be computable, the user could simply be assigned $f_i(x)$. This could be chosen at random. The center could in fact generate all required functions from a pseudo-random function and a single seed.

Alternatively, instead of using completely random functions one can use $\log k$-wise independent functions such as degree $\log k$ polynomials. The results regarding the probabilistic construction remain more or less true. The advantage is that there is a succinct representation for the functions now. Storing such function representations in the user decryption devices is not much more expensive than storing the keys required in the above schemes.

Reducing Storage. Suppose that we are interested in limiting the number of keys that a user must store (at the the expense of the number of keys that the center must broadcast). We can get a certain tradeoff: instead of hashing to a range of size $2k^2$ we hash to range of size $m = a \cdot k^2$. The results that we get in this case are that the memory requirements are smaller by a $\log a$ factor and the broadcast requirements are larger by a factor of a. This is true for both k-resilient schemes and for (k,p)-random-resilient schemes.

We now describe yet another tradeoff that may reduce storage requirements. Every $R(i,j)$ scheme above deals with a subset of the users. If we assume that the f_i functions can be computed by anyone (e.g., k-wise independent functions as described above), then the $R(i,j)$ 1-resilient schemes can be devised so as to deal with the true number of users associated with the scheme, depending on the underlying 1-resilient scheme, this leads to a saving in the memory requirements described in the scheme, at the expense of some additional computation.

Adversary Limitations and Resiliency. A k-resilient scheme is resilient to any coalition of size k, this means that irrespective of how the adversary goes about choosing the coalition, no coalition of size smaller than k will be of any use to the adversary. However, the scheme is resilient to many sets of size much larger than k.

The adversary may capture devices at random, in this case the random resiliency measure is directly applicable. Given a $(V, 1/2)$ randomized resilient scheme, the expected number of devices that the adversary must capture to break the scheme is at least $V/2$.

A possibly legitimate assumption is that the user of the decryption device does not even know his unique index amongst all users. For example, the user index and all user secrets could be stored on a (relatively) secure smartcard, such a smartcard is probably vulnerable but not to a casual user. Thus, if user indices are assigned at random any set of devices captured will be a random set irrespective of the adversary strategy used.

The definition of (k,p) random resiliency is somewhat problematic for two reasons:

1. The probability p is an absolute probability, this does not make sense if the underlying one resilient schemes we are using can be themselves broken with relatively high probability (e.g., by guessing the short secret keys).
2. The assignment of users ids (index numbers) to users is assumed to be random and secret. But, it may be possible to learn the user identification by monitoring transmissions and user behavior.

To avoid both these problems we define a new notion of resiliency and say that a scheme is *(k,p)-immune* if for any adversary choosing adaptively a subset S of at most k users and a disjoint subset T we have: the probability that the adversary (knowing all the secrets associated with S) guesses the value the center broadcasts to T is larger by at most (additive) p than the probability the adversary would have guessed it without knowing the secrets of S.

If we assume that the functions f are kept secret then the results we can get for (k,p)-immune schemes are very similar to the results for (k,p)-random-resilient schemes. However, we do not know whether this holds in general for all random-resilient schemes. This is true since the random constructions for both single level schemes and multi level schemes (described in the next section), the analysis fixes the subset S and evaluates the probability that it is good for a random construction. Since the adversary does not know the values of the hash functions (f_i for single level schemes) when adding a user to S, any choice of S has the same probability of being bad.

For completeness, we note that yet another attack is theoretically possible, although it may be rather difficult in practice. The adversary may attempt to actively subvert the system by publishing a solicitation for dishonest users that meet certain criteria. Specifically, it would be very useful for the adversary to capture pairs of devices that belong to the same 1-resilient $R(i,j)$ scheme described above, if he captures ℓ pairs (a_i, b_i) such that $f_i(a_i) = f_i(b_i)$ then he has corrupted our scheme above. In this case, a true k-resilient scheme is the only prevention. If k is sufficiently large and the number of traitors does not exceed k then the scheme is secure.

3.3 Multi-Level Schemes

We now describe a general multi-level scheme that converts a scheme with small resiliency to one with large resiliency. Consider a family of functions $f_1, \ldots, f_l$, $f_i : U \mapsto \{1, \ldots, m\}$ and a collection of sets of schemes,

$$\{R(i,j) | 1 \leq i \leq l, 1 \leq j \leq m\},$$

where each $R(i,j)$ consists of l' schemes labeled $R(i,j,1), \ldots, R(i,j,l')$. These functions and schemes obey the following condition: For every subset $S \subset U$ of size k, there exists some $1 \leq i \leq l$ such that for all $1 \leq j \leq m$ there exists some $1 \leq r_j \leq l'$ such that the scheme $R(i,j,r_j)$ is resilient to the set $\{x \in S | f_i(x) = j\}$.

We claim that such a structure can be used to obtain a k-resilient scheme: Generate independently chosen keys for all schemes $R(i,j,r)$. A user $x \in U$ receives for every $1 \le i \le l$ and every $1 \le r \le l'$ the keys associated with x in scheme $R(i, f_i(x), r)$. Given a subset $T \subset U$ and a secret message M, the center generates:

- Strings $M^1, \ldots, M^l$ such that $\bigoplus_{i=1}^{l} M^i = M$ and $M^1, \ldots, M^{l-1}$ are chosen at random.
- For every $1 \le i \le l$, and $1 \le j \le m$ random strings $M_1^{(i,j)}, \ldots, M_{l'}^{(i,j)}$, such that $\bigoplus_{t=1}^{l'} M_t^{(i,j)} = M^i$.

The center broadcasts for all $1 \le i \le \ell$ and $1 \le j \le m$ and $1 \le r \le l'$ the message $M_r^{(i,j)}$ to the privileged subset $\{x \in T | f_i(x) = j\}$ using scheme $R(i,j,r)$. Every user $x \in T$ can obtain for all $1 \le i \le \ell$ and $1 \le r \le l'$ messages $M_r^{(i,f_i(x))}$. To reconstruct the message M, the user $x \in T$ takes the bitwise exclusive or of all messages transmitted to the user in all schemes to which the user belongs, i.e., in all schemes $R(i,j,r)$ such that $f_i(x) = j$.

The number of keys associated with user x is therefore the number of keys associated with a scheme $R(i,j,r)$ times $l \times l'$. The length of a broadcast is equal to the number of messages transmitted in an $R(i,j,r)$ scheme times $l \times m \times l'$.

Claim 7 *The scheme described above is a k-resilient scheme.*

We now describe a concrete two level scheme using this method. Set $\ell = 2k \log n$, $m = k/\log k$, $t = 2e \log k$ and $l' = \log k + 1$. The first level consists of a family of ℓ functions $f_1, \ldots, f_l$, $f_i : U \mapsto \{1, \ldots, m\}$. At the second level we have function $g_r^{(i,j)} : U \mapsto \{1, \ldots 2t^2\}$ for all $1 \le i \le \ell$, $1 \le j \le m$ and $1 \le r \le l'$. Every such (i,j,r) and $1 \le h \le 2t^2$ defines a 1-resilient scheme $R(i,j,r,h)$ as in the scheme of Section 3.1. Every user x receives the keys of schemes $R(i, f_i(x), r, g_r^{(i,f_i(x))}(x))$ for all $1 \le i \le \ell$ and $1 \le r \le l'$.

For a set $S \subset U$ of size k we say that i is *good* if for all $1 \le j \le m$

1. $|\{x \in S | f_i(x) = j\}| \le t$.
2. there exists $1 \le r \le l'$ such that $g_r^{(i,j)}$ is 1-1 on $\{x \in S | f_i(x) = j\}$.

By Claim 7 we can show that if for every set $S \in U$ of size k there is a good i, then the scheme is k-resilient.

We prove that randomly chosen f_i and $g_r^{(i,j)}$ constitute a good scheme with reasonably high probability.

Fix a subset $S \subset U$ of size k and $j \in \{1 \ldots m\}$. The probability that Condition 1 above is not satisfied is at most

$$\binom{k}{t} \cdot \left(\frac{1}{m}\right)^t \le \left(\frac{ek}{2e \log k}\right)^{2e \log k} \cdot \left(\frac{\log k}{k}\right)^{2e \log k} = \left(\frac{1}{2}\right)^{2e \log k} = \frac{1}{k^{2e}}$$

Suppose that condition 1 is satisfied, then for any $1 \le r \le l'$ the probability that $g_r^{(i,j)}$ is 1-1 on $\{x \in S | f_i(x) = j\}$ is at least $1 - t\frac{1}{2t} = \frac{1}{2}$. Hence the probability that condition 2 is not satisfied is at most $1/2^{l'} = 1/2k$ and therefore the probability that Conditions 1 and 2 are both satisfied for every $1 \le j \le m$ is at least $1/2$. The probability that *no* i is good for S is at most $1/2^{\ell} = 1/n^{2k}$. Hence the probability that all subsets $S \subset U$ of size k have a good i is at least $1 - \binom{n}{k} \cdot \frac{1}{n^{2k}} \geq 1 - \frac{1}{n^k}$. We therefore conclude

Theorem 8. *There exists a k-resilient scheme that requires each user to store $O(k \log k \log n \cdot w)$ keys and the center to broadcast $O(k^2 \log^2 k \log n)$ messages. Moreover, the scheme can be constructed effectively with high probability.*

Corollary 9. *For any $1 \leq k \leq n$ and $0 \leq p \leq 1$ there exists a (k,p)-random-resilient scheme with the property that the number of keys each user should store is $O(\log k \log(1/p) \cdot w)$ and the center should broadcast $O(k \log^2 k \log(1/p))$ messages. Moreover, the scheme can be constructed effectively with high probability.*

4 An Example and Practical Considerations

The schemes described in this paper are valid for all possible values of the parameters. However, if random resiliency suffices, and if one seeks a solution to a concrete example then other considerations creep in.

Say we've got a user group of one billion subscribers. Also, assume that our goal is that to discourage any possible pirate box manufacturer, and thus the expectation should be that he is required to capture $k = 100,000$ devices before seeing any return on his investment.

Basing our 1-resilient scheme on the number theoretic scheme, and using our randomized $(100000, 1/2)$-resilient scheme, the number of keys stored in every subscriber decryption device is less than 20, and the length of a broadcast enabling transmission is on the order of two million keys. (Vs., one billion keys transmitted for standard schemes).

However, there is a major problem, with the set identification transmission. It seems that all subscribers will have to listen to one billion bits of set identification transmission without making a single error. In fact, the subscriber is apathic to the presence or absence of most of the users. It is only users that belong to the same underlying 1-resilient schemes that he belongs to that matter. Thus, there are advantages to splitting up users into independent broadcast encryption schemes, determining what user gets assigned to what scheme at random. By appropriately resynchronizing and labeling schemes, the decryption device will only have to deal with the set identification transmission dealing with one (smaller) scheme.

There is a tradeoff between error control issues and security. If the number of broadcast encryption schemes gets too large, and the resiliency gets too small, then the (multiple) birthday paradox enters into consideration. (We say such a scheme is broken if any of it's component broadcast encryption schemes is broken).

Say we split the billion users above into randomly assigned broadcast encryption groups of 1000 users. We use a non-random 5-resilient broadcast encryption scheme which requires about 10 keys stored per user, and 100 keys transmission per broadcast encryption scheme, for a total of 10^8 key transmissions. The total random resiliency is approximately $1,000,000^{5/6} = 100,000$. (The adversary must randomly select devices until he has 5 different devices from the same broadcast encryption scheme). Transmissions are 50 times longer than before, but still significantly shorter than individual transmissions. This is a practical scheme since there is no longer any serious error control problem.

Another advantage of the scheme presented in this section is that if the adversary is in fact successful, after collecting 100,000 decryption devices, and if we have captured one of the adversary eavesdropping devices, all is not lost. It is still a relatively simple matter to disable all adversary devices by disabling one group of 1000 users, splitting these users amongst other groups, the adversary effort has been in vain.

References

1. N. Alon and J. Spencer, **The Probabilistic Method**, Wiley, 1992.
2. J. L. Carter and M. N. Wegman, *Universal Classes of Hash Functions*, Journal of Computer and System Sciences 18 (1979), pp. 143-154.
3. W. Diffie and M. Hellman, *New Directions in Cryptography* , IEEE Trans. on Information Theory, vol. IT-22, 6 (1976), pp. 644-654.
4. M.L. Fredman, J. Komlós and E. Szemerédi, *Storing a Sparse Table with $O(1)$ Worst Case Access Time*, Journal of the ACM, Vol 31, 1984, pp. 538–544.
5. O. Goldreich, S. Goldwasser and S. Micali, *How to Construct Random Functions* Journal of the ACM **33**, 1986.
6. R. Impagliazzo, L. Levin and M. Luby, *Pseudo-random Generation given from a One-way Function*, Proc. of the 20th ACM Symp. on Theory of Computing, 1989.
7. K. Mehlhorn, **Data Structures and Algorithms: Sorting and Searching**, Springer-Verlag, Berlin Heidelberg, 1984.
8. R. Rivest, A. Shamir and L. Adleman, *A Method for Obtaining Digital Signature and Public Key Cryptosystems*, Comm. of ACM, 21 (1978), pp. 120-126.
9. A. Shamir, *On the Generation of Cryptographically Strong Pseudo-Random Number Sequences*, ACM Trans. Comput. Sys., 1 (1983), pp. 38-44.
10. M. N. Wegman and J. L. Carter, *New Hash Functions and Their Use in Authentication and Set Equality*, Journal of Computer and System Sciences 22, pp. 265-279 (1981).

Index

Lecture Notes in Computer Science

For information about Vols. 1–693
please contact your bookseller or Springer-Verlag

Vol. 694: A. Bode, M. Reeve, G. Wolf (Eds.), PARLE '93. Parallel Architectures and Languages Europe. Proceedings, 1993. XVII, 770 pages. 1993.

Vol. 695: E. P. Klement, W. Slany (Eds.), Fuzzy Logic in Artificial Intelligence. Proceedings, 1993. VIII, 192 pages. 1993. (Subseries LNAI).

Vol. 696: M. Worboys, A. F. Grundy (Eds.), Advances in Databases. Proceedings, 1993. X, 276 pages. 1993.

Vol. 697: C. Courcoubetis (Ed.), Computer Aided Verification. Proceedings, 1993. IX, 504 pages. 1993.

Vol. 698: A. Voronkov (Ed.), Logic Programming and Automated Reasoning. Proceedings, 1993. XIII, 386 pages. 1993. (Subseries LNAI).

Vol. 699: G. W. Mineau, B. Moulin, J. F. Sowa (Eds.), Conceptual Graphs for Knowledge Representation. Proceedings, 1993. IX, 451 pages. 1993. (Subseries LNAI).

Vol. 700: A. Lingas, R. Karlsson, S. Carlsson (Eds.), Automata, Languages and Programming. Proceedings, 1993. XII, 697 pages. 1993.

Vol. 701: P. Atzeni (Ed.), LOGIDATA+: Deductive Databases with Complex Objects. VIII, 273 pages. 1993.

Vol. 702: E. Börger, G. Jäger, H. Kleine Büning, S. Martini, M. M. Richter (Eds.), Computer Science Logic. Proceedings, 1992. VIII, 439 pages. 1993.

Vol. 703: M. de Berg, Ray Shooting, Depth Orders and Hidden Surface Removal. X, 201 pages. 1993.

Vol. 704: F. N. Paulisch, The Design of an Extendible Graph Editor. XV, 184 pages. 1993.

Vol. 705: H. Grünbacher, R. W. Hartenstein (Eds.), Field-Programmable Gate Arrays. Proceedings, 1992. VIII, 218 pages. 1993.

Vol. 706: H. D. Rombach, V. R. Basili, R. W. Selby (Eds.), Experimental Software Engineering Issues. Proceedings, 1992. XVIII, 261 pages. 1993.

Vol. 707: O. M. Nierstrasz (Ed.), ECOOP '93 – Object-Oriented Programming. Proceedings, 1993. XI, 531 pages. 1993.

Vol. 708: C. Laugier (Ed.), Geometric Reasoning for Perception and Action. Proceedings, 1991. VIII, 281 pages. 1993.

Vol. 709: F. Dehne, J.-R. Sack, N. Santoro, S. Whitesides (Eds.), Algorithms and Data Structures. Proceedings, 1993. XII, 634 pages. 1993.

Vol. 710: Z. Ésik (Ed.), Fundamentals of Computation Theory. Proceedings, 1993. IX, 471 pages. 1993.

Vol. 711: A. M. Borzyszkowski, S. Sokołowski (Eds.), Mathematical Foundations of Computer Science 1993. Proceedings, 1993. XIII, 782 pages. 1993.

Vol. 712: P. V. Rangan (Ed.), Network and Operating System Support for Digital Audio and Video. Proceedings, 1992. X, 416 pages. 1993.

Vol. 713: G. Gottlob, A. Leitsch, D. Mundici (Eds.), Computational Logic and Proof Theory. Proceedings, 1993. XI, 348 pages. 1993.

Vol. 714: M. Bruynooghe, J. Penjam (Eds.), Programming Language Implementation and Logic Programming. Proceedings, 1993. XI, 421 pages. 1993.

Vol. 715: E. Best (Ed.), CONCUR'93. Proceedings, 1993. IX, 541 pages. 1993.

Vol. 716: A. U. Frank, I. Campari (Eds.), Spatial Information Theory. Proceedings, 1993. XI, 478 pages. 1993.

Vol. 717: I. Sommerville, M. Paul (Eds.), Software Engineering – ESEC '93. Proceedings, 1993. XII, 516 pages. 1993.

Vol. 718: J. Seberry, Y. Zheng (Eds.), Advances in Cryptology – AUSCRYPT '92. Proceedings, 1992. XIII, 543 pages. 1993.

Vol. 719: D. Chetverikov, W.G. Kropatsch (Eds.), Computer Analysis of Images and Patterns. Proceedings, 1993. XVI, 857 pages. 1993.

Vol. 720: V.Mařík, J. Lažanský, R.R. Wagner (Eds.), Database and Expert Systems Applications. Proceedings, 1993. XV, 768 pages. 1993.

Vol. 721: J. Fitch (Ed.), Design and Implementation of Symbolic Computation Systems. Proceedings, 1992. VIII, 215 pages. 1993.

Vol. 722: A. Miola (Ed.), Design and Implementation of Symbolic Computation Systems. Proceedings, 1993. XII, 384 pages. 1993.

Vol. 723: N. Aussenac, G. Boy, B. Gaines, M. Linster, J.-G. Ganascia, Y. Kodratoff (Eds.), Knowledge Acquisition for Knowledge-Based Systems. Proceedings, 1993. XIII, 446 pages. 1993. (Subseries LNAI).

Vol. 724: P. Cousot, M. Falaschi, G. Filè, A. Rauzy (Eds.), Static Analysis. Proceedings, 1993. IX, 283 pages. 1993.

Vol. 725: A. Schiper (Ed.), Distributed Algorithms. Proceedings, 1993. VIII, 325 pages. 1993.

Vol. 726: T. Lengauer (Ed.), Algorithms – ESA '93. Proceedings, 1993. IX, 419 pages. 1993

Vol. 727: M. Filgueiras, L. Damas (Eds.), Progress in Artificial Intelligence. Proceedings, 1993. X, 362 pages. 1993. (Subseries LNAI).

Vol. 728: P. Torasso (Ed.), Advances in Artificial Intelligence. Proceedings, 1993. XI, 336 pages. 1993. (Subseries LNAI).

Vol. 729: L. Donatiello, R. Nelson (Eds.), Performance Evaluation of Computer and Communication Systems. Proceedings, 1993. VIII, 675 pages. 1993.

Vol. 730: D. B. Lomet (Ed.), Foundations of Data Organization and Algorithms. Proceedings, 1993. XII, 412 pages. 1993.

Vol. 731: A. Schill (Ed.), DCE – The OSF Distributed Computing Environment. Proceedings, 1993. VIII, 285 pages. 1993.

Vol. 732: A. Bode, M. Dal Cin (Eds.), Parallel Computer Architectures. IX, 311 pages. 1993.

Vol. 733: Th. Grechenig, M. Tscheligi (Eds.), Human Computer Interaction. Proceedings, 1993. XIV, 450 pages. 1993.

Vol. 734: J. Volkert (Ed.), Parallel Computation. Proceedings, 1993. VIII, 248 pages. 1993.

Vol. 735: D. Bjørner, M. Broy, I. V. Pottosin (Eds.), Formal Methods in Programming and Their Applications. Proceedings, 1993. IX, 434 pages. 1993.

Vol. 736: R. L. Grossman, A. Nerode, A. P. Ravn, H. Rischel (Eds.), Hybrid Systems. VIII, 474 pages. 1993.

Vol. 737: J. Calmet, J. A. Campbell (Eds.), Artificial Intelligence and Symbolic Mathematical Computing. Proceedings, 1992. VIII, 305 pages. 1993.

Vol. 738: M. Weber, M. Simons, Ch. Lafontaine, The Generic Development Language Deva. XI, 246 pages. 1993.

Vol. 739: H. Imai, R. L. Rivest, T. Matsumoto (Eds.), Advances in Cryptology – ASIACRYPT '91. X, 499 pages. 1993.

Vol. 740: E. F. Brickell (Ed.), Advances in Cryptology – CRYPTO '92. Proceedings, 1992. X, 593 pages. 1993.

Vol. 741: B. Preneel, R. Govaerts, J. Vandewalle (Eds.), Computer Security and Industrial Cryptography. Proceedings, 1991. VIII, 275 pages. 1993.

Vol. 742: S. Nishio, A. Yonezawa (Eds.), Object Technologies for Advanced Software. Proceedings, 1993. X, 543 pages. 1993.

Vol. 743: S. Doshita, K. Furukawa, K. P. Jantke, T. Nishida (Eds.), Algorithmic Learning Theory. Proceedings, 1992. X, 260 pages. 1993. (Subseries LNAI)

Vol. 744: K. P. Jantke, T. Yokomori, S. Kobayashi, E. Tomita (Eds.), Algorithmic Learning Theory. Proceedings, 1993. XI, 423 pages. 1993. (Subseries LNAI)

Vol. 745: V. Roberto (Ed.), Intelligent Perceptual Systems. VIII, 378 pages. 1993. (Subseries LNAI)

Vol. 746: A. S. Tanguiane, Artificial Perception and Music Recognition. XV, 210 pages. 1993. (Subseries LNAI).

Vol. 747: M. Clarke, R. Kruse, S. Moral (Eds.), Symbolic and Quantitative Approaches to Reasoning and Uncertainty. Proceedings, 1993. X, 390 pages. 1993.

Vol. 748: R. H. Halstead Jr., T. Ito (Eds.), Parallel Symbolic Computing: Languages, Systems, and Applications. Proceedings, 1992. X, 419 pages. 1993.

Vol. 749: P. A. Fritzson (Ed.), Automated and Algorithmic Debugging. Proceedings, 1993. VIII, 369 pages. 1993.

Vol. 750: J. L. Díaz-Herrera (Ed.), Software Engineering Education. Proceedings, 1994. XII, 601 pages. 1994.

Vol. 751: B. Jähne, Spatio-Temporal Image Processing. XII, 208 pages. 1993.

Vol. 752: T. W. Finin, C. K. Nicholas, Y. Yesha (Eds.), Information and Knowledge Management. Proceedings, 1992. VII, 142 pages. 1993.

Vol. 753: L. J. Bass, J. Gornostaev, C. Unger (Eds.), Human-Computer Interaction. Proceedings, 1993. X, 388 pages. 1993.

Vol. 754: H. D. Pfeiffer, T. E. Nagle (Eds.), Conceptual Structures: Theory and Implementation. Proceedings, 1992. IX, 327 pages. 1993. (Subseries LNAI).

Vol. 755: B. Möller, H. Partsch, S. Schuman (Eds.), Formal Program Development. Proceedings. VII, 371 pages. 1993.

Vol. 756: J. Pieprzyk, B. Sadeghiyan, Design of Hashing Algorithms. XV, 194 pages. 1993.

Vol. 757: U. Banerjee, D. Gelernter, A. Nicolau, D. Padua (Eds.), Languages and Compilers for Parallel Computing. Proceedings, 1992. X, 576 pages. 1993.

Vol. 758: M. Teillaud, Towards Dynamic Randomized Algorithms in Computational Geometry. IX, 157 pages. 1993.

Vol. 759: N. R. Adam, B. K. Bhargava (Eds.), Advanced Database Systems. XV, 451 pages. 1993.

Vol. 760: S. Ceri, K. Tanaka, S. Tsur (Eds.), Deductive and Object-Oriented Databases. Proceedings, 1993. XII, 488 pages. 1993.

Vol. 761: R. K. Shyamasundar (Ed.), Foundations of Software Technology and Theoretical Computer Science. Proceedings, 1993. XIV, 456 pages. 1993.

Vol. 762: K. W. Ng, P. Raghavan, N. V. Balasubramanian, F. Y. L. Chin (Eds.), Algorithms and Computation. Proceedings, 1993. XIII, 542 pages. 1993.

Vol. 763: F. Pichler, R. Moreno Díaz (Eds.), Computer Aided Systems Theory – EUROCAST '93. Proceedings, 1993. IX, 451 pages. 1994.

Vol. 764: G. Wagner, Vivid Logic. XII, 148 pages. 1994. (Subseries LNAI).

Vol. 765: T. Helleseth (Ed.), Advances in Cryptology – EUROCRYPT '93. Proceedings, 1993. X, 467 pages. 1994.

Vol. 766: P. R. Van Loocke, The Dynamics of Concepts. XI, 340 pages. 1994. (Subseries LNAI).

Vol. 767: M. Gogolla, An Extended Entity-Relationship Model. X, 136 pages. 1994.

Vol. 768: U. Banerjee, D. Gelernter, A. Nicolau, D. Padua (Eds.), Languages and Compilers for Parallel Computing. Proceedings, 1993. XI, 655 pages. 1994.

Vol. 769: J. L. Nazareth, The Newton-Cauchy Framework. XII, 101 pages. 1994.

Vol. 770: P. Haddawy (Representing Plans Under Uncertainty. X, 129 pages. 1994. (Subseries LNAI).

Vol. 771: G. Tomas, C. W. Ueberhuber, Visualization of Scientific Parallel Programs. XI, 310 pages. 1994.

Vol. 772: B. C. Warboys (Ed.),Software Process Technology. Proceedings, 1994. IX, 275 pages. 1994.

Vol. 773: D. R. Stinson (Ed.), Advances in Cryptology – CRYPTO '93. Proceedings, 1993. X, 492 pages. 1994.